D0151519

Stability and Stabilization

Stability and Stabilization

An Introduction

William J. Terrell

PRINCETON UNIVERSITY PRESS
PRINCETON AND OXFORD

Published by Princeton University Press, 41 William Street, Princeton, New Jersey 08540

In the United Kingdom: Princeton University Press, 6 Oxford Street, Woodstock, Oxfordshire OX20 1TW

ISBN-13: 978-0-691-13444-4
Library of Congress Control Number: 2008926510

British Library Cataloging-in-Publication Data is available
Printed on acid-free paper. ∞
press.princeton.edu
Typeset by S R Nova Pvt Ltd, Bangalore, India
Printed in the United States of America

10 9 8 7 6 5 4 3 2 1

Contents

List of Figures

Preface

Thanks for turning to the Preface.

This book is a text on stability theory and applications for systems of ordinary differential equations. It covers a portion of the core of mathematical control theory, including the concepts of linear systems theory and Lyapunov stability theory for nonlinear systems, with applications to feedback stabilization of control systems.

The book is written as an introduction for beginning students who want to learn about the mathematics of control through a study of one of its major problems: the problem of stability and feedback stabilization of equilibria. Readers can then explore the concepts within their own areas of scientific, engineering, or mathematical interest.

Previous exposure to control theory is not required. The minimal prerequisite for reading the book is a working knowledge of elementary ordinary differential equations and elementary linear algebra. Introductory courses in each of these areas meet this requirement. Some exposure to undergraduate analysis (advanced calculus) beyond the traditional three-semester calculus sequence will be helpful as the reader progresses through the book, but it is not a strict prerequisite for beginning.

It may be helpful to mention that one or two of the following courses probably provide more than sufficient background, due to the mathematical maturity required by such courses:

- a course in undergraduate analysis (advanced calculus) that covers mappings from $\mathbf{R}^n$ to $\mathbf{R}^m$, continuity, and differentiability
- a course in the theory of ordinary differential equations that covers existence and uniqueness, continuation (extension) of solutions, and continuous dependence of solutions, including discussion of both linear and nonlinear systems
- a second course in linear algebra or matrix theory
- a senior undergraduate or first-year graduate course in numerical analysis or numerical linear algebra

At this point I should note some important omissions. Numerical issues of control analysis and design are not discussed in the book. It is also important to mention right away that the book deals with systems in the state space framework only, with no discussion of transfer function analysis of linear systems. No experience with transfer functions is required, and no properties of transfer functions are invoked. I have included some references for these important areas as pointers to the literature. In addition, there is no systematic coverage of linear optimal control; however, there is a section on the algebraic Riccati equation and its connection with the linear quadratic regulator problem. The book includes an introduction to the state space framework of linear systems theory in Chapters 3–7. Chapters 8–16 are on nonlinear systems. (Detailed chapter descriptions appear in the introductory chapter.)

This book emphasizes basic system concepts, stability, and feedback stabilization for autonomous systems of ordinary differential equations. Thus, it covers a portion of the core of mathematical control theory, and attempts to show the cohesiveness and unity of the mathematical ideas involved.

For deeper foundations and a broader perspective on the wider field of mathematical control theory, as well as extensions and further applications of the ideas, I recommend additional reading. Suggested reading for this purpose appears at the end of each chapter in a Notes and References section and in a brief Further Reading chapter at the end of the book.

Exercise sets appear in separate sections at the end of each chapter, and in total there are over 190 exercises. In general, the exercises are low-to-intermediate hurdles, although there may be a few exceptions to this rule. There are exercises that require showing how something is done, and exercises that require showing why something is true. There are also exercises that ask for an example or counterexample to illustrate certain points. Occasionally, when all the required tools are available, the proof of a labeled result in the text is left as an exercise. Computations in the exercises usually require only a modest amount of calculation by hand, and some standard computations such as matrix rank or phase portraits can be done by software.

The book provides enough material for two academic semesters of coursework for begininng students. It can serve as a text for a second course in ordinary differential equations which provides an introduction to a core area of mathematical control theory. The material is presented at an intermediate level suitable for advanced undergraduates or beginning graduate students, depending on instructor choices and student backgrounds. When teaching this material I have usually covered most of the material of the Mathematical Background chapter (Chapter 2) in detail. Occasionally, I have covered some material from the Appendices; however, the Appendices might be assigned for self-study.

The text is also suitable for self-study by well-motivated readers with experience in ordinary differential equations and linear algebra. In fact, I hope that the book will provide some stimulation for readers to learn more about basic analysis and the theory of ordinary differential equations with the help of a motivating core problem of scientific, engineering, and mathematical significance.

ACKNOWLEDGMENTS . The results of this book originate with others. The selection of topics and the organization have been guided by a careful reading of some key papers, texts, and monographs. It is a pleasure to acknowledge my debt to all the authors of works in the Bibliography, Notes and References sections, and Further Reading. I have made selections from a very large literature and attempted to tie the topics together in a coherent way, while remaining aware that other choices might have been made. After sincere efforts to produce a clear and accurate text, any errors that have crept in are my own reponsibility, and I will remain grateful to any readers kind enough to send their comments, suggestions, or notices of correction.

Special thanks go to Stephen L. Campbell at North Carolina State University and Vickie Kearn at Princeton University Press for their interest and encouragement in this project, and to Jan C. Willems at Katholieke Universiteit Leuven for his willingness to read, comment and offer suggestions on many chapters of a draft manuscript. My thanks also to Robert E. Terrell for reading even earlier versions of many chapters and for offering illustration ideas and graphics files.

In addition, I want to thank all the anonymous reviewers of a draft manuscript for their interest and willingness to review portions of the text as it developed.

I also thank the Office of the Dean, College of Humanities and Sciences, Virginia Commonwealth University, for their interest in the project and for financial support during the summer of 2007 toward completion of the book, and the VCU Department of Mathematics and Applied Mathematics for the opportunity to teach topics courses on this material from 2003 to 2007. I also thank the students who participated with interest and patience in those topics courses.

This book is dedicated to: my father, Edgar A. Terrell, Jr., to the memory of my mother, Emmie J. Terrell, and my brothers, Robert E. Terrell and Jon A. Terrell.

William J. Terrell
Richmond, Virginia

Chapter One

Introduction

In this short introductory chapter, we introduce the main problem of stability and stabilization of equilibria, and indicate briefly the central role it plays in mathematical control theory. The presentation here is mostly informal. Precise definitions are given later. The chapter serves to give some perspective while stating the primary theme of the text.

We start with a discussion of simple equations from an elementary differential equations course in order to contrast open loop control and feedback control. These examples lead us to a statement of the main problem considered in the book, followed by an indication of the central importance of stability and stabilization in mathematical control theory. We then note a few important omissions. A separate section gives a complete chapter-by-chapter description of the book. The final section of the chapter is a list of suggested collateral reading.

1.1 OPEN LOOP CONTROL

Students of elementary differential equations already have experience with open loop controls. These controls appear as a given time-dependent forcing term in the second order linear equations that are covered in the first course on the subject. A couple of simple examples will serve to illustrate the notion of open loop control and allow us to set the stage for a discussion of feedback control in the next section.

The Forced Harmonic Oscillator. Consider the nonhomogeneous linear mass-spring equation with unit mass and unit spring constant,

$$\ddot{y} + y = u(t).$$

We use $\dot{y}$ and $\ddot{y}$ to denote the first and second derivatives of $y(t)$ with respect to time. The equation involves a known right-hand side, which can be viewed as a preprogrammed, or *open loop*, control defined by $u(t)$. The general real-valued solution for such equations is considered in differential equations courses, and it takes the form

$$y(t) = y_h(t) + y_p(t),$$

where $y_p(t)$ is any particular solution of the nonhomogeneous equation and $y_h(t)$ denotes the general solution of the homogeneous equation, $\ddot{y} + y = 0$.

For this mass-spring equation, we have

$$y_h(t) = c_1 \cos t + c_2 \sin t,$$

where the constants c_1 and c_2 are uniquely determined by initial conditions for $y(0)$ and $\dot{y}(0)$.

Suppose the input signal is $u(t) = \sin t$. This would not be an effective control, for example, if our purpose is to damp out the motion asymptotically or to regulate the motion to track a specified position or velocity trajectory. Since the frequency of the input signal equals the natural frequency of the unforced harmonic oscillator, $\ddot{y} + y = 0$, the sine input creates a resonance that produces unbounded motion of the mass.

On the other hand, the decaying input $u(t) = e^{-t}$ yields a particular solution given by $y_p(t) = \frac{1}{2}e^{-t}$. In this case, every solution approaches a periodic response as $t \to \infty$, given by $y_h(t)$, which depends on the initial conditions $y(0)$ and $\dot{y}(0)$, but not on the input signal.

Suppose we wanted to apply a continuous input signal which would guarantee that all solutions approach the origin defined by zero position and zero velocity. It is not difficult to see that we cannot do this with a continuous open loop control. The theory for second-order linear equations implies that there is no continuous open loop control $u(t)$ such that each solution of $\ddot{y} + y = u(t)$ approaches the origin as $t \to \infty$, independently of initial conditions.

THE DOUBLE INTEGRATOR. An even simpler equation is $\ddot{y} = u(t)$. The general solution has the form $y(t) = c_1 + c_2 t + y_p(t)$, where $y_p(t)$ is a particular solution that depends on $u(t)$. Again, there is no continuous control $u(t)$ that will guarantee that the solutions will approach the origin defined by zero position and zero velocity, independently of initial conditions.

Open loop, or preprogrammed, control does not respond to the state of the system it controls during operation. A standard feature of engineering design involves the idea of injecting a signal into a system to determine the response to an impulse, step, or ramp input signal. Recent work on the active approach to the design of signals for failure detection uses open loop controls as test signals to detect abnormal behavior [22]; an understanding of such open loop controls may enable more autonomous operation of equipment and condition-based maintenance, resulting in less costly or safer operation.

The main focus of this book is on principles of stability and feedback stabilization of an equilibrium of a dynamical system. The next section explains this terminology and gives a general statement of this core problem of dynamics and control.

1.2 THE FEEDBACK STABILIZATION PROBLEM

The main theme of stability and stabilization is focused by an emphasis on time invariant (autonomous) systems of the form

$$\dot{x} = f(x),$$

where $f : \mathcal{D} \subset \mathbf{R}^n \to \mathbf{R}^n$ is a continuously differentiable mapping (a smooth vector field on an open set $\mathcal{D} \subset \mathbf{R}^n$) and $\dot{x} := \frac{dx}{dt}$. If f is continuously differentiable, then f satisfies a local Lipschitz continuity condition in a neighborhood of each point in its domain. From the theory of ordinary differential equations, the condition of local Lipschitz continuity of f guarantees the existence and uniqueness of solutions of initial value problems

$$\dot{x} = f(x), \quad x(0) = x_0,$$

where x_0 is a given point of $\mathcal{D}$.

The state of the system at time t is described by the vector x. Assuming that $f(0) = 0$, so that the origin is an equilibrium (constant) solution of the system, the core problem is to determine the stability properties of the equilibrium. The main emphasis is on conditions for asymptotic stability of the equilibrium. A precise definition of the term *asymptotic stability* of $x = 0$ is given later. For the moment, we simply state its intuitive meaning: Solutions $x(t)$ with initial condition close to the origin are defined for all forward time $t \geq 0$ and remain close to $x = 0$ for all $t \geq 0$; moreover, initial conditions sufficiently close to the equilibrium yield solutions that approach the equilibrium asymptotically as $t \to \infty$.

We can now discuss the meaning of *feedback stabilization* of an equilibrium. Let $f : \mathbf{R}^n \times \mathbf{R}^m \to \mathbf{R}^n$ be a continuously differentiable function of $(x, u) \in \mathbf{R}^n \times \mathbf{R}^m$. The introduction of a feedback control models the more complicated process of actually measuring the system state and employing some mechanism to feed the measured state back into the system as a real time control on system operation. The feedback stabilization problems in this book involve autonomous systems with control u, given by

$$\dot{x} = f(x, u).$$

In this framework, the introduction of a smooth (continuously differentiable) state feedback control $u = k(x)$ results in the *closed loop system*

$$\dot{x} = f(x, k(x)),$$

which is autonomous as well. If $f(0,0) = 0$, then the origin $x_0 = 0$ is an equilibrium of the unforced system, $\dot{x} = f(x, 0)$. If the feedback satisfies $k(0) = 0$, then it preserves the equilibrium; that is, the closed loop system also has an equilibrium at the origin.

We apply stability theory in several different settings to study the question of existence of a continuously differentiable feedback $u = k(x)$ such that the origin $x_0 = 0$ is an asymptotically stable equilibrium of the closed loop system. For certain system classes and conditions, explicit stabilizing feedback controls are constructed. The system classes we consider are not chosen arbitrarily; they are motivated by (i) their relevance in the research activity on stabilization of recent decades, and (ii) their accessibility in an introductory text.

FEEDBACK IN THE HARMONIC OSCILLATOR AND DOUBLE INTEGRATOR SYSTEMS. The system corresponding to the undamped and unforced harmonic oscillator, obtained by writing $x_1 = y$ and $x_2 = \dot{y}$, and setting $u = 0$, is given by

$$\begin{aligned} \dot{x}_1 &= x_2, \\ \dot{x}_2 &= -x_1. \end{aligned}$$

This system does not have an asymptotically stable equilibrium at the origin $(x_1, x_2) = (0, 0)$. If we had both state components available for feedback, we could define a feedback control of the form $u = k_1 x_1 + k_2 x_2$, producing the closed loop system

$$\begin{aligned} \dot{x}_1 &= x_2, \\ \dot{x}_2 &= (k_1 - 1)x_1 + k_2 x_2. \end{aligned}$$

If we can measure only the position variable x_1 and use it for feedback, say $u = k_1 x_1$, then we are not able to make the origin $(0, 0)$ asymptotically stable, no matter what the value of the real coefficient k_1 may be. However, using only feedback from the velocity, if available, say $u = k_2 x_2$, it is possible to make the origin an asymptotically stable equilibrium of the closed loop system. Verification of these facts is straightforward, and to accomplish it, we can even use the second order form for the closed loop system; for position feedback only, $\ddot{y} + (1 - k_1)y = 0$; for velocity feedback only, $\ddot{y} - k_2\dot{y} + y = 0$. For position feedback, the characteristic equation is $r^2 + (1 - k_1) = 0$, and the general real-valued solution for $t \geq 0$ is (i) periodic for $k_1 < 1$, (ii) the sum of an increasing exponential term and a decreasing exponential term for $k_1 > 1$, and (iii) a constant plus an unbounded linear term for $k_1 = 1$. For velocity feedback, choosing $k_2 < 0$ ensures that all solutions that start close to the origin at time $t = 0$ remain close to the origin for all $t \geq 0$, and also satisfy $(x_1(t), x_2(t)) \to (0, 0)$ as $t \to \infty$.

For the simpler double integrator equation, $\ddot{y} = u(t)$, or its equivalent system,

$$\begin{aligned} \dot{x}_1 &= x_2, \\ \dot{x}_2 &= u, \end{aligned}$$

one can check that neither position feedback, $u = k_1 x_1$, nor velocity feedback, $u = k_2 x_2$, can make all solutions approach the origin as $t \to \infty$. However, feedback using both position and velocity, $u = k_1 x_1 + k_2 x_2$, will accomplish this if $k_1 > 0$ and $k_2 > 0$.

The study of stability and stabilization of equilibria for ordinary differential equations (ODEs) is a vast area of applications-oriented mathematics. The restriction to smooth feedback still leaves a huge area of results. This area will be explored in selected directions in the pages of this introductory text.

The restriction to smooth feedback avoids some technical issues that arise with discontinuous feedback, or even with merely continuous feedback.

Discontinuous feedback is mathematically interesting and relevant in many applications. For example, the solutions of many optimal control problems (not discussed in this book) involve discontinuous feedback. However, a systematic study of such feedback requires a reconsideration of the type of system under study and the meaning of solution. These questions fall essentially outside the scope of the present book.

Although we consider primarily smooth feedback controls, at several places in the text the admissible open loop controls are piecewise continuous, or at least integrable on any finite interval, that is, *locally integrable.*

The Importance of the Subject

Stability theory provides core techniques for the analysis of dynamical systems, and it has done so for well over a hundred years, at least since the 1892 work of A. M. Lyapunov; see [73]. An earlier feedback control study of a steam engine governor, by J. Clerk Maxwell, was probably the first modern analysis of a control system and its stability. Stability concepts have always been a central concern in the study of dynamical control systems and their applications. The problem of feedback stabilization of equilibria is a core problem of mathematical control theory. Possibly the most important point to make here is that many other issues and problems of control theory depend on concepts and techniques of stability and stabilization for their mathematical foundation and expression. Some of these areas are indicated in the end-of-chapter Notes and References sections.

Some Important Omissions

There are many important topics of stability, stabilization, and, more generally, mathematical control theory which are not addressed in this book. In particular, as mentioned in the Preface, there is no discussion of transfer function analysis for linear time invariant systems, and transfer functions are not used in the text. Also, there is no systematic coverage of optimal control beyond the single section on the algebraic Riccati equation. Since there is no coverage of numerical computation issues in this text, readers interested specifically in numerical methods should be aware of the text by B. N. Datta, *Numerical Methods for Linear Control Systems*, Elsevier Academic Press, London, 2004.

The end-of-chapter Notes and References sections have resources for a few other areas not covered in the text.

1.3 CHAPTER AND APPENDIX DESCRIPTIONS

In general, the chapters follow a natural progression. It may be helpful to mention that readers with a background in the state space framework of linear system theory and a primary interest in nonlinear systems might proceed with Chapter 8 (Stability Theory) after the introductory material of Chapter 2 (Mathematical Background) and Chapter 3 (Linear Systems and Stability). Definitions and examples of stability and instability appear in Chapter 3. For such readers, Chapters 4–7 could be used for reference as needed.

CHAPTER 2. The Mathematical Background chapter includes material mainly from linear algebra and differential equations. For basic analysis we reference Appendix B or the text [7]. The section on linear and matrix algebra includes some basic notation, linear independence and rank, similarity of matrices, invariant subspaces, and the primary decomposition theorem. The section on matrix analysis surveys ifferentiation and integration of matrix functions, inner products and norms, sequences and series of functions, and quadratic forms. A section on ordinary differential equations states the existence and uniqueness theorem for locally Lipschitz vector fields and defines the Jacobian linearization of a system at a point. The final section has examples of linear and nonlinear mass-spring systems, pendulum systems, circuits, and population dynamics in the phase plane. These examples are familiar from a first course in differential equations. The intention of the chapter is to present only enough to push ahead to the first chapter on linear systems.

CHAPTER 3. This chapter develops the basic facts for linear systems of ordinary differential equations. It includes existence and uniqueness of solutions for linear systems, the stability definitions that apply throughout the book, stability results for linear systems, and some theory of Lyapunov equations. Jordan forms are introduced as a source of examples and insight into the structure of linear systems. The chapter also includes the Cayley-Hamilton theorem. A few basic facts on linear time varying systems are included as well.

The next four chapters, Chapters 4–7, provide an introduction to the four fundamental structural concepts of linear system theory: controllability, observability, stabilizability, and detectability. We discuss the invariance (or preservation) of these properties under linear coordinate change and certain feedback transformations. All four properties are related to the study of stability and stabilization throughout these four chapters. While there is some focus on single-input single-output (SISO) systems in the examples, we include basic results for multi-input multi-output (MIMO) systems as well. Throughout Chapters 4–7, Jordan form systems are used as examples to help develop insight into each of the four fundamental concepts.

CHAPTER 4. Controllability deals with the input-to-state interaction of the system. This chapter covers controllability for linear time invariant systems. Single-input controllable systems are equivalent to systems in a special companion form (controller form). Controllability is a strong sufficient condition for stabilization by linear feedback, and it ensures the solvability of transfer-of-state control problems. The chapter includes the eigenvalue placement theorem for both SISO and MIMO systems, a controllability normal form (for uncontrollable systems) and the PBH controllability test. (The PBH controllability test and related tests for observability, stabilizability, and detectability are so designated in recognition of the work of V. M. Popov, V. Belevitch, and M. L. J. Hautus.)

Chapter 5. Observability deals with the state-to-output interaction of the system. The chapter covers the standard rank criteria for observability and the fundamental duality between observability and controllability. Lyapunov equations are considered under some special hypotheses. The chapter includes an observability normal form (for unobservable systems), and a brief discussion of output feedback versus full-state feedback.

Chapter 6. This chapter on stabilizability begins with a couple of standard stabilizing feedback constructions for controllable systems, namely, linear feedback stabilization using the controllability Gramian and Ackermann's formula. We characterize stabilizability with the help of the controllability normal form and note the general limitations on eigenvalue placement by feedback when the system is not controllable. The chapter also includes the PBH stabilizability test and some discussion on the construction of the controllability and observability normal forms.

Chapter 7. Detectability is a weaker condition than observability, but it guarantees that the system output is effective in distinguishing trajectories asymptotically, and this makes the property useful, in particular, in stabilization studies. The chapter begins with an example of an observer system for asymptotic state estimation. We define the detectability property, and establish the PBH detectability test and the duality of detectability and stabilizability. We discuss the role of detectability and stabilizability in defining observer systems, the role of observer systems in observer-based dynamic stabilization, and general linear dynamic controllers and stabilization. The final section provides a brief look at the algebraic Riccati equation, its connection with the linear quadratic regulator problem, and its role in generating stabilizing linear feedback controls.

Chapter 8. Chapter 8 presents the basic concepts and most important Lyapunov theorems on stability in the context of nonlinear systems. We discuss the use of linearization for determining asymptotic stability and instability of equilibria, and we define critical problems of stability and smooth stabilization. We state Brockett's necessary condition for smooth stabilization. This chapter also develops basic properties of limit sets and includes the invariance theorem. There is a discussion of scalar equations which is useful for examples, a section on the basin of attraction for asymptotically stable equilibria, and a statement of converse Lyapunov theorems.

CHAPTER 9. Chapter 9 develops the stability properties of equilibria for cascade systems. The assumptions are strengthened gradually through the chapter, yielding results on Lyapunov stability, local asymptotic stability, and global asymptotic stability. Two foundational results lead to the main stability results: first, the theorem on total stability of an asymptotically stable equilibrium under a class of perturbations; second, a theorem establishing that the boundedness of certain driven trajectories in a cascade implies the convergence of those trajectories to equilibrium. Cascade systems play a central role in control studies; they arise in control problems directly by design or as a result of attempts to decompose, or to transform, a system for purposes of analysis. The final section shows that cascade forms may also be obtained by appropriate aggregation of state components.

CHAPTER 10. Center manifold theory provides tools for the study of critical problems of stability, the problems for which Jacobian linearization cannot decide the issue. Many critical problems can be addressed by the theorems of Chapter 8 or Chapter 9. However, center manifold theory is an effective general approach. The chapter begins with examples to show the value of the center manifold concept and the significance of dynamic behavior on a center manifold. Then we state the main results of the theory: (i) the existence of a center manifold; (ii) the reduction of stability analysis to the behavior on a center manifold; and (iii) the approximation of a center manifold to an order sufficient to accomplish the analysis in (ii). Two applications of these ideas are given in this chapter: the preservation of smooth stabilizability when a stabilizable system is augmented by an integrator, and a center manifold proof of a result on asymptotic stability in cascades with a linear driving system. Another application, on the design of a center manifold, appears in Chapter 11.

CHAPTER 11. In this chapter we consider single-input single-output systems and the zero dynamics concept. We define the relative degree at a point, the normal form, the zero dynamics manifold, and the zero dynamics subsystem on that manifold. Next, we consider asymptotic stabilization by an analysis of the zero dynamics subsystem, including critical cases. A simple model problem of aircraft control helps in contrasting linear and nonlinear problems and their stability analysis. The concept of vector relative degree for multi-input multi-output systems is defined, although it is used within the text only for the discussion of passive systems with uniform relative degree one. (Further developments on MIMO systems with vector relative degree, or on systems without a well-defined relative degree, are available through resources in the Notes and References.) The chapter ends with two applications: the design of a center manifold for the airplane example, and the computation of zero dynamics for low-dimensional controllable linear systems which is useful in Chapter 15.

CHAPTER 12. We consider feedback linearization only for single-input single-output systems. A single-input control-affine nonlinear system is locally equivalent, under coordinate change and regular feedback transformation in a neighborhood of the origin, to a linear controllable system, if and only if the relative degree at the origin is equal to the dimension of the state space. Feedback linearizable systems are characterized by geometric conditions that involve the defining vector fields of the system. The proof of the main theorem involves a special case of the Frobenius theorem, which appears in an Appendix. Despite the lack of robustness of feedback linearization, there are important areas, for example mechanical systems, where feedback linearization has achieved successes. Most important, the ideas of feedback linearization have played an important role in the development of nonlinear geometric control.

CHAPTER 13. In the first section of this chapter, we present a theorem on the global stabilization of a special class of nonlinear systems using the feedback construction known as damping control (also known as L_gV *control*, or *Jurdjevic-Quinn feedback*). This theorem provides an opportunity to contrast the strong connections among Lie brackets, controllability, and stabilization for linear systems, with the very different situation of nonlinear systems. Thus, the second section shows that the Lie bracket-based generalization of the controllability rank condition does not imply a local controllability property of the nonlinear system, and even global controllability does not imply stabilizability by smooth feedback. (The definition of controllability used here is the same one used for linear systems: any point can be reached from any other point in finite time along a trajectory corresponding to some admissible open loop control.) We give references for more information on controllability ideas and their application.

CHAPTER 14. The passivity concept has roots in the study of passive circuit elements and circuit networks. Passivity is defined as an input-output property, but passive systems can be characterized in state-space terms by the KYP property. (The KYP property is so designated in recognition of the work of R. E. Kalman, V. A. Yakubovich, and V. M. Popov.) This chapter develops the stability and stabilization properties of passive systems. It is an exploration of systems having relative degree at the opposite extreme from the feedback linearizable systems: passive systems having a smooth storage function have uniform relative degree one. Moreover, systems that are feedback passive, that is, passive with a smooth positive definite storage function after application of smooth feedback, are characterized by two conditions: they have uniform relative degree one and Lyapunov stable zero dynamics in a neighborhood of the origin. Passivity plays an important role in the feedback stabilization of cascades in Chapter 15.

Chapter 15. Chapter 15 returns to cascade systems. Partial-state feedback, which uses only the states of the driving system, is sufficient for local asymptotic stabilization of a cascade. In general, however, partial-state feedback cannot guarantee global asymptotic stabilization without restrictive growth assumptions on the interconnection term in the driven system of the cascade. This chapter considers an important situation in which global stabilization is assured using full-state feedback. We assume that the driving system is feedback passive with an output function that appears in the interconnection term in an appropriate factored form; global asymptotic stabilization is then achieved with a constructible feedback control.

Chapter 16. This chapter motivates the input-to-state stability concept based on earlier considerations in the text. In particular, input-to-state stability (ISS) addresses the need for a condition on the driven system of a cascade that guarantees not only (i) bounded solutions in response to bounded inputs, but also (ii) converging solutions from converging inputs. This material requires an introduction to the properties of comparison functions from Appendix E. The comparison functions are used in a proof of the basic Lyapunov theorems on stability and asymptotic stability. We give the definition of ISS Lyapunov function and present the main result concerning them: a system is ISS if and only if an ISS Lyapunov function exists for it. This result is applied to establish the ISS property for several examples. We state a result on the use of input-to-state stability in cascade systems and provide some further references.

Chapter 17. This brief chapter collects some additional notes on further reading.

Appendix A. This brief key to notation provides a convenient reference.

Appendix B. This appendix provides a quick reference for essential facts from basic analysis in $\mathbf{R}$ and $\mathbf{R}^n$.

Appendix C. This material on ordinary differential equations is self-contained and includes proofs of basic results on existence and uniqueness of solutions, continuation of solutions, and continuous dependence of solutions on initial conditions and on right-hand sides.

Appendix D. This material on manifolds and the preimage theorem is useful background for the center manifold chapter (Chapter 10) as well as Chapters 11 and 12, which deal with some aspects of geometric nonlinear control. The material on distributions and the Frobenius theorem supports Chapters 11–12 specifically.

APPENDIX E. The comparison functions are standard tools for the study of ordinary differential equations; they provide a convenient language for expressing basic inequality estimates in stability theory. This material is used explicitly in the text only in Chapter 16 (and in a brief appearance in the proof of Theorem 10.2 (d) on center manifold reduction).

APPENDIX F. Some hints and answers to selected exercises are included here.

1.4 NOTES AND REFERENCES

For some review of a first course in differential equations, see [21] and [84]. For additional recommended reading in differential equations to accompany this text, see [15] and [40]. In addition, see the text by V. I. Arnold, *Ordinary Differential Equations*, MIT Press, Cambridge, MA, 1973, for its geometric and qualitative emphasis.

The texts [9] and [72] have many examples of control systems described by ordinary differential equations. The material in these books is accessible to an audience having a strong background in upper level undergraduate mathematics. The same is true of the texts [53] and [80]. A senior level course in control engineering is contained in K. Ogata, *Modern Control Engineering*, Prentice-Hall, Upper Saddle River, NJ, third edition, 1997.

For the mathematical foundations of control theory for linear and nonlinear systems, see [91], which is a comprehensive mathematical control theory text on deterministic finite-dimensional systems. It includes material on a variety of model types: continuous and discrete, time invariant and time varying, linear and nonlinear. The presentation in [91] also includes many bibliographic notes and references on stabilization and its development, as well as three chapters on optimal control.

For an interesting and mostly informal article on feedback, see [60].

Chapter Two

Mathematical Background

This chapter contains material from linear algebra and differential equations. It is assumed that the reader has some previous experience with most of these ideas. Thus, in this chapter, in order to allow for ease of reference for items of the most basic importance, as well as a relatively quick reading experience, some terms are defined in labeled definitions and others within the text itself. Careful attention to the notation, definitions, results, and exercises in the background presented here should provide for relatively easy reading in the first chapter on linear systems.

2.1 ANALYSIS PRELIMINARIES

The basic topological concepts in Euclidean space are assumed, as are the concepts of continuity and differentiability of functions mapping a subset of $\mathbf{R}^n$ to $\mathbf{R}^m$. For more background, see Appendix B, or any of the excellent analysis texts available. Most often, we cite the text [7].

2.2 LINEAR ALGEBRA AND MATRIX ALGEBRA

This section includes some basic notation, material on linear independence and rank, similarity of matrices, eigenvalues and eigenvectors, invariant subspaces, and the Primary Decomposition Theorem.

Assumptions and Basic Notation

We assume that the reader is familiar with the algebraic concept of a *field*, and with the axioms that define the concept of a *vector space over a field*. Some experience with the concepts of subspace, basis, and dimension, and how these concepts summarize the solution of systems of linear algebraic equations, is essential. The concepts of eigenvalues and eigenvectors for a linear mapping (or operator) of a vector space V to itself should be familiar, as well as the fact that a linear mapping from a vector space V of dimension n to another vector space W of dimension m (over the same field) may be represented, after a choice of a basis for V and a basis for W, by an $m \times n$ matrix having entries in the common field of scalars.

We write $\mathbf{R}$ for the set of real numbers, $\mathbf{R}_+ := [0, \infty)$, and $\mathbf{R}^n$ for Euclidean n-dimensional space, which is a vector space over the real field $\mathbf{R}$. Vectors are usually column vectors, although they may be written in n-tuple form; when component detail is needed, we may write a vector x in $\mathbf{R}^n$ in any of the following forms:

$$x = \begin{bmatrix} x_1 \\ \vdots \\ x_n \end{bmatrix}, \quad \begin{bmatrix} x_1 & \cdots & x_n \end{bmatrix}^T, \quad (x_1, \ldots, x_n).$$

Similarly, we write $\mathbf{C}$ for the field of complex numbers, and $\mathbf{C}^n$ for complex n-dimensional space, which is a vector space over $\mathbf{C}$.

The systems of differential equations in this book involve real vector fields. In particular, linear systems or Jacobian linearizations of nonlinear systems involve real coefficient matrices. However, eigenvalues and corresponding eigenvectors of these matrices may be complex, so we must work with matrices with either real or complex number entries. We write $\mathbf{R}^{n\times n}$ for the set of $n \times n$ matrices with real entries. Elements of $\mathbf{R}^{n\times n}$ are also called real $n \times n$ matrices. $\mathbf{R}^{n\times n}$ is a vector space over $\mathbf{R}$. We write $\mathbf{C}^{n\times n}$ for the set of $n \times n$ matrices with complex entries. Elements of $\mathbf{C}^{n\times n}$ are also called complex $n \times n$ matrices. $\mathbf{C}^{n\times n}$ is a vector space over $\mathbf{C}$. As sets, $\mathbf{R}^{n\times n} \subset \mathbf{C}^{n\times n}$, so every real matrix is a complex matrix of the same size.

Linear Independence and Rank

We first recall the definition of linear independence of vectors in a vector space.

Definition 2.1 *Let V be a vector space over the field $\mathbf{F}$ ($\mathbf{F} = \mathbf{R}$ or $\mathbf{F} = \mathbf{C}$).*

(a) *The vectors $v_1, \ldots, v_k$ in V are* linearly dependent *over $\mathbf{F}$ if at least one of these vectors can be written as a linear combination of the others. That is, there are a j and scalars $c_1, c_2, \ldots, c_{j-1}, c_{j+1}, \ldots, c_k$ such that $v_j = c_1 v_1 + \cdots + c_{j-1}v_{j-1} + c_{j+1}v_{j+1} + \cdots + c_k v_k$.*

(b) *The vectors $v_1, \ldots, v_k$ in V are* linearly independent *over $\mathbf{F}$ if they are not linearly dependent over $\mathbf{F}$.*

Equivalently, $v_1, \ldots, v_k$ are linearly independent if and only if none of the vectors can be written as a linear combination of the remaining vectors in the collection, which is equivalent to saying that for scalars $c_1, \ldots, c_k$,

$$c_1 v_1 + c_2 v_2 + \cdots + c_k v_k = 0 \quad \Longrightarrow \quad c_1 = c_2 = \cdots = c_k = 0.$$

The $n\times n$ identity matrix, I_n, is the matrix with ones on the main diagonal and zeros elsewhere. Equivalently, we write $I = I_n = [e_1 \cdots e_n]$, where the j-th column e_j is the j-th *standard basis vector*. A matrix $A \in \mathbf{C}^{n\times n}$ is *invertible* if there exists a matrix $B \in \mathbf{C}^{n\times n}$ such that $BA = AB = I_n$.

Then B, if it exists, is unique and we write B as A^{-1} for the inverse of A. An invertible matrix is also called a *nonsingular* matrix.

If A is real and invertible, then A^{-1} is also real. To see this, write the inverse as $B = X + iY$ with X and Y real. Then $I = AB = AX + iAY$, and therefore $AY = 0$; hence, $Y = 0$ and $B = X$ is real. The idea of invertibility is to solve $Ax = y$ in the form $x = A^{-1}y$. This solution can be carried out simultaneously for the standard basis vectors in place of y, by the row reduction of the augmented matrix $[A \quad I_n]$ to the form $[I_n \quad A^{-1}]$, assuming that A is indeed invertible. Since elementary row operations preserve the property of nonzero value for the determinant of a square matrix, we have the criterion that a matrix A is invertible if and only if the determinant of A is nonzero, $\det A \neq 0$.

A *subspace* of a vector space V is a nonempty subset X of V that is closed under vector addition and scalar multiplication. That is, for any vectors $v, w \in X$ and any scalar $c \in \mathbf{F}$, we have $v + w \in X$ and $cv \in X$. A *basis* for a subspace X is a set $\{v_1, \ldots, v_k\}$ such that $v_1, \ldots, v_k$ are linearly independent and $X = \text{span}\,\{v_1, \ldots, v_k\}$, where $\text{span}\,\{v_1, \ldots, v_k\}$ is the *set of all linear combinations* of $v_1, \ldots, v_k$. A basis for X is a minimal spanning set for X and a maximal linearly independent subset of X. All bases for X must have the same number of elements, and the *dimension* of the subspace X, $\dim X$, is defined to be the number of elements in any basis for X. Two useful subspaces associated with an $m \times n$ matrix A are the range space and the nullspace. The *range* of A is denoted $R(A)$ and is the span of the columns of A. (If A is real and considered as a linear operator on $\mathbf{R}^n$, then this span is defined using real scalars, and if A is complex and considered as a linear operator on $\mathbf{C}^n$, then the span is defined using complex scalars.) The *nullspace* of A is denoted $N(A)$ or $\ker(A)$; it is the solution space of the linear system $Ax = 0$. (If A is real, we are usually interested in $N(A)$ as a subspace of $\mathbf{R}^n$.)

The *rank* of a matrix $A \in \mathbf{C}^{m\times n}$ is the dimension of the column space or range $R(A)$, and is written $\text{rank}\,A$, so $\text{rank}\,A = \dim R(A)$. (The rank is also equal to the number of basic columns in the row echelon form of the matrix. The basic columns are the columns which contain the pivots used in the row reduction.) The rank is always the rank over the field of complex numbers. In particular, the rank of a real matrix $A \in \mathbf{R}^{m\times n}$ is the rank over $\mathbf{C}$, but this is also the rank of A over the field of real numbers $\mathbf{R}$. (See Exercise 2.2) The *nullity* of A equals $\dim N(A)$. For any $A \in \mathbf{C}^{m\times n}$, we have

$$\dim R(A) + \dim N(A) = n.$$

Similarity of Matrices

Let $A, B \in \mathbf{C}^{n\times n}$. A is *similar* to B if there exists a nonsingular matrix $S \in \mathbf{C}^{n\times n}$ such that $S^{-1}AS = B$. This relation of similarity of matrices is an equivalence relation on $\mathbf{C}^{n\times n}$. An important similarity invariant of a matrix

A is the *characteristic polynomial* of A, defined by $p(\lambda) = \det(\lambda I - A)$. For later reference it is worth noting that, by this definition, the characteristic polynomial of a matrix is a *monic polynomial*, meaning that the coefficient of the highest power of λ in $\det(\lambda I - A)$ is equal to one. To see the invariance of this polynomial under a similarity transformation S, first recall that $\det(AB) = \det A \det B$ for square matrices A, B of the same size. Then compute

$$\det(\lambda I - S^{-1}AS) = \det S^{-1}(\lambda I - A)S = \det S^{-1} \det(\lambda I - A) \det S,$$

and finally, use the fact that $\det S^{-1} \det S = \det I_n = 1$, to conclude that $\det(\lambda I - S^{-1}AS) = \det(\lambda I - A)$. The set of eigenvalues of A is also a similarity invariant, since the *eigenvalues* are the roots of the *characteristic equation* of A, which is $p(\lambda) = \det(\lambda I - A) = 0$. The relation of similarity is also an equivalence relation on $\mathbf{R}^{n \times n}$. Let $A, B \in \mathbf{R}^{n \times n}$. We say that A and B are *similar via a real similarity* if and only if there exists a nonsingular matrix $S \in \mathbf{R}^{n \times n}$ such that $S^{-1}AS = B$. As we will see later in this chapter, the complex Jordan form of a (real or complex) $n \times n$ matrix A is an especially simple representative of the similarity equivalence class of A using complex similarities. The real Jordan form of a *real* matrix A is an especially simple representative of the similarity equivalence class of A using real similarities.

An $n \times n$ matrix $A = [a_{ij}]$ is *diagonal* if every entry off the main diagonal is zero. Thus, A is diagonal if and only if $a_{ij} = 0$ for $i \neq j$. An $n \times n$ matrix A is *diagonalizable* if A is similar to a diagonal matrix; that is, there exist a diagonal matrix, denoted $D = \mathrm{diag}\,[d_1, \ldots, d_n]$, where the d_i are the diagonal entries, and a nonsingular matrix S such that $S^{-1}AS = D$. The diagonal entries $d_1, \ldots, d_n$ are necessarily the eigenvalues of A, because the eigenvalues of a diagonal matrix are the main diagonal entries.

Invariant Subspaces

Let V be a vector space over $\mathbf{C}$. A linear mapping $A : V \to V$ is a function such that $A(v + w) = A(v) + A(w)$ for all $v, w \in V$, and $A(cv) = cA(v)$ for all $v \in V$ and $c \in \mathbf{C}$. We usually write $Av := A(v)$ when A is linear. A subspace $W \subset V$ is an *invariant subspace* for a linear mapping $A : V \to V$ if $A(W) \subset W$, that is, $Aw \in W$ for all $w \in W$. We also say that the subspace W is invariant under A.

Important examples of invariant subspaces of $\mathbf{C}^n$ are given by the eigenspaces associated with the eigenvalues of a matrix $A \in \mathbf{C}^{n \times n}$. First, we recall the definitions of eigenvalues and eigenvectors of a square matrix.

Definition 2.2 (Eigenvalues and Eigenvectors of $A \in \mathbf{C}^{n \times n}$)
Let $A \in \mathbf{C}^{n \times n}$. The complex number λ is an eigenvalue *of A if there exists a nonzero vector v such that $Av = \lambda v$. Such a vector v is called an* eigenvector *for the eigenvalue λ.*

Thus, if $W = N(A - \lambda I)$, where λ is an eigenvalue of A, then W is the *eigenspace* asssociated with λ, and W is invariant under A. Note that $v, w \in W$ implies $A(v + w) = Av + Aw = \lambda v + \lambda w = \lambda(v + w)$, hence $v + w \in W$. Also, $v \in W$ and $c \in \mathbf{C}$ imply $A(cv) = cAv = c\lambda v = \lambda(cv)$, so $cv \in W$. By definition, the nonzero vectors in $W = N(A - \lambda I)$ are the eigenvectors of A associated with λ. The *geometric multiplicity* of an eigenvalue λ is the dimension of $N(A - \lambda I)$, that is, the number of linearly independent eigenvectors for λ.

Additional examples of invariant subspaces of A are provided by the following construction. If $f(t)$ is a polynomial, then we write $f(A)$ for the polynomial expression obtained by substituting A for t in $f(t)$. If $f(t)$ is a polynomial, let $W := N(f(A))$. Then W is invariant under A. For if $w \in W$, then $f(A)w = 0$, and therefore $f(A)Aw = Af(A)w = A0 = 0$, which says that $Aw \in W$. In particular, let $f(t) = (t - \lambda)^r$, where λ is an eigenvalue of A and $r \geq 1$. The subspace $N(f(A)) = N((A - \lambda I)^r)$ is invariant under A, and, when $r \geq 2$, this subspace may contain vectors which are not true eigenvectors of A, that is, vectors not in $N(A - \lambda I)$. Any vector in $N((A - \lambda I)^r)$ for some r is called a *generalized eigenvector* of A associated with the eigenvalue λ. It is a consequence of the Primary Decomposition Theorem below that we do not generate any new generalized eigenvectors by considering $N((A - \lambda I)^r)$ for r greater than the algebraic multiplicity of λ. The *algebraic multiplicity* of an eigenvalue λ is the algebraic multiplicity of λ as a root of the characteristic polynomial of A. The subspace $N((A - \lambda I)^m)$, where m is the algebraic multiplicity of the eigenvalue λ, is called the *generalized eigenspace* associated with λ. This generalized eigenspace includes all true eigenvectors for λ as well as all generalized eigenvectors (and the zero vector). The algebraic multiplicity of an eigenvalue must be greater than or equal to the geometric multiplicity.

Example 2.1 The matrix

$$A = \begin{bmatrix} 2 & 0 & 0 \\ 0 & 3 & 1 \\ 0 & 0 & 3 \end{bmatrix}$$

has eigenvalues $\lambda_1 = 2$ and $\lambda_2 = 3$, as is easily seen from the triangular structure of A. The eigenspace for $\lambda_1 = 2$, which is the solution space of $(A - 2I)v = 0$, is spanned by the vector $e_1 = [1\ 0\ 0]^T$, the first standard basis vector. The eigenspace for $\lambda_2 = 3$ is spanned by the vector $e_2 = [0\ 1\ 0]^T$. Thus, A has only two linearly independent eigenvectors and is therefore not diagonalizable. However, we can find a nonzero solution of the system $(A - 3I)^2 v = 0$, which is linearly independent of the eigenvectors e_1 and e_2. The vector $e_3 = [0\ 0\ 1]^T$ satisfies $(A - 3I)e_3 = e_2$; therefore $(A - 3I)^2 e_3 = (A - 3I)e_2 = 0$. Thus, the generalized eigenspace for $\lambda_2 = 3$ is span $\{e_2, e_3\}$. $\triangle$

The Primary Decomposition Theorem

The Primary Decomposition Theorem is used later in the text to deduce a general solution formula for systems of linear differential equations. The resulting solution formula is then used to deduce norm bounds for the solutions.

In order to state the Primary Decomposition Theorem, we need the next definition.

Definition 2.3 *Let V be a vector space, and let $W_1, \dots, W_k$ be subspaces of V. We say that V is the* direct sum *of the subspaces W_i, $i = 1, \dots, k$ if every vector v in V can be written uniquely as a sum*

$$v = w_1 + \cdots + w_k$$

with $w_i \in W_i$, $i = 1, \dots, k$.

When V is a direct sum of subspaces W_i, we write

$$V = W_1 \oplus \cdots \oplus W_k .$$

If V is the direct sum of the subspaces W_i, and $\mathcal{B}_i$ is a basis for W_i, $1 \le i \le k$, then a basis $\mathcal{B}$ for V is obtained by setting

$$\mathcal{B} = \bigcup_{i=1}^{k} \mathcal{B}_i .$$

We can now state the Primary Decomposition Theorem.

Theorem 2.1 (The Primary Decomposition Theorem)
Let A be an $n \times n$ real or complex matrix, and write $p(\lambda) = \prod_{i=1}^{k}(\lambda - \lambda_i)^{m_i}$ for the characteristic polynomial of A, where $\lambda_1, \dots, \lambda_k$ are the distinct eigenvalues of A, having algebraic multiplicities $m_1, \dots, m_k$, respectively.

Then C^n is the direct sum of the generalized eigenspaces of A, and the dimension of each generalized eigenspace equals the algebraic multiplicity of the corresponding eigenvalue. That is,

$$C^n = N((A - \lambda_1 I)^{m_1}) \oplus \cdots \oplus N((A - \lambda_k I)^{m_k}).$$

Proof. See [15] (pp. 278–279) or [40] (pp. 331–333). □

2.3 MATRIX ANALYSIS

In a quick survey, this section covers differentiation and integration of matrix functions, and norms, sequences and series of functions, and quadratic forms.

Differentiation and Integration of Matrix Functions

If $A(t) \in \mathbf{C}^{n\times n}$ for each t in a real interval $\mathcal{I}$, then we say that $A(t)$ is a *matrix function* on $\mathcal{I}$. A matrix function $A(t) = [a_{ij}(t)]$ is *continuous* on $\mathcal{I}$ if each entry $a_{ij}(t)$ is a continuous function on $\mathcal{I}$. Similarly, $A(t)$ is *differentiable* (or *smooth*) of class C^k if each entry $a_{ij}(t)$ is differentiable (or smooth) of class C^k on $\mathcal{I}$ (has k continuous derivatives on $\mathcal{I}$). If $A(t) = [a_{ij}(t)]$ is a C^1 matrix function, we write $\dot{A}(t)$ for the matrix with, ij-entry equal to $\frac{da_{ij}}{dt}(t)$.

Example 2.2 Given that

$$A(t) = \begin{bmatrix} t & t^2 \\ 2 & \frac{1}{t} \end{bmatrix},$$

then $A(t)$ is smooth of class C^∞ on any interval not containing $t = 0$, and, in particular, for $t \neq 0$,

$$\dot{A}(t) = \begin{bmatrix} 1 & 2t \\ 0 & -\frac{1}{t^2} \end{bmatrix}.$$

△

See Exercise 2.1 for a few important facts about differentiation.

A matrix function $A(t) = [a_{ij}(t)]$ is *integrable* on the closed and bounded interval $\mathcal{I} = [\alpha, \beta]$ if each entry is integrable on $\mathcal{I}$. If $A(t)$ is integrable on $\mathcal{I} = [\alpha, \beta]$, then we define

$$\int_\alpha^\beta A(t)\,dt := \Big[\int_\alpha^\beta a_{ij}(t)\,dt\Big].$$

Example 2.3 For the 2×2 $A(t)$ considered in Example 2.2, we have

$$\int_\alpha^\beta A(t)\,dt = \begin{bmatrix} \displaystyle\int_\alpha^\beta t\,dt & \displaystyle\int_\alpha^\beta t^2\,dt \\ \displaystyle\int_\alpha^\beta 2\,dt & \displaystyle\int_\alpha^\beta \frac{1}{t}\,dt \end{bmatrix} = \begin{bmatrix} \dfrac{\beta^2 - \alpha^2}{2} & \dfrac{\beta^3 - \alpha^3}{3} \\ 2(\beta - \alpha) & \ln|\beta| - \ln|\alpha| \end{bmatrix},$$

if the interval $[\alpha, \beta]$ does not contain zero. On the other hand, if $0 \in [\alpha, \beta]$, then the definite integral of this $A(t)$ from α to β is an improper matrix integral because of the lower right entry. In this case, the improper matrix integral does not converge because the improper real integral in the lower right entry does not converge. △

Inner Products and Norms

Suppose that $A \in \mathbf{C}^{m\times n}$ with $A = [a_{ij}]$, $1 \le i \le m$, $1 \le j \le n$. The *transpose* of A is written A^T and is the matrix with i, j-entry equal to a_{ji}, for $1 \le j \le n$, $1 \le i \le m$. That is, the rows of A^T are the columns of A, taken in the same order.

Example 2.4 Let

$$A = \begin{bmatrix} 2 & 3 \\ -3i & 4i \end{bmatrix} \quad \text{and} \quad B = \begin{bmatrix} 2 & 3 \\ 1 & 5 \end{bmatrix}.$$

Then the transposes of these matrices are given by

$$A^T = \begin{bmatrix} 2 & -3i \\ 3 & 4i \end{bmatrix} \quad \text{and} \quad B^T = \begin{bmatrix} 2 & 1 \\ 3 & 5 \end{bmatrix}. \qquad \triangle$$

The dot product of vectors in $\mathbf{R}^n$ is a pairing of vectors which is bilinear in each vector argument. Formally, the dot product is a mapping $B : \mathbf{R}^n \times \mathbf{R}^n \to \mathbf{R}$ often written $B(u, v) := u \cdot v$. It is an *inner product* according to the following definition.

Definition 2.4 *Let V be a vector space over $\mathbf{R}$. A function $B : V \times V \mapsto \mathbf{R}$ is a* real inner product *on V if*

(a) $B(u, u) > 0$ *for all* $u \neq 0 \in V$, *and* $B(0, 0) = 0$;
(b) $B(v, u) = B(u, v)$ *for all* $u, v \in V$;
(c) $B(u, \alpha v + \beta w) = \alpha B(u, v) + \beta B(u, w)$ *for all* $u, v, w \in V$ *and all* $\alpha, \beta \in \mathbf{R}$.

Definition 2.5 *The* standard inner product (*Euclidean inner product*) *on $\mathbf{R}^n$ is defined by*

$$u^T v = u_1 v_1 + \cdots + u_n v_n,$$

for any vectors u and v in $\mathbf{R}^n$.

The verification that the product $(u, v) \mapsto u^T v$ satisfies Definition 2.4 is left as an exercise.

We also need complex-valued inner products. The formal definition only requires a change in item (b) of Definition 2.4. First, recall that the *conjugate* of a complex number, written in standard form $z = a + ib$ with a, b real, is $\bar{z} := a - ib$.

Definition 2.6 *Let V be a vector space over $\mathbf{C}$. A function $H : V \times V \mapsto \mathbf{C}$ is a* Hermitian (complex) inner product *on V if*

(a) $B(u, u) > 0$ *for all* $u \neq 0 \in V$, *and* $B(0, 0) = 0$;
(b) $B(v, u) = \overline{B(u, v)}$ *for all* $u, v \in V$;
(c) $B(u, \alpha v + \beta w) = \alpha B(u, v) + \beta B(u, w)$ *for all* $u, v, w \in V$ *and all* $\alpha, \beta \in \mathbf{C}$.

Note that (b) and (c) of Definition 2.6 imply that $B(\alpha u + \beta v, w) = \bar{\alpha} B(u, w) + \bar{\beta} B(v, w)$ for all choices of the arguments.

The bar notation is also used to indicate the componentwise conjugate of a vector $v = [v_1 \cdots v_n]^T$ in $\mathbf{C}^n$; thus, $\bar{v} = [\bar{v}_1 \cdots \bar{v}_n]^T$. The *conjugate transpose* of the column vector v is the row vector $v^* := [\bar{v}_1 \cdots \bar{v}_n]$.

The next definition features the most important complex inner product for our purposes.

Definition 2.7 *The* standard complex inner product (*Hermitian inner product*) *on* $\mathbf{C}^n$ *is defined by*

$$u^* v = \bar{u}_1 v_1 + \cdots + \bar{u}_n v_n,$$

for any vectors $u = [u_1 \ \cdots \ u_n]^T$, $v = [v_1 \ \cdots \ v_n]^T$ *in* $\mathbf{C}^n$.

The verification that the product $(u, v) \mapsto u^* v$ satisfies Definition 2.6 is left as an exercise. The conjugation of one of the factors guarantees that $u^* u \geq 0$. We use conjugation on the left-hand factor so that we can use the conjugate transpose operation on the left-hand vector; this choice is consistent with [75]. Without the conjugation, we may have $u^T u < 0$ for a complex vector u. For example, if $u = [i \ \ 0]^T$, then $u^T u = -1$.

Suppose that $A \in \mathbf{C}^{m \times n}$ and $A = [a_{ij}]$, $1 \leq i \leq m$, $1 \leq j \leq n$. The *conjugate* of A is given by

$$\bar{A} := [\bar{a}_{ij}], \quad 1 \leq i \leq m, 1 \leq j \leq n.$$

The *conjugate transpose* of A is written A^* and is given by

$$A^* := (\bar{A})^T.$$

Note that we also have $A^* = \overline{(A^T)}$. (The operations of conjugation and transposition commute.) Matrix A^* is also called the *Hermitian transpose* of A. If the matrix product AB is defined, then $(AB)^* = B^* A^*$. If A and B are in $\mathbf{R}^{m \times n}$, then the last property reads $(AB)^T = B^T A^T$. The conjugate transpose A^* has the property that

$$(Ax)^* y = x^* A^* y$$

for all vectors x and y in $\mathbf{C}^n$. In general, $(Ax)^* y \neq x^*(Ay)$, unless $A^* = A$, that is, A is a *Hermitian* matrix.

Example 2.5 Suppose that

$$A = \begin{bmatrix} i & 1 \\ 3 & i \end{bmatrix}, \quad x = \begin{bmatrix} 2 \\ i \end{bmatrix}, \quad y = \begin{bmatrix} i \\ 2i \end{bmatrix}.$$

Then we have

$$x^* A^* y = [2 \ \ -i] \begin{bmatrix} -i & 3 \\ 1 & -i \end{bmatrix} \begin{bmatrix} i \\ 2i \end{bmatrix} = [2 \ \ -i] \begin{bmatrix} 1 + 6i \\ 2 + i \end{bmatrix} = 3 + 10i,$$

and

$$x^*(Ay) = [2 \ \ -i] \begin{bmatrix} i & 1 \\ 3 & i \end{bmatrix} \begin{bmatrix} i \\ 2i \end{bmatrix} = [2 \ \ -i] \begin{bmatrix} -1 + 2i \\ -2 + 3i \end{bmatrix} = 1 + 6i. \qquad \triangle$$

The next result is a useful fact that emerges from a careful development of the Gaussian elimination process.

Lemma 2.1 (Transposes and Rank)
If $A \in \mathbf{C}^{m\times n}$, then

$$\operatorname{rank} A = \operatorname{rank} A^T \quad \textit{and} \quad \operatorname{rank} A = \operatorname{rank} A^*.$$

Proof. See [75] (page 139). □

If V is a subspace of $\mathbf{C}^n$, then the *orthogonal complement* of V in $\mathbf{C}^n$ is the subspace defined by

$$V^\perp := \{w \in \mathbf{C}^n \ : \ w^* v = 0 \text{ for all } v \in V\}.$$

Similarly, if V is a subspace of $\mathbf{R}^n$, then the *orthogonal complement* of V in $\mathbf{R}^n$ is the subspace defined by

$$V^\perp := \{w \in \mathbf{R}^n \ : \ w^T v = 0 \text{ for all } v \in V\}.$$

In either context, the set $V^\perp$ is closed under vector addition and appropriate scalar multiplication, so $V^\perp$ is indeed a subspace. And, in either context, we have $(V^\perp)^\perp = V$.

The next theorem is often called the Fundamental Theorem of Linear Algebra.

Theorem 2.2 (Fundamental Theorem of Linear Algebra)
Let A be in $\mathbf{R}^{m\times n}$ with $\operatorname{rank} A = r$. Then the following statements are true:

- $\dim R(A) = r$;
- $\dim N(A) = n - r$;
- $\dim R(A^T) = r$;
- $\dim N(A^T) = m - r$;
- $R(A) = N(A^T)^\perp$: *equivalently*, $N(A^T) = R(A)^\perp$;
- $R(A^T) = N(A)^\perp$: *equivalently*, $N(A) = R(A^T)^\perp$.

If A is in $\mathbf{C}^{m\times n}$, with $\operatorname{rank} A = r$, then the statements of Theorem 2.2 remain true when A^T is replaced in each instance with A^*.

We now discuss vector norms.

Definition 2.8 *Let V be a vector space. A function $\nu : V \mapsto \mathbf{R}$ is a* norm *on V, written $\|v\| := \nu(v)$, if*

(a) $\|v\| \geq 0$ *and* $\|v\| = 0$ *only if* $v = 0$;
(b) $\|cv\| = |c|\,\|v\|$ *for every v in V and scalar c*;
(c) $\|v + w\| \leq \|v\| + \|w\|$ *for every v, w in V.*

An example is the vector space of real numbers, which is normed by the absolute value function. An inner product $B(u, v)$ always induces a norm defined by $\|x\| := B(x, x)^{\frac{1}{2}}$. The Euclidean norm on $\mathbf{R}^n$ is induced in this way by the standard (Euclidean) inner product: $\|x\|_2^2 = x^T x$ for $x \in \mathbf{R}^n$. In $\mathbf{C}^n$ we have the standard norm defined by $\|x\|_2^2 = x^* x$ for $x \in \mathbf{C}^n$.

Definition 2.9 *Let V be a normed vector space, with norm $\|\cdot\|$. A sequence of vectors v_n in V is a* Cauchy sequence *if for every $\epsilon > 0$ there is an $N(\epsilon) > 0$ such that*

$$m, n \geq N(\epsilon) \implies \|v_m - v_n\| < \epsilon.$$

A sequence of vectors v_n in V converges *with limit $w \in V$ if for every $\epsilon > 0$ there is an $N(\epsilon) > 0$ such that*

$$n \geq N(\epsilon) \implies \|v_n - w\| < \epsilon.$$

It is a good exercise to show that a convergent sequence (i) has a unique limit and (ii) must be a Cauchy sequence.

Definition 2.10 *A normed vector space V is* complete *if every Cauchy sequence in V converges to a vector in V. A complete normed vector space is also called a* Banach space.

A closed subset of a Banach space need not be a subspace (a vector space in its own right); however, a closed subset of a Banach space is itself complete in the sense that it contains the limit of every Cauchy sequence of elements from the subset. This follows from the definition of closed set, since, by definition, a closed set contains all its accumulation points.

The most important examples of complete normed spaces for this book are the spaces $\mathbf{R}^n$ and $\mathbf{C}^n$. Their completeness depends on the completeness of the set of real numbers $\mathbf{R}$: every Cauchy sequence of real numbers converges to a real number. In contrast, the field Q of rational numbers is a vector space over Q, but there are Cauchy sequences of rationals which do not converge to a rational number. The completeness of $\mathbf{R}$ follows from the least upper bound property of the ordered field $\mathbf{R}$. For more information on the foundations of the real numbers and their completeness, see Appendix B or [7].

The next lemma states that a norm on a vector space V is a continuous function on that space.

Lemma 2.2 (Continuity of a Norm)
Let $\|\cdot\|$ be a norm on a vector space V. If x_n is a sequence in V that converges to x in V, then

$$\lim_{n\to\infty} \|x_n\| = \|\lim_{n\to\infty} x_n\| = \|x\|.$$

Proof. From the triangle inequality, we have

$$\Big| \|v\| - \|w\| \Big| \leq \|v - w\|$$

for any vectors v, w in V. Thus,

$$\Big| \, \|x_n\| - \|x\| \, \Big| \le \|x_n - x\|$$

for all n. Letting $n \to \infty$, the lemma follows. □

It is useful to know that all norms on a finite-dimensional vector space V are equivalent in the sense that they define the same notion of convergence of sequences in V. Thus, a sequence converges with respect to one norm if and only if it converges with respect to every other possible norm that might be defined on V. Formally, two norms $\|\cdot\|_a$ and $\|\cdot\|_b$ are *equivalent* if there exist numbers $\alpha > 0$ and $\beta > 0$ such that

$$\alpha\|x\|_a \le \|x\|_b \le \beta\|x\|_a \quad \text{for all } x \in V.$$

It is a good exercise to check that this defines an equivalence relation. In $\mathbf{C}^n$, $\|x\|_\infty := \max_{1\le k\le n} |x_k|$ gives a norm which satisfies

$$\max_k |x_k| \le \sqrt{|x_1|^2 + \cdots + |x_n|^2} \le \sqrt{n}\, \max_k |x_k| \; ;$$

that is,

$$\|x\|_\infty \le \|x\|_2 \le \sqrt{n}\, \|x\|_\infty \; ,$$

where $\|x\|_2$ is the Euclidean norm on $\mathbf{R}^n$. The proof of the next proposition shows that any norm $\|\cdot\| : \mathbf{R}^n \to \mathbf{R}$ is continuous in the Euclidean norm on $\mathbf{R}^n$, and the equivalence of any two norms on $\mathbf{R}^n$ follows from this fact.

Proposition 2.1 *Any two norms on* $\mathbf{R}^n$ *are equivalent.*

Proof. We show that an arbitrary norm, denoted $\|\cdot\|$, is equivalent to the Euclidean norm, $\|x\|_2 = (\sum_{i=1}^n |x_i|^2)^{\frac{1}{2}}$. Let e_i be the i-th standard basis vector. For each vector x there exist unique real numbers x_i such that $x = \sum_{i=1}^n x_i e_i$. Then

$$\|x\| \le \sum_{i=1}^n \|x_i e_i\| = \sum_{i=1}^n |x_i| \, \|e_i\| \le \|x\|_2 \sum_{i=1}^n \|e_i\| = \beta\|x\|_2,$$

where $\beta = \sum_{i=1}^n \|e_i\|$. This inequality shows that the norm $\|\cdot\|$ is continuous in the Euclidean norm, since

$$\Big| \, \|x\| - \|y\| \, \Big| \le \|x - y\| \le \beta\|x - y\|_2$$

for any two vectors x and y. Let S be the unit sphere in $\mathbf{R}^n$ defined by the Euclidean norm: $S = \{x : \|x\|_2 = 1\}$. Since S is compact, the continuous function $x \mapsto \|x\|$ achieves its minimum value on S at some point $x_0 \in S$. Then we have

$$\|x\| \ge \|x_0\| =: \alpha > 0 \quad \text{for all } x \in S.$$

Any nonzero vector x can be written in the form $x = cu$ for some $u \in S$ and $c = \|x\|_2$. It follows that

$$\|x\| = c\,\|u\| \geq c\alpha = \alpha\|x\|_2.$$

This completes the proof of the equivalence of $\|\cdot\|$ and $\|\cdot\|_2$. □

If D is a nonempty subset of $\mathbf{R}^n$, and $\|\cdot\|$ is a norm on $\mathbf{R}^n$, the distance from a point x to the set D is given by

$$\operatorname{dist}(x, D) := \inf\{\|x - a\| \,:\, a \in D\}.$$

The infimum, or greatest lower bound, exists because the set $\{\|x - a\| : a \in D\}$ is bounded below (by zero, for instance). The distance from x to D depends on the norm. A curve defined by a function $x : [0, \infty) \to \mathbf{R}^n$ *approaches* the set D if

$$\lim_{t\to\infty} \operatorname{dist}(x(t), D) = 0.$$

This concept is well defined, being independent of the norm used. If $\epsilon > 0$, we define the *ϵ-neighborhood* of a nonempty set $D \subset \mathbf{R}^n$ by

$$B_\epsilon(D) := \{y \in \mathbf{R}^n : \operatorname{dist}(y, D) < \epsilon\}.$$

For a single point $a \in \mathbf{R}^n$, we simply write $B_\epsilon(a)$ rather than $B_\epsilon(\{a\})$; thus,

$$B_\epsilon(a) := \{x \,:\, \|x - a\| < \epsilon\}$$

is the open ball of radius ϵ centered at the point a.

A *matrix norm* is a function $\|\cdot\|$ on the vector space of $n \times n$ matrices which satisfies, in addition to the properties of Definition 2.8, the property that

(d) $\|AB\| \leq \|A\|\,\|B\|$ for any two $n \times n$ matrices A and B.

A *matrix norm compatible with a given vector norm* $\|x\|$ is a matrix norm $\|\cdot\|$ on the $n \times n$ matrices which satisfies, in addition, the compatibility property

(e) $\|Ax\| \leq \|A\|\,\|x\|$ for any $n \times n$ matrix A and vector x.

The *matrix norm induced by a given vector norm* is defined as follows. Let $\|x\|$ denote a given vector norm. This norm induces a matrix norm on $\mathbf{C}^{n\times n}$, given by

$$\|A\| := \max_{\|x\|\leq 1} \|Ax\|, \tag{2.1}$$

where $\|Ax\|$ is the given vector norm of the image vector Ax. This does indeed define a norm on the space $\mathbf{C}^{n\times n}$ compatible with the given vector norm. For a matrix norm, it is straightforward to show by induction that $\|A^k\| \leq \|A\|^k$ for every positive integer k.

THE ABSOLUTE SUM NORMS. For some estimates needed later on, we choose to work with the vector norm defined by

$$\|x\| = \sum_{j=1}^{n} |x_j| = |x_1| + \cdots + |x_n|. \tag{2.2}$$

In some references this norm is denoted $\|x\|_1$, but we will not use the subscript. We will also work with the matrix norm defined by

$$\|A\| = \sum_{i,j=1}^{n} |a_{ij}|, \quad \text{where } A = [a_{ij}]. \tag{2.3}$$

We leave as an exercise the verification that (2.2) defines a vector norm and (2.3) defines a matrix norm. The matrix norm (2.3) is not induced by the absolute sum norm (2.2); however, we now show the compatibility of (2.3) with (2.2). We have

$$\begin{aligned}
\|Ax\| &= \sum_{i=1}^{n} |(Ax)_i| \\
&= \sum_{i=1}^{n} \left| \sum_{j=1}^{n} a_{ij} x_j \right| \\
&\leq \sum_{i=1}^{n} \sum_{j=1}^{n} |a_{ij}| \, |x_j|.
\end{aligned}$$

Since $|x_j| \leq \|x\|$ for each j, we have $\|Ax\| \leq \|A\| \, \|x\|$, as we wanted to show.

Finally, we note that, using the absolute sum norms for vector functions and matrix functions, for any real interval $[a, b]$ we have

$$\left\| \int_a^b f(t)\, dt \right\| \leq \int_a^b \|f(t)\|\, dt$$

for a vector function $f : [a, b] \to \mathbf{R}^n$, and

$$\left\| \int_a^b A(t)\, dt \right\| \leq \int_a^b \|A(t)\|\, dt$$

for a matrix function $A : [a, b] \to \mathbf{R}^n$. These estimates can be useful in establishing the convergence of certain improper integrals of matrix functions.

Sequences and Series of Functions

The Cauchy-Schwartz inequality states that $|v^T w| \leq \|v\|_2 \, \|w\|_2$ for any vectors $v, w \in \mathbf{R}^n$. If e_i denotes the i-th standard basis vector in $\mathbf{R}^n$, and $A = [a_{ij}]$ for $A \in \mathbf{R}^{n \times n}$, then

$$|a_{ij}| = |e_i^T A e_j| \leq \|e_i\|_2 \, \|Ae_j\|_2 \leq \|A\|_2, \tag{2.4}$$

where $\|A\|_2$ is the matrix norm induced by the Euclidean vector norm. It follows from (2.4) that every Cauchy sequence of matrices A_k in $\mathbf{R}^{n\times n}$ must converge to a matrix $A \in \mathbf{R}^{n\times n}$. This last fact is also transparent from the matrix norm in (2.3). That is, if $A_k \in \mathbf{R}^{n\times n}$ is a sequence with the property that for every $\epsilon > 0$, there exists an $N(\epsilon) > 0$ such that $m, n > N(\epsilon)$ implies $\|A_m - A_n\| < \epsilon$, then there is a matrix $A \in \mathbf{R}^{n\times n}$ such that $\lim_{k\to\infty} A_k = A$, that is, $\lim_{k\to\infty} \|A_k - A\| = 0$. Thus, with any matrix norm, the space $\mathbf{R}^{n\times n}$ is a complete normed vector space.

Definition 2.11 (Absolute Convergence)
Let V be a normed vector space with norm $\|\cdot\|$. The infinite series $\sum_{k=1}^{\infty} a_k$, $a_k \in V$, is absolutely convergent *if the series $\sum_{k=1}^{\infty} \|a_k\|$ of nonnegative real numbers converges.*

The next lemma is used later on in the discussion of the matrix exponential.

Lemma 2.3 (Absolute Convergence Implies Convergence)
Let V be a complete normed vector space. If the infinite series $\sum_{k=1}^{\infty} a_k$, $a_k \in V$, is absolutely convergent, then it converges in the norm on V to a limit $s \in V$, that is,

$$\sum_{k=1}^{\infty} a_k = \lim_{n\to\infty} \sum_{k=1}^{n} a_k = s.$$

Proof. See Exercise 2.7. □

It is useful to recall the definition of uniform convergence of sequences and series of functions. This property is important because it allows us to interchange limit processes of calculus in certain situations.

Definition 2.12 (Uniform Convergence of Sequences and Series)
Let the functions s_j, f_j, $j \geq 1$, be defined on a common domain $D \subseteq \mathbf{R}^n$, and suppose that s_j, f_j all take values in $\mathbf{R}^m$.

(a) *The sequence s_j* converges uniformly *on D to a function $s : D \to \mathbf{R}^m$ if, given any $\epsilon > 0$, there is an $N(\epsilon) > 0$ such that*

$$j \geq N(\epsilon) \implies \|s_j(x) - s(x)\| < \epsilon \quad \text{for all } x \in D.$$

(b) *The series $\sum_{j=1}^{\infty} f_j$* converges uniformly *on D to a function $f : D \to \mathbf{R}^m$ if the sequence of partial sums*

$$s_n := \sum_{j=1}^{n} f_j$$

converges uniformly on D to f.

The Weierstrass M-test is useful in showing that a series of functions converges uniformly, when the terms in the series satisfy appropriate bounds.

Lemma 2.4 (Weierstrass M-Test for Uniform Convergence)
Let the sequence of functions f_j, $j \geq 1$, be defined on a common domain D in $\mathbf{R}^n$, with common range space $\mathbf{R}^m$. Suppose that each f_j satisfies a bound of the form

$$\|f_j(x)\| \leq M_j \quad \text{for all } x \in D,$$

where the M_j are fixed numbers. If the series of the M_j converges, that is,

$$\sum_{j=1}^{\infty} M_j < \infty,$$

then the series of functions

$$\sum_{j=1}^{\infty} f_j(x)$$

converges uniformly on D.

Proof. By the boundedness hypothesis, we can invoke Lemma 2.3 and conclude that the series $\sum_{j=1}^{\infty} f_j$ converges pointwise to a limit function f which is defined on D. It remains to show that the series converges uniformly to f on D. Define the sequence S_n of partial sums of the series by

$$S_n(x) = \sum_{j=1}^{n} f_j(x), \quad x \in D.$$

We want to show that the sequence S_n converges uniformly on D, and to do so we may work with any norm on the range space $\mathbf{R}^m$. Let T_n be the sequence of partial sums of the series $\sum_{j=1}^{\infty} M_j$, and note that $T_n \geq 0$ for each n. Given $\epsilon > 0$, there is a number $N(\epsilon) > 0$ such that

$$m > n > N(\epsilon) \implies \sum_{j=n+1}^{m} M_j = T_m - T_n < \frac{\epsilon}{2}.$$

Thus, for $m > n > N(\epsilon)$ and all $x \in D$, the partial sums $S_n(x)$ satisfy

$$\begin{aligned}
\|S_m(x) - S_n(x)\| &= \left\| \sum_{j=n+1}^{m} f_j(x) \right\| \\
&\leq \sum_{j=n+1}^{m} \|f_j(x)\| \\
&\leq \sum_{j=n+1}^{m} M_j \\
&< \frac{\epsilon}{2}.
\end{aligned}$$

Fix $x \in D$ and let $m \to \infty$. By the continuity of the norm (Lemma 2.2), for any fixed $n > N(\epsilon)$ we have

$$\lim_{m\to\infty} \|S_m(x) - S_n(x)\| = \|f(x) - S_n(x)\| \le \frac{\epsilon}{2} < \epsilon.$$

Since x in D was fixed but arbitrary, we conclude that if $n > N(\epsilon)$, then $\|f(x) - S_n(x)\| < \epsilon$ for all $x \in D$. Thus the sequence S_n converges uniformly to f on D. This completes the proof. □

Quadratic Forms

The Euclidean norm $\|x\|_2$ is associated with the quadratic form $x_1^2 + \cdots + x_n^2$, because

$$\|x\|_2 = (x_1^2 + \cdots + x_n^2)^{1/2}.$$

Vector norms defined by more general positive definite quadratic forms are very convenient in discussions of stability of equilibria.

Definition 2.13 *Let P be a symmetric matrix in $\mathbf{R}^{n\times n}$, that is, $P^T = P$.*

(a) *P is* positive definite *if $x^T Px > 0$ for every nonzero vector x in $\mathbf{R}^n$.*
(b) *P is* positive semidefinite *if $x^T Px \ge 0$ for every x in $\mathbf{R}^n$.*
(c) *P is* negative definite *if $x^T Px < 0$ for every nonzero vector x in $\mathbf{R}^n$.*
(d) *P is* negative semidefinite *if $x^T Px \le 0$ for every x in $\mathbf{R}^n$.*
(e) *P is* indefinite *if none of the conditions (a)–(d) hold.*

Note that a matrix Q is negative definite if and only if $-Q$ is positive definite, and that Q is negative semidefinite if and only if $-Q$ is positive semidefinite.

Suppose P is symmetric positive definite, and define

$$\|x\|_P := (x^T Px)^{\frac{1}{2}}.$$

It is straightforward to show that this defines a norm on $\mathbf{R}^n$. In fact, $B(u, v) = u^T Pv$ defines an inner product on $\mathbf{R}^n$. There is no loss of generality in specifying symmetric matrices in Definition 2.13, when defining quadratic forms; see Exercise 2.4.

Recall that every real symmetric matrix P can be diagonalized by a real *orthogonal* matrix S, that is $S^T S = I$, hence $S^T = S^{-1}$, and

$$S^T PS = \text{diag}\,[\lambda_1, \dots, \lambda_n],$$

where λ_i, $1 \le i \le n$, are the eigenvalues of P. (See [75] (page 549).) The diagonalizing transformation S is given by $S = [v_1 \cdots v_n]$, where $\{v_1, \dots, v_n\}$ is an *orthornormal basis* of $\mathbf{R}^n$ consisting of eigenvectors of P. By definition of orthonormal basis, $v_1, \dots, v_n$ are linearly independent and satisfy $\|v_i\|_2^2 = v_i^T v_i = 1$, $1 \le i \le n$, and $v_i^T v_j = 0$ for $i \ne j$.

There are two criteria for positive definiteness that are useful to remember, especially when dealing with small size matrices. Recall that the *leading principal minors* of an $n \times n$ matrix P are the determinants of the $k \times k$ upper left submatrices of P, for $k = 1, \ldots, n$.

Proposition 2.2 *Let P be a real symmetric matrix. The following are equivalent:*

(a) *P is positive definite.*
(b) *All eigenvalues of P are positive.*
(c) *All leading principal minors of P are positive.*

Proof. We prove only the equivalence of (a) and (b). There is a real orthogonal S such that $S^{-1}PS = \text{diag}\,[\lambda_1, \ldots, \lambda_n]$, with the eigenvalues of P, which are necessarily real, on the main diagonal. If $x = Sz$, then the quadratic form $x^T Px$ is, in z coordinates, $\sum_k \lambda_k z_k^2$, and the equivalence of (a) and (b) follows. A proof of the equivalence of (a) and (c) may be found in [75] (pages 558–559), or in K. Hoffman and R. Kunze, *Linear Algebra*, Prentice-Hall, Englewood Cliffs, NJ, second edition, 1971 (pages 328–329). □

Example 2.6 Consider the matrix

$$P = \begin{bmatrix} 3 & 2 & 0 \\ 2 & 4 & -2 \\ 0 & -2 & 5 \end{bmatrix}.$$

The three principal minors of P are

$$\det[3] = 3 > 0, \quad \det \begin{bmatrix} 3 & 2 \\ 2 & 4 \end{bmatrix} = 8 > 0, \quad \det P = 3(16) - 2(10) = 28 > 0,$$

so P is positive definite. On the other hand, the matrix

$$Q = \begin{bmatrix} 3 & 0 & 0 \\ 0 & -4 & 2 \\ 0 & 2 & -5 \end{bmatrix}$$

has a negative 2×2 principal minor, so Q is not positive definite. Notice that Q cannot be negative definite either, because of the first principal minor. Q is indefinite. As an alternative argument, we might observe that Q has eigenvalues 3, λ_2, and λ_3, with $\lambda_2 \lambda_3 = 16$ and $\lambda_2 + \lambda_3 = -9$. Therefore the symmetric Q must have a negative eigenvalue, and therefore cannot be positive definite. △

It is useful to know that a real symmetric matrix P is positive semidefinite if and only if all eigenvalues of P are nonnegative. Suppose that R is a real $n \times n$ matrix. Then $R^T R$ is symmetric positive semidefinite: For every real vector x, $x^T R^T Rx = \|Rx\|^2 \geq 0$. The next lemma provides a converse.

Lemma 2.5 *Let $Q \in \mathbf{R}^{n\times n}$ be symmetric positive semidefinite. Then there exists a positive semidefinite matrix $R \in \mathbf{R}^{n\times n}$ such that*

$$Q = R^T R.$$

If Q is positive definite, then R can be chosen positive definite.

Proof. Since Q is symmetric, it is diagonalizable by a real orthogonal matrix S. Write $S^T Q S = D = \operatorname{diag}[\lambda_1, \ldots, \lambda_n]$, and define

$$R = S \operatorname{diag}[\sqrt{\lambda_1}, \ldots, \sqrt{\lambda_n}]\, S^T =: S D^{\frac{1}{2}} S^T.$$

Then R is symmetric and

$$\begin{aligned} R^T R &= (S D^{\frac{1}{2}} S^T)(S D^{\frac{1}{2}} S^T) \\ &= S D S^T \\ &= Q. \end{aligned}$$

Clearly, $z^T R z \geq 0$ for all vectors z, so R is positive semidefinite. By this construction, it is clear that if Q is positive definite then so is R. □

A matrix $Q \geq 0$ can have other factorizations $Q = C^T C$, where C need not be positive semidefinite. For example, consider the factorization

$$Q = \begin{bmatrix} 1 & 0 \\ 0 & 0 \end{bmatrix} = \begin{bmatrix} 1 \\ 0 \end{bmatrix} \begin{bmatrix} 1 & 0 \end{bmatrix} =: C^T C.$$

2.4 ORDINARY DIFFERENTIAL EQUATIONS

In this section we state the basic theorem on the existence and uniqueness of solutions of initial value problems for ordinary differential equations and give examples to illustrate the theorem. We define the Jacobian linearization of a system at a point. The section concludes with examples of linear and nonlinear systems in the plane.

Existence and Uniqueness Theorem

Consider the initial value problem for a system of ordinary differential equations,

$$\dot{x}(t) = f(t, x(t)), \quad x(t_0) = x_0, \tag{2.5}$$

where $f : D \mapsto \mathbf{R}^n$ is a C^1 (continuously differentiable) vector field defined on an open set $D \subset \mathbf{R}^{n+1}$ and $\dot{x} = \frac{dx}{dt}$. A *solution* is a differentiable function $x(t)$ that satisfies (2.5) on some real interval J containing t_0. We say that system (2.5) is *autonomous* or *time invariant* if the vector field f does not depend explicitly on time t; otherwise the system is nonautonomous (time varying). For autonomous systems we usually take the initial time to be $t_0 = 0$.

In order to guarantee that initial-value problems have a unique solution, some local growth restriction must be imposed on the vector field f in (2.5).

Definition 2.14 (Locally Lipschitz Vector Field)
Let D be an open set in $\mathbf{R}^{n+1}$. A function $f : D \mapsto \mathbf{R}^n$, denoted $f(t,x)$ with $t \in \mathbf{R}$ and $x \in \mathbf{R}^n$, is locally Lipschitz *in x on D if for any point $(t_0, x_0) \in D$, there are an open ball $B_r(t_0, x_0)$ about (t_0, x_0) and a number L such that*

$$\|f(t, x_1) - f(t, x_2)\| \leq L\|x_1 - x_2\|$$

for all (t, x_1), (t, x_2) in $B_r(t_0, x_0) \cap D$.

If f has continuous first order partial derivatives with respect to all components of x at every point of $D \subseteq \mathbf{R}^{n+1}$, then f is locally Lipschitz in x on D. (See Theorem B.5 and the discussion that follows it.)

Theorem 2.3 (Existence and Uniqueness)
Let D be an open set in $\mathbf{R}^{n+1}$. If $f : D \mapsto \mathbf{R}^n$ is locally Lipschitz on D, then, given any $x_0 \in D$ and any $t_0 \in \mathbf{R}$, there exists a $\delta > 0$ such that the initial value problem (2.5) *has a unique solution $x(t, t_0, x_0)$ defined on the interval $[t_0 - \delta, t_0 + \delta]$; that is, if $z(t) := x(t, t_0, x_0)$, then*

$$\frac{d}{dt} z(t) = f(t, z(t)), \quad \text{for } t \in [t_0 - \delta, t_0 + \delta],$$

and $z(t_0) = x(t_0, t_0, x_0) = x_0$.

Proof. See Appendix C on ordinary differential equations. □

Solutions of locally Lipschitz systems can always be extended to a maximal interval of existence $(t_{\min}(x_0), t_{\max}(x_0))$, which depends on x_0 and f. In some cases, a final extension may be made to include one or both endpoints $t_{\min}(x_0)$, $t_{\max}(x_0)$, when these are finite numbers. A solution will normally mean the unique solution determined by an initial condition, with domain given by the maximal interval assured by the extension of solutions. (See the discussion of extension of solutions and Theorem C.3 in the Appendix.) If all solutions exist for all forward times $t \geq 0$, then the system is said to be *forward complete*. If all solutions are defined for $t \in (-\infty, \infty)$, then the system is *complete*.

We consider a few scalar autonomous differential equations in order to illustrate Theorem 2.3. We often write $\phi_t(x_0) = x(t, t_0, x_0)$ for the solution of an initial value problem.

Example 2.7 The linear initial value problem $\dot{x} = ax$, $x(0) = x_0 \in \mathbf{R}$, has unique solution $x(t) = e^{at}x_0$ defined on the whole real line. △

Example 2.8 Consider the initial value problem $\dot{x} = x^2$, $x(0) = x_0 \in \mathbf{R}$. After separating the variables, a direct integration gives the following unique and maximally defined solutions:

$$\phi_t(x_0) = \frac{x_0}{1 - x_0 t} \quad \text{for} \quad -\infty < t < \frac{1}{x_0} \quad \text{if} x_0 > 0,$$

and

$$\phi_t(x_0) = \frac{x_0}{1 - x_0 t} \quad \text{for} \quad \frac{1}{x_0} < t < \infty \quad \text{if } x_0 < 0,$$

as well as $x(t) \equiv 0$ (for all real t) if $x(0) = 0$. In this case, we always have $\dot{x} > 0$ for $x_0 \neq 0$, and the vector field grows rapidly enough as $|x|$ increases that all nonzero solutions exhibit a finite escape time, either in positive time or negative time. Therefore the differential equation is not complete. $\triangle$

Example 2.9 The equation $\dot{x} = x^{\frac{3}{2}}$ has the constant solution $x = 0$, and real nonconstant solutions only for $x_0 > 0$. We take $t_0 = 0$ and write $x(0) = x_0$. The general solution is given by

$$\phi_t(x_0) = \left(\frac{2\sqrt{x_0}}{2 - \sqrt{x_0}\, t} \right)^2,$$

and the maximal interval of existence, given $x_0 > 0$, is $(-\infty, \frac{2}{\sqrt{x_0}})$. This equation is not forward complete. The equation is defined on the *open* set $D = \{x : x > 0\}$, but the initial value problem with $x_0 = 0$ has the equilibrium solution $x(t) \equiv 0$ as well. $\triangle$

Example 2.10 The equation $\dot{x} = x^{\frac{1}{3}}$ has a solution for any real $x_0 = x(0)$. For $x_0 \neq 0$, separation of variables yields the unique solution

$$\phi_t(x_0) = \left(\frac{2}{3}t + x_0^{\frac{2}{3}} \right)^{\frac{3}{2}}.$$

Note that $f(x) = x^{\frac{1}{3}}$ is locally Lipschitz for $x \neq 0$. When $x_0 = 0$, it is easy to check that the preceding formula does give a solution $(\frac{2}{3}t)^{\frac{3}{2}}$ of the initial value problem; however, it is not the only solution when $x_0 = 0$. There are infinitely many solutions of the initial value problem when $x_0 = 0$. In fact, for any $\sigma > 0$, the function

$$x(t) = \begin{cases} 0 & \text{if } t < \sigma \\ (\frac{2}{3}(t - \sigma))^{\frac{3}{2}} & \text{if } t \geq \sigma \end{cases},$$

satisfies the differential equation and the initial condition $x(0) = 0$. Note that $f(x) = x^{\frac{1}{3}}$ does not satisfy a Lipschitz condition in any neighborhood of $x = 0$. (See Exercise 2.9.) $\triangle$

Extension of Solutions: An Example

For stability studies it is important to determine if the solutions of a system of differential equations exist for all $t \geq 0$. Sometimes we wish to determine this only for initial conditions in some subset of the domain. As an example, we consider the following application of Theorem C.3 on the extension of solutions.

Example 2.11 A model of the population dynamics of two species in competitive interaction is given by the equations

$$\dot{x}_1 = x_1(3 - x_1 - 2x_2),$$
$$\dot{x}_2 = x_2(4 - 3x_1 - x_2).$$

This system is defined on the whole plane, although the interest of the system as a model of population dynamics means that the closed first quadrant is of primary interest. It is straightforward to check that each nonnegative axis is forward invariant. Consequently, the entire first quadrant, including the nonnegative coordinate axes, is forward invariant. (By uniqueness of solutions, no initial condition in the open first quadrant corresponds to a trajectory that intersects a coordinate axis; moreover, continuity of solutions rules out jumping an axis.) There are four equilibrium points, at $A = (0,0)$, $B = (3,0)$, $C = (1,1)$, and $D = (0,4)$. Solutions on a positive axis with sufficiently large initial condition must decrease along that axis in forward time. Consider any closed square region K_s in the first quadrant, with one vertex at the origin, and with sides on the axes and side length $s > 4$. Then the vector field points inward on the boundary segments of K_s lying in the interior of the first quadrant. Thus, K_s is an invariant region, meaning that any trajectory that starts at $t = 0$ within the square K_s must remain in K_s as long as it is defined. (See Figure 2.1.) Since K_s is a compact invariant region, Theorem C.3 implies that any such trajectory is extendable to all $t \geq 0$. Since this is true for a square of any side length $s > 4$, Theorem C.3 implies the forward completeness of this system on the entire closed first quadrant. △

Jacobian Linearization of a Vector Field

Suppose we want to study the local behavior of solutions of the system $\dot{x} = f(x)$ near a point $x = a$. One of the main tools for such a study is the Jacobian linearization of the system at $x = a$.

Definition 2.15 *Let D be an open set in $\mathbf{R}^n$, $a \in D$, and $f : D \to \mathbf{R}^n$ a C^1 vector field. The Jacobian linearization of the system $\dot{x} = f(x)$ at the point $x = a$ is the linear system*

$$\dot{z} = Df(a)z,$$

where $Df(a)$ is the Jacobian matrix of f at $x = a$, defined by

$$Df(a) := \begin{bmatrix} \dfrac{\partial f_1}{\partial x_1}(a) & \cdots & \dfrac{\partial f_1}{\partial x_n}(a) \\ \vdots & & \vdots \\ \dfrac{\partial f_n}{\partial x_1}(a) & \cdots & \dfrac{\partial f_n}{\partial x_n}(a) \end{bmatrix}.$$

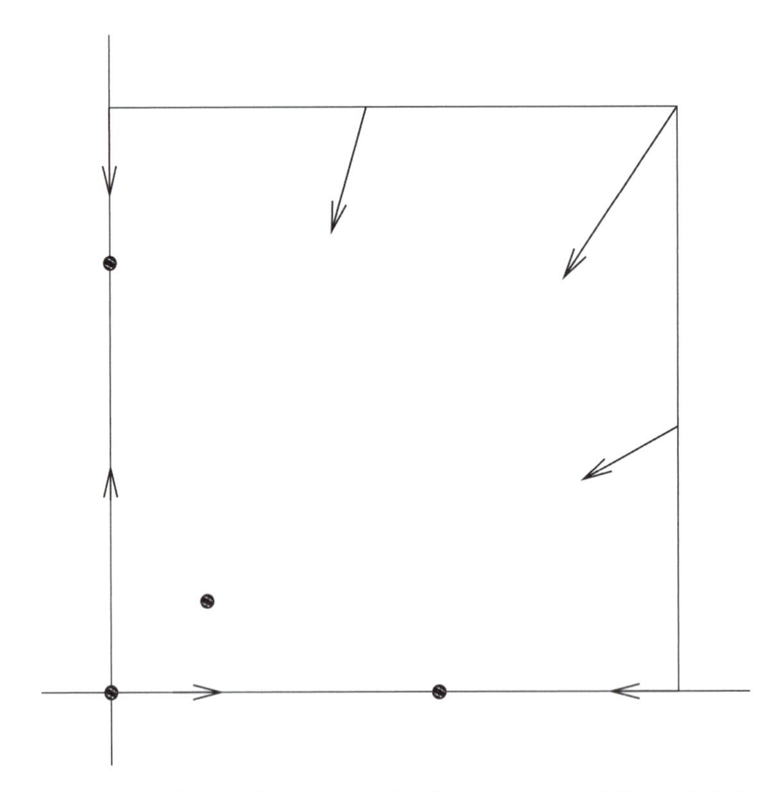

Figure 2.1 A forward invariant region for the system of Example 2.11.

We think of the variable z in the linearization as $z = x - a$ in the Taylor expansion of f about $x = a$. Of course, the Taylor expansion begins with $\dot{x} = f(a) + Df(a)(x-a) + \cdots$, and the terms $Df(a)z = Df(a)(x-a)$ give only the first order terms (the linear terms) in the Taylor expansion centered at a. These linear terms may carry a great deal of information about the local behavior of f in a neighborhood of $z = 0$, that is, in a neighborhood of $x = a$. In particular, information about eigenvalues of $Df(a)$ allows us in many (but not all) cases to deduce local stability properties of an equilibrium at $x = a$ for the nonlinear system $\dot{x} = f(x)$. Note that there is an equilibrium (that is, constant) solution given by $x \equiv a$ if and only if $f(a) = 0$.

As an example, consider again the vector field in Example 2.11.

Example 2.12 The function

$$f(x_1, x_2) = \begin{bmatrix} x_1(3 - x_1 - 2x_2) \\ x_2(4 - 3x_1 - x_2) \end{bmatrix}$$

has Jacobian matrix

$$Df(a) = \begin{bmatrix} \dfrac{\partial f_1}{\partial x_1} & \dfrac{\partial f_1}{\partial x_2} \\ \dfrac{\partial f_2}{\partial x_1} & \dfrac{\partial f_2}{\partial x_2} \end{bmatrix}_{x=a} = \begin{bmatrix} 3 - 2x_1 - 2x_2 & -2x_1 \\ -3x_2 & 4 - 3x_1 - 2x_2 \end{bmatrix}_{x=a}.$$

In particular, the Jacobian linearization at the equilibrium point $a = (1, 1)$ is given by

$$\dot{z} = Df(a)z = \begin{bmatrix} -1 & -2 \\ -3 & -1 \end{bmatrix} z. \qquad \triangle$$

No confusion should arise if we abuse the notation slightly by using the same state variable x for the system linearization as for the original system; thus, we may write $\dot{x} = Df(a)x$ for the Jacobian linearization of $\dot{x} = f(x)$ at the point a.

2.4.1 Phase Plane Examples: Linear and Nonlinear

Examples like the ones discussed here are probably familiar from elementary differential equations. These examples illustrate some of the ideas discussed earlier in this chapter, and they are used later on to help motivate stability definitions.

REMARKS ON STANDARD NOTATION. We often compute the quantity

$$(\nabla V(x))^T f(x) = \nabla V(x) \cdot f(x)$$

when we differentiate a real-valued function $V(x)$ along a solution of the system $\dot{x} = f(x)$. More precisely, if $x(t)$ is the solution with initial condition $x(0) = x$, then

$$\frac{dV}{dt}(x(t))\Big|_{t=0} = (\nabla V(x))^T f(x)$$

by the chain rule and evaluation at $t = 0$. It is important to realize that the right-hand side depends only on the point x; at each x, it gives the directional derivative of V in the direction of the vector $f(x)$. There are two other standard and accepted notations for the same quantity:

$$\dot{V}(x) := (\nabla V(x))^T f(x) =: L_f V(x).$$

While any of these may be used, and the reader should be familiar with them, some caution is advised at the beginning. The notation $\dot{V}(x)$ carries an implicit understanding of the vector field f, and despite the dot notation it does not depend explicitly on t when f is autonomous. The notation $L_f V(x)$ can be read as the *Lie derivative of V with respect to f*. The Lie derivative notation has some advantages later on when certain higher order derivatives are needed.

LINEAR SYSTEMS: TWO-DIMENSIONAL EXAMPLES. We begin with an example which is easy to understand geometrically.

Example 2.13 Consider the system

$$\begin{bmatrix} \dot{x}_1 \\ \dot{x}_2 \end{bmatrix} = \begin{bmatrix} 0 & 9 \\ -1 & 0 \end{bmatrix} \begin{bmatrix} x_1 \\ x_2 \end{bmatrix} =: A \begin{bmatrix} x_1 \\ x_2 \end{bmatrix}.$$

The characteristic equation of the coefficient matrix is $\lambda^2 + 9 = 0$, so the eigenvalues are $\lambda = \pm 3i$. These equations imply that $\frac{dx_2}{dx_1} = -\frac{x_1}{9x_2}$, which is separable, and a direct integration shows that solutions must satisfy the constraint

$$\frac{x_1^2}{9} + x_2^2 = \text{constant}.$$

The solution trajectories form a family of ellipses surrounding the origin, corresponding to periodic solutions. (One can argue that each solution may be extended to the time domain $(-\infty, \infty)$; moreover, there is a lower bound on the speed at which the curve is traversed.) A simple sketch shows that the Euclidean norm $\|x(t)\|$ is an increasing function of t over part of the orbit, and it is a decreasing function of t over part of the orbit. (See Figure 2.2.) We compute that

$$\frac{d}{dt}\|x\|^2 = \frac{d}{dt}x^T x = (Ax)^T x + x^T Ax = x^T(A^T + A)x = x^T \begin{bmatrix} 0 & 8 \\ 8 & 0 \end{bmatrix} x,$$

and observe that $A^T + A$ is neither negative definite nor positive definite; it is indefinite. On the other hand, if we use the norm defined by

$$\|x\|_P^2 = x^T P x := x^T \begin{bmatrix} \frac{1}{9} & 0 \\ 0 & 1 \end{bmatrix} x = \frac{1}{9}x_1^2 + x_2^2,$$

then we have

$$\frac{d}{dt}\|x\|_P^2 = \frac{2}{9}x_1\dot{x}_1 + 2x_2\dot{x}_2 = 2x_1x_2 - 2x_2x_1 = 0.$$

This is not too surprising, since it is a restatement of the fact that each solution traces out an ellipse within the family indicated above. If we think of $\|x\|_P$ as a measure of the energy of the system in the state x, then the $\|\cdot\|_P$ energy is conserved. △

The system in Example 2.13 is a two-dimensional *Hamiltonian system*. By definition, this means that the system has the form

$$\dot{x}_1 = \frac{\partial H}{\partial x_2}(x_1, x_2),$$
$$\dot{x}_2 = -\frac{\partial H}{\partial x_1}(x_1, x_2) \ ,$$

where $H(x_1, x_2)$ is a continuously differentiable real-valued function, called the *Hamiltonian*, or *Hamiltonian energy function*. For Example 2.13, it is

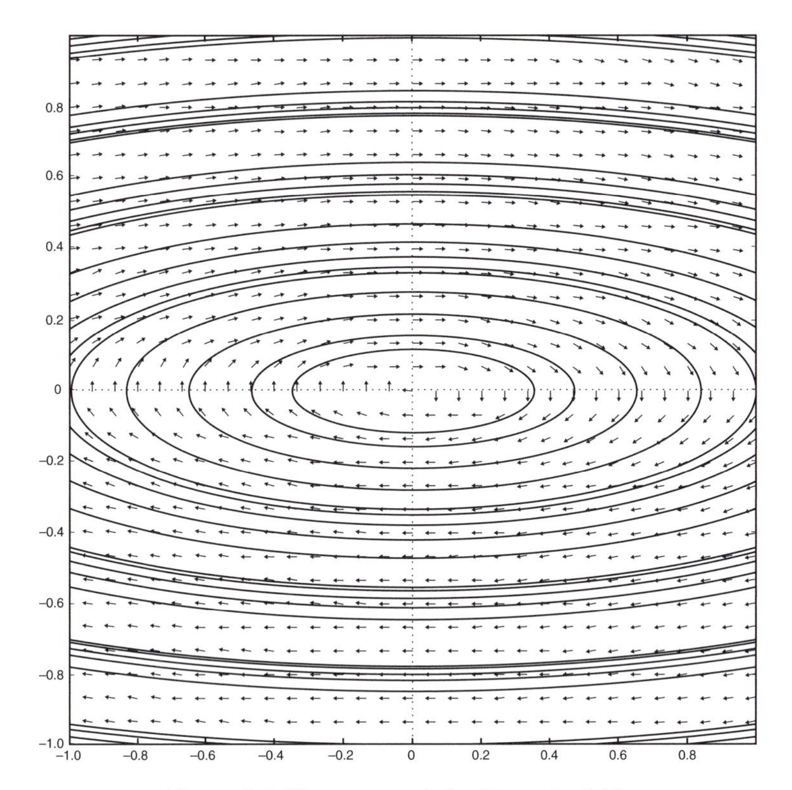

Figure 2.2 Phase portrait for Example 2.13.

straightforward to check that we may take

$$H(x) = H(x_1, x_2) = \frac{1}{2}x_1^2 + \frac{9}{2}x_2^2 = \frac{9}{2}\|x\|_P^2$$

as the Hamiltonian. The (Hamiltonian) energy function is constant along each solution of a Hamiltonian system. To see this, let $(x_1(t), x_2(t))$ be a solution, and use the chain rule to compute that

$$\frac{dH}{dt}(x_1, x_2) = \frac{\partial H}{\partial x_1}(x_1, x_2)\frac{\partial H}{\partial x_2}(x_1, x_2) + \frac{\partial H}{\partial x_2}(x_1, x_2)\left(-\frac{\partial H}{\partial x_1}(x_1, x_2)\right) = 0.$$

This identity in t implies that $H(x_1(t), x_2(t)) = H(x_1(0), x_2(0))$ for all t for which the solution is defined. The value of the energy is determined by the initial conditions.

Example 2.14 Consider the system

$$\dot{x} = \begin{bmatrix} -1 & 9 \\ -1 & -1 \end{bmatrix} x =: Ax.$$

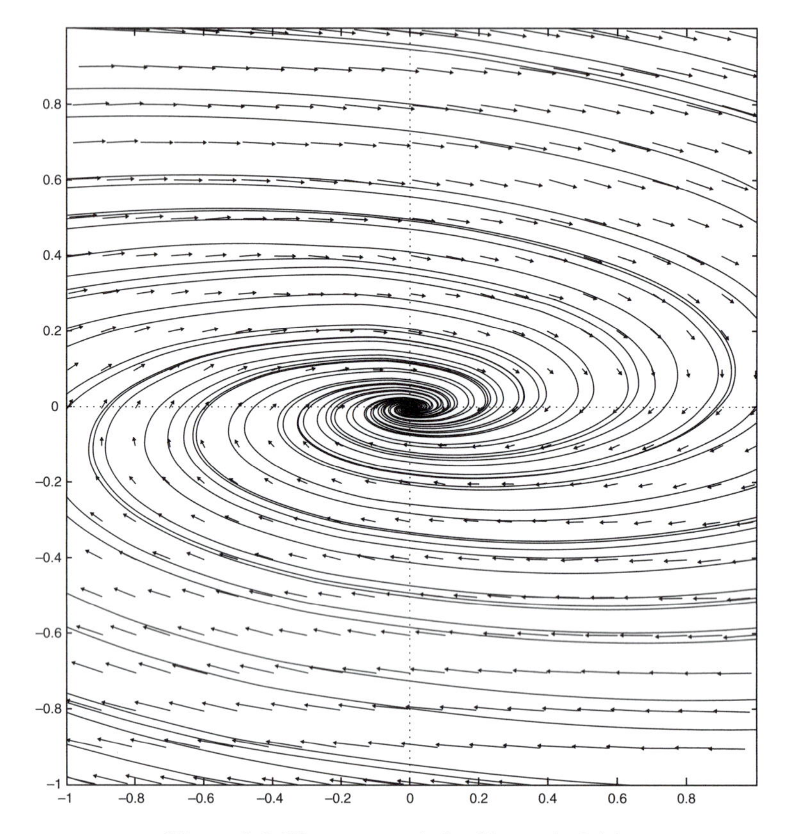

Figure 2.3 Phase portrait for Example 2.14.

The characteristic equation of A is $(\lambda + 1)^2 + 9 = 0$, so the eigenvalues are $\lambda = -1 \pm 3i$. As in Example 2.13, suppose we calculate

$$A^T + A = \begin{bmatrix} -2 & 8 \\ 8 & -2 \end{bmatrix}.$$

The eigenvalues of $A^T + A$ are 6 and -10, so $A^T + A$ is indefinite. We conclude that along a nonzero solution $x(t)$ the Euclidean norm is sometimes decreasing and sometimes increasing. But a phase portrait suggests that all solutions approach the origin as $t \to \infty$. (See Figure 2.3.) For a definite conclusion, without knowledge of later theorems, one could diagonalize A by a nonsingular complex matrix S, such that

$$S^{-1}AS = \begin{bmatrix} -1 + 3i & 0 \\ 0 & -1 - 3i \end{bmatrix},$$

and argue that all complex-valued solutions approach the origin as $t \to \infty$. Note that this is so precisely because the eigenvalues $\lambda = -1 \pm 3i$ have negative real part. Given this understanding, it is natural to expect that there must be some norm, or *energy measure*, which decreases along real solution trajectories as t increases. Not all norms have the form $\|x\|_P = (x^T P x)^{\frac{1}{2}}$ for a positive definite P, and not all norms will be decreasing along nonzero solutions, but it is possible to find such a norm here. Note that

$$\frac{d}{dt}(x^T P x) = x^T(A^T P + PA)x < 0 \quad \text{for } x \neq 0$$

if and only if $A^T P + PA$ is negative definite. Define

$$P = \begin{bmatrix} 3/5 & 1/5 \\ 1/5 & 23/10 \end{bmatrix}, \qquad \|x\|_P^2 = x^T P x.$$

It is straightforward to verify that $A^T P + PA = -I$. Thus we have

$$\frac{d}{dt}\|x\|_P^2 = x^T(A^T P + PA)x = -x^T x < 0 \quad \text{for } x \neq 0.$$

Therefore $\|x(t)\|_P$ strictly decreases along nonzero solutions. It can be shown that $\|x(t)\|_P \to 0$ as $t \to \infty$, and consequently $x(t)$ approaches zero. A precise argument for this convergence could be an interesting exercise here; however, a detailed argument is given later in Theorem 8.1. $\triangle$

The previous examples involved energy-like functions to help track the solutions, in the sense that a decrease in energy means that the solutions are crossing the level curves of the energy function in a certain direction. When such energy functions are used to study stability properties of equilibria, then (at least in the context of this book) they will generally be continuously differentiable (C^1) and positive-valued except at the equilibrium, where, like a norm, they have the value zero. These energy functions are called *Lyapunov functions*; a formal definition appears later. Sometimes the assumption of positive definiteness of an energy function can be relaxed. Useful energy functions may have a strict decrease along solutions, as in Example 2.14, or not, as in Example 2.13.

For two-dimensional linear systems, conclusions about asymptotic behavior might be verified, at least in some cases, by explicit knowledge of solution formulas. For nonlinear systems, however, the study of asymptotic behavior usually cannot rely on explicit solution formulas. The study of stability of equilibria in the nonlinear case often involves the search for an energy function that is nonincreasing along solutions.

Before giving some nonlinear examples, we summarize linear planar systems by giving an *algebraic catalog* of them that complements the *geometric* phase plane catalog with which many readers are familiar.

Example 2.15 (Algebraic Catalog of Linear Planar Systems)
Consider a two-dimensional linear system $\dot{x} = Ax$ with real coefficient matrix

$$A = \begin{bmatrix} a & b \\ c & d \end{bmatrix}.$$

The characteristic equation is $\lambda^2 - (a+d)\lambda + (ad - bc) = 0$. Recall the definitions of the trace of A, $\operatorname{tr} A$, and the determinant of A, $\det A$:

$$\operatorname{tr} A := a + d, \quad \det A := ad - bc.$$

If we write the roots of the characteristic equation (which are the eigenvalues of A) as λ_1 and λ_2, then the characteristic polynomial is

$$(\lambda - \lambda_1)(\lambda - \lambda_2) = \lambda^2 - (\lambda_1 + \lambda_2)\lambda + \lambda_1\lambda_2,$$

and we conclude that

$$\operatorname{tr} A = a + d = \lambda_1 + \lambda_2 \quad \text{and} \quad \det A = ad - bc = \lambda_1\lambda_2.$$

The algebraic catalog is described in terms of the trace and determinant of A, so we simply write the characteristic equation of A as

$$\lambda^2 - \operatorname{tr} A\lambda + \det A = 0.$$

The eigenvalues are given by

$$\lambda = \frac{1}{2}\operatorname{tr} A \pm \frac{1}{2}\sqrt{(\operatorname{tr} A)^2 - 4\det A}. \tag{2.6}$$

In the classification that follows, we use certain standard names in *italics* for the phase portrait classification of the equilibrium at the origin. These names can be taken as definitions here, although we assume the reader has some familiarity with this terminology. These names are often used to describe the whole phase portrait as well as the type of the equilibrium $x = 0$ itself. The algebraic catalog follows, by direct reference to (2.6):

(a) If $\det A < 0$, then the eigenvalues of A are real and have opposite sign, hence the equilibrium $x = 0$ is a *saddle* point.

(b) If $\det A = 0$, (but $A \neq 0$), then at least one of the eigenvalues of A is zero, and there is a line of equilibrium points including the origin. The full range of geometric phase portraits can be described using the two possible real Jordan forms for A, displayed in these systems:

$$\dot{x} = \begin{bmatrix} 0 & 0 \\ 0 & \lambda \end{bmatrix} x \quad \text{and} \quad \dot{x} = \begin{bmatrix} 0 & 1 \\ 0 & 0 \end{bmatrix} x.$$

(c) If $\det A > 0$, then the classification proceeds according to the sign of the *discriminant* $D := (\operatorname{tr} A)^2 - 4\det A$. If $D \geq 0$, then the eigenvalues are real and have the same sign, and we have a *node* at the origin. The special case $D = 0$ corresponds to a repeated real eigenvalue. If $D < 0$, then the eigenvalues are complex conjugates, and we have a *spiral* point at the origin; in these phase portraits, solution curves spiral inward toward the origin as t increases if and only if the real part of the eigenvalues is negative. These cases (nodes and spirals) are distinguished dynamically by the sign of $\operatorname{tr} A$. Observe that $\operatorname{tr} A < 0$

implies that the solutions are asymptotic to the origin as $t \to \infty$, $\operatorname{tr} A > 0$ implies that they are asymptotic to the origin as $t \to -\infty$, while $\operatorname{tr} A = 0$ implies a *center* point at $x = 0$. For example, the elliptic trajectories in Example 2.13 surround the center at $x = 0$.

A comparison of this algebraic catalog with the geometric phase portrait catalog in a favorite differential equations book can be a useful exercise and review. △

NONLINEAR SYSTEMS: TWO-DIMENSIONAL EXAMPLES. Here are two examples of nonlinear systems with energy functions that help to reveal asymptotic solution behavior. The purpose of these examples at this point is the same as the purpose of the linear system examples given above, and it is twofold: first, to indicate some models assumed as background from a differential equations course, and second, to motivate the need for the general stability theorems given later on.

Example 2.16 Consider the motion of a pendulum arm of length l, suspended vertically, with attached mass m subject only to a constant gravitational force g; that is, there is no damping force. Suppose we consider the motion of small oscillations with the mass near the natural (vertical down) equilibrium position. Write y for the angular displacement from the equilibrium at $y = 0$. The situation is modeled by the differential equation

$$\ddot{y} + \frac{g}{l} \sin y = 0.$$

Writing $x_1 = y$ and $x_2 = \dot{y}$, the equivalent first order system is the Hamiltonian system

$$\begin{aligned} \dot{x}_1 &= x_2 = \frac{\partial H}{\partial x_2}(x_1, x_2), \\ \dot{x}_2 &= -\frac{g}{l} \sin x_1 = -\frac{\partial H}{\partial x_1}(x_1, x_2), \end{aligned}$$

where the Hamiltonian energy function H is given by

$$H(x_1, x_2) = \frac{1}{2} x_2^2 + \frac{g}{l}(1 - \cos x_1).$$

The function $H(x_1, x_2)$ is the sum of a kinetic energy term that depends on velocity x_2 and a potential energy term that depends on position x_1. Note that $H(x_1, x_2) > 0$ for nonzero (x_1, x_2) in a neighborhood of $(0, 0)$, and $H(0, 0) = 0$; we say that H is positive definite in a neighborhood of the origin. Along any solution (x_1, x_2), the rate of change of energy with respect to t is given by

$$\frac{d}{dt} H(x_1, x_2) = \frac{\partial H}{\partial x_1}(x_1, x_2)\dot{x}_1 + \frac{\partial H}{\partial x_2}(x_1, x_2)\dot{x}_2 \equiv 0,$$

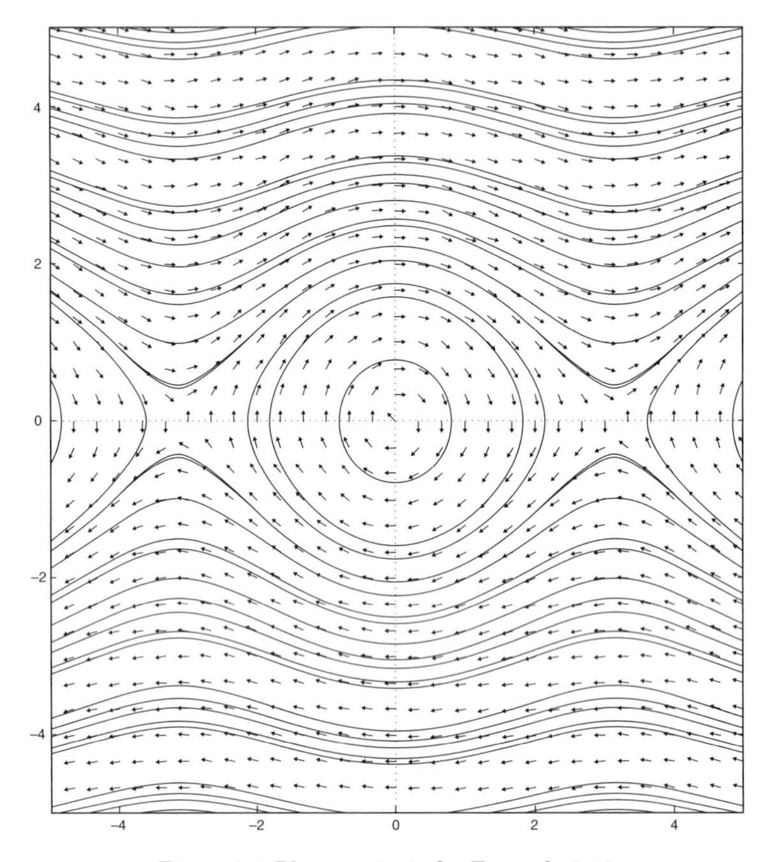

Figure 2.4 Phase portrait for Example 2.16.

an identity in t. Thus, solutions that start within the level set defined by $H = h$, for sufficiently small $h > 0$, must remain in that level set for as long as the solution is defined, in this case for $-\infty < t < \infty$. Thus, the solutions are periodic. For sufficiently small $h > 0$, the level set $H = h$ is a closed curve surrounding the origin. (See Figure 2.4.) In the absence of damping, an initial condition (x_{10}, x_{20}), having a small displacement $\|(x_{10}, x_{20})\|$ from the equilibrium at $(0, 0)$, determines an oscillatory motion of small amplitude (norm) about the origin. △

We now consider an example with damping, or friction, that dissipates energy.

Example 2.17 Suppose that we include in the pendulum model of Example 2.16 a damping force proportional to the velocity. Then the equation is

$$\ddot{y} + c\dot{y} + \frac{g}{l}\sin y = 0,$$

where c is constant. The equivalent system for $x_1 = y$ and $x_2 = \dot{y}$ is

$$\begin{aligned}\dot{x}_1 &= x_2, \\ \dot{x}_2 &= -\frac{g}{l}\sin x_1 - c\,x_2.\end{aligned}$$

Using the same total energy function as before (we call it $V(x_1, x_2)$ this time, because now it is not Hamiltonian energy), we have

$$V(x_1, x_2) = \frac{1}{2}x_2^2 + \frac{g}{l}(1 - \cos x_1)$$

and

$$\dot{V}(x_1, x_2) = \frac{\partial V}{\partial x_1}(x_1, x_2)\dot{x}_1 + \frac{\partial V}{\partial x_2}(x_1, x_2)\dot{x}_2 = -c\,x_2^2 \leq 0.$$

Therefore the total energy is nonincreasing. In fact, it can be shown that the energy V is strictly decreasing along nonzero solution curves, because the solution curves do indeed cut across the level curves of V, rather than remaining within an energy level curve for some nontrivial time interval. To see this, notice that $\dot{V}(x_1, x_2)$ is zero only at points of the x_1-axis (where $x_2 = 0$). At points of the form $x = (x_1, 0)$ sufficiently close to the origin, the velocity vector is $(0, -\frac{g}{l}\sin x_1)$, which cuts across the set where $\dot{V}(x_1, x_2) = 0$. Thus, nonzero solutions near the origin never become trapped within positive-energy level curves. Since $V(x_1(t), x_2(t))$ is strictly decreasing as a function of t, the solution $(x_1(t), x_2(t))$ remains bounded in forward time and therefore exists for all $t \geq 0$. Moreover, it can be shown that $V(x_1(t), x_2(t))$ must asymptotically approach zero, and thus $(x_1(t), x_2(t))$ must approach the origin. (See Figure 2.5.) Once more, a precise argument for this convergence might be an interesting exercise here; however, a detailed argument is given later. △

In Example 2.17, the stated asymptotic behavior is plausible on physical grounds, since we know that the oscillations of the pendulum arm will die out eventually. But a rigorous argument is needed to establish this behavior for the model itself, and thus to confirm that the model does indeed reflect appropriately the asymptotic behavior of a real pendulum. The main technical point of the analysis, when the solutions cut across the level curves of energy, is to show that for sufficiently small $\|x_0\|$, the strict decrease of $V(x(t))$ implies that $V(x(t)) \to 0$ as $t \to \infty$. Hence, $x(t) \to 0$ as $t \to \infty$, for if not, then we have a contradiction of $V(x(t)) \to 0$. Observe that the limit

$$V_0 := \lim_{t \to \infty} V(x(t))$$

must exist, since $V(x(t))$ is a decreasing function which is bounded below (by zero). The details of the argument that $V(x(t)) \to 0$ as $t \to \infty$ are covered in Chapter 8.

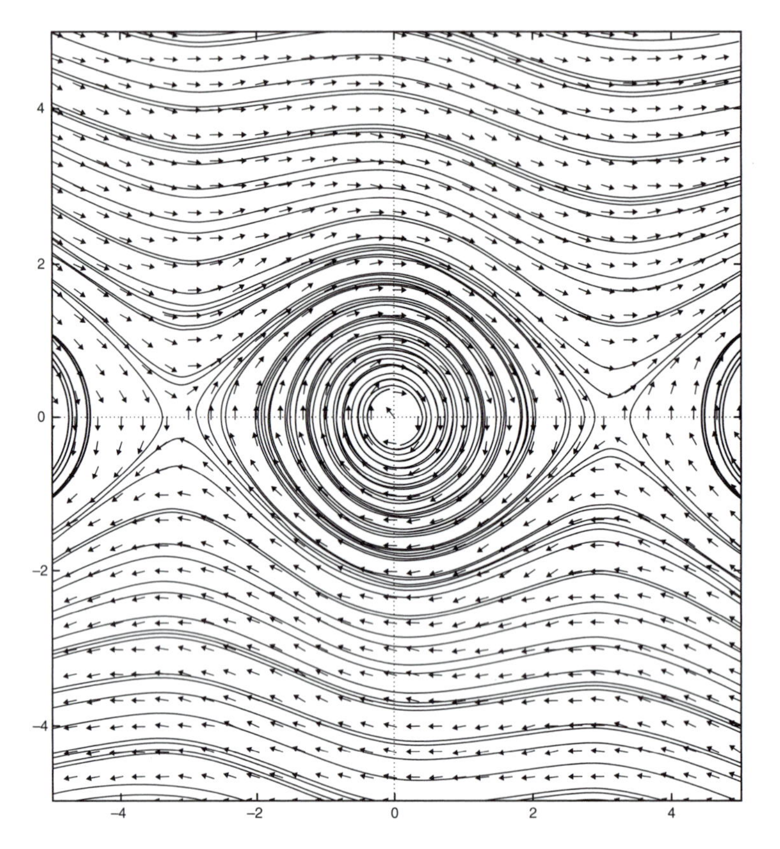

Figure 2.5 Phase portrait for Example 2.17.

2.5 EXERCISES

Exercise 2.1 Prove the following facts:

(a) If $A(t)$ and $B(t)$ are differentiable matrix functions on an interval $\mathcal{I}$, and the product $A(t)B(t)$ is defined, then

$$\frac{d}{dt}[A(t)B(t)] = \frac{dA(t)}{dt}B(t) + A(t)\frac{dB(t)}{dt}, \quad t \in \mathcal{I}.$$

(b) If $A(t)$ is differentiable and invertible for each $t \in \mathcal{I}$, then

$$\frac{d}{dt}[A^{-1}(t)] = -A^{-1}(t)\frac{dA(t)}{dt}A^{-1}(t), \quad t \in \mathcal{I}.$$

(c) If $x(t)$ is a differentiable vector function, and $\|\cdot\|_2$ denotes the Euclidean norm, then

$$\frac{d}{dt}\|x(t)\|_2 = \frac{\dot{x}^T(t)x(t)}{\|x(t)\|_2}.$$

Exercise 2.2 *On Linear Independence*
Let $v_1, \ldots, v_n$ be in $\mathbf{R}^m$.

(a) Show that the real vectors $v_1, \ldots, v_n$ are linearly independent over $\mathbf{C}$ if and only if they are linearly independent over $\mathbf{R}$.
(b) Let A be the $m \times n$ matrix $[v_1 \; v_2 \; \cdots \; v_n]$. Show that the rank of A over $\mathbf{C}$ equals the rank of A over $\mathbf{R}$.

Exercise 2.3 *Invariant Subspaces*
Let S be a subspace of $\mathbf{R}^n$ and $A \in \mathbf{R}^{n\times n}$. Show that $A(S) \subseteq S$ if and only if $A^T(S^\perp) \subseteq S^\perp$.

Exercise 2.4 *Matrix of a Quadratic Form*
There is no loss of generality in specifying symmetric matrices in Definition 2.13. Let P be an $n \times n$ real matrix.

(a) Verify that P may be written as

$$P = \frac{1}{2}(P + P^T) + \frac{1}{2}(P - P^T) = P_{\text{sym}} + P_{\text{skew}},$$

where $P_{\text{sym}} := \frac{1}{2}(P+P^T)$ and $P_{\text{skew}} := \frac{1}{2}(P - P^T)$ are the *symmetric* and *skew-symmetric* parts of P, respectively. Verify that $P_{\text{sym}}^T = P_{\text{sym}}$ and $P_{\text{skew}}^T = -P_{\text{skew}}$.
(b) Show that

$$x^T P x = x^T P_{\text{sym}} x \quad \text{for all } x.$$

Exercise 2.5 *More on Matrices in Quadratic Forms*

(a) If P is a symmetric $n \times n$ matrix, then $\nabla(x^T P x)$ is given by

$$\nabla(x^T P x) = 2Px.$$

(b) For any $n \times n$ matrix B,

$$2x^T B x = x^T(B + B^T)x.$$

Hint: See Exercise 2.4.
(c) Suppose $Q^T = Q$ is (positive or negative) semidefinite. Show that $x^T Q x = 0$ implies $Qx = 0$. *Hint*: Use Lemma 2.5.

Exercise 2.6 *Positive Definite Matrices*
Suppose $P \in \mathbf{R}^{n\times n}$ is positive definite (Definition 2.13).

(a) Show that $||x||_P = (x^T P x)^{\frac{1}{2}}$ defines a norm on $\mathbf{R}^n$.
(b) Show that $w^* P w > 0$ for every nonzero complex vector $w \in \mathbf{C}^n$. *Hint*: Write $w = u + iv$ with $u, v \in \mathbf{R}^n$, and expand the quadratic form to show that $w^* P w = u^T P u + v^T P v$.

Exercise 2.7 *Absolute Convergence in a Normed Space*
Let V be a complete normed vector space. Show that if the infinite series $\sum_{k=1}^{\infty} a_k$, $a_k \in V$, is absolutely convergent, then it is convergent. *Hint*: Let $s_n = \sum_{k=1}^{n} a_k$ and $S_n = \sum_{k=1}^{n} \|a_k\|$. Show that if $\{S_n\}$ is a Cauchy sequence in $\mathbf{R}$, then $\{s_n\}$ is a Cauchy sequence in V. Then invoke the completeness of V.

Exercise 2.8 *Geometric Series*
Let x be a real or complex variable. Show that the domain of convergence of the series $\sum_{k=0}^{\infty} x^k$ is $|x| < 1$. What is the sum of this series? Show that this series converges uniformly on the domain $|x| \le 1 - \delta$ for any δ with $0 < \delta < 1$.

Exercise 2.9 *Local Lipschitz Condition*

(a) Verify that in each of Examples 2.7, 2.8, 2.9, the right-hand side of the differential equation is locally Lipschitz on $\mathbf{R}$.
(b) Consider the equation $\dot{x} = x^{\frac{1}{3}}$ of Example 2.10. Show that $f(x) = x^{\frac{1}{3}}$ does not satisfy a Lipschitz condition in any neighborhood of $x = 0$.

Exercise 2.10 Consider the nonlinear system

$$\begin{aligned}\dot{x}_1 &= -x_2 + \epsilon x_1(r-1),\\ \dot{x}_2 &= x_1 + \epsilon x_2(r-1),\end{aligned}$$

where $r = (x_1^2 + x_2^2)^{\frac{1}{2}}$ is the radial polar coordinate.

(a) Verify that, for $\epsilon = 0$, the Jacobian linearization at the origin has a *center* phase portrait for which all solutions are periodic. Classify the phase portrait of the linearization at the origin when $\epsilon \neq 0$.
(b) Show that, in polar coordinate form, the nonlinear system is

$$\begin{aligned}\dot{r} &= \epsilon r(r-1),\\ \dot{\theta} &= 1.\end{aligned}$$

(c) Show that, if $\epsilon < 0$, then the origin is an unstable spiral (in the sense that $r(t)$ increases with time); and if $\epsilon > 0$, then the origin is a stable spiral (in the sense that $r(t)$ decreases with increasing time).

Exercise 2.11 *Some Periodic Orbits*
There are cases where the linearization of a planar system has pure imaginary eigenvalues, and the periodic solution behavior of the linearization is preserved *locally* near equilibrium in the nonlinear system itself. This situation does not always occur; however, show that this phenomenon occurs in the following systems:

(a) the nonlinear undamped pendulum in Example 2.16;

(b) the predator-prey system

$$\dot{x}_1 = -x_1 + x_1 x_2,$$
$$\dot{x}_2 = x_2 - x_1 x_2.$$

Hint: See [40] (pp. 258–262) or [41] (pp. 239–243) if needed.

Exercise 2.12 *Pure Imaginary Eigenvalues Again*
Consider the system

$$\dot{x}_1 = x_1^3 - x_2,$$
$$\dot{x}_2 = x_1 + x_2^3.$$

(a) Show that the origin is the only equilibrium of the system, and the eigenvalues of the Jacobian linearization at the origin are $\pm i$.
(b) Show that the Euclidean norm strictly increases along nonzero solutions of this system. Thus, show that a nonzero solution cannot be periodic.

Exercise 2.13 Consider the planar system

$$\dot{x}_1 = x_2(x_1^2 + x_2^2),$$
$$\dot{x}_2 = -x_1(x_1^2 + x_2^2)$$

(a) Show that the solution trajectories are circles in the plane centered at the origin. *Hint*: Consider the slope field given by $\frac{dx_2}{dx_1}$ and integrate.
(b) Explain how the solutions in (a) differ from the solutions of the undamped, linear harmonic oscillator system

$$\dot{x}_1 = x_2,$$
$$\dot{x}_2 = -x_1.$$

Exercise 2.14 *Find the Equilibria First*
This example indicates one of the reasons why it is important to identify the equilibrium points when determining the phase portrait of a planar autonomous system, rather than just integrating an expression for $\frac{dx_2}{dx_1}$ (when the integration is possible). Consider the system

$$\dot{x}_1 = x_1 x_2,$$
$$\dot{x}_2 = -x_1^2.$$

Integrate $\frac{dx_2}{dx_1} = -\frac{x_1^2}{x_1 x_2}$ for $x_1 x_2 \neq 0$ to determine trajectories in the plane. Then complete a sketch of the phase portrait by taking into account the equilibria of the system. How does this phase portrait differ from that of the undamped linear harmonic oscillator?

Exercise 2.15 *Cayley-Hamilton Theorem for 2×2 Matrices*
Let A be a complex 2×2 matrix, written as

$$A = \begin{bmatrix} a & b \\ c & d \end{bmatrix},$$

and let $p(\lambda)$ be the characteristic polynomial of A. Show by direct calculation that A satisfies its own characteristic equation, meaning that

$$p(A) = A^2 - (a+d)A + (ad - bc)I = 0.$$

2.6 NOTES AND REFERENCES

Reference [75] is an excellent resource for linear algebra. See also [67]. For proofs of the Primary Decomposition Theorem, see [15] or [40]. For characterizations of a basis for a vector space, see [75] (page 195). For the Fundamental Theorem of Linear Algebra, see [75] or G. Strang, *Introduction to Linear Algebra*, Wellesley-Cambridge Press, Wellesley, MA 1993, or the interesting paper by G. Strang, The fundamental theorem of linear algebra, *Amer. Math. Monthly*, 100, 848–855, 1993.

For two-dimensional autonomous systems, phase plane analysis, and applications, see [21], [40], or [41]. Exercise 2.12 was taken from [21].

For background in analysis, see [7]. For more on uniform convergence and the interchange of limit processes of calculus, including term-by-term differentiation and term-by-term integration of series of functions, see [7] (Chapter 9).

The coefficients $\det A$ and $\operatorname{tr} A$ in the characteristic polynomial of a 2×2 complex matrix A are the symmetric functions of order 2 and 1, respectively, of the eigenvalues of A. For an $n \times n$ complex matrix A with $n > 2$, $\det A = \prod_{j=1}^{n} \lambda_j$ is the symmetric function of order n, and $\operatorname{tr} A = \sum_{j=1}^{n} \lambda_j$ is again the symmetric function of order 1. The other coefficients of the characteristic polynomial of A are determined by the other symmetric functions of the eigenvalues, having orders 2 through $n-1$. For more on the symmetric functions of the eigenvalues, see [75] (page 494).

Chapter Three

Linear Systems and Stability

This chapter includes the basic results on the solutions of linear systems with emphasis on time invariant systems. Stability definitions are introduced and applied to linear time invariant systems. Some results on Lyapunov equations are presented, including both the direct and converse Lyapunov theorems for linear time invariant (LTI) systems.

3.1 THE MATRIX EXPONENTIAL

Recall that if a is a fixed real number, then the real exponential function e^{at}, $t \in \mathbf{R}$, is defined by the real infinite series of functions

$$e^{at} := 1 + at + \frac{1}{2!}(at)^2 + \cdots + \frac{1}{k!}(at)^k + \cdots = \sum_{k=0}^{\infty} \frac{t^k}{k!} a^k.$$

We know that $\frac{d}{dt} e^{ta} = a e^{ta}$. But it is useful to obtain this derivative formula using term-by-term differentiation of the defining series. For $|t| \le \beta$, we can estimate the general term of the exponential series by

$$\left| \frac{1}{k!}(at)^k \right| = \frac{|t|^k}{k!} |a|^k \le \frac{\beta^k}{k!} |a|^k =: m_k.$$

The series $\sum m_k$ converges by the ratio test. Thus, the Weierstrass test applies and we conclude that the series converges uniformly on the interval $|t| \le \beta$. A similar argument shows that the power series defined by the term-by-term differentiated series also converges uniformly on compact time intervals, and thus the differentiated series represents, for each real t, the derivative of e^{at}. (See, for example, [7] (Theorem 9.14, page 230) for the relevant general theorem on uniform convergence and differentiation of series of functions. For a statement specific to power series, see [7] (Theorem 9.23, page 236).) Thus, it is legitimate to write

$$\frac{d}{dt}\left(1 + at + \frac{1}{2!}(at)^2 + \cdots + \frac{1}{k!}(at)^k + \cdots \right) = a + at + a\frac{1}{2!}t^2 + \cdots = ae^{at},$$

and from this calculation we conclude immediately that $e^{at}x_0$, $-\infty < t < \infty$, is the unique solution of the initial value problem $\dot{x} = ax$, $x(0) = x_0$. Here is a direct proof of uniqueness: If $\phi(t)$ solves the same initial value problem, then the function $e^{-at}\phi(t)$ is a differentiable function with zero derivative for all t, hence $e^{-at}\phi(t)$ is a constant function. Evaluation at $t = 0$ shows that the constant must be x_0, and therefore $\phi(t) = e^{at}x_0$.

There is a similar matrix construction that produces the general solution of the initial value problem

$$\dot{x} = Ax, \qquad x(0) = x_0 \tag{3.1}$$

when A is a real $n \times n$ matrix. If $A \in \mathbf{R}^{n\times n}$, we consider the series

$$I + tA + \frac{1}{2!}(tA)^2 + \cdots + \frac{1}{k!}(tA)^k + \cdots = \sum_{k=0}^{\infty} \frac{t^k}{k!}A^k. \tag{3.2}$$

For $|t| \le \beta$, we can estimate the general term of this series by

$$\left\| \frac{1}{k!}(tA)^k \right\| = \frac{|t|^k}{k!}\|A^k\| \le \frac{\beta^k}{k!}\|A\|^k =: M_k.$$

The series $\sum M_k$ converges by the ratio test, and therefore, by the Weierstrass test, the series in (3.2) converges absolutely and uniformly on the interval $|t| \le \beta$ for any finite β. (We identify the space $\mathbf{R}^{n\times n}$ of real matrices with $\mathbf{R}^{n^2}$, and the argument for the Weierstrass M-test applies.)

Definition 3.1 *If $A \in \mathbf{R}^{n\times n}$, then the matrix exponential function e^{tA}, $t \in \mathbf{R}$, is defined by*

$$e^{tA} := I + tA + \frac{t^2}{2!}A^2 + \cdots + \frac{t^k}{k!}A^k + \cdots = \sum_{k=0}^{\infty} \frac{t^k}{k!}A^k.$$

The general solution of a linear system $\dot{x} = Ax$ can be expressed in terms of the matrix exponential function. Again by standard results on power series, the series for e^{tA} can be differentiated term-by-term with respect to t to obtain

$$\begin{aligned}\frac{d}{dt}e^{tA} &= A + tA^2 + \frac{t^2}{2!}A^3 + \cdots + \frac{t^k}{k!}A^{k+1} + \cdots \\ &= Ae^{tA}.\end{aligned}$$

Thus, e^{tA} is a matrix solution of the differential equation $\dot{x} = Ax$. Equivalently, each column of e^{tA} is a vector function that solves $\dot{x} = Ax$. We have the following fundamental theorem for linear time invariant systems.

Theorem 3.1 (Fundamental Theorem for LTI Systems)
Let $x_0 \in \mathbf{R}^n$. The unique solution of the initial value problem (3.1) *is given by $x(t) = e^{tA}x_0$ and this solution is defined for all real t.*

Proof. We showed above that $e^{tA}x_0$ is a solution of (3.1) for $-\infty < t < \infty$. Uniqueness follows from the general uniqueness result for solutions of initial value problems. □

A real or complex matrix solution $X(t)$ of $\dot{x} = Ax$ which is nonsingular for all t is called a *fundamental matrix* solution. Corollary 3.1 below shows that e^{tA} is a fundamental matrix solution as a consequence of the uniqueness of solutions for initial value problems. It is easy to check that if $X(t)$ is any fundamental matrix and C is any invertible matrix of the same size, then $X(t)C$ is also a fundamental matrix. From this fact and the uniqueness of solutions, we have

$$e^{tA} = X(t)X(0)^{-1}.$$

We only need to observe that each side is a matrix solution of $\dot{x} = Ax$, and each side equals the identity matrix when $t = 0$.

There are some special solutions of the system $\dot{x} = Ax$ that are easily identified. In particular, let λ be an eigenvalue of A and let v be a corresponding eigenvector, so that $Av = \lambda v$. Then the function $\phi(t) = e^{\lambda t}v$ must be a (possibly complex-valued) solution, since

$$\dot{\phi}(t) = \frac{d}{dt}[e^{\lambda t}v] = \left(\frac{d}{dt}e^{\lambda t}\right)v = \lambda e^{\lambda t}v = e^{\lambda t}Av = A\phi(t).$$

Suppose now that A is diagonalizable, with eigenvalues $\lambda_1, \dots, \lambda_n$. If the vectors $v_1, \dots, v_n$ form a basis of n linearly independent eigenvectors of A (which exist by the fact that A is diagonalizable) with $Av_j = \lambda_j v_j$, then we have the n special solutions

$$\phi_1(t) = e^{\lambda_1 t}v_1, \quad \dots \quad , \phi_n(t) = e^{\lambda_n t}v_n.$$

Place these solutions as the columns of a complex matrix

$$\Phi(t) = [\phi_1(t) \cdots \phi_n(t)].$$

Then $\Phi(t)$ is a fundamental matrix solution of $\dot{x} = Ax$, with $\Phi(0) = [v_1 \cdots v_n]$. It follows that

$$e^{tA} = \Phi(t)\Phi(0)^{-1} = \Phi(t)[v_1 \cdots v_n]^{-1}.$$

Example 3.1 Consider the system

$$\dot{x} = \begin{bmatrix} 3 & -2 \\ 1 & 1 \end{bmatrix} x.$$

We will construct the general solution and the matrix solution e^{tA}. The characteristic equation is $\det(\lambda I - A) = (\lambda - 3)(\lambda - 1) + 2 = \lambda^2 - 4\lambda + 5 = 0$, so the eigenvalues are $\lambda = 2 \pm i$. An eigenvector for $2 + i$ is

$$v = \begin{bmatrix} 1+i \\ 1 \end{bmatrix}.$$

An eigenvector for $\lambda = 2 - i$ is

$$v = \begin{bmatrix} 1-i \\ 1 \end{bmatrix}.$$

These two eigenvectors are linearly independent since

$$\det \begin{bmatrix} 1+i & 1-i \\ 1 & 1 \end{bmatrix} = 2i \neq 0.$$

Therefore two solutions are given by

$$\phi_1(t) = e^{(2+i)t} \begin{bmatrix} 1+i \\ 1 \end{bmatrix}, \qquad \phi_2(t) = e^{(2-i)t} \begin{bmatrix} 1-i \\ 1 \end{bmatrix}.$$

Thus,

$$\Phi(t) := \begin{bmatrix} e^{(2+i)t}(1+i) & e^{(2-i)t}(1-i) \\ e^{(2+i)t} & e^{(2-i)t} \end{bmatrix}$$

is a matrix solution. Observe that the Wronskian test applied at $t = 0$,

$$\det \Phi(0) = \det \begin{bmatrix} 1+i & 1-i \\ 1 & 1 \end{bmatrix} = 2i \neq 0,$$

confirms that $\phi_1(t)$, $\phi_2(t)$ are linearly independent solutions and therefore $\Phi(t)$ is a fundamental matrix. The general real-valued solution using e^{tA} is given by

$$x(t) = e^{tA}x_0 = \Phi(t)\Phi(0)^{-1}x_0,$$

and a straightforward calculation gives

$$e^{tA} = \begin{bmatrix} e^{2t}(\cos t + \sin t) & -2e^{2t}\sin t \\ e^{2t}\sin t & e^{2t}(\cos t - \sin t) \end{bmatrix}.$$

It is also possible to obtain e^{tA} with the help of the real Jordan form of A; this alternative calculation of the general solution for this system is carried out in Example 3.7 below. △

The next result gives some useful properties of e^{tA} that are corollaries of the uniqueness of solutions.

Corollary 3.1 (Properties of e^{tA})
If $A \in \mathbf{R}^{n\times n}$, then the following properties hold:

(a) *If $AB = BA$, then $e^{t(A+B)} = e^{tA}e^{tB}$ for all t.*
(b) *e^{tA} is nonsingular for each t, and $(e^{tA})^{-1} = e^{-tA}$.*
(c) *If S is nonsingular, then $S^{-1}e^{tA}S = e^{t(S^{-1}AS)}$ for all t.*

Proof. (a) We show that the function $X(t) := e^{tA}e^{tB}$ is a solution of $\dot{X} = (A+B)X$ satisfying $X(0) = I$; the result then follows by the uniqueness of solutions. Differentiate X, using the product rule to get

$$\dot{X}(t) = Ae^{tA}e^{tB} + e^{tA}Be^{tB}.$$

Note that $AB = BA$ implies that $e^{tA}B = Be^{tA}$ (consider the defining series). It follows easily that $\dot{X}(t) = (A+B)X(t)$, $X(0) = I$, hence $X(t) = e^{t(A+B)}$.

(b) Let $B = -A$ in statement (a). Then $I = e^{t(A-A)} = e^{tA}e^{-tA}$, and (b) follows immediately.

(c) This property can be deduced from a power series argument, but again we will use uniqueness of solutions. Let $\bar{A} = S^{-1}AS$. Let $X(t) = S^{-1}e^{tA}S$. Clearly, $X(0) = I$. We now show that $X(t)$ satisfies $\dot{X}(t) = \bar{A}X(t)$. Differentiate $X(t)$ to get

$$\dot{X}(t) = S^{-1}Ae^{tA}S = \bar{A}S^{-1}e^{tA}S = \bar{A}X(t).$$

By uniqueness, $X(t) = e^{t\bar{A}} = e^{t(S^{-1}AS)}$, as we wished to show. □

3.2 THE PRIMARY DECOMPOSITION AND SOLUTIONS OF LTI SYSTEMS

In this section we use the Primary Decomposition Theorem (see Theorem 2.1) to develop a representation of the general solution of a linear time invariant system. Useful bounds on the solutions follow from the resulting representation. In particular, we will produce some useful norm bounds on e^{tA} for any $n \times n$ matrix A.

Let A be a real or complex $n \times n$ matrix. Suppose A has $k \leq n$ distinct eigenvalues, denoted by $\lambda_1, \lambda_2, \ldots, \lambda_k$, with corresponding algebraic multiplicities $m_1, m_2, \ldots, m_k$, so that $m_1 + m_2 + \cdots + m_k = n$. Associated with eigenvalue λ_j is the system of linear equations

$$(A - \lambda_j I)^{m_j} x = 0. \tag{3.3}$$

For each $j = 1, \ldots, k$, the solution space of (3.3) is a linear subspace of $\mathbf{C}^n$, which we denote by X_j. It follows from the Primary Decomposition Theorem that, for each j, $\dim X_j = m_j$, that is, the system (3.3) has m_j linearly independent solutions. As stated in Theorem 2.1, $\mathbf{C}^n$ is the direct sum of the subspaces X_j:

$$\mathbf{C}^n = X_1 \oplus X_2 \oplus \cdots \oplus X_k.$$

Indeed, for every vector $x \in \mathbf{C}^n$ there exist unique vectors $x_j \in X_j$ for $j = 1, \ldots, k$, such that

$$x = x_1 + x_2 + \cdots + x_k\,. \tag{3.4}$$

Each subspace X_j is invariant for A, hence invariant for $(A - \lambda_j I)$. The linear transformation induced on X_j by $(A - \lambda_j I)$ is nilpotent with index at most m_j, as shown by the defining equation (3.3) for X_j. (A matrix N is called *nilpotent* if $N^k = 0$ for some positive integer k. The least such positive integer is called the index of N.)

Example 3.2 The index of the nilpotent matrix $A - \lambda_j I$ may be less than the algebraic multiplicity m_j of the eigenvalue λ_j. Consider the following

3×3 matrix with a single eigenvalue,

$$A = \begin{bmatrix} 2 & 1 & 0 \\ 0 & 2 & 0 \\ 0 & 0 & 2 \end{bmatrix}.$$

The algebraic multiplicity is 3, and the Cayley-Hamilton Theorem says that $(A - 2I)^3 = 0$. But we also have $(A - 2I)^2 = 0$, so $A - 2I$ is nilpotent with index two. △

In the special case where $k = n$, we have $m_j = 1$ for $j = 1, 2, \dots, n$, and then each of the vectors x_j in (3.4) is a constant multiple of an eigenvector of A. Moreover, the n independent eigenvectors of A span $\mathbf{C}^n$ and thus form a basis of $\mathbf{C}^n$. We have seen above, in the discussion following Theorem 3.1, that in this case we can produce a fundamental matrix solution in a straightforward way.

We now want to apply the primary decomposition (3.4) in order to represent any solution of $\dot{x} = Ax$ with $x(0) = v$. Instead of computing the matrix solution e^{tA} directly, the idea is to evaluate $e^{tA}v$ in a single expression. From (3.4), we have

$$v = v_1 + \cdots + v_k$$

for unique vectors $v_j \in X_j$, $j = 1, \dots, k$. By definition of X_j, the vector v_j is a solution of (3.3). Using linearity and properties of the exponential, we may write

$$\begin{aligned} e^{tA}v &= \sum_{j=1}^{k} e^{tA}v_j \\ &= \sum_{j=1}^{k} e^{\lambda_j t} e^{t(A-\lambda_j I)} v_j \\ &= \sum_{j=1}^{k} e^{\lambda_j t} \left[I + t(A - \lambda_j I) + \cdots + \frac{t^{m_j-1}}{(m_j-1)!}(A - \lambda_j I)^{m_j-1} \right] v_j \end{aligned} \tag{3.5}$$

for $-\infty < t < \infty$. Observe that (3.5) involves a finite sum for each j because v_j is a solution of (3.3). Thus, we may write the solution $e^{tA}v$ of the initial value problem $\dot{x} = Ax$, $x(0) = v$ as

$$e^{tA}v = \sum_{j=1}^{k} e^{\lambda_j t} \left[\sum_{i=0}^{m_j-1} \frac{t^i}{i!}(A - \lambda_j I)^i \right] v_j, \quad -\infty < t < \infty. \tag{3.6}$$

It is also interesting to note that each of the vector summands $e^{tA}v_j$ in (3.5) belongs to the subspace X_j, since X_j is invariant for $A - \lambda_j I$.

As an application of (3.6) we prove the following important norm estimate for the matrix exponential.

Theorem 3.2 *Let A be an $n \times n$ matrix with distinct eigenvalues $\lambda_1, \ldots, \lambda_k$ of algebraic multiplicities $m_1, \ldots, m_k$, respectively, so that $m_1 + \cdots + m_k = n$. If α is any number larger than the real part of each λ_j, $j = 1, \ldots, k$, that is,*

$$\alpha > \max\{\operatorname{Re}\lambda_j \, : \, j = 1, \ldots, k\}, \tag{3.7}$$

and $\|\cdot\|$ denotes a matrix norm, then there exists a constant $M > 0$ such that

$$\|e^{tA}\| \leq Me^{\alpha t} \quad \text{for } 0 \leq t < \infty\,. \tag{3.8}$$

Proof. By the equivalence of norms in $\mathbf{C}^{n\times n}$, we can work with the matrix norm given by the absolute sum of the entries.

By formula (3.6), every entry of e^{tA} has the form $\sum_{j=1}^{k} p_j(t)e^{\lambda_j t}$, where $p_j(t)$ is a polynomial of degree at most $(m_j - 1)$. If α satisfies (3.7), then, for each index pair (r, s), there is a constant M_{rs} such that the (r, s) entry of e^{tA} is bounded by

$$M_{rs}e^{\alpha t} \quad \text{for } t \geq 0.$$

Let $M := n^2 \max\{M_{rs} \, : \, 1 \leq r, s \leq n\}$. Then (3.8) holds. □

Before stating an important corollary of Theorem 3.2, we introduce the following terminology, which is standard and very convenient.

Definition 3.2 *A matrix A in $\mathbf{C}^{n\times n}$ is said to be* Hurwitz *if all of its eigenvalues have negative real part.*

Corollary 3.2 *If A is Hurwitz, then there exist constants $\rho > 0$ and $M > 0$ such that every solution $x(t)$ of $\dot{x} = Ax$ satisfies*

$$\|x(t)\| \leq M\|x(0)\|e^{-\rho t} \quad \text{for } 0 \leq t < \infty\,. \tag{3.9}$$

Thus, every solution $x(t)$ satisfies

$$\lim_{t\to\infty} x(t) = 0.$$

Proof. Using the fact that A is Hurwitz, define

$$-\rho := \frac{1}{2}\max\{\operatorname{Re}\lambda : \lambda \text{ is an eigenvalue of } A\}.$$

By Theorem 3.2, there exists $M > 0$ such that

$$\|x(t)\| = \|e^{tA}x(0)\| \leq \|e^{tA}\| \, \|x(0)\| \leq M\|x(0)\|e^{-\rho t}, \; t \geq 0,$$

which is (3.9). □

If all eigenvalues of A have real part either negative or zero, and the eigenvalues with zero real part have geometric multiplicity equal to their algebraic multiplicity, then a constant upper bound exists for $\|e^{tA}\|$ on the interval $t \geq 0$. Using this fact, bounds may be deduced for solutions of some nonhomogeneous linear systems in critical cases. (See Exercises 3.13 and 3.14.)

Formula (3.6) provides conceptual insight along with some qualitative consequences in Theorem 3.2 and Corollary 3.2. In addition, (3.6) can serve as the basis for computing solutions to specific systems, and we now give some examples of its use in that role.

Example 3.3 Consider $\dot{x} = Ax$, where

$$A = \begin{bmatrix} 1 & 1 \\ -1 & 3 \end{bmatrix}.$$

The characteristic equation is $\lambda^2 - 4\lambda + 4 = 0$, with single root $\lambda_1 = 2$, repeated with algebraic multiplicity 2. An eigenvector v is a solution of the linear system

$$(A - 2I)v = \begin{bmatrix} -1 & 1 \\ -1 & 1 \end{bmatrix} = 0.$$

Therefore v must have the form

$$v = \alpha \begin{bmatrix} 1 \\ 1 \end{bmatrix},$$

where α is a nonzero scalar. It is easy to verify that

$$(A - 2I)^2 = 0$$

and the generalized eigenspace for $\lambda_1 = 2$ must have dimension 2. We apply (3.6) to compute $e^{tA}v$ for an arbitrary vector $v = [v_1 \; v_2]^T$. By (3.6), we have

$$\begin{aligned} e^{tA}v &= e^{2t}[I + t(A - 2I)]v \\ &= e^{2t}\left[\begin{bmatrix} v_1 \\ v_2 \end{bmatrix} + \begin{bmatrix} t(-v_1 + v_2) \\ t(-v_1 + v_2) \end{bmatrix}\right] \end{aligned}$$

From this expression we can produce e^{tA}. We get the first column of e^{tA} by computing

$$e^{tA}e_1 = e^{2t} \begin{bmatrix} 1 - t \\ -t \end{bmatrix}.$$

The second column of e^{tA} is

$$e^{tA}e_2 = e^{2t} \begin{bmatrix} t \\ 1 + t \end{bmatrix}.$$

Thus,

$$e^{tA} = e^{2t} \begin{bmatrix} 1 - t & t \\ -t & 1 + t \end{bmatrix}.$$

△

Example 3.4 Consider the matrix

$$A = \begin{bmatrix} -9 & 5 & -1 \\ 0 & -2 & 4 \\ 0 & -1 & 2 \end{bmatrix}.$$

The characteristic polynomial of A is $\lambda^2(\lambda+9)$; thus, the eigenvalues, listed according to algebraic multiplicity, are $-9, 0, 0$. It is straightforward to verify that each of the two distinct eigenvalues has geometric multiplicity one. In fact, an eigenvector for $\lambda_1 = -9$ is $v_1 = [1 \quad 0 \quad 0]^T$, and an eigenvector for $\lambda_2 = 0$ is $v_2 = [1 \quad 2 \quad 1]^T$. In order to complete a basis in which A takes its Jordan form, we solve the equation $(A - \lambda_2)w = v_2$, or $Aw = [1 \quad 2 \quad 1]^T$, to obtain any particular solution, for example, $w = [\frac{1}{3} \quad 1 \quad 1]^T$. Thus, if we set

$$S = [v_1 \quad v_2 \quad w] = \begin{bmatrix} 1 & 1 & \frac{1}{3} \\ 0 & 2 & 1 \\ 0 & 1 & 1 \end{bmatrix},$$

then we have

$$S^{-1}AS = J := \begin{bmatrix} -9 & 0 & 0 \\ 0 & 0 & 1 \\ 0 & 0 & 0 \end{bmatrix}.$$

Note that it is straightforward to verify that $AS = SJ$, where J is defined as indicated. We can now compute e^{tA} since we have

$$A = SJS^{-1} \Longrightarrow e^{tA} = Se^{tJ}S^{-1} = S \begin{bmatrix} e^{-9t} & 0 & 0 \\ 0 & 1 & t \\ 0 & 0 & 1 \end{bmatrix} S^{-1}.$$

Since

$$S^{-1} = \begin{bmatrix} 1 & -\frac{2}{3} & \frac{1}{3} \\ 0 & 1 & -1 \\ 0 & -1 & 2 \end{bmatrix},$$

we have

$$\begin{aligned} e^{tA} &= \begin{bmatrix} 1 & 1 & \frac{1}{3} \\ 0 & 2 & 1 \\ 0 & 1 & 1 \end{bmatrix} \begin{bmatrix} e^{-9t} & 0 & 0 \\ 0 & 1 & t \\ 0 & 0 & 1 \end{bmatrix} \begin{bmatrix} 1 & -\frac{2}{3} & \frac{1}{3} \\ 0 & 1 & -1 \\ 0 & -1 & 2 \end{bmatrix} \\ &= \begin{bmatrix} e^{-9t} & (-\frac{2}{3}e^{-9t} + \frac{2}{3} - t) & (\frac{1}{3}e^{-9t} - \frac{1}{3} + 2t) \\ 0 & 1 - 2t & 4t \\ 0 & -t & 1 + 2t \end{bmatrix}, \end{aligned}$$

from which the general solution to $\dot{x} = Ax$ is easily obtained. △

Finally, note that the existence of unbounded entries in e^{tA} on the interval $0 \le t < \infty$ in the last example is due to the presence of a zero eigenvalue with geometric multiplicity less than its algebraic multiplicity.

3.3 JORDAN FORM AND MATRIX EXPONENTIALS

Jordan forms are a useful theoretical tool as they provide considerable insight into the structure of linear systems. In this section we discuss the *real Jordan form* for real $n \times n$ matrices. The real Jordan form will be used here to illustrate computations of e^{tA}. Given $A \in \mathbf{R}^{n\times n}$ there exists a nonsingular real S such that $S^{-1}AS = J$, where J is the real Jordan form of A. Matrix J is a block diagonal matrix which is unique up to a rearrangement of the blocks along the diagonal. It should be mentioned that Jordan forms can be difficult to compute accurately by any numerical algorithm. In this text, Jordan forms are used primarily for the insight they provide, and only small size examples are needed as illustrations.

3.3.1 Jordan Form of Two-Dimensional Systems

As motivation for the general result on the real Jordan form, we now consider the possible real Jordan forms for a 2×2 real matrix A. Our application context is the solution of two-dimensional linear systems, $\dot{x} = Ax$.

There are three cases to consider: (i) real eigenvalues and diagonalizable A, (ii) a repeated real eigenvalue with A nondiagonalizable, and (iii) a complex conjugate pair of eigenvalues.

REAL EIGENVALUES AND DIAGONALIZABLE. Suppose A has real eigenvalues and is diagonalizable. Let λ_1 and λ_2 be the eigenvalues of A (possibly repeated), with independent eigenvectors v_1 and v_2, respectively. Then the matrix $S = [v_1 \quad v_2]$ is nonsingular. Writing $x = Sz$, we have

$$AS = S \begin{bmatrix} \lambda_1 & 0 \\ 0 & \lambda_2 \end{bmatrix}.$$

In z coordinates we have the linear system

$$\dot{z} = \begin{bmatrix} \lambda_1 & 0 \\ 0 & \lambda_2 \end{bmatrix} z.$$

The real Jordan form in this case is the diagonal matrix, $J = \operatorname{diag}[\lambda_1, \lambda_2]$. Note that the asymptotic behavior as $t \to \infty$ is clear from this form. In particular, all solutions $z(t) \to 0$ as $t \to \infty$ if and only if both eigenvalues are negative. The same is true of solutions $x(t)$ in the original coordinates.

Example 3.5 As a specific example, consider the system

$$\dot{x} = \begin{bmatrix} 2 & 3 \\ 3 & 2 \end{bmatrix} x.$$

The eigenvalues are real and distinct, $\lambda_1 = -1$, $\lambda_2 = 5$, so A is diagonalizable. Corresponding eigenvectors are given by

$$v_1 = \begin{bmatrix} -1 \\ 1 \end{bmatrix}, \quad v_2 = \begin{bmatrix} 1 \\ 1 \end{bmatrix}.$$

Set $x = Sz$ with

$$S = \begin{bmatrix} -1 & 1 \\ 1 & 1 \end{bmatrix}.$$

The system in z coordinates is

$$\dot{z} = S^{-1}ASz = \begin{bmatrix} -1 & 0 \\ 0 & 5 \end{bmatrix} z.$$

The general solution of this system is

$$z(t) = \text{diag}\,[e^{-t}, e^{5t}]z(0) = \begin{bmatrix} e^{-t} & 0 \\ 0 & e^{5t} \end{bmatrix} z(0).$$

The transformation back to the general solution in the x coordinates is straightforward:

$$\begin{aligned} x(t) = e^{tA}x(0) &= S \begin{bmatrix} e^{-t} & 0 \\ 0 & e^{5t} \end{bmatrix} S^{-1}x(0) \\ &= \begin{bmatrix} -1 & 1 \\ 1 & 1 \end{bmatrix} \begin{bmatrix} e^{-t} & 0 \\ 0 & e^{5t} \end{bmatrix} \left(-\frac{1}{2}\right) \begin{bmatrix} 1 & -1 \\ -1 & -1 \end{bmatrix} x(0) \\ &= -\frac{1}{2} \begin{bmatrix} -e^{-t} - e^{5t} & e^{-t} - e^{5t} \\ e^{-t} - e^{5t} & -e^{-t} - e^{5t} \end{bmatrix} x(0) \\ &= \frac{1}{2} \begin{bmatrix} e^{-t} + e^{5t} & e^{5t} - e^{-t} \\ e^{5t} - e^{-t} & e^{-t} + e^{5t} \end{bmatrix} x(0). \end{aligned}$$

Written out explicitly by coordinates, the general solution is

$$\begin{aligned} x_1(t) &= \frac{1}{2}(e^{-t} + e^{5t})x_1(0) + \frac{1}{2}(e^{5t} - e^{-t})x_2(0) \\ x_2(t) &= \frac{1}{2}(e^{5t} - e^{-t})x_1(0) + \frac{1}{2}(e^{-t} + e^{5t})x_2(0). \end{aligned}$$

△

NONDIAGONALIZABLE WITH A REPEATED REAL EIGENVALUE. Suppose A has a repeated real eigenvalue λ, and there is only one independent eigenvector, v. We can always find a vector w such that v and w are linearly independent and $(A - \lambda I)w = v$; thus, $Aw = v + \lambda w$. If $S = [v \quad w]$, and $x = Sz$, then

$$\dot{z} = \begin{bmatrix} \lambda & 1 \\ 0 & \lambda \end{bmatrix} z =: Jz.$$

Note that $J = D + N = \text{diag}\,[\lambda, \lambda] + N$, with $DN = ND$. By Corollary 3.1, we obtain

$$e^{tJ} = e^{\lambda t}e^{tN} = e^{\lambda t} \begin{bmatrix} 1 & t \\ 0 & 1 \end{bmatrix} = \begin{bmatrix} e^{\lambda t} & te^{\lambda t} \\ 0 & e^{\lambda t} \end{bmatrix}.$$

Then we have $e^{tA} = Se^{tJ}S^{-1}$. In particular, we see that all solutions satisfy $\lim_{t\to\infty} x(t) = S \lim_{t\to\infty} z(t) = 0$ as $t \to \infty$, if and only if $\lambda < 0$.

Example 3.6 As a specific case, consider the system

$$\dot{x} = \begin{bmatrix} -1 & 1 \\ -1 & -3 \end{bmatrix} x.$$

The characteristic equation is $(\lambda+1)(\lambda+3)+1 = \lambda^2+4\lambda+4 = (\lambda+2)^2 = 0$, so the repeated eigenvalue is $\lambda = -2$. An eigenvector is

$$v = \begin{bmatrix} -1 \\ 1 \end{bmatrix}.$$

We solve $(A + 2I)w = v$ to find, for example,

$$w = \begin{bmatrix} 0 \\ -1 \end{bmatrix}.$$

Set $S = [v \; w]$ and $x = Sz$ to get

$$\begin{aligned} \dot{z} = S^{-1}ASz &= \begin{bmatrix} -1 & 0 \\ 1 & -1 \end{bmatrix}^{-1} \begin{bmatrix} -1 & 1 \\ -1 & -3 \end{bmatrix} \begin{bmatrix} -1 & 0 \\ 1 & -1 \end{bmatrix} z \\ &= \begin{bmatrix} -2 & 1 \\ 0 & -2 \end{bmatrix} z. \end{aligned}$$

It is clear that we have $\lim_{t\to\infty} z(t) = 0$ for all solutions. However, explicit computation gives

$$x(t) = e^{tA}x(0) = Se^{tJ}S^{-1}x(0) = \begin{bmatrix} (1+t)e^{-2t} & te^{-2t} \\ -te^{-2t} & (1-t)e^{-2t} \end{bmatrix} x(0),$$

from which the asymptotic behavior is also evident. △

Complex Conjugate Eigenvalues. Suppose the real matrix A has a conjugate pair of eigenvalues, $\lambda = a \pm ib$. Let w be a (necessarily complex) eigenvector, and write $w = u + iv$. Then v and u must be linearly independent over the real field (why?). Define $S = [v \quad u]$ and let $x = Sz$. The system in the z coordinates is given by

$$\dot{z} = S^{-1}ASz = \begin{bmatrix} a & -b \\ b & a \end{bmatrix} z.$$

This equation is obtained from the eigenvector equation, $Aw = \lambda w$, or $A(u + iv) = (a + ib)(u + iv)$, by explicitly writing the real and imaginary parts of each side, equating the corresponding parts, and using the ordered basis $\{v, u\}$. The matrix

$$D_{a,b} := \begin{bmatrix} a & -b \\ b & a \end{bmatrix}$$

is the real Jordan form of A. We may write

$$D_{a,b} = \begin{bmatrix} a & 0 \\ 0 & a \end{bmatrix} + \begin{bmatrix} 0 & -b \\ b & 0 \end{bmatrix},$$

and compute $e^{tD_{a,b}}$ using Corollary 3.1 (a) and a power series argument to get

$$e^{tD_{a,b}} = e^{at} \begin{bmatrix} \cos bt & -\sin bt \\ \sin bt & \cos bt \end{bmatrix}.$$

We can recover e^{tA}, if needed explicitly, as

$$e^{tA} = Se^{tD_{a,b}}S^{-1} = [v\ u]e^{at}\begin{bmatrix}\cos bt & -\sin bt\\ \sin bt & \cos bt\end{bmatrix}[v\ u]^{-1}.$$

Example 3.7 Consider again the system from Example 3.1,

$$\dot{x} = \begin{bmatrix}3 & -2\\ 1 & 1\end{bmatrix}x.$$

The characteristic equation is $(\lambda - 3)(\lambda - 1) + 2 = \lambda^2 - 4\lambda + 5 = 0$, so the eigenvalues are $\lambda = 2 \pm i$. An eigenvector for $2 + i$ is

$$w = \begin{bmatrix}1+i\\ 1\end{bmatrix} = \begin{bmatrix}1\\ 1\end{bmatrix} + i\begin{bmatrix}1\\ 0\end{bmatrix} =: u + iv.$$

Set $x = Sz$, with

$$S = [v \quad u] = \begin{bmatrix}1 & 1\\ 0 & 1\end{bmatrix} \quad \text{and} \quad S^{-1} = \begin{bmatrix}1 & -1\\ 0 & 1\end{bmatrix}.$$

Then

$$\dot{z} = D_{2,1}z := \begin{bmatrix}2 & -1\\ 1 & 2\end{bmatrix}z.$$

As indicated above for $e^{tD_{a,b}}$, we have

$$e^{tD_{2,1}} = e^{2t}\begin{bmatrix}\cos t & -\sin t\\ \sin t & \cos t\end{bmatrix}.$$

By Corollary 3.1, it follows that

$$\begin{aligned}e^{tA} = Se^{tD_{2,1}}S^{-1} &= \begin{bmatrix}1 & 1\\ 0 & 1\end{bmatrix}(e^{2t})\begin{bmatrix}\cos t & -\sin t\\ \sin t & \cos t\end{bmatrix}\begin{bmatrix}1 & -1\\ 0 & 1\end{bmatrix}\\ &= \begin{bmatrix}e^{2t}(\cos t + \sin t) & -2e^{2t}\sin t\\ e^{2t}\sin t & e^{2t}(\cos t - \sin t)\end{bmatrix}.\end{aligned}$$

Thus, the real-valued solution of $\dot{x} = Ax$, $x(0) = v \in \mathbf{R}^2$, is given by

$$e^{tA}v = \begin{bmatrix}e^{2t}(\cos t + \sin t) & -2e^{2t}\sin t\\ e^{2t}\sin t & e^{2t}(\cos t - \sin t)\end{bmatrix}\begin{bmatrix}v_1\\ v_2\end{bmatrix},$$

as promised at the end of Example 3.1. △

3.3.2 Jordan Form of n-Dimensional Systems

Theorem 3.3 below states that every real $n \times n$ matrix A is similar to a canonical form known as a real Jordan form matrix.

As shown by the two-dimensional examples above, property (c) of Corollary 3.1 is very convenient for use with real Jordan forms. If S transforms A to real Jordan form $J = S^{-1}AS$, then $e^{tJ} = e^{tS^{-1}AS} = S^{-1}e^{tA}S$, and therefore $e^{tA} = Se^{tJ}S^{-1}$. In principle, the exponential of a Jordan form can always be computed by writing $J = D + N$, where D is diagonal and N is nilpotent with $DN = ND$, so that Corollary 3.1 (a) applies. This procedure yields $e^{tJ} = e^{tD}e^{tN}$, and e^{tN} will be a finite sum.

A useful fact to remember is that, if A is block diagonal, say

$$A = \operatorname{diag}\,[A_1, A_2, \ldots, A_k],$$

then e^{tA} is also block diagonal:

$$e^{tA} = \operatorname{diag}\,[e^{tA_1}, e^{tA_2}, \ldots, e^{tA_k}].$$

We can now state the result on existence of the real Jordan form.

Theorem 3.3 (Real Jordan Form)
For every real $n \times n$ matrix A, there exists a real nonsingular matrix S such that $J := S^{-1}AS$ is a block diagonal matrix of the form

$$J = \begin{bmatrix} J_{\lambda_1} & & & & & & & \\ & J_{\lambda_2} & & & & & & \\ & & \ddots & & & & & \\ & & & J_{\lambda_r} & & & & \\ & & & & J_{a_1,b_1} & & & \\ & & & & & J_{a_2,b_2} & & \\ & & & & & & \ddots & \\ & & & & & & & J_{a_s,b_s} \end{bmatrix}, \tag{3.10}$$

where each Jordan block J_{λ_j} (for λ_j real) is of the form

$$J_{\lambda_j} = \begin{bmatrix} \lambda_j & 1 & & & \\ & \lambda_j & 1 & & \\ & & \ddots & & \\ & & & \ddots & 1 \\ & & & & \lambda_j \end{bmatrix}, \tag{3.11}$$

and each Jordan block J_{a_j,b_j} *(associated with a conjugate pair of eigenvalues* $\lambda_j, \bar{\lambda}_j = a_j \pm ib_j$*) is of the form*

$$J_{a_j,b_j} = \begin{bmatrix} D_{a_j,b_j} & I & & & \\ & D_{a_j,b_j} & I & & \\ & & \ddots & & \\ & & & \ddots & I \\ & & & & D_{a_j,b_j} \end{bmatrix}, \tag{3.12}$$

where

$$D_{a_j,b_j} = \begin{bmatrix} a_j & -b_j \\ b_j & a_j \end{bmatrix}, \quad I = \begin{bmatrix} 1 & 0 \\ 0 & 1 \end{bmatrix}. \tag{3.13}$$

Proof. A complete proof is available in [40]. □

Much more can be said about Jordan forms. For our purposes, only a few facts are especially important. There can be more than one Jordan block for a given eigenvalue, and a Jordan block may be of size 1×1. The number of Jordan blocks for each real eigenvalue equals the geometric multiplicity of the eigenvalue, that is, the number of linearly independent eigenvectors for that eigenvalue. Similarly, the number of Jordan blocks of the form J_{a_j,b_j} equals the geometric multiplicity of the eigenvalue $a_j + ib_j$, which is the same as the geometric multiplicity of the conjugate eigenvalue $a_j - ib_j$.

If the transformation S is allowed to be a complex matrix, then there exists a complex nonsingular S such that $J := S^{-1}AS$ is block diagonal with the diagonal blocks (in general of different size) all having the form J_{λ_j} as in (3.11) (including now any complex eigenvalues λ_j as well). The matrix J is then called the *complex Jordan form* of A. It is unique except for the ordering of the blocks. For more on Jordan forms, see [40], [67], or [75].

We now consider the exponential of a real Jordan form. By Theorem 3.3, it suffices to carry out the computation of the exponentials for the two types of Jordan blocks given by (3.11) and (3.12).

First, consider a block of the form (3.11), associated with a real eigenvalue λ, and write this block as

$$J_\lambda = \begin{bmatrix} \lambda & 1 & & & \\ & \lambda & 1 & & \\ & & \ddots & & \\ & & & \ddots & 1 \\ & & & & \lambda \end{bmatrix}.$$

We now write $J_\lambda = D + N$, where D is the diagonal matrix with λ's down the main diagonal, and N is the nilpotent matrix with ones on the superdiagonal and zeros elsewhere. By Corollary 3.1 (a), if the size of the block J_λ is $m \times m$, that is, $N^m = 0$ but $N^{m-1} \neq 0$, then we have

$$e^{tJ_\lambda} = e^{tD}e^{tN} = \operatorname{diag}[e^{\lambda t}, \dots, e^{\lambda t}] \left(I + tN + \frac{t^2}{2!}N^2 + \cdots + \frac{t^{(m-1)}}{(m-1)!}N^{m-1} \right)$$

$$= e^{\lambda t} \begin{bmatrix} 1 & t & \frac{t^2}{2!} & \cdots & \frac{t^{(m-1)}}{(m-1)!} \\ & 1 & t & & \vdots \\ & & & \ddots & \frac{t^2}{2!} \\ & & & \ddots & t \\ & & & & 1 \end{bmatrix}$$

$$= \begin{bmatrix} e^{\lambda t} & te^{\lambda t} & \frac{t^2}{2!}e^{\lambda t} & \cdots & \frac{t^{(m-1)}}{(m-1)!}e^{\lambda t} \\ & e^{\lambda t} & te^{\lambda t} & & \vdots \\ & & & \ddots & \frac{t^2}{2!}e^{\lambda t} \\ & & & \ddots & te^{\lambda t} \\ & & & & e^{\lambda t} \end{bmatrix}.$$

Now consider a block of the form (3.12), associated with a pair of complex conjugate eigenvalues $\lambda = a \pm ib$, and write this block as

$$J_{a,b} = \begin{bmatrix} D_{a,b} & I & & & \\ & D_{a,b} & I & & \\ & & \ddots & & \\ & & \ddots & I & \\ & & & & D_{a,b} \end{bmatrix}.$$

Now write $J_{a,b} = D + N$, where D is the block diagonal matrix with the blocks $D_{a,b}$ down the main diagonal, and N is the nilpotent matrix with 2×2 identity matrices I on the block superdiagonal and zeros elsewhere. Then $DN = ND$. Suppose that $J_{a,b}$ is $2j \times 2j$, so there are exactly j of the $D_{a,b}$ blocks on the main block diagonal of $J_{a,b}$. We have seen that

$$e^{tD_{a,b}} = e^{at} \begin{bmatrix} \cos bt & -\sin bt \\ \sin bt & \cos bt \end{bmatrix},$$

and finally

$$e^{tJ_{a,b}} = e^{tD}e^{tN} = \begin{bmatrix} e^{tD_{a,b}} & te^{tD_{a,b}} & \cdots & \cdots & \dfrac{t^{j-1}}{(j-1)!}e^{tD_{a,b}} \\ 0 & e^{tD_{a,b}} & te^{tD_{a,b}} & \cdots & \vdots \\ \vdots & \ddots & \ddots & \ddots & \vdots \\ \vdots & & \ddots & e^{tD_{a,b}} & te^{tD_{a,b}} \\ 0 & \cdots & \cdots & 0 & e^{tD_{a,b}} \end{bmatrix}.$$

Bounds on e^{tA} for Hurwitz A. We now apply the real Jordan form to obtain bounds on the solutions of $\dot{x} = Ax$ when A is Hurwitz. Given A, there is a nonsingular matrix S such that $A = SJS^{-1}$, and $e^{tA} = Se^{tJ}S^{-1}$, so we have

$$\|e^{tA}\| \le \|S\| \, \|e^{tJ}\| \, \|S^{-1}\|.$$

Assuming that A is Hurwitz, let $\alpha > 0$ such that

$$\max\{\mathrm{Re}\,(\lambda) \, : \, \lambda \text{ is an eigenvalue of } A\} < -\alpha < 0.$$

Factor out $e^{-\alpha t}$ from each entry of e^{tJ}, and define $M(t) = [m_{ij}(t)]$ by the equation

$$e^{tJ} =: e^{-\alpha t} M(t).$$

Then each entry $m_{ij}(t)$ of $M(t)$ is bounded on the interval $[0, \infty)$. Choose $m > 0$ such that

$$|m_{ij}(t)| \le m, \quad \text{for } 1 \le i, j \le n, \ t \ge 0.$$

Then $\|M(t)\| \le n^2 m$ for $t \ge 0$, and we have

$$\|e^{tJ}\| = e^{-\alpha t}\|M(t)\| \le n^2 m e^{-\alpha t}, \quad t \ge 0.$$

Recalling that $e^{tA} = Se^{tJ}S^{-1}$, we have

$$\|e^{tA}\| = \|Se^{tJ}S^{-1}\| \le Ke^{-\alpha t},$$

where $K := n^2 m \, \|S\| \, \|S^{-1}\|$. Observe that we have again reached the estimate given earlier in Theorem 3.2, but this time by an argument using the Jordan form of A. The constant α is easily related to the largest real part of an eigenvalue of A. For example, we may take α to be half of the largest real part of an eigenvalue of the Hurwitz matrix A. Again, this alternative approach (via the Jordan form) shows that if A is Hurwitz, then every solution of $\dot{x} = Ax$ satisfies $x(t) \to 0$ as $t \to \infty$. To complete the picture, we prove the following result.

Theorem 3.4 *If $A \in \mathbf{R}^{n \times n}$ has an eigenvalue with nonnegative real part, then there is a real nonzero solution $x(t)$ of $\dot{x} = Ax$ such that $x(t)$ does not tend to zero as $t \to \infty$.*

Proof. Suppose $\lambda = a+ib$ is an eigenvalue of A, with $a \geq 0$, and $w = u+iv$ is a corresponding eigenvector. Then $e^{(a+ib)t}(u+iv)$ is a complex solution. The real part of this solution, $x(t) := e^{at}(\cos(bt)u - \sin(bt)v)$, is a real-valued solution which is either (i) nonzero periodic, if $a = 0$, or (ii) unbounded, if $a > 0$. □

Taken together, Corollary 3.2 and Theorem 3.4 say that each solution of $\dot{x} = Ax$ converges asymptotically to the origin if and only if A is Hurwitz.

The next example shows that real Jordan forms can arise naturally when simple systems are interconnected by feedback.

Example 3.8 Consider two identical controlled harmonic oscillators, each described by the two-dimensional system

$$\dot{x} = \begin{bmatrix} 0 & 1 \\ -1 & 0 \end{bmatrix} x + \begin{bmatrix} 0 \\ 1 \end{bmatrix} u,$$
$$y = \begin{bmatrix} 1 & 0 \end{bmatrix} x.$$

The *series* connection of these two identical systems uses the output of one as a scalar input to the other, and is modeled by the four-dimensional system

$$\dot{z} = A_s z + b_s u = \begin{bmatrix} 0 & 1 & 0 & 0 \\ -1 & 0 & 0 & 0 \\ 0 & 0 & 0 & 1 \\ 1 & 0 & -1 & 0 \end{bmatrix} z + \begin{bmatrix} 0 \\ 1 \\ 0 \\ 0 \end{bmatrix} u,$$
$$y = \begin{bmatrix} 0 & 0 & 1 & 0 \end{bmatrix} z.$$

The *parallel* connection of the two systems uses the same input structure with the output being the sum of the outputs of each system. Thus the parallel connection is modeled by the four-dimensional system

$$\dot{z} = A_p z + b_p u = \begin{bmatrix} 0 & 1 & 0 & 0 \\ -1 & 0 & 0 & 0 \\ 0 & 0 & 0 & 1 \\ 0 & 0 & -1 & 0 \end{bmatrix} z + \begin{bmatrix} 0 \\ 1 \\ 0 \\ 1 \end{bmatrix} u,$$
$$y = \begin{bmatrix} 1 & 0 & 1 & 0 \end{bmatrix} z.$$

It is straighforward to check that the eigenvalues of A_s and A_p are the same, namely, $\pm i$, with each of these having algebraic multiplicity 2; the characteristic polynomial for each system matrix is $(\lambda^2 + 1)(\lambda^2 + 1)$. It is also easy to check that the eigenvalue i for A_p has geometric multiplicity 2, as does the eigenvalue $-i$. On the other hand, matrix A_s has only one independent eigenvector associated with the eigenvalue i. Thus, A_s and A_p have distinct real Jordan forms.

The unforced parallel connection has bounded solutions, because there are two decoupled subsystems, each subsystem being a harmonic oscillator.

On the other hand, the unforced series connection allows the component z_1 as an input that drives the second oscillator. Suppose $z_1 = \sin t$. The natural frequency of z_1 exactly matches the natural frequency of the unforced second oscillator, producing a resonance that allows unbounded solutions for z_3 and z_4. It is not necessary to compute the full system solution to see this. Observe that the equation $\ddot{y} + y = \sin t$ has unbounded solutions due to the exact match (the *resonance*) between the frequency of $\sin t$ and the natural frequency of the homogeneous system for z_3 and z_4. (See Exercise 3.5.) $\triangle$

3.4 THE CAYLEY-HAMILTON THEOREM

Suppose A is a diagonalizable $n \times n$ complex matrix. Then there exists a basis of $\mathbf{C}^n$ consisting of eigenvectors $v_1, \ldots, v_n$ of A. Let $Av_j = \lambda_j v_j$ for $1 \le j \le n$. We may write the characteristic polynomial of A as

$$p(\lambda) = \sum_{k=1}^{n} a_k \lambda^k = (\lambda - \lambda_1) \cdots (\lambda - \lambda_n).$$

The polynomial $p(\lambda)$ determines a mapping of $n \times n$ matrices, $B \mapsto p(B)$, with the matrix polynomial $p(B)$ defined by substitution of B for λ and the identity I for the scalar unit. In particular, we have

$$p(A) = \sum_{k=1}^{n} a_k A^k = (A - \lambda_1 I) \cdots (A - \lambda_n I)\,.$$

The linear factors in this operator polynomial may be arranged in any order, as these factors commute as linear operators, being polynomials in A. By applying $p(A)$ to each of the basis vectors v_i, $i = 1, \ldots, n$, we find that $p(A)$ must be the zero matrix. We say that A satisfies its own characteristic equation.

It is important to realize that not all square matrices are diagonalizable. However, it is true that every $n \times n$ complex matrix satisfies its own characteristic equation. This is the assertion of the Cayley-Hamilton Theorem (Theorem 3.5 below).

Example 3.9 The matrix

$$A = \begin{bmatrix} 2 & 1 \\ 0 & 2 \end{bmatrix}$$

is not diagonalizable. There is only a one-dimensional eigenspace associated with the eigenvalue $\lambda = 2$, so there is no basis of $\mathbf{C}^2$ consisting of eigenvectors of A. The characteristic equation of A is

$$\det(\lambda I - A) = (\lambda - 2)^2 = 0,$$

and it is easy to verify that A satisfies the matrix equation

$$(A - 2I)^2 = A^2 - 4A + 4I = 0$$

as guaranteed by the Cayley-Hamilton Theorem. △

We now state and prove the Cayley-Hamilton Theorem.

Theorem 3.5 (Cayley-Hamilton Theorem)
Let A be a real or complex $n \times n$ matrix, and

$$p(\lambda) = \det(\lambda I - A) = \sum_{k=0}^{n} a_k \lambda^k$$

the characteristic polynomial of A. Then A satisfies its own characteristic equation, that is,

$$p(A) = \sum_{k=0}^{n} a_k A^k = 0_{n\times n}.$$

Proof. Write $p(\lambda) = \det(\lambda I - A) = \prod_{j=1}^{k}(\lambda - \lambda_j)^{m_j}$, where $\lambda_1, \dots, \lambda_k$ are the distinct eigenvalues of A, having algebraic multiplicity m_j for $j = 1, \dots, k$. By the Primary Decomposition theorem, C^n is the direct sum of the generalized eigenspaces of A:

$$C^n = N((A - \lambda_1 I)^{m_1}) \oplus N((A - \lambda_2 I)^{m_2}) \cdots \oplus N((A - \lambda_k I)^{m_k}). \quad (3.14)$$

The matrix $p(A)$ annihilates each of these null spaces: Let $x \in N((A - \lambda_j I)^{m_j})$. Then, since the primary factors commute, $p(A)x$ may be evaluated by first evaluating $(A - \lambda_j I)^{m_j} x$, which gives the zero vector. Thus, $p(A)$ annihilates each of the null spaces in the primary decomposition (3.14); therefore $p(A)x = 0$ for all x. This completes the proof. □

There are other approaches to proving the Cayley-Hamilton Theorem.

3.5 LINEAR TIME VARYING SYSTEMS

In this section, we consider the initial value problem for a nonautonomous linear system,

$$\dot{x} = A(t)x, \quad x(t_0) = x_0. \quad (3.15)$$

If $A(t)$ is continuous on an interval J, then existence and uniqueness for the problem (3.15) *on the entire interval* J follows from Theorem 2.3 (existence and uniqueness) and Theorem C.3 (extension to the entire interval J) under a continuity assumption on the $n \times n$ matrix function $A(t)$, as stated in the following corollary.

Corollary 3.3 (Existence and Uniqueness for LTV Systems)
Suppose $A(t)$ is continuous for t in the interval J. If $(t_0, x_0) \in J \times \mathbf{R}^n$, then

there is a unique solution of the initial value problem (3.15) *defined on the entire interval* J.

The interval J in Corollary 3.3 may contain one or both or neither of its endpoints, and we may have $J = (-\infty, \infty)$. The derivative of a time function at an endpoint of an interval J is the appropriate one-sided derivative at that point.

A *fundamental matrix solution* of a linear system $\dot{x} = A(t)x$, $(t, x) \in J \times \mathbf{R}^n$, is an $n \times n$ matrix function $M(t)$ defined on J, such that $M(t)$ is nonsingular for all $t \in J$ and

$$\dot{M}(t) = A(t)M(t), \quad \text{for all } t \in J.$$

The next result states several important facts about the general solution of the system.

Theorem 3.6 (General Solution of LTV Systems)
Let $A(t)$ be a real $n \times n$ matrix function, $t \in J$. The following statements hold for solutions of the system $\dot{x} = A(t)x$.

(a) *If $x^1(t), \ldots, x^n(t)$ are (column vector) solutions of $\dot{x} = A(t)x$ and*

$$M(t) := [x^1(t) \ \cdots \ x^n(t)],$$

then $M(t)$ is a matrix solution of the system, that is, $M(t)$ satisfies the linear matrix differential equation

$$\dot{M}(t) = A(t)M(t).$$

(b) *If $x^1(t), \ldots, x^n(t)$ are solutions, $M(t)$ is defined as in (a), and $M(t_0)$ is nonsingular for some t_0, then each solution can be written as $x(t) = M(t)v$ for some constant vector v.*

(c) *If $M(t)$ is a matrix solution such that $M(t_0)$ is nonsingular for some t_0, then $M(t)$ is a fundamental matrix solution on J.*

Proof. (a) With $M(t)$ as defined, we have

$$\begin{aligned} \dot{M}(t) &= \begin{bmatrix} \dot{x}^1(t) & \cdots & \dot{x}^n(t) \end{bmatrix} \\ &= \begin{bmatrix} A(t)x^1(t) & \cdots & A(t)x^n(t) \end{bmatrix} \\ &= A(t)M(t). \end{aligned}$$

(b) If the indicated solution vectors are linearly independent at time t_0, and $x(t)$ is any other solution, then there is a unique solution for v in the system of linear algebraic equations $x(t_0) = M(t_0)v$. But then the functions $x(t)$ and $M(t)v$ are solutions of the system with the same value at time t_0. By uniqueness of solutions of initial value problems, we have $x(t) = M(t)v$ for all $t \in J$.

(c) Consider the contrapositive statement: If $M(t)$ is a matrix solution that is singular at some time t_0, then $M(t)$ is singular at all t. Assume that $M(t)$ is a matrix solution and $M(t_0)$ is singular for some t_0. Then $M(t_0)$

has a linear dependency among its columns, so there is a constant vector v such that $M(t_0)v = 0$. But then the function $M(t)v$ and the zero function are solutions with the same value at t_0. By uniqueness, we have $M(t)v = 0$ for all t, which shows that $M(t)$ is singular for all t. □

If $g(t)$ is continuous on J, and $M(t)$ is a fundamental matrix solution for $\dot{x} = A(t)x$ on J, then the general solution of the nonhomogeneous equation $\dot{x} = A(t)x + g(t)$, $t \in J$, is given by the *variation of parameters* formula,

$$x(t) = M(t)\left(M^{-1}(t_0)x_0 + \int_{t_0}^{t} M^{-1}(s)g(s)\,ds\right), \quad \text{where } x(t_0) = x_0.$$

If A is constant, and we take $M(t) = e^{tA}$, then the variation of parameters formula reads

$$x(t) = e^{tA}x_0 + e^{tA}\int_0^t e^{-sA}g(s)\,ds.$$

The stability analysis of linear time varying systems is not discussed in detail in this book. We offer a few general remarks to serve as motivation for this interesting and important area. And we give an example as a warning that additional concepts are needed for a detailed stability analysis.

Observe that if all eigenvalues of $A^T(t) + A(t)$ are negative real, then the function $V(x) = x^T x$ strictly decreases along nonzero solutions of $\dot{x} = A(t)x$. It follows that every solution of $\dot{x} = A(t)x$ must approach the origin as $t \to \infty$; the reader might try this as an exercise, or see [82] (Theorem 8.2, page 132) for the details.

The next example is a good one to remember, as it shows that negative real parts for all eigenvalues is not a sufficient condition for asymptotic convergence of all solutions to the origin. (Compare this with Corollary 3.2 on time invariant linear systems.)

Example 3.10 Consider the system

$$\dot{x} = \begin{bmatrix} -1 & e^{2t} \\ 0 & -1 \end{bmatrix} x, \quad t \in (-\infty, \infty).$$

For each t, the coefficient matrix has -1 as a repeated eigenvalue. The solution for x_2 is $x_2(t) = e^{-t}x_{20}$. If we substitute this into the equation for x_1, then variation of parameters gives the solution

$$\begin{aligned} x_1(t) &= e^{-t}x_{10} + e^{-t}\left(\int_0^t e^{3s}x_2(s)\,ds\right) \\ &= e^{-t}x_{10} + e^{-t}\left(\int_0^t e^{2s}x_{20}\,ds\right) \\ &= e^{-t}x_{10} + e^{-t}\frac{1}{2}\left(e^{2t}x_{20} - x_{20}\right) \\ &= e^{-t}x_{10} + \frac{1}{2}e^{t}x_{20} - \frac{1}{2}e^{-t}x_{20}. \end{aligned}$$

Because of the exponential growth term, if $x_{20} \neq 0$ then $x_1(t) \to \infty$ as $t \to \infty$. Thus, negative real parts for all eigenvalues is not a sufficient condition for asymptotic convergence of all solutions to the origin in a linear time-varying system. △

Despite cases such as Example 3.10, eigenvalues can indeed exert influence on the asymptotic behavior of solutions of linear time varying systems. If A is Hurwitz, then the origin of a system of the form

$$\dot{x} = (A + B(t))x$$

will be globally attractive, all solutions approaching the origin as $t \to \infty$, if the perturbation $B(t)$ satisfies

$$\lim_{t \to \infty} B(t) = 0,$$

or even if $B(t)$ is suitably bounded for large t. See [15] (Theorem 4.2, Corollary to Theorem 4.2, and Exercise 15, pp. 156-158).

3.6 THE STABILITY DEFINITIONS

The stability definitions given here apply throughout the text. We begin with the definitions of Lyapunov stability and asymptotic stability of an equilibrium for a system of ordinary differential equations. Then we study in some detail the case of linear time invariant systems, including the role of Lyapunov equations.

3.6.1 Motivations and Stability Definitions

Our main interest in this text is in time invariant (autonomous) systems

$$\dot{x} = f(x), \tag{3.16}$$

where $f : D \to \mathbf{R}^n$ and D is an open set in $\mathbf{R}^n$ containing the origin. The basic definitions for stability will be stated only for (3.16).

An *equilibrium solution* is a constant solution. The next lemma is immediate.

Lemma 3.1 *The vector x_0 is an equilibrium (constant) solution of* (3.16) *if and only if $f(x_0) = 0$.*

We assume that solutions are uniquely determined by the initial condition x in D, and we write $\phi(t, x) = \phi_t(x)$ for the solution with $\phi(0, x) = \phi_0(x) = x$. The domain of a solution is the maximal interval of existence.

TRANSLATION OF EQUILIBRIUM. The Jacobian linearization of system (3.16) at the origin is the linear time invariant system

$$\dot{x} = Ax, \tag{3.17}$$

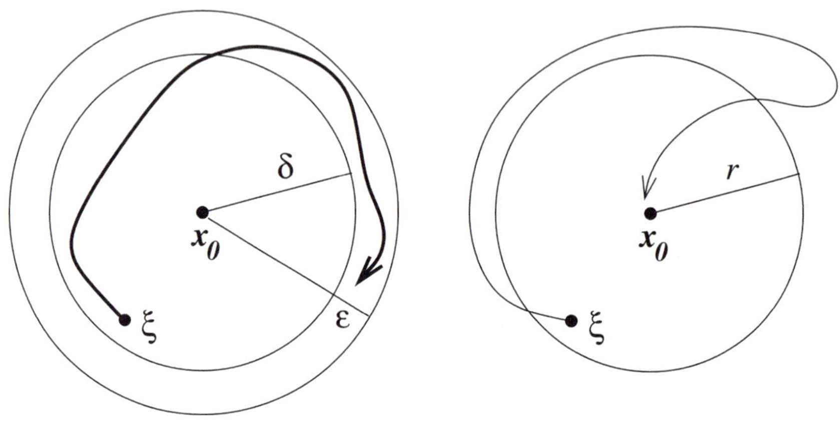

Figure 3.1 Stability of Equilibrium—Definition 3.3 (a).

Figure 3.2 Asymptotic stability of equilibrium—Definition 3.3 (b).

where $A = Df(0)$ is the Jacobian matrix of the vector field f, evaluated at $x = 0$. If $f(x_0) = 0$ and we want to study the stability of a nonzero equilibrium x_0, then, by a translation coordinate change, $z = x - x_0$, we can consider the equilibrium $z = 0$ of the transformed nonlinear system, $\bar{f}(z) := f(z + x_0)$, for which $D\bar{f}(0) = Df(x_0)$.

This is a good time to recall Examples 2.13 and 2.14, as they provide motivation for the developments of this section. The next example is instructive as well.

Example 3.11 The two-dimensional time invariant system given in polar coordinates by

$$\dot{r} = r(1 - r),$$
$$\dot{\theta} = \sin^2(\theta/2)$$

has a unique solution for any initial condition in the plane. There is an equilibrium at $(x_{10}, x_{20}) = (1, 0)$ (equivalently, at $(r_0, \theta_0) = (1, 0)$). An analysis of the phase portrait shows that $\lim_{t\to\infty}(x_1(t), x_2(t)) = (1, 0)$ for every nonzero solution $(x_1(t), x_2(t))$. However, it cannot be said that solutions with initial condition close to $(1, 0)$ stay close to $(1, 0)$ for all $t \geq 0$. We do not wish to call such an equilibrium stable, but it is reasonable to call it attractive. The point is that for nonlinear systems, an equilibrium can be attractive and yet nearby initial conditions have trajectories that make large excursions away from the equilibrium. (See Exercise 3.11) △

Stability concepts refer to properties of individual solutions of a system. Our interest is in equilibrium solutions, and we generally assume that the equilibrium of interest is $x_0 = 0$, but we state the definitions for an arbitrary equilibrium of a smooth vector field. Here are the definitions of stability and asymptotic stability of equilibria. (See also Figure 3.1 and Figure 3.2.)

Definition 3.3 *Suppose that x_0 is an equilibrium solution of* (3.16)*, where $f : D \to \mathbf{R}^n$. Then*

(a) x_0 *is* (Lyapunov) stable *if, for any $\epsilon > 0$ there exists a $\delta(\epsilon) > 0$ such that*

$$\|\xi - x_0\| < \delta(\epsilon) \implies \|\phi(t, \xi) - x_0\| < \epsilon \quad \textit{for all } t \geq 0.$$

(b) x_0 *is* asymptotically stable *if it is Lyapunov stable and there exists a number $r > 0$ such that*

$$\|\xi - x_0\| < r \implies \lim_{t \to \infty} \phi(t, \xi) = x_0.$$

(c) x_0 *is* globally asymptotically stable in D *if it is Lyapunov stable, and for every initial condition $\xi \in D$, we have*

$$\lim_{t \to \infty} \phi(t, \xi) = x_0.$$

(This term usually refers to the case where $D = \mathbf{R}^n$.)

(d) x_0 *is* unstable *if it is not (Lyapunov) stable.*

Let $f(x)$ be an autonomous C^1 vector field on an open set $D \subset \mathbf{R}^n$, with $f(p) = 0$ for some $p \in D$. We view the domain of f as $\mathbf{R} \times D$, since the vector field is time invariant. It is not hard to see that a nonconstant solution $x(t)$ of $\dot{x} = f(x)$, with maximal forward interval $[0, \hat{t})$, can satisfy $x(t) \to p$ as $t \to \hat{t}$ only if the approach is asymptotic, that is, $\hat{t} = \infty$. If $\hat{t}$ were finite, then $x(t)$ could be extended to $[0, \hat{t}]$ by defining $x(\hat{t}) = p$. But this contradicts uniqueness of solutions. Hence, a nonconstant solution cannot reach an equilibrium in finite time.

If there exists a number $r > 0$ such that

$$\|z - x_0\| < r \quad \implies \quad \lim_{t \to \infty} \phi(t, z) = x_0,$$

then the equilibrium x_0 is called *attractive*. Thus, an equilibrium is asymptotically stable if and only if it is attractive and Lyapunov stable. Example 3.11 shows that attractivity does not imply Lyapunov stability, and therefore attractivity does not imply asymptotic stability.

We have seen that linear systems (3.17) with an attractive equilibrium at the origin are characterized by the fact that all eigenvalues of the matrix A have negative real part. Moreover, the norm estimate of Corollary 3.2 shows that for linear systems, attractivity implies Lyapunov stability. Hence, for linear systems, attractivity implies asymptotic stability. In summary, for linear systems (3.17), attractivity of the origin is equivalent to (global) asymptotic stability, which is equivalent to the matrix A being Hurwitz.

3.6.2 Lyapunov Theory for Linear Systems

We present the basic Lyapunov equations in some detail. The study of these equations leads to another characterization of asymptotic stability for linear time invariant systems.

Example 2.13 and Example 2.14 suggest that stability questions can be related to the behavior of a positive definite function along solution curves. In particular, in Example 2.14, we found a positive definite P such that

$$A^T P + PA = -I,$$

which gave a strict decrease of the function $V(x) = x^T Px$ along nonzero solutions. A systematic study of these questions involves the basic Lyapunov equation

$$A^T P + PA + W = 0, \tag{3.18}$$

where W is a given positive definite matrix, and we are interested in a positive definite solution P of (3.18).

The next proposition says that if (3.18) has a positive definite solution P for some given positive definite W, then A must be Hurwitz.

Proposition 3.1 *Let $A \in \mathbf{R}^{n\times n}$, and suppose $W \in \mathbf{R}^{n\times n}$ is positive definite. If P is positive definite and satisfies* (3.18), *then A is Hurwitz.*

Proof. Let λ be an eigenvalue of A. We will show that $\text{Re}\,(\lambda) < 0$. Let $v \neq 0$ be an eigenvector for λ. Then

$$v^*(A^T P + PA)v = (Av)^* Pv + v^* P(Av) = \bar{\lambda} v^* Pv + \lambda v^* Pv = (\bar{\lambda} + \lambda) v^* Pv.$$

By hypothesis, $A^T P + PA$ is negative definite and P is positive definite, so we must have $\bar{\lambda} + \lambda = 2\text{Re}\,(\lambda) < 0$. Since λ was an arbitrary eigenvalue of A, A must be Hurwitz. □

With P as in Proposition 3.1, we conclude that the norm defined by $||x||_P^2 := x^T Px$ strictly decreases along nonzero solutions of $\dot{x} = Ax$. In fact,

$$\frac{d}{dt}||x||_P^2 = x^T(A^T P + PA)x = -x^T Wx < 0 \quad \text{for } x \neq 0.$$

Since A is Hurwitz, Corollary 3.2 implies that $||x||_P \to 0$ as $t \to \infty$ for each solution $x(t)$ of $\dot{x} = Ax$. Thus, Proposition 3.1 gives a sufficient condition for asymptotic stability of the origin of $\dot{x} = Ax$. Before moving ahead, we compute a solution for a simple case of (3.18).

Example 3.12 Let

$$A = \begin{bmatrix} -1 & 9 \\ -1 & -1 \end{bmatrix},$$

which is the coefficient matrix in Example 2.14. Let $W = I$. Then W is positive definite. We will find a symmetric solution P of (3.18). Write

$$P = \begin{bmatrix} a & b \\ b & c \end{bmatrix}.$$

By direct calculation, the Lyapuov equation (3.18) is equivalent to the system of algebraic equations

$$\begin{aligned} -2a - 2b &= -1, \\ 9a - 2b - c &= 0, \\ 18b - 2c &= -1. \end{aligned}$$

There is a unique solution given by $a = \frac{3}{10}$, $b = \frac{1}{5}$, and $c = \frac{23}{10}$. Therefore the unique solution P of (3.18) is

$$P = \begin{bmatrix} \frac{3}{10} & \frac{1}{5} \\ \frac{1}{5} & \frac{23}{10} \end{bmatrix}.$$

It is straightforward to check that P is positive definite. Proposition 3.1 implies that A is Hurwitz; this can be verified, of course, by knowledge of the eigenvalues. This solution of (3.18) explains our choice of P in Example 2.14. $\triangle$

The next proposition summarizes the most important facts about the Lyapunov equation (3.18), when A is Hurwitz.

Proposition 3.2 *Suppose $A \in \mathbf{R}^{n\times n}$ is Hurwitz. Then the following statements are true:*

(a) *For every matrix $W \in \mathbf{R}^{n\times n}$, there is a unique solution $P \in \mathbf{R}^{n\times n}$ of (3.18), and P is given by*

$$P = \int_0^\infty e^{tA^T} W e^{tA}\, dt. \tag{3.19}$$

(b) *If W is positive semidefinite, then P is positive semidefinite.*
(c) *If W is positive definite, then P is positive definite.*

Proof. We first show that P as defined in (3.19) is a solution of (3.18). From the proof of Theorem 3.2, there are positive constants α, M such that $||e^{tA}|| \le Me^{-\alpha t}$ for $t \ge 0$. Therefore the integral in (3.19) converges absolutely, hence P is well-defined. Thus, we may write

$$A^T P + PA = \int_0^\infty \left[A^T e^{tA^T} W e^{tA} + e^{tA^T} W e^{tA} A\right] dt.$$

Recalling that

$$e^{tA} A = \frac{d(e^{tA})}{dt} \quad \text{and} \quad A^T e^{tA^T} = \frac{d(e^{tA^T})}{dt},$$

we have

$$\begin{aligned} A^T P + PA &= \int_0^\infty \left[\frac{d(e^{tA^T})}{dt} W e^{tA} + e^{tA^T} W \frac{d(e^{tA})}{dt} \right] dt \\ &= \int_0^\infty \frac{d}{dt}(e^{tA^T} W e^{tA})\, dt \\ &= \lim_{b\to\infty} \int_0^b \frac{d}{dt}(e^{tA^T} W e^{tA})\, dt \\ &= \lim_{b\to\infty} (e^{bA^T} W e^{bA} - W) = -W, \end{aligned}$$

where we used the facts that $\lim_{b\to\infty} e^{bA} = 0$ and $\lim_{b\to\infty} e^{bA^T} = 0$. Therefore P solves equation (3.18).

We now show that (3.18) has a unique solution. Suppose $\bar{P}$ is a solution. Let $b > 0$, and compute that

$$\begin{aligned} \int_0^b e^{tA^T} W e^{tA}\, dt &= -\int_0^b e^{tA^T} (A^T \bar{P} + \bar{P} A) e^{tA}\, dt \\ &= -\int_0^b \frac{d}{dt}(e^{tA^T} \bar{P} e^{tA})\, dt \\ &= \bar{P} - e^{bA^T} \bar{P} e^{bA}. \end{aligned}$$

Now take the limit as $b \to \infty$ on both sides to obtain

$$\int_0^\infty e^{tA^T} W e^{tA}\, dt = \bar{P} - \lim_{b\to\infty} e^{bA^T} \bar{P} e^{bA} = \bar{P}.$$

Therefore the solution of (3.18) is unique.

Finally, knowing that the integral in (3.19) exists, it follows from a standard limit argument that if W is symmetric then P is symmetric. Statements (b) and (c) now follow easily. (See Exercise 3.15.) □

The analytic formula (3.19) is primarily a useful theoretical tool, rather than a computational method to find P.

The hypothesis that A is Hurwitz in Proposition 3.2 is stronger than is actually needed to deduce solvability of (3.18). Suppose $\lambda_1, \ldots, \lambda_n$ are the eigenvalues of A. It can be shown that the Lyapunov equation

$$A^T P + PA + Q = 0$$

has a unique solution P for any given Q if and only if $\lambda_i + \lambda_j \neq 0$ for any pair i, j. For a simple example, see Exercise 3.16, where, in addition, a more general Lyapunov equation is considered.

Theorem 3.7 *Let $A \in \mathbf{R}^{n\times n}$. The following statements are equivalent:*

(a) *The origin of the system $\dot{x} = Ax$ is asymptotically stable.*

(b) *There exist positive definite matrices P and W that satisfy the Lyapunov equation* (3.18).

(c) *For any positive definite matrix W, the Lyapunov equation* (3.18) *has a unique solution P which is positive definite.*

(d) *A is similar to a matrix B that satisfies $B + B^T < 0$. In other words, the system $\dot{x} = Ax$ is equivalent by a linear change of coordinates to a system $\dot{z} = Bz$ for which the Euclidean norm is strictly decreasing along nonzero solutions.*

Proof. Proposition 3.2 (part (b)) shows that (a) implies (c). Proposition 3.1 shows that (b) implies (a). Therefore (b) implies (c). It is clear that (c) implies (b). Thus, statements (a), (b), and (c) are equivalent.

We now concentrate on showing that (a) is equivalent to (d). Similar matrices have the same set of eigenvalues, hence (d) implies (a). For the converse, suppose A is Hurwitz. Since we know (c) holds, if we set $W = I$ then there is a positive definite solution P of the equation

$$A^T P + PA + I = 0. \tag{3.20}$$

There is a positive definite matrix S such that $S^2 = P$; it is natural to write $S = P^{\frac{1}{2}}$ and call it the positive square root of P. Matrix $P^{\frac{1}{2}}$ is invertible, and we write $P^{-\frac{1}{2}} := (P^{\frac{1}{2}})^{-1}$. Now multiply (3.20) on the right and on the left by $P^{-\frac{1}{2}}$, and rearrange to obtain

$$P^{-\frac{1}{2}} A^T P^{\frac{1}{2}} + P^{\frac{1}{2}} A P^{-\frac{1}{2}} = -P^{-1}.$$

Note that the right hand side of the resulting equation is negative definite. Now with $B := P^{\frac{1}{2}} A P^{-\frac{1}{2}}$, we see that A is similar to B, and $B + B^T$ is negative definite. Thus (a) implies (d). This completes the proof. □

3.7 EXERCISES

Exercise 3.1 *A Lyapunov Equation*
Consider the matrix A of the system in Example 2.13.

(a) Show that there is no positive definite matrix P such that

$$A^T P + PA + I = 0.$$

(b) Suppose Q is a 2×2 positive definite matrix. Show that there is no positive definite matrix P such that

$$A^T P + PA + Q = 0.$$

Exercise 3.2 *Solving a Lyapunov Equation*
Consider the system matrix A in Example 2.14. Find a symmetric solution P of the Lyapunov equation

$$A^T P + PA = -5I.$$

Is the solution unique? Is it positive definite?

Exercise 3.3 *Stability versus Strict Norm Decrease*
With A and W as in Example 3.12, illustrate Theorem 3.7 (d) by finding a matrix B similar to A such that

$$B^T + B < 0.$$

Hint: See the proof of Theorem 3.7 for the construction of such a B.

Exercise 3.4 Consider the planar system

$$\begin{aligned}\dot{x} &= xy - x^3, \\ \dot{y} &= 0.\end{aligned}$$

(a) Find all the equilibrium points of the system.
(b) Show directly from the definitions that the origin is stable but not asymptotically stable.
(c) Show that small perturbations of the vector field, which preserve an equilibrium at the origin, can make the origin either asymptotically stable or unstable.

Exercise 3.5 *Series and Parallel Oscillator Connections*
Verify the statements made in the discussion of Example 3.8. In particular, show that the equation $\ddot{y} + y = \sin t$ has unbounded solutions. *Hint*: Try a particular solution of the form $y_p(t) = At \cos t + Bt \sin t$.

Exercise 3.6 *Working with the Characteristic Polynomial*
Consider a linear system $\dot{x} = Ax$, where A is a 3×3 real matrix with

$$\det(\lambda I - A) = (\lambda + 2)(\lambda^2 + 4).$$

Let x_0 be an arbitrary point of $\mathbf{R}^3$, and let $x(t)$ be the solution with $x(0) = x_0$.

(a) Show that $x(t)$ approaches a fixed plane in $\mathbf{R}^3$, independently of the initial condition x_0. *Hint*: Analyze the system using coordinates in which the system coefficient matrix is in real Jordan form.
(b) Show that all solutions $x(t)$ must approach either the origin or a periodic orbit of the system.

Exercise 3.7 *Solution Types*
What is the smallest positive integer n for which there is a linear differential equation with real coefficients,

$$y^{(n)} + a_1 y^{(n-1)} + \cdots + a_n y = 0,$$

having among its solutions the functions

$$3te^{2t}, \quad t \sin t, \quad -2e^{-t} \quad ?$$

Find the coefficients a_i in the differential equation.

Exercise 3.8 *Eigenvalues of a Nilpotent Matrix*
A matrix N is called *nilpotent* if $N^k = 0$ for some positive integer k. Show that N is nilpotent if and only if every eigenvalue of N is zero.

Exercise 3.9 *On Corollary* 3.2
Find an $M > 0$ and $\rho > 0$ as in Corollary 3.2 for each matrix below:

$$(a)\ \begin{bmatrix} -2 & 1 \\ 0 & -2 \end{bmatrix} \qquad (b)\ \begin{bmatrix} -2 & 50 \\ 0 & -2 \end{bmatrix} \qquad (c)\ \begin{bmatrix} -2 & -\sqrt{5} \\ \sqrt{5} & -2 \end{bmatrix}.$$

Exercise 3.10 Let A be a real or complex square matrix. Show that, in a Jordan form of A, any and all nonzero off-diagonal entries (which are ones by definition) may be taken to have an arbitrarily given positive value $\gamma > 0$. More precisely, there exists $S(\gamma)$ such that $AS(\gamma) = S(\gamma)J(\gamma)$, where $J(\gamma)$ is a Jordan form except that any and all off-diagonal unit entries are replaced by γ. *Hint*: It suffices to show this for a single Jordan block. Consider this simple case:

$$\begin{bmatrix} \lambda & 1 & 0 \\ 0 & \lambda & 1 \\ 0 & 0 & \lambda \end{bmatrix} \begin{bmatrix} 1 & 0 & 0 \\ 0 & \gamma & 0 \\ 0 & 0 & \gamma^2 \end{bmatrix} = \begin{bmatrix} 1 & 0 & 0 \\ 0 & \gamma & 0 \\ 0 & 0 & \gamma^2 \end{bmatrix} \begin{bmatrix} \lambda & \gamma & 0 \\ 0 & \lambda & \gamma \\ 0 & 0 & \lambda \end{bmatrix} = \begin{bmatrix} \lambda & \gamma & 0 \\ 0 & \gamma\lambda & \gamma^2 \\ 0 & 0 & \gamma^2\lambda \end{bmatrix}.$$

Exercise 3.11 *An Attractive Equilibrium That is not Asymptotically Stable*
Consider the two-dimensional system given in polar coordinates by

$$\dot{r} = r(1 - r),$$
$$\dot{\theta} = \sin^2\left(\frac{\theta}{2}\right).$$

This system has a unique solution for each initial condition in the plane.

(a) Show that there are two equilibria, one at $(r_0, \theta_0) = (1, 0)$ and the other at the origin.
(b) Show that for any nonzero initial condition, the solution approaches the equilibrium at $(r_0, \theta_0) = (1, 0)$, which is therefore attractive.
(c) Explain why the equilibrium $(r_0, \theta_0) = (1, 0)$ is not asymptotically stable.

Exercise 3.12 *Lyapunov Equation for a Stable System*
Suppose that the origin of the linear system $\dot{x} = Ax$ is stable. Show that there exists a positive definite matrix P such that $A^T P + PA \leq 0$.

Exercise 3.13 *Critical Cases for Linear Systems*
Consider the system $\dot{x} = Ax$, and suppose that all eigenvalues of A have nonpositive real part, and those with zero real part have geometric

multiplicity equal to their algebraic multiplicity. Show that the following statements are true.

(a) There exists a constant $M > 0$ such that

$$\|e^{tA}\| \leq M \quad \text{for } 0 \leq t < \infty.$$

(b) Every solution of $\dot{x} = Ax$ is bounded for $0 \leq t < \infty$.
(c) The equilibrium at the origin is Lyapunov stable.

Exercise 3.14 *More on Critical Cases for Linear Systems*
Suppose, as in Exercise 3.13, that all eigenvalues of A have nonpositive real part, and those with zero real part have geometric multiplicity equal to their algebraic multiplicity. Consider the system $\dot{x} = Ax + g(t)$, where $g(t)$ is continuous on $0 \leq t < \infty$ and

$$\int_0^\infty \|g(s)\|\, ds < \infty.$$

Use variation of parameters to show that every solution of $\dot{x} = Ax + g(t)$ is bounded on $0 \leq t < \infty$.

Exercise 3.15 Complete the details of the proof of Proposition 3.2, parts (b) and (c).

Exercise 3.16 *More Lyapunov Equations*
Using the concept of Kronecker product of matrices, it can be shown that the Lyapunov equation

$$BX + XA + Q = 0$$

has a unique solution X for any given Q if and only if $\lambda_i + \mu_j \neq 0$ for all pairs i, j, where the λ_i are the eigenvalues of A and μ_j are the eigenvalues of B. (See [9] pages 38–39).

(a) Find the unique solution of the Lyapunov equation $A^T P + PA + Q = 0$ if

$$A = \begin{bmatrix} -1 & 3 \\ 0 & 5 \end{bmatrix}, \qquad Q = I.$$

Note that a unique solution is guaranteed if $\lambda_1 + \lambda_2 \neq 0$, where λ_1, λ_2 are the eigenvalues of A, since A and A^T have the same eigenvalues. *Hint*: The solution is symmetric.

(b) Find the unique solution of the Lyapunov equation $BP + PA + Q = 0$ if

$$A = \begin{bmatrix} -1 & 3 \\ 0 & 5 \end{bmatrix}, \qquad B = \begin{bmatrix} 2 & 0 \\ 3 & 4 \end{bmatrix}, \qquad Q = I.$$

Hint: The solution is *not* symmetric.

Exercise 3.17 *A Margin of Stability*
Let $\sigma > 0$ be a fixed number, Q a positive definite matrix, and A a matrix of the same size as Q. Show that every eigenvalue λ of A satisfies $\text{Re}\,\lambda < -\sigma$ if and only if there exists a positive definite matrix P such that

$$A^T P + PA + 2\sigma P = -Q.$$

3.8 NOTES AND REFERENCES

For differential equations, see [15], [21], [40], [41].

For linear algebra and matrix analysis, including Jordan forms, see [75], [40], [42], or [67]. The use of the primary decomposition theorem, as well as the presentation of the 2×2 real Jordan forms, is influenced by [15]. Example 3.8 is influenced by a discussion of connected oscillators in [58]. Jordan forms and linear differential equations are discussed in [40]. For background in analysis, see [7].

The basic results on the Lyapunov equation and stability results for linear time invariant systems are available from many sources. The presentation of Lyapunov equations in this chapter closely follows [26] (pages 29–32), with some influence from the discussion of norms in [40]. More detail on the solvability of Lyapunov matrix equations is available in [43].

The example in Exercise 3.11 is taken from [12] (page 59).

The statement and proof of Theorem 3.6 follow [81] (pages 100–101). More on the asymptotic behavior of linear time varying systems, and stability concepts for them, appears in [15] (pages 156–158), [58] (Section 4.6), and [82].

Chapter Four

Controllability of Linear Time Invariant Systems

The central concept of this chapter is *controllability*. We begin with a question about the equivalence between n-dimensional first order linear systems and n-th order scalar linear equations. This approach allows us to focus on the companion form which is one of the greatest advantages of single-input controllable systems. Afterwards, we define and discuss the analytical, control-theoretic meaning of controllability.

4.1 INTRODUCTION

For purposes of analysis it is standard to convert a single n-th order non-homogeneous linear differential equation to a system of n first order linear differential equations. Specifically, an n-th order linear equation

$$y^{(n)} + k_1 y^{(n-1)} + \cdots + k_{n-1}\dot{y} + k_n y = u(t), \tag{4.1}$$

with real constant coefficients k_i, is equivalent, via the standard definition of the vector variable $z = [y \quad \dot{y} \quad \ddot{y} \quad \ldots \quad y^{(n-1)}]^T$, to the linear system

$$\dot{z} = Pz + du(t), \tag{4.2}$$

where

$$P = \begin{bmatrix} 0 & 1 & 0 & \cdots & 0 \\ 0 & 0 & 1 & \cdots & \\ \cdots & \cdots & \cdots & \ddots & \\ 0 & \cdots & 0 & \cdots & 1 \\ -k_n & -k_{n-1} & -k_{n-2} & \cdots & -k_1 \end{bmatrix} \tag{4.3}$$

is a companion matrix, the k_i are as in (4.1), and

$$d = [0 \quad 0 \quad \ldots \quad 1]^T \tag{4.4}$$

is the n-th standard basis vector.

System (4.2) is very special. We call it a *companion system* because P in (4.3) is a *companion matrix* defined by the characteristic polynomial

$$p(\lambda) = \lambda^n + k_1\lambda^{n-1} + k_2\lambda^{n-2} + \cdots + k_{n-1}\lambda + k_n;$$

note that the last row of P is completely determined by the coefficients of this polynomial, and matrix P has characteristic polynomial given by $p(\lambda)$. The special form (4.2) is also called a *controller form*.

The form (4.2) has advantages, for example, in the problem of feedback stabilization. Note that feedback in (4.2) of the form $u = -\alpha_n z_1 - \alpha_{n-1} z_2 - \cdots - \alpha_1 z_n$, with the α_j real, produces a closed loop coefficient matrix also in companion form with characteristic polynomial

$$\lambda^n + (k_1 + \alpha_1)\lambda^{n-1} + \cdots + (k_{n-1} + \alpha_{n-1})\lambda + (k_n + \alpha_n).$$

The numbers $k_j + \alpha_j$, $1 \le j \le n$, determine the entries on the last row of this new coefficient matrix. These coefficients may be assigned by appropriate choice of the α_j, yielding any desired set of closed loop eigenvalues, subject only to the restriction that complex eigenvalues must occur in conjugate pairs, since the coefficients are real.

We have seen that (4.1) can be written as a linear first order system. What about the converse? When can a constant coefficient linear system

$$\dot{x} = Ax + bu(t), \tag{4.5}$$

where A is $n \times n$ and b is $n \times 1$, be transformed to the special form (4.2) by a nonsingular linear transformation $z = Tx$, where T is a constant matrix? Since $\dot{z} = T\dot{x} = (TAT^{-1})(Tx) + Tbu = (TAT^{-1})z + (Tb)u$, we are led to ask: When is there a nonsingular T such that TAT^{-1} is a companion matrix and Tb is the n-th standard basis vector? This is a very natural mathematical question; in advanced courses in differential equations, the subject of normal forms achieved by coordinate transformations is an important topic. And, as indicated above, the converse is relevant for the stabilization question as well.

A linear transformation of the state x that transforms (4.5) to (4.2) is not always possible, as can be seen by considering the diagonal system

$$\begin{bmatrix} \dot{x}_1 \\ \dot{x}_2 \end{bmatrix} = \begin{bmatrix} 1 & 0 \\ 0 & 1 \end{bmatrix} \begin{bmatrix} x_1 \\ x_2 \end{bmatrix} + \begin{bmatrix} b_1 \\ b_2 \end{bmatrix} u, \tag{4.6}$$

or the system

$$\begin{bmatrix} \dot{x}_1 \\ \dot{x}_2 \end{bmatrix} = \begin{bmatrix} 0 & 1 \\ 0 & 0 \end{bmatrix} \begin{bmatrix} x_1 \\ x_2 \end{bmatrix} + \begin{bmatrix} b_1 \\ 0 \end{bmatrix} u. \tag{4.7}$$

For these simple examples, easy calculations show that we cannot reach the form (4.2) by $z = Tx$ with any nonsingular T. (See Exercise 4.1.)

The question of a transformation from (4.5) to (4.2) is not usually addressed systematically in differential equations texts. However, alternative system representations are always of interest. The answer to the converse question posed above will lead us to characterize a *controllability* property originally motivated by control engineering problems. The controllability property guarantees the solvability of some basic transfer-of-state control problems. Moreover, it will be seen (by means of the eigenvalue placement theorem) that controllability is a strong sufficient condition for stabilization by linear feedback control.

The chapter proceeds as follows. First, we clarify the relation between (4.5) and (4.2) and derive a necessary and sufficient condition for equivalence, the Kalman rank criterion. Second, we explore the rank criterion by explaining its meaning as a *controllability* condition. Next, we show the importance of these developments to the questions of eigenvalue placement and feedback stabilization. Finally, we consider controllability in the case of diagonal and Jordan form systems. These forms provide additional insight and they lead us to an alternative rank condition for controllability, the PBH controllability test.

4.2 LINEAR EQUIVALENCE OF LINEAR SYSTEMS

A SIMPLE EXAMPLE. We begin by transforming a simple example, and afterwards, we consider a precise definition of *linear equivalence* of systems.

Example 4.1 Consider the system

$$\dot{x}_1 = -2x_1 + 2x_2 + u, \tag{4.8}$$

$$\dot{x}_2 = x_1 - x_2, \tag{4.9}$$

which has the form (4.5) with

$$A = \begin{bmatrix} -2 & 2 \\ 1 & -1 \end{bmatrix}, \quad b = \begin{bmatrix} 1 \\ 0 \end{bmatrix}.$$

Differentiate (4.9) and substitute from (4.8) to obtain $\ddot{x}_2 + 3\dot{x}_2 = u(t)$. This second order equation for x_2 has the form (4.1) for $n = 2$. On the other hand, a solution of $\ddot{x}_2 + 3\dot{x}_2 = u(t)$ for $x_2(t)$ determines a function $x_1(t)$, using $x_1 = \dot{x}_2 + x_2$ from (4.9), so that we get a solution of the system. Thus, system (4.8)–(4.9) can reasonably be said to be equivalent to the second order equation $\ddot{y} + 3\dot{y} = u$.

Is there another second order equation of the form $\ddot{y} + k_1\dot{y} + k_0 y = u$ that is also equivalent to (4.8)–(4.9)? For example, we might try to get a second order equation for x_1. This question is handled using the precise definition of equivalence (Definition 4.1 below). Note that the equation $\ddot{y} + 3\dot{y} = u$ has the form

$$\dot{z} = \begin{bmatrix} 0 & 1 \\ 0 & -3 \end{bmatrix} z + \begin{bmatrix} 0 \\ 1 \end{bmatrix} u \tag{4.10}$$

when we write $z_1 = y$ and $z_2 = \dot{y}$. We expect that there is a transformation from $\mathbf{R}^2$ to itself that transforms our original system (4.8)–(4.9) to the form (4.10). Since the differential equations are linear, the transformation is linear, say $z = Tx$, such that the differential equation in the z-coordinates, namely $\dot{z} = TAT^{-1}z + Tbu$, is exactly (4.10). In fact, direct calculation verifies that the unique T which does this is given by

$$T = \begin{bmatrix} 0 & 1 \\ 1 & -1 \end{bmatrix}.$$

We obtained a second order equation for the variable x_2. If we set $z_1 = x_2$ and $z_2 = \dot{x}_2$, then $z_2 = x_1 - x_2$, which yields the transformation $z = Tx$ with matrix T. △

Here is the precise definition of linear equivalence of two linear time invariant systems.

Definition 4.1 *The system* $\dot{x} = Ax + bu$ *is* linearly equivalent *to the system* $\dot{z} = Ez + fu$ *if there exists a nonsingular matrix* T *such that*

$$TAT^{-1} = E \quad \text{and} \quad Tb = f. \tag{4.11}$$

Thus, (4.5) is equivalent to (4.2) if and only if there is a nonsingular T such that $TAT^{-1} = P$ and $Tb = d$, where P and d are given in (4.3) and (4.4).

The natural questions concerning existence, uniqueness, and computation of T arise. Before proceeding to answer these questions it is instructive to try to transform the following example to the form (4.2).

Example 4.2 Let A be as in Example 4.1, but replace the input matrix there with the new input matrix $b = [1 \quad 1]^T$. Then the resulting system cannot be transformed to the form (4.2). (See Exercise 4.2.) △

EQUIVALENCE AND THE COMPANION MATRIX P. We have seen by examples that system (4.5) may or may not be equivalent to a companion system. It is convenient to make the following definition.

Definition 4.2 *The vector* b *is a* cyclic vector *for the* $n \times n$ *matrix* A *if the* n *vectors* b, Ab, $\ldots$, $A^{n-1}b$ *are linearly independent.*

In (4.2), vector d is a cyclic vector for P, as can be seen by observing that the matrix

$$\begin{bmatrix} d & Pd & \cdots & P^{n-1}d \end{bmatrix} = \begin{bmatrix} 0 & \cdot & \cdot & \cdot & \cdot & 0 & 1 \\ 0 & \cdot & \cdot & \cdot & 0 & 1 & * \\ 0 & \cdot & \cdot & 0 & 1 & * & * \\ & & & \vdots & & & \\ 0 & 1 & * & * & * & * & * \\ 1 & * & * & * & * & * & * \end{bmatrix}$$

must be nonsingular. The existence of a cyclic vector is a similarity invariant. If A is similar to P in (4.3) and $TAT^{-1} = P$, then A has a cyclic vector given by $T^{-1}d$.

The next proposition gives a useful condition that is equivalent to similarity between A and a companion form like P in (4.3).

Proposition 4.1 *The matrix A is similar to the companion matrix P of its characteristic polynomial if and only if the minimal and characteristic polynomials of A are identical.*

Proof. The characteristic polynomial and the minimal polynomial are similarity invariants. The minimal polynomial of a companion matrix P is the same as its characteristic polynomial [42] (page 147, Theorem 3.3.14). Thus, if A is similar to P in (4.3), then the minimal polynomial and the characteristic polynomial of A must be identical.

On the other hand, if the minimal polynomial and characteristic polynomial of A are identical, then the Jordan form of A must contain exactly one Jordan block for each distinct eigenvalue; the size of each Jordan block is equal to the multiplicity of the corresponding eigenvalue as a root of the characteristic (minimal) polynomial of A. But the same statement is true for P, because for a companion matrix the characteristic polynomial and the minimal polynomial are the same. Thus, the Jordan form of the companion matrix P has the same Jordan block structure as the Jordan form of A, hence A must be similar to P. □

Proposition 4.1 makes it easy to construct examples of matrices that have (or do not have) cyclic vectors. The similarity condition holds in Examples 4.1–4.2, where the characteristic and minimal polynomial for A is $\lambda(\lambda+3)$. We conclude that there is some other obstruction in Example 4.2 to an equivalence with system (4.2), and the obstruction must involve the b vector. Thus, the problem with transforming Example 4.2 is related to the way the forcing function u enters the equations. We pursue this observation later on.

Uniqueness of the Transformation T. Assume that we have a nonsingular T such that $TAT^{-1} = P$ and $Tb = d$. Then $TAT^{-1}d = TAb$, and $TA^k b = TA^k T^{-1}d = (TAT^{-1})^k d = P^k d$ for all $k \geq 0$. Nonsingularity of T implies that

$$n = \operatorname{rank}\begin{bmatrix} d & Pd & \cdots & P^{n-1}d \end{bmatrix} = \operatorname{rank}\begin{bmatrix} b & Ab & \cdots & A^{n-1}b \end{bmatrix}.$$

Moreover, T is uniquely determined by its action on the basis defined by the vectors b, Ab, ..., $A^{n-1}b$. Thus, we have the following uniqueness result and necessary condition for equivalence.

Proposition 4.2 *There is at most one nonsingular linear transformation, $z = Tx$, taking (4.5) to the companion form (4.2). Such a T exists only if*

$$\operatorname{rank}\begin{bmatrix} b & Ab & \cdots & A^{n-1}b \end{bmatrix} = n. \tag{4.12}$$

Example 4.2 is explained by this result, because in that example we have

$$[b \;\; Ab] = \begin{bmatrix} 1 & 0 \\ 1 & 0 \end{bmatrix}.$$

We return to Example 4.2 later for additional insight. Proposition 4.2 also explains why we cannot get a second order equation (4.1) for the variable x_1 in Example 4.1; the second order equation for $y = z_1$ must be a *unique* linear combination of the components of x, namely, the linear combination using the coefficients in the first row of a transformation T to (4.2).

We now show that the rank condition (4.12) implies that there is a nonsingular T such that the coordinate change $z = Tx$ transforms (4.5) to (4.2). We also show how to construct T by a simple direct method.

THE RANK CONDITION IS SUFFICIENT FOR EQUIVALENCE. Referring back to Example 4.1, the key in transforming (4.5) to (4.2) is to identify the variable z_1 that satisfies an equivalent n-th order equation (4.1). Let $z = Tx$ be the transformation, and let τ denote the first row of T. Note that $z_1 = \tau \cdot x$. Since $Tb = d = [0 \; 0 \; \dots \; 0 \; 1]^T$ and $TA^k b = P^k d$, we must have $\tau \cdot b = 0$, $\tau \cdot Ab = 0$, ... , $\tau \cdot A^{n-2} b = 0$, and $\tau \cdot A^{n-1} b = 1$. Write this as

$$\tau \begin{bmatrix} b & Ab & \cdots & A^{n-1}b \end{bmatrix} = \begin{bmatrix} 0 & \cdots & 0 & 1 \end{bmatrix} = d^T. \tag{4.13}$$

If we assume that $\operatorname{rank} [b \;\; Ab \;\; \cdots \;\; A^{n-1}b] = n$, then there is a unique solution for τ in (4.13). (Again, the crucial z_1 variable must be a unique linear combination of the x components.) What about the rest of T? The form of the companion system requires that

$$z_2 = \dot{z}_1 = \tau \cdot \dot{x} = \tau \cdot (Ax + bu) = \tau Ax + \tau bu = \tau Ax;$$

therefore the second row of T is τA. Continuing in this way, the equations defining τ and the form of the z system imply that

$$T = \begin{bmatrix} \tau \\ \tau A \\ \tau A^2 \\ \vdots \\ \tau A^{n-1} \end{bmatrix}. \tag{4.14}$$

We combine this argument with Proposition 4.2 as follows.

Theorem 4.1 *The system $\dot{x} = Ax + bu(t)$ with $x \in \mathbf{R}^n$ can be transformed to the companion system, $\dot{z} = Pz + du(t)$, by a nonsingular linear transformation, $z = Tx$, if and only if*

$$\operatorname{rank} \begin{bmatrix} b & Ab & \cdots & A^{n-1}b \end{bmatrix} = n.$$

When this rank condition holds, T is uniquely defined by (4.14), *where τ is the unique solution of* (4.13).

Theorem 4.1 answers our original question. If the algebraic fact concerning the existence of a cyclic vector for the companion matrix P is understood, then the situation regarding equivalence between (4.5) and (4.2) becomes more transparent.

Our original question got us to this point. We now know more about the structure of systems equivalent to companion form systems. But there is much more involved here. Think about varying the nonhomogeneous term in (4.5). What if we apply different input functions $u(t)$? To what extent can this affect the solutions of the system? We consider the question of varying the input in the next section. By doing so, we obtain an analytic, control-theoretic meaning of the rank condition in Theorem 4.1.

4.3 CONTROLLABILITY WITH SCALAR INPUT

System (4.5) is called a *single-input* system because the input function u is scalar-valued rather than vector-valued. We show in this section that a natural concept of *controllability* for the single-input system (4.5) coincides with b being a cyclic vector for A.

In an elementary differential equations course the nonhomogeneous term in (4.1) is usually a specified function of t. But we now ask: What happens with the system dynamics as we change u? To what extent can the motion of the state vector $x(t)$ be influenced, starting from an initial state x_0 and using fairly arbitrary inputs, $u(t)$? Before stating the definition of controllability, we must specify a set of admissible input functions. The solutions of linear time invariant systems are defined on the entire real line. However, for some questions, the inputs are restricted to an interval $[t_0, t_f]$. Thus, with an appropriate restriction of domain when necessary, we could consider several vector spaces of functions for the set $\mathcal{U}$ of admissible controls, including piecewise constant, continuous, or locally integrable functions. A real-valued function $u(t)$ is *locally integrable* if, for any $t_1 < t_2$, we have

$$\int_{t_1}^{t_2} |u(s)|\, ds < \infty.$$

An $\mathbf{R}^m$-valued function $u(t) = [u_1(t) \ \cdots \ u_m(t)]^T$ is locally integrable if each component function is locally integrable.

For an m-input system, the set of admissible control functions is (considering forward time only)

$$\mathcal{U} = \{u : [0, \infty) \mapsto \mathbf{R}^m \ : u \text{ is locally integrable}\}. \tag{4.15}$$

The interval $[0, \infty)$ may be replaced by any real interval J as needed.

We can now state the definition of controllability.

Definition 4.3 *The linear system $\dot{x} = Ax + Bu$, where A is $n \times n$ and B is $n \times m$, is* controllable *if, given any x_0, $x_f \in \mathbf{R}^n$, there exist a $t_f > 0$ and an* *admissible control function $u(t)$, defined for $0 \le t \le t_f$, such that the solution of $\dot{x} = Ax + Bu(t)$ with initial condition $x(0) = x_0$ satisfies $x(t_f) = x_f$.*

For the moment, we emphasize single-input systems; and we write $B = b$ when the control u is a scalar input. The solution of the single-input system (4.5) with $x(0) = x_0$ is given by the variation of parameters formula,

$$x(t) = e^{tA}\left(x_0 + \int_0^t e^{-sA} bu(s)\, ds\right). \tag{4.16}$$

The linear system (4.5) is controllable if, for any given x_0, x_f, there are some t_f and some locally integrable function u on $0 \le t \le t_f$ such that

$$x_f = e^{t_f A}\left(x_0 + \int_0^{t_f} e^{-sA} bu(s)\, ds\right). \tag{4.17}$$

It may be surprising that the solvability of (4.17) for arbitrary x_0, x_f is determined by a purely algebraic criterion. The explanation lies with the Cayley-Hamilton Theorem, which says that the matrix A satisfies $p(A) = 0$, where $p(\lambda)$ is the characteristic polynomial of A. The rank condition (4.12) is known as the *controllability rank condition*, and the matrix $[b\ Ab \cdots A^{n-1}b]$ is called the *controllability matrix*, because of Theorem 4.2 below.

It may be helpful to note that the construction in the proof of the *if* part of Theorem 4.2 can be motivated in Exercise 4.10 using a scalar linear system.

Theorem 4.2 *The linear system $\dot{x} = Ax + bu$ is controllable if and only if*

$$\text{rank}\begin{bmatrix} b & Ab & \cdots & A^{n-1}b \end{bmatrix} = n. \tag{4.18}$$

Proof. By the Cayley-Hamilton theorem, for each $k \ge n$, A^k can be expressed as a linear combination of $I, A, A^2, \ldots, A^{n-1}$. Recall that the column space (range) of the controllability matrix, denoted $R([b\ \ Ab\ \ \cdots\ \ A^{n-1}b])$, is a closed subspace of $\mathbf{R}^n$. From the definition of the matrix exponential, we may then conclude that the range of $e^{-sA}b$ must lie in $R([b\ \ Ab\ \ \cdots\ \ A^{n-1}b])$ for every s; thus, the integral on the right side of (4.16) must also lie in $R([b\ \ Ab\ \ \cdots\ \ A^{n-1}b])$ for all t. Let x_0 be given. The states that are reachable from x_0 in finite time by means of some input $u(t)$ must all lie within $R([b\ \ Ab\ \ \cdots\ \ A^{n-1}b])$. Thus, if the rank condition does *not* hold, then the system is *not* controllable because there are states that cannot be reached from x_0. This establishes the *only if* part, that controllability implies the rank condition (4.18).

Conversely, suppose that (4.18) holds. We must now show that the system $\dot{x} = Ax + bu$ is controllable. Choose any finite time $t_f > 0$, and consider the symmetric $n \times n$ matrix

$$M = \int_0^{t_f} e^{-sA} bb^T e^{-sA^T} \, ds.$$

We first show that M is nonsingular, and then we show that nonsingularity of M implies controllability. Suppose that $Mv = 0$ for some v; then also $v^T M v = 0$, and therefore

$$0 = v^T M v = \int_0^{t_f} v^T e^{-sA} bb^T e^{-sA^T} v \, ds = \int_0^{t_f} (\psi(s))^2 \, ds,$$

where $\psi(s) := v^T e^{-sA} b$. Since $(\psi(s))^2$ is continuous and nonnegative, we conclude that $\psi(s) \equiv 0$. Thus,

$$\psi(0) = v^T b = 0, \quad \dot{\psi}(0) = -v^T A b = 0, \quad \ldots, \quad \psi^{(n-1)}(0) = \pm v^T A^{n-1} b = 0.$$

Therefore v is perpendicular to $R([b \;\; Ab \;\; \cdots \;\; A^{n-1}b])$. By the rank assumption, we must have $v = 0$, and therefore M is nonsingular. Now take any two points x_0, x_f in $\mathbf{R}^n$, and define the control $u(s) = b^T e^{-sA^T} x$ for $0 \le s \le t_f$, where x is to be chosen later. The solution $x(t)$ with input u and initial condition x_0 has final point x_f at time t_f provided that x can be chosen so that

$$x_f = e^{t_f A} \left(x_0 + \left(\int_0^{t_f} e^{-sA} bb^T e^{-sA^T} \, ds \right) x \right) = e^{t_f A}(x_0 + Mx).$$

But $e^{t_f A}$ is nonsingular, with $(e^{t_f A})^{-1} = e^{-t_f A}$, and M is nonsingular, so $x = M^{-1}(e^{-t_f A} x_f - x_0)$. Thus, any x_0 can be steered to any x_f in time t_f by an admissible control, so the system is controllable. □

It is often convenient to write

$$\mathcal{C}(A, b) := [b \;\; Ab \;\; \cdots \;\; A^{n-1}b]$$

for the controllability matrix determined by the matrix pair (A, b). We say that (A, b) is a *controllable pair* if (4.18) holds, and an *uncontrollable pair* if (4.18) does not hold.

We can illustrate Theorem 4.2 and the idea of controllability by returning to Example 4.2.

Example 4.3 The system is

$$\dot{x} = \begin{bmatrix} -2 & 2 \\ 1 & -1 \end{bmatrix} x + \begin{bmatrix} 1 \\ 1 \end{bmatrix} u.$$

Note that $\lambda = 0$ is an eigenvalue of A, and $b = [1 \;\; 1]^T$ is a corresponding eigenvector, so the controllability rank condition does not hold. However, A is similar to its companion matrix. Using the T computed before and

$z = Tx$ we have the system

$$\dot{z} = \begin{bmatrix} 0 & 1 \\ 0 & -3 \end{bmatrix} z + \begin{bmatrix} 1 \\ 0 \end{bmatrix} u.$$

Differentiation of the z_1 equation and direct substitutions produce a second order equation for z_1:

$$\ddot{z}_1 + 3\dot{z}_1 = 3u + \dot{u}.$$

This equation does not match the form of (4.1) due to the $\dot{u}$ term. One integration produces the first order equation

$$\dot{z}_1 + 3z_1 = 3 \int u(s)\, ds + u(t),$$

which shows that arbitrary inputs u can affect the dynamics only in a one-dimensional space. Recall from the calculation of T that $z_1 = x_2$ and $z_2 = x_1 - x_2$. The original x equations might lead us to think that u can fully affect both x_1 and x_2; however, note that u cannot influence the dynamics of the difference $x_1 - x_2 = z_2$. Only when the initial condition for z involves $z_2(0) = 0$ can u be used to control a trajectory. That is, the inputs completely control only the states that lie in the subspace $R([b\ Ab]) = \text{span}\,\{b\} = \text{span}\,\{[1\ \ 1]^T\}$. Solutions starting with $x_1(0) = x_2(0)$ satisfy $x_1(t) = x_2(t) = \int_0^t u(s)\, ds + x_1(0)$. One can steer along the line $x_1 = x_2$ from any initial point to any final point $x_1(t_f) = x_2(t_f)$ at any finite time t_f by appropriate choice of $u(t)$. On the other hand, if the initial condition lies off the line $x_1 = x_2$, then the difference $z_2 = x_1 - x_2$ decays exponentially so there is no chance of steering to an arbitrarily given final state in finite time. △

For a controllable system, there are generally many control functions that can implement a state transfer from x_0 to x_f in finite time t_f. (Note also that, according to the proof of Theorem 4.2, this transfer can be done, in principle if not in practice, in *any* specified finite time.) This flexibility can be exploited in some applications to optimize the behavior of the system in some way, for example by minimizing a measure of the cost of carrying out the transfer. In particular, if the cost of the control action is measured by the integral

$$\int_{t_0}^{t_f} |u(s)|^2\, ds,$$

then a control that minimizes this cost can be determined. For additional details on this problem for linear systems, see [9] (pp. 102–105). Linear optimal control theory is a well-developed branch of mathematical control theory concerned with optimizing various performance indices of system (4.5) by appropriate choice of the control.

Example 4.4 A simple linear model of airplane flight from [91] (p. 90) is given by the linear equations

$$\begin{aligned}\dot{\alpha} &= a(\phi - \alpha),\\ \ddot{\phi} &= -\omega^2(\phi - \alpha - bu),\\ \dot{h} &= c\alpha.\end{aligned}$$

These equations assume that the body of the plane is inclined at an angle ϕ with the horizontal, that the flight path is along a straight line at angle α with the horizontal, and that the plane flies at a constant nonzero ground speed of c meters per second. The altitude of the plane is given by h in meters. Constants a and b are positive, and ω is a natural oscillation frequency of the pitch angle ϕ. The control input u supplies a control force using the elevators at the tail of the plane. These equations are intended to model the plane's flight only for small angles ϕ and α, with $\alpha > 0$ for ascending flight and $\alpha < 0$ for descent. Using the variables $x_1 = \alpha$, $x_2 = \phi$, $x_3 = \dot{\phi}$, and $x_4 = h$, we have a four-dimensional linear system, which can be shown to be controllable using the single input u. (See Exercise 4.4.) △

4.4 EIGENVALUE PLACEMENT WITH SINGLE INPUT

A major theme in control theory is the use of feedback control to modify the system dynamics to achieve some desired behavior, for example, to stabilize an otherwise unstable equilibrium point. In this section we indicate some advantages of an equivalence with the companion system (4.2) with regard to such issues.

Recall that A is Hurwitz if all eigenvalues of A lie in the open left half-plane, and that the origin of $\dot{x} = Ax$ is asymptotically stable if and only if A is Hurwitz.

Definition 4.4 *In the linear system $\dot{x} = Ax + Bu$, where A is $n \times n$ and B is $n \times m$, a* linear state feedback *is given by $u = Kx$, where K is an $m \times n$ matrix. The corresponding* closed loop system *is $\dot{x} = (A + BK)x$. (We may write $B = b$ when the control u is a scalar input.)*

Consider the companion system (4.2). Using linear state feedback, $u = Kx$, it is possible to assign eigenvalues arbitrarily to the resulting closed loop system, subject only to the restriction that the complex eigenvalues of $A + bK$ occur in conjugate pairs. Specifically, by setting $u = Kx = [-\alpha_n \;\; -\alpha_{n-1} \;\; \cdots \;\; -\alpha_1]\,x$ in (4.2) we get the closed loop system $\dot{z} = \hat{P}z$, where $\hat{P}$ has the same form as P in (4.3) except the last row is now $[-(k_n + \alpha_n) \;\; -(k_{n-1} + \alpha_{n-1}) \;\; \cdots \;\; -(k_1 + \alpha_1)]$. Suppose that $m_1, m_2, \ldots, m_n$ are the desired coefficients of the characteristic polynomial of the closed loop. With the k_i known and the m_i specified, we have $\alpha_i = m_i + k_i$. Thus, the coefficients of the characteristic polynomial of $A + bK$ may be chosen so that all the roots lie in the open left half-plane, that is, $A + bK$ is Hurwitz. Moreover, the exponential rate of convergence of $z(t)$ to the origin can be increased by shifting the roots to the left in the complex plane.

A linear system that is not Hurwitz might be made Hurwitz if modified by appropriate linear feedback. There is no difficulty in defining stabilizability for vector-input systems.

Definition 4.5 *The linear system $\dot{x} = Ax + Bu$, where A is $n \times n$ and B is $n \times m$, is* stabilizable *if there exists an $m \times n$ matrix K such that $A + BK$ is Hurwitz.*

For the moment, we continue to emphasize single-input systems. The next result shows the strength of controllability in modifying the system dynamics by linear feedback. It says that if (A, b) is controllable, then the eigenvalues of $A + bK$ can be assigned arbitrarily by appropriate choice of the real feedback matrix K, subject only to the restriction that complex eigenvalues occur in conjugate pairs.

Theorem 4.3 (Eigenvalue Placement Theorem: Single Input)
If the single-input system $\dot{x} = Ax + bu$ is controllable, then it is stabilizable. Moreover, if $p(\lambda)$ is any monic real polynomial, then there exists a $1 \times n$ matrix K such that $p(\lambda) = \det(\lambda I - A - bK)$.

Proof. We have discussed the proof only for the special case of a companion system. By controllability, there is a nonsingular T with $z = Tx$ such that $\dot{z} = TAT^{-1}z + Tbu$ is a controller (companion) system. Therefore, as we have seen, the eigenvalues of $TAT^{-1} + Tb\hat{K}$ may be arbitrarily assigned, as described, by appropriate choice of $\hat{K}$. Now observe that the similarity

$$TAT^{-1} + Tb\hat{K} = T(A + b\hat{K}T)T^{-1} = T(A + bK)T^{-1}, \quad K := \hat{K}T,$$

shows that the eigenvalues of $A + bK$ can be arbitrarily assigned by appropriate choice of feedback $u = Kx$. $\square$

Controllability guarantees strong stabilization properties of a linear system, since the exponential rate of decay in the direction of each eigenvector can be made as fast as desired. Later on, we consider weaker conditions that guarantee stabilizability as in Definition 4.5.

The considerations in this section help to indicate that much can be accomplished with linear feedback in linear controllable systems. Consequently, it is of interest to have an extension of the solution of the equivalence problem involving systems (4.2) and (4.5) to the case where (4.5) is replaced by a single-input *nonlinear* system. The extended equivalence problem is the feedback linearization problem considered in detail in a later chapter.

4.5 CONTROLLABILITY WITH VECTOR INPUT

This section is a convenient place to discuss some straightforward results for vector-input linear systems. We consider here an extension of the controllability characterization in Theorem 4.2. We also prove the normal form Theorem 4.5 for uncontrollable systems, which eventually leads to an alternative controllability test in a later section, the PBH test. Finally, we state Proposition 4.3 and Proposition 4.4 on invariance properties of controllability. The extension of the eigenvalue placement Theorem 4.3 is a bit more involved and is done in a later section.

Consider a linear system with multivariable input,

$$\dot{x} = Ax + Bu, \tag{4.19}$$

where $u \in \mathbf{R}^m$, hence B is $n \times m$. The admissible control functions are the locally integrable $\mathbf{R}^m$-valued functions $u(t)$.

THE KALMAN RANK CRITERION. The controllability characterization in Theorem 4.2 extends directly to (4.19) with no change in the statement except to replace b by B. In this case, the m-input *controllability matrix*

$$\mathcal{C}(A,B) := [B \;\; AB \;\; \cdots \;\; A^{n-1}B]$$

is $n \times nm$. Here is the extension of Theorem 4.2.

Theorem 4.4 *The multi-input system* (4.19) *is controllable if and only if*

$$\operatorname{rank} \mathcal{C}(A,B) = \operatorname{rank}\,[B \;\; AB \;\; \cdots \;\; A^{n-1}B] = n,$$

that is, if and only if $\mathcal{C}(A,B)$ has full row rank.

Proof Outline. The proof proceeds as in the proof of Theorem 4.2. With careful attention to the dimensions involved, the same proof carries through. The M matrix is now

$$M := \int_0^{t_f} e^{-sA} B B^T e^{-sA^T}\, ds,$$

which is $n \times n$, while $\psi := v^T e^{-sA} B$ is now $1 \times m$. The details of the proof are left to Exercise 4.11. □

We say that the matrix pair (A,B) is a *controllable pair*, or that (A,B) is *controllable*, if $\operatorname{rank} \mathcal{C}(A,B) = n$. Controllability implies stabilizability in the vector-input case; this will follow from the general eigenvalue placement theorem given later (see Theorem 4.6).

The next theorem presents a *controllability normal form* for systems where the pair (A,B) is uncontrollable. This result depends only on the fact that $R(\mathcal{C}(A,B))$ is an invariant subspace for A.

Theorem 4.5 *Suppose the controllability matrix $\mathcal{C}(A, B)$ for system (4.19) has rank $r < n$. Then there is a real nonsingular matrix T such that the transformation $x = Tz$ produces the transformed system in normal form*

$$\dot{z} = T^{-1}ATz + T^{-1}Bu =: \begin{bmatrix} A_{11} & A_{12} \\ 0 & A_{22} \end{bmatrix} z + \begin{bmatrix} B_1 \\ 0 \end{bmatrix} u, \tag{4.20}$$

where A_{11} is $r \times r$ and (A_{11}, B_1) is controllable.

Proof. Define T so that the first r columns form a basis for $R(\mathcal{C}(A, B))$. The zero entries in $T^{-1}AT$ and $T^{-1}B$ are due to the invariance of $R(\mathcal{C}(A, B))$ under A, and A_{11} is $r \times r$ by construction of the mentioned basis. To see that (A_{11}, B_1) is controllable, note that

$$T^{-1}\mathcal{C}(A, B) = \begin{bmatrix} B_1 & A_{11}B_1 & \cdots & A_{11}^{n-1}B_1 \\ 0 & 0 & 0 & 0 \end{bmatrix}.$$

Since $\operatorname{rank} \mathcal{C}(A, B) = r$, it follows that

$$\operatorname{rank} \begin{bmatrix} B_1 & A_{11}B_1 & \cdots & A_{11}^{n-1}B_1 \end{bmatrix} = r.$$

Since A_{11} is $r \times r$ and B_1 is $r \times m$, the Cayley-Hamilton theorem implies that

$$\operatorname{rank}[B_1 \ A_{11}B_1 \ \cdots \ A_{11}^{r-1}B_1] = r.$$

Thus, the pair (A_11, B_1) is controllable. □

The normal form of Theorem 4.5 will be useful in several ways later on, in this chapter and in later chapters on linear systems.

We now consider two important invariance properties of controllability. The next proposition states that controllability is invariant under linear state coordinate transformations.

Proposition 4.3 *If (A, B) is controllable and T is nonsingular, then the pair $(T^{-1}AT, T^{-1}B)$ is also controllable. In other words, controllability is invariant under linear state coordinate transformations.*

Proof. See Exercise 4.5. □

In many control problems, a preliminary state feedback Kx is used to simplify the system, and then a new reference input term βv is available through the same channel for further control purposes. This situation is modeled by using the regular feedback transformation $u = Kx + \beta v$, where β is assumed to be a nonsingular $m \times m$ matrix, thus preserving the number of independent inputs. The next proposition says that such feedback transformations do indeed preserve controllability.

Proposition 4.4 *If (A, B) is controllable and β is an $m \times m$ nonsingular matrix, then for any $m \times n$ matrix K, the pair $(A + BK, B\beta)$ is controllable.*

That is, controllability is invariant under regular feedback transformations of the form $u = Kx + \beta v$.

Proof. See Exercise 4.5. □

4.6 EIGENVALUE PLACEMENT WITH VECTOR INPUT

The extension of Theorem 4.3 on eigenvalue placement to the case of m-input controllable systems can be based on the single-input result; the key step is Lemma 4.2 below. Eigenvalue placement with vector input is not as transparent as in the case of single-input controllable systems, where a companion form is available and the input enters only through the final equation. First, controllability of the pair (A, B) does not imply that A is similar to a companion matrix, because any A can be paired with some B to obtain a controllable pair (take $B = I$, for example). Thus, matrix A need not have a cyclic vector. Even if A is similar to a companion form, the similarity transformation does not guarantee any particularly simple structure for the transformed input matrix.

We begin with a lemma about the discrete linear system defined by a controllable pair (A, B).

Lemma 4.1 *If (A, B) is controllable and b is a nonzero column of B, then there exist vectors $u_0, \ldots, u_{n-2}$ such that the vectors $x_0, \ldots, x_{n-1}$ defined by*

$$x_0 := b, \quad x_{j+1} := Ax_j + Bu_j \quad \text{for } j = 0, \ldots, n-2 \tag{4.21}$$

are linearly independent.

Proof. We show by induction that there exist vectors $u_0, \ldots, u_{n-2}$ and linearly independent, vectors $x_0, \ldots, x_{n-1}$, with $x_0 = b$, satisfying (4.21). To start the induction, note that $x_0 := b \neq 0$. Now suppose that $k < n-1$, $x_0, \ldots, x_k$ are linearly independent, and $u_0, \ldots, u_{k-1}$ are such that (4.21) is satisfied. Let $\mathcal{S} := \text{span}\,\{x_0, \ldots, x_k\}$; thus, $\dim \mathcal{S} = k+1$.

We must show that we can choose a vector u_k such that $x_{k+1} := Ax_k + Bu_k \notin \mathcal{S}$. Suppose, for purposes of contradiction, that such a choice of u_k is not possible; then $Ax_k + Bu \in \mathcal{S}$ for all $u \in \mathbf{R}^m$. With $u = 0$ we see that $Ax_k \in \mathcal{S}$, and therefore $Bu \in \mathcal{S}$ for all u. Thus, $\mathcal{R}(B) \subset \mathcal{S}$. For $j < k$ we have $Ax_j = x_{j+1} - Bu_j \in \mathcal{S}$, by definition of $\mathcal{S}$ and the fact that $Ax_k \in \mathcal{S}$. Therefore $\mathcal{S}$ is invariant under A, that is, $A(\mathcal{S}) \subset \mathcal{S}$. Controllability of (A, B) implies that $\mathcal{S} = \mathbf{R}^n$, because the range of the controllability matrix is the smallest subspace which contains $\mathcal{R}(B)$ and is invariant under A. (See Exercise 4.12.) But this contradicts $\dim \mathcal{S} = k+1 < n$. Hence, we can find a vector u_k as desired.

This completes the proof of Lemma 4.1 by induction. □

The next lemma says that linear feedback can always be chosen to transform a vector-input controllable system into a single-input controllable system.

Lemma 4.2 *Let B be $n \times m$ and suppose (A, B) is controllable. If b is any nonzero column of B, then there exists an $m \times n$ matrix K such that $(A + BK, b)$ is controllable.*

Proof. By Lemma 4.1 there are vectors $u_0, \ldots, u_{n-2}$ and linearly independent vectors $x_0, \ldots, x_{n-1}$, with $x_0 = b$, satisfying (4.21). We may choose a vector u_{n-1} arbitrarily. Define the matrix K by requiring that $Kx_j = u_j$ for $j = 0, \ldots, n-1$. Then b is a cyclic vector for $A + BK$, since the vectors x_j are linearly independent and we have $(A+BK)^{j-1}b = x_j$ for $j = 0, \ldots, n-1$. Therefore $(A + BK, b)$ is controllable. □

We can now prove the main result of this section.

Theorem 4.6 (General Eigenvalue Placement Theorem)
Suppose B is $n \times m$ and (A, B) is controllable. If $p(\lambda)$ is any monic real polynomial, then there exists a real $m \times n$ matrix G such that $p(\lambda) = \det(\lambda I - A - BG)$. In particular, (A, B) is stabilizable.

Proof. By Lemma 4.2, if b is any nonzero column of B, then there exists K such that the pair $(A + BK, b)$ is controllable. Let β be such that $b = B\beta$. Define the feedback transformation $u = Kx + \beta v$, where v is a new scalar reference input. Substitution of this input for u in $\dot{x} = Ax + Bu$ gives $\dot{x} = (A + BK)x + B\beta v = (A + BK)x + bv$. By Theorem 4.3, we may place the eigenvalues of this single-input system arbitrarily by appropriate choice of linear feedback $v = Fx$, where F is $1 \times n$; indeed, any set of eigenvalues symmetric with respect to the real axis may be realized as the eigenvalues of the real matrix $A + BK + bF$, for appropriate F. In summary, then, the eigenvalues of $A+BK+B\beta F = A+B(K+\beta F)$ can be placed by appropriate choice of the $m \times n$ feedback matrix $G := K + \beta F$. This completes the proof. □

In the proof of Theorem 4.6, note that K depends on $b = B\beta$, whereas F is chosen to place the eigenvalues of the controllable pair $(A + BK, b)$.

The next example illustrates the construction of an appropriate feedback K as in Lemma 4.2.

Example 4.5 Consider the pair (A, B), where

$$A = \begin{bmatrix} 2 & 1 & 0 \\ 0 & 2 & 0 \\ 0 & 0 & 2 \end{bmatrix}, \quad B = \begin{bmatrix} 0 & 0 \\ 1 & 0 \\ 0 & 1 \end{bmatrix} := \begin{bmatrix} b_0 & b_1 \end{bmatrix}.$$

The column notation for B will be useful in a moment. The characteristic polynomial of A is $\det(\lambda I - A) = (\lambda - 2)^3$ and the minimal polynomial is $(\lambda - 2)^2$. Therefore A is not similar to a companion matrix (Proposition 4.1), hence A has no cyclic vector. It is straightforward to check that the first three columns of the controllability matrix are linearly independent, so the pair (A, B) is controllable by Theorem 4.4. Write a candidate feedback matrix K as

$$K = \begin{bmatrix} k_{11} & k_{12} & k_{13} \\ k_{21} & k_{22} & k_{23} \end{bmatrix}.$$

Let $b = b_0 = x_0$. Then we may take

$$\begin{aligned} x_1 &= (A + BK)x_0 = Ax_0 + BKx_0 \quad \text{with} \quad Kx_0 = u_0 := [1\ 0]^T, \\ x_2 &= (A + BK)x_1 = Ax_1 + BKx_1 \quad \text{with} \quad Kx_1 = u_1 := [0\ 1]^T, \end{aligned}$$

where we are free to choose both u_0 and u_1. The choice for u_2 is arbitrary as well, and we choose to set $Kx_2 = u_2 := u_1$. Therefore the equation for K is

$$K \begin{bmatrix} x_0 & x_1 & x_2 \end{bmatrix} = \begin{bmatrix} u_0 & u_1 & u_2 \end{bmatrix},$$

where the u_j have been chosen and the x_j are to be computed. We find that

$$x_1 = Ab_0 + Bu_0 = Ab_0 + b_0 = \begin{bmatrix} 1 \\ 3 \\ 0 \end{bmatrix},$$

and

$$x_2 = Ax_1 + Bu_1 = \begin{bmatrix} 5 \\ 6 \\ 1 \end{bmatrix}.$$

Written out in detail, the equation for K is

$$\begin{bmatrix} k_{11} & k_{12} & k_{13} \\ k_{21} & k_{22} & k_{23} \end{bmatrix} \begin{bmatrix} 0 & 1 & 5 \\ 1 & 3 & 6 \\ 0 & 0 & 1 \end{bmatrix} = \begin{bmatrix} 1 & 0 & 0 \\ 0 & 1 & 1 \end{bmatrix},$$

and it is straightforward to compute that

$$K = \begin{bmatrix} -3 & 1 & 9 \\ 1 & 0 & -4 \end{bmatrix}.$$

Thus, we have

$$A + BK = \begin{bmatrix} 2 & 1 & 0 \\ 0 & 2 & 0 \\ 0 & 0 & 2 \end{bmatrix} + \begin{bmatrix} 0 & 0 \\ 1 & 0 \\ 0 & 1 \end{bmatrix} \begin{bmatrix} -3 & 1 & 9 \\ 1 & 0 & -4 \end{bmatrix} = \begin{bmatrix} 2 & 1 & 0 \\ -3 & 3 & 9 \\ 1 & 0 & -2 \end{bmatrix},$$

and, by construction, $A + BK$ has the cyclic vector b. Indeed,

$$\begin{bmatrix} b & (A+BK)b & (A+BK)^2 b \end{bmatrix} = \begin{bmatrix} 0 & 1 & 5 \\ 1 & 3 & 6 \\ 0 & 0 & 1 \end{bmatrix} = \begin{bmatrix} x_0 & x_1 & x_2 \end{bmatrix},$$

which is invertible. Therefore $(A + BK, b)$ is controllable. △

Using Theorem 4.6, eigenvalue placement for the system in Example 4.5 can now be carried out; see Exercise 4.13.

4.7 THE PBH CONTROLLABILITY TEST

We begin this section with some examples of systems $\dot{x} = Ax + Bu$ with the A matrix in diagonal form or in Jordan form. The purpose of these examples is to help in gaining further insight into the controllability concept and to motivate another very useful test for controllability, the PBH controllability test.

We consider first a diagonal form system.

Example 4.6 Consider a two-dimensional diagonal system

$$\dot{x} = Ax + bu := \begin{bmatrix} \lambda_1 & 0 \\ 0 & \lambda_2 \end{bmatrix} x + \begin{bmatrix} b_1 \\ b_2 \end{bmatrix} u.$$

Any two-dimensional system with diagonalizable A matrix can be transformed to this special form by a linear coordinate transformation. The controllability matrix is

$$\mathcal{C}(A,b) = \begin{bmatrix} b_1 & \lambda_1 b_1 \\ b_2 & \lambda_2 b_2 \end{bmatrix},$$

and it is easy to see that the system is controllable if and only if each $b_j \neq 0$ and $\lambda_1 \neq \lambda_2$. It may be useful to recall here the diagonal example which appeared at the beginning of this chapter. △

We move on to a three-dimensional diagonal system.

Example 4.7 Consider the system

$$\dot{x} = Ax + bu := \begin{bmatrix} \lambda_1 & 0 & 0 \\ 0 & \lambda_2 & 0 \\ 0 & 0 & \lambda_3 \end{bmatrix} x + \begin{bmatrix} b_1 \\ b_2 \\ b_3 \end{bmatrix} u,$$

with eigenvalues λ_j. The controllability matrix is

$$\mathcal{C}(A,b) = \begin{bmatrix} b_1 & \lambda_1 b_1 & \lambda_1^2 b_1 \\ b_2 & \lambda_2 b_2 & \lambda_2^2 b_2 \\ b_3 & \lambda_3 b_3 & \lambda_3^2 b_3 \end{bmatrix}.$$

As with the two-dimensional diagonal system, it is still easy to see that controllability requires that each $b_j \neq 0$ (so there are no zero rows) and that $\lambda_i \neq \lambda_j$ for $i \neq j$ (otherwise there are dependent rows). Thus, necessary conditions for controllability are that all components of the input vector b are nonzero and that all the eigenvalues are distinct.

Are these two necessary conditions, taken together, also sufficient for controllability of this diagonal system? It turns out that they are, since these conditions do imply that $\mathcal{C}(A, b)$ is nonsingular. In addition, it might be an interesting exercise here to verify that the two necessary conditions taken together are equivalent to the statement that rank $[A - \lambda I \quad b] = 3$ for every eigenvalue of A. △

In the next example, we consider the controllability of a system with a single Jordan block as coefficient matrix.

Example 4.8 Consider the system

$$\dot{x} = Ax + bu := \begin{bmatrix} 3 & 1 & 0 \\ 0 & 3 & 1 \\ 0 & 0 & 3 \end{bmatrix} x + \begin{bmatrix} b_1 \\ b_2 \\ b_3 \end{bmatrix} u.$$

A direct computation of $\mathcal{C}(A, b)$ shows that a necessary condition for controllability is that $b_3 \neq 0$; this is also clear by the fact that the last component equation is decoupled from the first two equations. Suppose for the moment that we have $b_3 = 0$; then the system is in a controllability normal form (see Theorem 4.5). One can consider different cases. For example, if $b_2 \neq 0$, then the system is already in normal form with A_{11} the upper left 2×2 block and $A_{22} = [3]$. On the other hand, if $b_2 = 0$ as well, then the system is in normal form with $A_{11} = [3]$ and A_{22} is the lower right 2×2 block.

Is this three-dimensional diagonal system controllable? In fact, the condition $b_3 \neq 0$ is also sufficient for controllability, but this fact may be difficult to see by a direct analysis of the 3×3 controllability matrix. △

These examples suggest that an alternative controllability test could be helpful. The normal form of Theorem 4.5 will play a key role in developing an alternative test. But first, we give one more example, which shows how the result stated in Example 4.8 might be used.

Example 4.9 Consider the four-dimensional system,

$$\dot{x} = Ax + bu := \begin{bmatrix} 2 & 0 & 0 & 0 \\ 0 & 3 & 1 & 0 \\ 0 & 0 & 3 & 1 \\ 0 & 0 & 0 & 3 \end{bmatrix} x + \begin{bmatrix} b_1 \\ b_2 \\ b_3 \\ b_4 \end{bmatrix} u,$$

which has a single Jordan block associated with each of the distinct eigenvalues, 2 and 3. Assuming the result stated in Example 4.8, it is not difficult to see that this system is controllable if and only if $b_1 \neq 0$ and $b_4 \neq 0$. △

Multiple Jordan blocks associated with the same eigenvalue present a different situation; such single-input systems cannot be controllable. (See Exercise 4.7 for a simple case.)

Theorem 4.5 on the controllability normal form helps in the next step toward the desired alternative controllability test. Recall that $R(\mathcal{C}(A,B))$ is invariant under A; thus, the orthogonal complement of this space is invariant under A^T (Exercise 2.3). Consider the eigenvalues of A^T restricted to $R(\mathcal{C}(A,b))^{\perp}$; these are precisely the eigenvalues of the matrix A_{22} in the controllability normal form (4.20). Note that $w^*A = \lambda w^*$ if and only if w is an eigenvector of A^T for the eigenvalue $\bar{\lambda}$. (The vector w is often called a *left eigenvector* for A associated with the eigenvalue λ of A, terminology which is potentially confusing if *left* is not emphasized. Remember that A and A^T have the same eigenvalues, but, in general, different eigenvectors.) These are the basic ideas behind the next result.

Theorem 4.7 *The linear system $\dot{x} = Ax + Bu$ is controllable if and only if for every complex λ the only $n \times 1$ vector w that satisfies*

$$w^*A = \lambda w^* \quad \text{and} \quad w^*B = 0 \tag{4.22}$$

is the zero vector, $w = 0$.

Proof. Suppose that (4.22) holds for some λ and a nonzero w; we must show that the system is not controllable. With the stated assumption, repeated use of (4.22) implies that

$$\begin{aligned} w^*\begin{bmatrix} B & AB & \cdots & A^{n-1}B \end{bmatrix} &= \begin{bmatrix} w^*B & w^*AB & \cdots & w^*A^{n-1}B \end{bmatrix} \\ &= \begin{bmatrix} w^*B & \lambda w^*B & \cdots & \lambda^{n-1}w^*B \end{bmatrix} \\ &= 0_{1\times nm}. \end{aligned}$$

This is equivalent to saying that the controllability matrix has linearly dependent rows. Therefore (A,B) is not controllable.

Now suppose that (A,B) is not controllable; we must show that (4.22) holds for some λ and a nonzero w. With the stated assumption, Theorem 4.5 applies. Let T be a nonsingular matrix such that the transformation $x = Tz$ produces the normal form

$$\dot{z} = \begin{bmatrix} A_{11} & A_{12} \\ 0 & A_{22} \end{bmatrix} z + \begin{bmatrix} B_1 \\ 0 \end{bmatrix} u, \tag{4.23}$$

where A_{11} is $r \times r$, $r < n$, and (A_{11}, B_1) is controllable. Let w_{uc} be an eigenvector of A_{22}^T with eigenvalue $\bar{\lambda}$; then $A_{22}^T w_{uc} = \bar{\lambda} w_{uc}$, and therefore $w_{uc}^* A_{22} = \lambda w_{uc}^*$. Since A_{22} is real, both λ and $\bar{\lambda}$ are eigenvalues of A_{22}, and therefore also eigenvalues of A. Write $w^* := [0_{1\times r} \quad w_{uc}^*]T^{-1}$. Since w_{uc} is an eigenvector for A_{22}^T, w is nonzero, and we have

$$w^*B = [0_{1\times r} \quad w_{uc}^*]T^{-1}T\begin{bmatrix} B_1 \\ 0 \end{bmatrix} = [0_{1\times r} \quad w_{uc}^*]\begin{bmatrix} B_1 \\ 0 \end{bmatrix} = 0.$$

We also have

$$w^*A = [0_{1\times r} \quad w_{uc}^*]T^{-1}T \begin{bmatrix} A_{11} & A_{12} \\ 0 & A_{22} \end{bmatrix} T^{-1} = [0_{1\times r} \quad w_{uc}^* A_{22}]T^{-1},$$

and therefore $w^*A = \bar{\lambda}[0_{1\times r} \quad w_{uc}^*]T^{-1} = \bar{\lambda}w^*$. Hence, the nonzero w^* and eigenvalue $\bar{\lambda}$ satisfy (4.22). □

Note that the eigenvalues of the restriction of A^T to its invariant subspace $R(\mathcal{C}(A,b))^{\perp}$ are precisely the eigenvalues of matrix A_{22} in (4.23). If $\operatorname{rank} \mathcal{C}(A,B) = r$, there are exactly $n-r$ of these eigenvalues, counting multiplicities.

We can now establish the PBH rank test for controllability. This test and the later PBH tests (for observability, stabilizability, and detectability of linear time invariant systems) are so designated to credit work by V. M. Popov, V. Belevitch, and M. L. J. Hautus.

Theorem 4.8 (PBH Controllability Test)
The matrix pair (A, B) is controllable if and only if

$$\operatorname{rank} \begin{bmatrix} A - \lambda I & B \end{bmatrix} = n \tag{4.24}$$

for every eigenvalue λ of A.

Proof. Suppose that (A, B) is not controllable. Then, by Theorem 4.7, there are an eigenvalue λ for A and a vector w such that $A^T w = \lambda w$ and $B^T w = 0$; hence, we have

$$\begin{bmatrix} A^T - \lambda I \\ B^T \end{bmatrix} w = 0, \tag{4.25}$$

with w nonzero. Therefore $\operatorname{rank} [A - \lambda I \quad B] < n$ and (4.24) fails to hold. Now suppose that (4.24) does not hold. Then there exist a number λ and a nonzero vector w such that $w^*[A - \lambda I \quad B] = 0$; hence, equation (4.25) holds for a nonzero vector w, and therefore, by Theorem 4.7, (A, B) is not controllable. □

Returning to Example 4.8, we observe that the PBH test establishes the controllability criterion, $b_3 \neq 0$, quite easily. The PBH test can make controllability results for Jordan form systems more transparent, and the study of such systems adds insight into controllability. Consider the case of a single Jordan block system, as in Example 4.8. In that example, if we are able to control the last state variable, then (intuitively) we expect that the system is controllable since the last component feeds into the previous equations by back substitution. For controllability, if $x(t_0)$ is given, we only need enough freedom in the choice of $u(t)$ to achieve any three assigned components of a final state vector $x(t_f)$. The PBH test confirms this intuition for such cases. See also Exercise 4.6.

TERMINOLOGY FOR SOME EIGENPAIRS. Here is a formal definition which may be convenient.

Definition 4.6 *The pair* (λ, w) *is an* uncontrollable eigenpair *of* (A, B), *or simply an* uncontrollable pair *of* (A, B), *if* $w \neq 0$ *and we have*

$$w^* A = \lambda w^* \quad \text{and} \quad w^* B = 0.$$

The proof of Theorem 4.7 shows that, given an eigenpair (λ, w), it is an uncontrollable pair if and only if $w \in \mathcal{R}(\mathcal{C}(A, B))^\perp$. The statement in Definition 4.6 is a simple form that can help to reveal *uncontrollable modes*, defined here as vectors in $\mathcal{R}(\mathcal{C}(A, B))^\perp$, and corresponding eigenvalues.

Note that the eigenvalues of A_{11} in (4.23) correspond to controllable modes. The next definition uses invariance of the subspace $\mathcal{R}(\mathcal{C}(A, B))$ under A to describe these modes.

Definition 4.7 *The pair* (λ, v) *is a* controllable eigenpair *of* (A, B), *or simply a* controllable pair *of* (A, B), *if* $v \neq 0$ *and we have*

$$Av = \lambda v \quad \text{and} \quad v \in \mathcal{R}(\mathcal{C}(A, B)).$$

Using Definition 4.6, a restatement of Theorem 4.7 is that *a system is controllable if and only if there are no uncontrollable eigenpairs.* And using Definition 4.7, we can say that *a system is controllable if and only if all eigenpairs are controllable.*

Example 4.10 Consider the system

$$\dot{x} = \begin{bmatrix} 2 & 1 & 0 \\ 0 & 2 & 0 \\ 0 & 0 & 2 \end{bmatrix} x + \begin{bmatrix} 0 \\ 0 \\ 1 \end{bmatrix} u.$$

The only eigenvalue is $\lambda = 2$, and this eigenvalue corresponds to both controllable and uncontrollable modes. (Some authors use the terms *controllable eigenvalue* and *uncontrollable eigenvalue*. As this example shows, these two terms are not mutually exclusive. See also Example 5.6.) $\triangle$

This completes our study of controllability of linear time invariant systems. The next section offers a few comments on linear time varying systems as motivation for further reading.

4.8 LINEAR TIME VARYING SYSTEMS: AN EXAMPLE

Even for systems that are modeled using time invariant physical laws, there are situations where an appropriate model might be time varying. For example, satellite motion might be studied using a rotating coordinate frame where the rotation is time dependent. And when a nonlinear time invariant system is linearized along a *nonconstant* solution trajectory, the resulting Jacobian linearization is a linear time varying system.

For linear time varying systems, additional technical issues arise in any extension of controllability, although several extensions have been accomplished. For example, several useful definitions of controllability for time varying systems are possible, all of which coalesce in the linear time invariant case to describe the same concept. These definitions may involve the initial time t_0, the particular initial state x_0 considered, and the time interval over which control action is to take place. The reader interested in these issues is invited to explore the references. Here we merely give an example to show that controllability of time varying linear systems requires some additional ideas.

Example 4.11 Consider the system

$$x' = \begin{bmatrix} 1 & 0 \\ 0 & 2 \end{bmatrix} x + \begin{bmatrix} b_1(t) \\ b_2(t) \end{bmatrix} u.$$

Write $x_{10} := x_1(0)$ and $x_{20} := x_2(0)$. Then the general solution is

$$x(t) = \begin{bmatrix} x_1(t) \\ x_2(t) \end{bmatrix} = \begin{bmatrix} e^t(x_{10} + \int_0^t e^{-s} b_1(s) u(s)\, ds) \\ e^{2t}(x_{20} + \int_0^t e^{-2s} b_2(s) u(s)\, ds) \end{bmatrix}.$$

If b_1 and b_2 are constant, then Theorem 4.2 ensures that the system is completely controllable. Suppose now that $b_1(t) = e^t$ and $b_2(t) = e^{2t}$; then we have

$$x(t) = \begin{bmatrix} x_1(t) \\ x_2(t) \end{bmatrix} = \begin{bmatrix} e^t(x_{10} + \int_0^t u(s)\, ds) \\ e^{2t}(x_{20} + \int_0^t u(s)\, ds) \end{bmatrix}.$$

Thus, solutions that start on the line $x_2 = x_1$ at $t = 0$ always satisfy the condition $x_2(t) = e^t x_1(t)$. If $x_{10} = x_{20}$, then the motion is confined to the first or third quadrant since x_1 and x_2 must have the same sign. In particular, the set of points reachable from the origin lies within those two quadrants. Therefore the system is not controllable according to Definition 4.3. If we consider the controllability matrix and controllability rank condition in a pointwise manner, that is, if we consider the matrix function

$$\begin{bmatrix} B(t) & AB(t) \end{bmatrix} = \begin{bmatrix} e^t & e^t \\ e^{2t} & 2e^{2t} \end{bmatrix},$$

we do obtain a nonsingular matrix. This example shows that a pointwise interpretation of the controllability rank condition of Theorem 4.2 does not lead to a criterion for controllability (Definition 4.3) of linear time varying systems. △

4.9 EXERCISES

Exercise 4.1 *First Examples*

(a) Show that (4.6) cannot be transformed to the form (4.2) via the linear transformation $z = Tx$ with invertible T. *Hint*: If the transformation is possible, then the entries in P are known.

(b) Do the same for (4.7).

Exercise 4.2 Let A be as in Example 4.1, and let $b = [1\ 1]^T$. Show that the resulting system $\dot{x} = Ax + bu$ cannot be transformed to the form (4.2).

Exercise 4.3 *A Transfer-of-State Problem*

For the system

$$\begin{bmatrix} \dot{x}_1 \\ \dot{x}_2 \end{bmatrix} = \begin{bmatrix} 2 & 1 \\ 0 & 2 \end{bmatrix} \begin{bmatrix} x_1 \\ x_2 \end{bmatrix} + \begin{bmatrix} 0 & 1 \\ 1 & 0 \end{bmatrix} u,$$

find a control function $u(t) = [u_1(t)\ u_2(t)]^T$ such that the solution $x(t) = [x_1(t)\ x_2(t)]^T$ with initial condition $x(0) = [0\ 0]^T$ reaches the final position $x_f = x(2) = [1\ 3]^T$ at time $t_f = 2$. *Hint*: Start with the variation of parameters formula for $x_2(t)$, and try a constant control $u_1 = \alpha$ for $0 \le t \le 2$: the condition $x_2(2) = 3$ uniquely determines α. Then proceed with variation of parameters for $x_1(t)$. Is the problem more difficult with a single scalar input, with input vector $b = [0\ 1]^T$?

Exercise 4.4 *The Linear Airplane*

Write the linear airplane equations in Example 4.4 as a four-dimensional linear system using the variables $x_1 = \alpha$, $x_2 = \phi$, $x_3 = \dot{\phi}$, and $x_4 = h$, and show that the system is controllable.

Exercise 4.5 *Invariance Properties of Controllability*

Prove Proposition 4.3 and Proposition 4.4.

Exercise 4.6 *Diagonal Systems and Controllability*

(a) In Example 4.7, verify that the statement

$$b_j \neq 0, \quad j = 1, 2, 3 \quad \text{and} \quad \lambda_i \neq \lambda_j \quad \text{for all } i \neq j$$

is equivalent to the following statement:

$$\text{rank} \begin{bmatrix} A - \lambda I & b \end{bmatrix} = n \quad \text{for all eigenvalues } \lambda \text{ of } A.$$

(b) Show, using the PBH test, that an n-dimensional diagonal system $\dot{x} = Ax + bu$ is controllable if and only if all components of the input vector b are nonzero and the eigenvalues of A are distinct.

Exercise 4.7 *Controllability and Jordan Forms*

(a) Show that the system

$$\dot{x} = \begin{bmatrix} \lambda & 1 & 0 \\ 0 & \lambda & 0 \\ 0 & 0 & \lambda \end{bmatrix} x + \begin{bmatrix} b_1 \\ b_2 \\ b_3 \end{bmatrix} u$$

cannot be controllable. Argue that a single-input system with multiple Jordan blocks associated with the same eigenvalue cannot be controllable.

(b) Suppose

$$A = \begin{bmatrix} \mu & 0 & 0 \\ 0 & \lambda & 1 \\ 0 & 0 & \lambda \end{bmatrix} \quad \text{and} \quad b = \begin{bmatrix} b_1 \\ b_2 \\ b_3 \end{bmatrix}.$$

Find necessary and sufficient conditions on b for controllability of the pair (A, b). Does it matter whether $\mu \neq \lambda$ or $\mu = \lambda$?

Exercise 4.8 *More on Controllability and Jordan Forms*

(a) Show that this system is controllable:

$$\dot{x} = Ax + Bu := \begin{bmatrix} 2 & 1 & 0 & 0 \\ 0 & 2 & 0 & 0 \\ 0 & 0 & 3 & 1 \\ 0 & 0 & 0 & 3 \end{bmatrix} x + \begin{bmatrix} 0 & 0 \\ 0 & 1 \\ 0 & 0 \\ 1 & 1 \end{bmatrix} u.$$

(b) Let $u = [u_1 \; u_2]^T$ be the input in (a). If the system in (a) is augmented by the equation $\dot{z} = 2z + u_2$, is the augmented system controllable?

Exercise 4.9 *Assignable Characteristic Polynomials*

Show that arbitrary assignability of monic real characteristic polynomials by state feedback implies controllability. (*Hint*: Theorem 4.5.)

Exercise 4.10 *Motivating a Transfer-of-State Control*

This exercise motivates the choice of control input that accomplishes the transfer of state in the proof of Theorem 4.2. Consider the scalar equation with scalar control, $\dot{x} = ax + bu$, with $b \neq 0$. It can be seen that this is a controllable system, as follows. Let x_0, x_f be in $\mathbf{R}$, and let $t_f > 0$.

(a) Let $M := \int_0^{t_f} e^{-sa} bbe^{-sa}\, ds$, and observe that $M > 0$.

(b) Let $u(s) := be^{-sa}x$ for $0 \leq s \leq t_f$, where x is to be chosen. Verify that, with input $u(t)$, the solution $x(t)$ having initial condition $x(0) = x_0$ has final point x_f at time t_f given by

$$x_f = e^{at_f}\left(x_0 + \int_0^{t_f} (e^{-sa} bbe^{-sa} ds)\, x \right) = e^{at_f}(x_0 + Mx).$$

(c) Show that in (b) there is a unique solution for x which makes $x(t_f) = x_f$.

Exercise 4.11 Complete the details in the proof of Theorem 4.4.

Exercise 4.12 *Range Space of the Controllability Matrix*
Show that $R(\mathcal{C}(A, B))$ is the smallest invariant subspace of A that contains $R(B)$. That is, if S is any invariant subspace for A that contains $R(B)$, then $R(\mathcal{C}(A, B)) \subseteq S$.

Exercise 4.13 *Eigenvalue Placement with Vector-Input*
Consider the system in Example 4.5, with vector $b = b_0$ and feedback matrix K constructed there.

(a) Find a 1×3 feedback matrix F such that $A + BK + bF$ has characteristic polynomial $p(\lambda) = (\lambda + 1)^3 = \lambda^3 + 3\lambda^2 + 3\lambda + 1$.
(b) Using the result of (a), find the corresponding feedback matrix G as in Theorem 4.6 such that $p(\lambda) = \det(\lambda I - A - BG)$.

Exercise 4.14 *Controllability and Jordan Form*
Consider the matrix pair (A, B), where

$$A = \begin{bmatrix} 2 & 0 & 0 \\ 0 & 2 & 1 \\ 0 & 0 & 2 \end{bmatrix}, \qquad B = \begin{bmatrix} 1 & 0 \\ 0 & 0 \\ 0 & 1 \end{bmatrix} = \begin{bmatrix} b_0 & b_1 \end{bmatrix}.$$

Verify that (A, B) is controllable, and find K and β such that $A + BK$ is the Jordan block of size 3 for $\lambda = 2$, while $B\beta = b_1$. Show that $(A + BK, b_1)$ is controllable. Note that this procedure is to channel all input through the second of the original input vectors, so an appropriate choice of K should provide an interconnection term for the first state component; such an interconnection term is necessary for controllability when using only b_1 as the input matrix.

Exercise 4.15 *Bounds on Controls*
The characterizations of the controllability property in this chapter depend on the availability of controls of arbitrarily large norm. In practice there is always a maximum possible norm for controls. Such restrictions on the controls require some restrictions on a linear system in order to be controllable. This exercise explores a necessary condition for controllability with bounded scalar controls.

Consider a single-input single-output linear system $\dot{x} = Ax + bu$ with controls u restricted to take values only within the interval $-M < u < M$ for some $M > 0$.

(a) Show that a necessary condition for controllability is that every eigenvalue of A have real part equal to zero. *Hint.* Consider a single Jordan block in the real Jordan form of A, corresponding to an eigenvalue $\lambda = \alpha + i\beta$, and show that $\alpha \neq 0$ implies that the system is not controllable. Consider the different possible Jordan blocks depending on whether β is zero or nonzero.

(b) Explain why controllability as in Definition 4.3 will almost never be possible in practice when there are bounds on the size of controls.

4.10 NOTES AND REFERENCES

The main resources for the preparation of this chapter were [53] and [82]. The concept of controllability is due to R. E. Kalman, see for example [54] or [56]. Another early paper is [29]. Reference [82] is a comprehensive text on linear systems, with much material on linear time varying systems; Example 4.11 was taken from there. Other helpful text references are [9], [16], [104]. Reference [91] is an excellent resource for theoretical foundations of both linear and nonlinear systems. For linear algebra and matrix analysis, see [75].

Proposition 4.1 is taken from [42] (page 147, Theorem 3.3.15). The proof of Theorem 4.2 follows [80]. The proof of Theorem 4.6 using Lemma 4.1 and Lemma 4.2 follows [35]. The development of the PBH controllability test was guided by [82]. For additional discussion on Jordan forms and controllability, see [10] (Chapter 8).

Exercise 4.15 was suggested by [51] (Theorem 5, page 138), where further information can be found.

Chapter Five

Observability and Duality

In this chapter we introduce the concept of *observability* and establish the algebraic duality of controllability and observability for linear time invariant systems. We derive an observability normal form, discuss invariance properties, and establish the PBH test for observability. We also revisit some Lyapunov equations.

5.1 OBSERVABILITY, DUALITY, AND A NORMAL FORM

Suppose we have a multi-input multi-output (MIMO) system $\dot{x} = Ax + Bu$ for which certain linear combinations of the state components x_i are directly measured. These measurements might be accomplished, for example, by some convenient combination of sensors. We write the system and its measured outputs as

$$\dot{x} = Ax + Bu, \tag{5.1}$$

$$y = Cx, \tag{5.2}$$

where B is $n \times m$ and C is $p \times n$. Given a state solution $x(t)$, the function $y(t) = Cx(t)$ is the known output. We would like to know whether the full system state can be determined from the limited direct knowledge of y, without integrating the differential equations.

Definition 5.1 *The MIMO system* (5.1)–(5.2) *is* observable *if, for any* $x_0 = x(0)$*, there is a finite time* $t_f > 0$ *such that knowledge of the input* $u(t)$ *and output* $y(t)$ *on* $[0, t_f]$ *uniquely determines* x_0.

We will see in a moment that Definition 5.1 could be restated without loss of generality by considering only the zero input, $u \equiv 0$.

Observability and Duality for SISO Systems

In order to see how a determination of x_0 might be made from knowledge of the output, consider for simplicity the case of a single-input single-output (SISO) system

$$\dot{x} = Ax + bu, \tag{5.3}$$

$$y = cx, \tag{5.4}$$

where b is $n \times 1$ and c is $1 \times n$. We may differentiate the output equation $(n-1)$ times, using substitutions for $\dot{x}$ from the differential equation, and then set $t = 0$ to get

$$\begin{bmatrix} y(0) \\ \dot{y}(0) \\ \cdots \\ y^{(n-1)}(0) \end{bmatrix} = \begin{bmatrix} c \\ cA \\ \cdots \\ cA^{n-1} \end{bmatrix} x_0 + \text{(terms dependent on u)}$$

$$=: \mathcal{O}(c, A)x_0 + \text{(terms dependent on u)}. \tag{5.5}$$

If we assume that $\operatorname{rank} \mathcal{O}(c, A) = n$, then the coefficient of x_0 in (5.5) is nonsingular and we can solve for x_0 in terms of y, u and their derivatives.

Example 5.1 Consider the system

$$\begin{aligned} \dot{x}_1 &= -2x_1 + 2x_2 + u, \\ \dot{x}_2 &= x_1 - x_2 + u, \\ y &= x_1. \end{aligned}$$

In this case, the output and its first derivative give the equations

$$\begin{bmatrix} y(0) \\ \dot{y}(0) \end{bmatrix} = \begin{bmatrix} 1 & 0 \\ -2 & 2 \end{bmatrix} x_0 + \begin{bmatrix} 0 \\ 1 \end{bmatrix} u.$$

Since the coefficient of x_0 is nonsingular, the system is observable. $\triangle$

Theorem 5.1 *The system* (5.3)–(5.4) *is* observable *if and only if*

$$\operatorname{rank} \mathcal{O}(c, A) = \operatorname{rank} \begin{bmatrix} c \\ cA \\ \cdots \\ cA^{n-1} \end{bmatrix} = n.$$

Proof. We have already shown the sufficiency of the rank condition. Now assume that observability holds; we must show that the rank condition holds. Suppose also, for the purpose of contradiction, that $\operatorname{rank} \mathcal{O}(c, A) < n$. Then there is a nonzero vector v such that

$$\begin{bmatrix} c \\ cA \\ \cdots \\ cA^{n-1} \end{bmatrix} v = 0. \tag{5.6}$$

Take $x_0 = v$ and consider the output $y = ce^{tA}v$ corresponding to the input $u \equiv 0$. By (5.6) and the definition of the matrix exponential, the Taylor series expansion for $y(t)$ must have all coefficients equal to zero. Thus, $y \equiv 0$; but this is also the output when $x_0 = 0$ under zero input. This contradicts the observability assumption, because there are two initial conditions which cannot be distinguished by the output when the zero input is applied. $\square$

The rank condition of Theorem 5.1 is known as the *observability rank condition* and the matrix $\mathcal{O}(c, A)$ appearing there is called the *observability matrix* for (5.3)–(5.4). As we have noted, the rank condition implies that the system state x can be reconstructed from knowledge of y, u, and their derivatives.

We can now rephrase the equivalence problem of the previous chapter. We consider the SISO system (5.3)–(5.4). We now ask, when is c the first row of a transformation $z = Tx$ to the controller form (4.2), where y is the dependent variable in the n-th order scalar equation (4.1)? We know that if such a T exists, then T must have the form (4.14) with $\tau = c$. We must also have $Tb = d = [0 \quad \cdots \quad 0 \quad 1]^T$. Thus,

$$\text{rank } T = \text{rank} \begin{bmatrix} c \\ cA \\ \cdots \\ cA^{n-1} \end{bmatrix} = n. \tag{5.7}$$

Moreover, we may write

$$\begin{bmatrix} c \\ cA \\ \cdots \\ cA^{n-1} \end{bmatrix} b = \begin{bmatrix} 0 \\ 0 \\ \cdots \\ 1 \end{bmatrix} = \begin{bmatrix} b^T \\ b^T A^T \\ \cdots \\ b^T (A^{n-1})^T \end{bmatrix} c^T, \tag{5.8}$$

where the last equality displays a useful symmetry, or duality. We will explain the significance of this duality in a moment. One can check that the differential equation for $z := [y \quad \dot{y} \quad \cdots \quad y^{(n-1)}]^T$ really is the controller form (4.2), by using the fact that A satisfies its own characteristic polynomial:

$$A^n + k_1 A^{n-1} + \cdots + k_{n-1} A + k_n I = 0.$$

We have proved the following result.

Proposition 5.1 *There exists a nonsingular linear transformation $z = Tx$ that transforms* (5.3)–(5.4) *to controller form* (4.2) *with $z_1 = y = cx$, if and only if the rank condition* (5.7) *holds and* (5.8) *is satisfied. Then T is uniquely determined and is the matrix in* (5.7).

Note that the matrix $\mathcal{O}(c, A)$ has the same rank as its transpose given by

$$\left[\; c^T \quad A^T c^T \quad \cdots \quad (A^T)^{n-1} c^T \; \right].$$

Therefore, according to Proposition 5.1, we can say that $y = cx$ satisfies the n-th order scalar equation (4.1) if and only if the system

$$\dot{z} = A^T z + c^T u \tag{5.9}$$

is controllable (Theorem 4.2). Moreover, (5.8) shows that b^T is the first row of the transformation matrix that takes (5.9) to controller form; this transformation matrix can be seen on the right-hand side of (5.8).

The connection of Proposition 5.1 with system (5.9) indicates a fundamental duality between controllability and observability. We define the *dual system* of (5.3)–(5.4) as

$$\dot{z} = A^T z + c^T u, \tag{5.10}$$

$$y = b^T z. \tag{5.11}$$

Then the dual of the dual of a system is the original system, that is, the dual of (5.10)–(5.11) is (5.3)–(5.4). This definition of dual system allows us to encapsulate the discussion thus far with the following classical duality statement.

Theorem 5.2 *The single-input single-output system* (5.3)–(5.4) *is observable if and only if system* (5.10) *is controllable. The system* (5.3) *is controllable if and only if the dual system* (5.10)–(5.11) *is observable.*

Note that the observability property is independent of the input matrix, and the controllability property is independent of the output matrix. Thus, if the pair (A, b) is controllable then every triple of matrices (c, A, b) yields a controllable system; similarly, if the pair (c, A) is observable then every triple of matrices (c, A, b) yields an observable system.

Observability and Duality for MIMO Systems

We now extend the Kalman rank criterion (Theorem 5.1) and the duality between observability and controllability (Theorem 5.2) to the multi-input multi-output systems (5.1)–(5.2), shown here for convenience:

$$\dot{x} = Ax + Bu, \tag{5.12}$$

$$y = Cx, \tag{5.13}$$

where B is $n \times m$ and C is $p \times n$. The definition of the single-output observability matrix $\mathcal{O}(c, A)$ in (5.5) has a straightforward extension for (5.12)–(5.13), given by the matrix

$$\mathcal{O}(C, A) := \begin{bmatrix} C \\ CA \\ \cdots \\ CA^{n-1} \end{bmatrix},$$

and motivated in the same way, by differentiation of the output and substitution from the differential equation. Matrix $\mathcal{O}(C, A)$ is called the *observability matrix* for (5.12)–(5.13), or the observability matrix of the pair (C, A). The observability matrix is now $pn \times n$, and we have the following rank characterization of observability.

Theorem 5.3 *System* (5.12)–(5.13) *is observable if and only if the observability matrix* $\mathcal{O}(C, A)$ *has full column rank, that is, if and only if*

$$\operatorname{rank} \mathcal{O}(C, A) = n.$$

A careful trace through the proof of Theorem 5.1 shows that the argument extends in a straightforward way to give a proof of Theorem 5.3. When (5.12)–(5.13) is observable we also say that the pair (C, A) is observable.

The *dual system* of (5.12)–(5.13) is defined by

$$\dot{z} = A^T z + C^T u, \tag{5.14}$$

$$y = B^T z, \tag{5.15}$$

with matrix dimensions determined by the original system. This definition of dual system allows us to state the fundamental duality between observability and controllability in the next theorem.

Theorem 5.4 *System* (5.12)–(5.13) *is observable if and only if system* (5.14) *is controllable. System* (5.12) *is controllable if and only if the dual system* (5.14)–(5.15) *is observable.*

Again we note that the pair (A, B) is controllable if and only if every triple of system matrices (C, A, B) is controllable; similarly, (C, A) is observable if and only if every triple of system matrices (C, A, B) is observable.

An Observability Normal Form

An observability normal form can be developed for systems with an *unobservable pair* (C, A), that is, a pair such that $\operatorname{rank} \mathcal{O}(C, A) < n$. Such a normal form is the dual version of the controllability normal form.

Theorem 5.5 *Consider a MIMO linear system* $\dot{x} = Ax + Bu$, $y = Cx$, *where* C *is* $p \times n$. *If* $\operatorname{rank} \mathcal{O}(C, A) = \rho < n$, *then there is a nonsingular matrix* T *such that the transformation* $x = Tz$ *produces the normal form*

$$\begin{bmatrix} \dot{z}_1 \\ \dot{z}_2 \end{bmatrix} = \begin{bmatrix} A_{11} & 0 \\ A_{21} & A_{22} \end{bmatrix} \begin{bmatrix} z_1 \\ z_2 \end{bmatrix} + \begin{bmatrix} B_1 \\ B_2 \end{bmatrix} u,$$

$$y = \begin{bmatrix} C_1 & 0 \end{bmatrix} z,$$

where A_{11} *is* $\rho \times \rho$ *and* (C_1, A_{11}) *is observable.*

Proof. Use Theorem 4.5 and a duality argument. □

The special system form in Theorem 5.5 is called an *observability normal form*. The construction of an observability normal form is considered further in the next chapter.

When $u = 0$, Theorem 5.5 implies that for any initial conditions with $z_1(0) = 0$, the solution trajectory $z(t)$ will evolve within the subspace defined by $z_1 = 0$, and this trajectory will be determined by the dynamics of the system $\dot{z}_2 = A_{22} z_2$. We may refer to the states z_1 (or components thereof) as *observable modes*, and we say that the states z_2 (or components thereof) are *unobservable modes*.

Invariance Properties of Observability

We proceed to consider some invariance properties of observability. In particular, the observability property is invariant under linear coordinate transformations.

Proposition 5.2 *Let A and T be $n \times n$ and C be $p \times n$. If (C, A) is observable and T is nonsingular, then $(T^{-1}AT, CT)$ is observable. That is, observability is invariant under linear coordinate transformations.*

Proof. The proof is left to Exercise 5.1. □

In many applications it is not realistic to assume that the full state vector x is available for measurement and subsequent feedback. Suppose there are $p < n$ scalar-valued measurements available, each measurement being a linear combination of the state components, and only these outputs may be used in a feedback control. This brings us to the definition of output feedback and the resulting closed loop systems.

Definition 5.2 *In system* (5.12)–(5.13), static output feedback *is defined by $u = Ky = KCx$ where K is a real $m \times p$ matrix. The corresponding* closed loop system *is $\dot{x} = (A + BKC)x$.*

The term *static* is used in Definition 5.2 to emphasize the expression of the feedback directly in terms of the state x. Of course, full state feedback given by $u = Kx$ is a special case of static feedback (taking $C = I$). A more general framework of dynamic output feedback will be discussed in some detail later. To state the difference between these types of feedback very briefly, we can say that static output feedback depends on the state without involving any integrations of the state vector, whereas dynamic output feedback processes the state through integration operations before yielding the control term that modifies the original dynamics.

We now consider two questions about the invariance of observability under feedback transformations:

- If (C, A) is observable, is $(C, A + BKC)$ observable for arbitrary $m \times p$ matrices K? In other words, is observability invariant under static output feedback transformations?
- If (C, A) is observable, is $(C, A + BK)$ observable for arbitrary $m \times n$ matrices K? In other words, is observability invariant under full state feedback $u = Kx$, where the whole state x is considered available for feedback?

The next proposition gives a positive answer to the first of these invariance questions.

Proposition 5.3 *Suppose that A and T are $n \times n$, B is $n \times m$, and C is $p \times n$. If (C, A) is observable, then for any $m \times n$ matrix K, the pair $(C, A+BKC)$ is observable. That is, observability is invariant under static output feedback transformations.*

Proof. The proposition follows from the set equality

$$\ker \mathcal{O}(C, A) = \ker \mathcal{O}(C, A + BKC).$$

The details are a recommended exercise (Exercise 5.2). □

Regarding the second of the invariance questions mentioned above, the larger class of full state feedback transformations does not generally preserve observability. Here is a single-input single-output example that illustrates this fact.

Example 5.2 Consider the system

$$\dot{x} = \begin{bmatrix} 0 & 1 \\ 1 & 0 \end{bmatrix} x + \begin{bmatrix} 1 \\ 0 \end{bmatrix} u,$$
$$y = \begin{bmatrix} 1 & 0 \end{bmatrix} x.$$

It is easy to check that it is observable by the rank criterion. Let $K = [k_1 \; k_2] = [k_1 \;\; -1]$. Then we have

$$A + BK = \begin{bmatrix} k_1 & 1 + k_2 \\ 1 & 0 \end{bmatrix} = \begin{bmatrix} k_1 & 0 \\ 1 & 0 \end{bmatrix},$$

and the observability matrix for the pair $(C, A + BK)$ is

$$\begin{bmatrix} C \\ C(A + BK) \end{bmatrix} = \begin{bmatrix} 1 & 0 \\ k_1 & 0 \end{bmatrix},$$

which has rank one. Thus, state feedback may not preserve observability. Note that the feedback produced a zero eigenvalue for the matrix $A + BK$, with a corresponding eigenvector contained in $\ker C$. It is worth observing that this cannot happen with static output feedback, which has the form $u = ky = kx_1$ where k is a scalar. △

Output Feedback versus Full State Feedback

We now define the concept of static output feedback stabilizability.

Definition 5.3 *System* (5.12)–(5.13) *is* stabilizable by static output feedback *if there exists an $m \times p$ matrix K such that, with $u = Ky = KCx$, the* *closed loop linear system $\dot{x} = (A + BKC)x$ is asymptotically stable, that is, $A + BKC$ is Hurwitz.*

We might naturally ask whether observability implies any particular eigenvalue placement properties, or at least stabilizability by output feedback. However, general expectations along these lines should not be high, as the next example shows.

Example 5.3 An undamped linear pendulum with position output is modeled by the system

$$\begin{aligned} \dot{x}_1 &= x_2, \\ \dot{x}_2 &= -x_1 + u, \\ y &= x_1. \end{aligned}$$

This system is clearly observable by Theorem 5.1. However, output feedback $u = ky = kx_1$, with k scalar, leads to the system

$$\dot{x} = (A + bkc)x = \begin{bmatrix} 0 & 1 \\ k - 1 & 0 \end{bmatrix} x.$$

The eigenvalues are the roots of $\lambda^2 = k - 1$. If $k < 1$ then we have a conjugate pair of imaginary eigenvalues. If $k = 1$, then there is a double zero eigenvalue. And if $k > 1$, then the origin is a saddle point. Therefore the system cannot be stabilized by static output feedback. Intuitively, we realize that output feedback fails to stabilize the system because there is no information available about the velocity x_2 and no damping occurs through output feedback.

On the other hand, if the output is the velocity $y = x_2$, then it is easily checked that the system with this new output is stabilizable by static output feedback. If $u = ky = kx_2$, then the closed loop matrix is

$$A + bkc = \begin{bmatrix} 0 & 1 \\ -1 & k \end{bmatrix},$$

and any choice of k such that $k < 0$ will make $A + bkc$ Hurwitz. $\triangle$

According to Definition 5.3, a stabilizing static output feedback, if one exists, must have the factorized form $u = KCx$. This can be a very restrictive condition, as Example 5.3 shows.

The output feedback stabilization problem for linear time-invariant systems is a difficult one. (A relatively easier case concerns a class of passive linear systems later in the text; see "Passivity and Output Feedback Stabilization.") Some resources for more information on output feedback stabilization are indicated in the Notes and References for this chapter.

5.2 LYAPUNOV EQUATIONS AND HURWITZ MATRICES

When a system is observable, the conditions that characterize a Hurwitz coefficient matrix A by the solvability of a Lyapunov equation can be simplified to some extent as compared with Theorem 3.7. The simplified characterization is in the next proposition.

Proposition 5.4 *Suppose (C, A) is an observable pair. Then the following statements are equivalent:*

(a) *The matrix A is Hurwitz.*
(b) *There exists a positive definite solution P of the Lyapunov equation*

$$A^T P + PA + C^T C = 0. \tag{5.16}$$

Proof. (a) implies (b): Let $W = C^T C$. Since A is Hurwitz, Proposition 3.2 applies, and therefore (5.16) has a unique solution P. Matrix W is positive semidefinite, since $x^T W x = x^T C^T C x = \|Cx\|^2 \geq 0$ for all x; hence P is positive semidefinite. In fact, we have

$$x^T P x = x^T \left(\int_0^\infty e^{tA^T} C^T C e^{tA} \, dt \right) x = \int_0^\infty x^T e^{tA^T} C^T C e^{tA} x \, dt \geq 0.$$

Observability of the pair (C, A) implies that, if $x \neq 0$, then $Ce^{tA}x$ cannot be identically zero, and therefore $x^T P x > 0$. Therefore the unique solution of (5.16) is positive definite. (Note that this is true even though $W = C^T C$ is not necessarily positive definite; in fact $C^T C$ is positive definite if and only if C has full column rank.)

(b) implies (a): We assume that (5.16) has a positive definite solution P. We must show that every eigenvalue of A has negative real part. Let λ be an eigenvalue of A, with eigenvector $v \neq 0$. Using the eigenvector v, we have (for real C and A)

$$\begin{aligned} -v^* C^T C v &= v^*(A^T P + PA)v \\ &= v^* A^T P v + v^* P A v \\ &= (\lambda v)^* P v + v^* P (\lambda v) \\ &= \bar{\lambda} v^* P v + \lambda v^* P v \\ &= (\bar{\lambda} + \lambda) v^* P v. \end{aligned}$$

Thus, $-\|Cv\|^2 = 2\text{Re}\,(\lambda) v^* P v$. Since $v^* P v > 0$, we must have $\text{Re}\,(\lambda) \leq 0$. If $\text{Re}\,(\lambda) = 0$, then $\|Cv\| = 0$, hence $Cv = 0$, and consequently $CA^k v = \lambda^k C v = 0$ for all nonnegative integers k. Therefore v is in the nullspace of the observability matrix. By the observability hypothesis, $v = 0$, and this contradicts the fact that v is an eigenvector. Therefore $\text{Re}\,(\lambda) < 0$. Since this is true for every eigenvalue of A, A is Hurwitz. $\square$

In Proposition 5.4, a structural condition (observability) gave a new characterization of Hurwitz A in terms of the solvability of a specific Lyapunov equation by a positive definite matrix. Duality gives the next result.

Proposition 5.5 *Suppose (A, B) is a controllable pair. Then the following statements are equivalent:*

(a) *The matrix A is Hurwitz.*
(b) *There exists a positive definite solution Q of the Lyapunov equation*

$$AQ + QA^T + BB^T = 0. \tag{5.17}$$

Proof. See Exercise 5.3. □

Note that the appearance of the unknown Q in (5.17) is reversed as compared with equation (5.16).

5.3 THE PBH OBSERVABILITY TEST

In this section we establish the PBH rank test for observability. We apply the test to some diagonal and Jordan form systems to provide additional insight into the test and the observability concept.

First, we have the dual version of Theorem 4.7.

Theorem 5.6 *The linear system $\dot{x} = Ax$ with output $y = Cx$ is observable if and only if for every complex λ the only $n \times 1$ vector v that satisfies*

$$Av = \lambda v \quad \text{and} \quad Cv = 0 \tag{5.18}$$

is the zero vector, $v = 0$.

Proof. Apply Theorem 4.7 to the dual system defined by the pair (A^T, C^T), and the theorem follows. □

Note that Theorem 5.6 is also clear from the definition of observability, and it is clear as well from the fact that $N(\mathcal{O}(C, A))$ is an invariant subspace for A.

We have the following important characterization of observability.

Theorem 5.7 (PBH Observability Test)
The pair (C, A) is observable if and only if

$$\text{rank} \begin{bmatrix} A - \lambda I \\ C \end{bmatrix} = n \tag{5.19}$$

for every eigenvalue λ of A.

Proof. The statement follows from the PBH controllability test by a duality argument. Alternatively, it follows directly from Theorem 5.6. □

To gain additional insight into this observability test, we consider some diagonal and Jordan form systems. The first example is a system with a diagonal coefficient matrix.

Example 5.4 Suppose that $\lambda_1, \lambda_2, \lambda_3$ are distinct real numbers. By Theorem 5.7, the diagonal system

$$\dot{x} = \begin{bmatrix} \lambda_1 & 0 & 0 \\ 0 & \lambda_2 & 0 \\ 0 & 0 & \lambda_3 \end{bmatrix} x,$$
$$y = \begin{bmatrix} c_1 & c_2 & c_3 \end{bmatrix} x$$

is observable if and only if $c_j \neq 0$ for $1 \leq j \leq 3$. See also Exercise 5.4. △

The next example has a Jordan block A matrix.

Example 5.5 For the system

$$\begin{aligned} \dot{x} &= \begin{bmatrix} \lambda & 1 & 0 \\ 0 & \lambda & 1 \\ 0 & 0 & \lambda \end{bmatrix} x, \\ y &= \begin{bmatrix} c_1 & c_2 & c_3 \end{bmatrix} x, \end{aligned} \tag{5.20}$$

we have

$$\begin{bmatrix} A - \lambda I \\ C \end{bmatrix} = \begin{bmatrix} 0 & 1 & 0 \\ 0 & 0 & 1 \\ 0 & 0 & 0 \\ c_1 & c_2 & c_3 \end{bmatrix}.$$

This matrix has full rank equal to 3 if and only if $c_1 \neq 0$. By the PBH observability test, the system is observable if and only if $c_1 \neq 0$. Alternatively, we could attempt to check the rank condition of Theorem 5.1, or we could attempt to compute the null space of the observability matrix, but the simple criterion $c_1 \neq 0$ does not emerge so easily that way. △

Since observability is a property invariant under linear state transformations, Example 5.5 gives some insight into why the PBH test works in general.

TERMINOLOGY FOR SOME EIGENPAIRS. Given a matrix pair (C, A), recall that $N(\mathcal{O}(C, A))$ is an invariant subspace for A; that is, $A(N(\mathcal{O}(C, A))) \subset N(\mathcal{O}(C, A))$. Certain eigenvectors of A can be identified as unobservable modes, as follows.

Definition 5.4 *The pair* (λ, v) *is an* unobservable eigenpair, *or simply an* unobservable pair, *if* $v \neq 0$ *and*

$$Cv = 0 \quad \textit{and} \quad Av = \lambda v.$$

The terminology in this definition is natural, because if (λ, v) is an unobservable pair of (C, A), then $x(t) = e^{\lambda t}v$ is a solution of $\dot{x} = Ax$ and $y = Cx(t) = Ce^{tA}v = Ce^{\lambda t}v = 0$ for all t. But we obtain identically zero output also for the zero solution $x(t) = 0$ determined by the zero initial condition $x(0) = 0$. Thus, the output trajectory $y(t)$ alone cannot distinguish between the zero solution and the simple nonzero solution $e^{\lambda t}v$ associated with an unobservable pair (λ, v). Note also that (λ, v) is an unobservable pair if and only if λ is an eigenvalue of A restricted to its invariant subspace $N(\mathcal{O}(C, A))$.

Recall that the subspace $(N(\mathcal{O}(C, A)))^{\perp}$ is invariant under A^T. Certain eigenvectors of A^T can be identified as observable modes, as follows.

Definition 5.5 *The pair (λ, v) is an* observable eigenpair, *or simply an* observable pair, *if we have*

$$A^T v = \lambda v \quad \textit{and} \quad v \in (N(\mathcal{O}(C, A)))^{\perp}.$$

Note that (λ, v) is an observable pair if and only if λ is an eigenvalue of A^T restricted to its invariant subspace $(N(\mathcal{O}(C, A)))^{\perp}$.

Example 5.6 Consider the system

$$\dot{x} = Ax := \begin{bmatrix} 2 & 1 & 0 \\ 0 & 2 & 0 \\ 0 & 0 & 2 \end{bmatrix} x.$$

The only eigenvalue of A is $\lambda = 2$, and the only eigenvectors of A are the standard basis vectors e_1 and e_3. On the other hand, the only eigenvalue of

$$A^T = \begin{bmatrix} 2 & 0 & 0 \\ 1 & 2 & 0 \\ 0 & 0 & 2 \end{bmatrix}$$

is $\lambda = 2$, and the only eigenvectors of A^T are the standard basis vectors e_2 and e_3. We illustrate Definition 5.4 and Definition 5.5 for several different choices of output.

If the output of the system is $y = x_3 = [0 \quad 0 \quad 1]x$, then we find that $N(\mathcal{O}(c, A)) = \text{span}\,\{e_1, e_2\}$, and therefore $(N(\mathcal{O}(c, A)))^{\perp} = \text{span}\,\{e_3\}$. It is straightforward to check that $(2, e_1)$ and $(2, e_2)$ are unobservable eigenpairs, whereas $(2, e_3)$ is an observable eigenpair.

If the output is $y = x_1 = [1 \quad 0 \quad 0]x$, then $N(\mathcal{O}(c, A)) = \text{span}\,\{e_3\}$, whereas $(N(\mathcal{O}(c, A)))^{\perp} = \text{span}\,\{e_1, e_2\}$. In this case, $(2, e_3)$ is an unobservable eigenpair, whereas $(2, e_2)$ is an observable eigenpair.

If $y = x_2 = [0 \quad 1 \quad 0]x$, then $N(\mathcal{O}(c, A)) = \text{span}\,\{e_1, e_3\}$ and $(N(\mathcal{O} \times (c, A)))^{\perp} = \text{span}\,\{e_2\}$. Then $(2, e_1)$ and $(2, e_3)$ are unobservable eigenpairs, and $(2, e_2)$ is an observable eigenpair. $\triangle$

Example 5.6 illustrates the fact that an eigenvalue can be associated with both observable modes and unobservable modes. (Some authors use the terms *observable eigenvalue* and *unobservable eigenvalue*. As the example shows, these two terms are not mutually exclusive. See also Example 4.10.)

According to Definition 5.4, we can say that *a system is observable if and only if there are no unobservable eigenpairs.* On the other hand, if $C \neq 0$ and the pair (C, A) is not observable, then, taking into account Definition 5.5, we can say that there must exist unobservable eigenpairs and also observable eigenpairs, as we saw in Example 5.6. These facts are clear also from the observability normal form.

5.4 EXERCISES

Exercise 5.1 Prove Proposition 5.2. Do this in two ways: (a) directly from the definition of the observability matrix, and (b) by duality, using Proposition 4.3.

Exercise 5.2 Supply the details in the proof of Proposition 5.3.

Exercise 5.3 Use Proposition 5.4 and duality to prove Proposition 5.5.

Exercise 5.4 *PBH Observability Test and Jordan Block Systems*
Generalize the result of Example 5.4 to the case of an n-dimensional diagonal system with n distinct real eigenvalues. Generalize the result of Example 5.5 to the case of an n-dimensional system with a single Jordan block as coefficient matrix.

Exercise 5.5 *Careful with Block Submatrices*

(a) Consider a matrix pair (C, A) with matrices of the form

$$A = \begin{bmatrix} A_1 & 0 \\ 0 & A_2 \end{bmatrix}, \quad C = \begin{bmatrix} C_1 & C_2 \end{bmatrix}.$$

Where is the flaw in the following statement? The pair (C, A) is observable if and only if each of the pairs (C_1, A_1), (C_2, A_2) is observable.

(b) Consider a matrix pair (A, B) with matrices of the form

$$A = \begin{bmatrix} A_1 & 0 \\ 0 & A_2 \end{bmatrix}, \quad B = \begin{bmatrix} B_1 \\ B_2 \end{bmatrix}.$$

Where is the flaw in the following statement? The pair (A, B) is controllable if and only if each of the pairs (A_1, B_1), (A_2, B_2) is controllable.

Exercise 5.6 *Observability and Asymptotic Stability*
Consider the system $\dot{x} = Ax$, where A is $n \times n$. Suppose there exist a positive definite matrix P and a $p \times n$ matrix C such that $A^T P + PA = -C^T C$. Show that if (C, A) is observable, then every solution $x(t)$ satisfies $x(t) \to 0$ as $t \to \infty$.

Exercise 5.7 *More on Observability and Asymptotic Stability*

(a) Apply the result of Exercise 5.6 to find a quadratic function $V(x) = x^T Px$ such that $V(x(t))$ decreases along nonzero solutions $x(t)$ of the system

$$\begin{aligned} \dot{x}_1 &= x_2, \\ \dot{x}_2 &= -x_1 - \sigma x_2, \\ y &= x_1, \end{aligned}$$

where $\sigma > 0$ is a damping constant.

(b) With $V(x)$ as in (a), are there any nonzero solutions $x(t)$ that satisfy $\frac{dV}{dt}(x(t)) \equiv 0$? Explain your answer in terms of the observability property. Then explain directly in terms of the differential equations in (a).

(c) With reference to the system of (a), consider the rate of change of $W(x) = \|x\|_2^2$ along solutions. Are there any nonzero solutions $x(t)$ that satisfy $\frac{dW}{dt}(x(t)) \equiv 0$? Explain your answer by analyzing the differential equations.

Exercise 5.8 Use a duality argument to deduce the PBH controllability test (Theorem 4.8) from the PBH observability test (Theorem 5.7).

Exercise 5.9 This exercise suggests alternative characterizations of observability and controllability of a system $\dot{x} = Ax + Bu$, $y = Cx$ when A is a Jordan form matrix. *Hint*: Use the PBH tests.

(a) Consider the pair (A, B), where

$$A = \begin{bmatrix} \lambda & 1 & 0 & 0 \\ 0 & \lambda & 1 & 0 \\ 0 & 0 & \lambda & 0 \\ 0 & 0 & 0 & \lambda \end{bmatrix}, \qquad B = \begin{bmatrix} b_{11} & b_{12} \\ b_{21} & b_{22} \\ b_{31} & b_{32} \\ b_{41} & b_{42} \end{bmatrix}.$$

Show that (A, B) is controllable if and only if the row vectors $[b_{31} \quad b_{32}]$ and $[b_{41} \quad b_{42}]$ are linearly independent.

(b) Consider the pair (C, A), where

$$C = \begin{bmatrix} c_{11} & c_{12} & c_{13} & c_{14} \\ c_{21} & c_{22} & c_{23} & c_{24} \end{bmatrix}, \qquad A = \begin{bmatrix} \lambda_1 & 1 & 0 & 0 \\ 0 & \lambda_1 & 0 & 0 \\ 0 & 0 & \lambda_1 & 1 \\ 0 & 0 & 0 & \lambda_1 \end{bmatrix}.$$

Show that (C, A) is observable if and only if the column vectors $[c_{11} \quad c_{21}]^T$ and $[c_{13} \quad c_{23}]^T$ are linearly independent.

(c) Formulate alternative characterizations of controllability and observability for (C, A, B) when A is a Jordan form matrix.

5.5 NOTES AND REFERENCES

The main resources for this chapter were [53] and [82]. The term *observability* is due to R. E. Kalman. Three major themes in mathematical control theory involve (i) the input-to-state interaction: *controllability*, (ii) the state-to-output interaction: *observability*, and (iii) transitions between different representations of a dynamical system. We have tried to illustrate those themes in the last two chapters. Helpful text references on systems and control are [9], [16], [82], [91], [104]. For linear algebra background, see [75].

As mentioned earlier for controllability concepts, linear time varying systems present additional technical issues in any extension of the observability concept presented here; again, see [82]. For more on static output feedback stabilization, [82] (note 14.5, pages 262–263) includes references where results up to 1991 may be found.

For a SISO linear time invariant system, the ratio of Laplace-transformed output to Laplace-transformed input defines the system *transfer function*. (A very quick look at this definition can be seen in Exercise 11.3.) For transfer function representations for linear time invariant systems (the *frequency domain* approach), one can consult [53] for a wealth of examples. The comprehensive text [80] on the behavioral approach to systems theory also has much material on transfer functions. The subject of *realization theory* deals with the existence and construction of state space models for a given input-output behavior (given, for example, by a transfer function description of a system). For the essentials of realization theory for linear time invariant systems, see [28] (Sections 2.5–2.6); and, for linear time varying systems with time invariant systems as a special case, see [16] (Chapter 2) and [82] (Chapter 10).

Chapter Six

Stabilizability of LTI Systems

In this chapter we study the stabilizability of linear time invariant systems by linear state feedback. For controllable systems, we show some standard ways that stabilizing feedback can be defined. Then we characterize stabilizability and describe the general limitations on eigenvalue placement by feedback for systems that are not controllable.

6.1 STABILIZING FEEDBACKS FOR CONTROLLABLE SYSTEMS

This section is devoted to the determination of stabilizing feedback matrices in the following situations:

1. Vector input controllable linear systems, $\dot{x} = Ax + Bu$, using the controllability Gramian
2. Single-input controllable linear systems, $\dot{x} = Ax + bu$, using Ackermann's formula

In each of these settings, there is a general construction of a stabilizing feedback. In the first setting, the controllability Gramian matrix can be used to define a stabilizing feedback, and in the second setting, Ackermann's formula can be used for arbitrary eigenvalue placement. (Another standard way of defining stabilizing feedbacks is by means of positive semidefinite solutions of the matrix algebraic Riccati equation; this is discussed in Section 7.5, "LQR and the Algebraic Riccati Equation.")

Stabilizing Feedback by Controllability Gramian

Assume that (A, B) is controllable, let $\beta > 0$ be an arbitrary positive number, and consider the matrix

$$Q_\beta := \int_0^\beta e^{-tA} BB^T e^{-tA^T} \, dt. \tag{6.1}$$

This matrix is called the finite time *controllability Gramian* over the interval $[0, \beta]$ defined by the matrix pair $(-A, B)$. Note that Q_β depends on β, but to keep the notation as simple as possible, we will simply write Q in the argument that follows. Since (A, B) is controllable, Q is a positive definite matrix (Exercise 6.1).

Our main goal is to show that the feedback matrix $K := -B^TQ^{-1}$ produces a Hurwitz closed loop matrix $A + BK = A - BB^TQ^{-1}$. Recall that the derivative of e^{tA} is Ae^{tA}, from which we derive

$$\begin{aligned} AQ + QA^T &= \int_0^\beta [Ae^{-tA}BB^Te^{-tA^T} + e^{-tA}BB^Te^{-tA^T}A^T]\,dt \\ &= \int_0^\beta -\frac{d(e^{-tA}BB^Te^{-tA^T})}{dt}\,dt \\ &= BB^T - e^{-\beta A}BB^Te^{-\beta A^T}. \end{aligned}$$

Set $K := -B^TQ^{-1}$, and use the previous expression to write

$$\begin{aligned} (A + BK)Q + Q(A + BK)^T &= (A - BB^TQ^{-1})Q + Q(A - BB^TQ^{-1})^T \\ &= AQ + QA^T - 2BB^T \\ &= -BB^T - e^{-\beta A}BB^Te^{-\beta A^T}. \end{aligned} \tag{6.2}$$

Define $\hat{B} := [B \quad e^{-\beta A}B]$, and note that the right hand side of (6.2) may be written

$$-\hat{B}\hat{B}^T = -\begin{bmatrix} B & e^{-\beta A}B \end{bmatrix} \begin{bmatrix} B^T \\ B^Te^{-\beta A^T} \end{bmatrix}.$$

Therefore (6.2) is exactly the Lyapunov equation

$$(A - BB^TQ^{-1})Q + Q(A - BB^TQ^{-1})^T + \hat{B}\hat{B}^T = 0.$$

Since (A, B) is controllable, $(A - BB^TQ^{-1}, B)$ is also controllable. From the definition of $\hat{B}$ it follows easily that $(A - BB^TQ^{-1}, \hat{B})$ is also controllable. Now apply Proposition 5.5; since Q is positive definite, we conclude that $A + BK = A - BB^TQ^{-1}$ is Hurwitz. Note that this result holds for any choice of the number $\beta > 0$ in the definition (6.1) of the controllability Gramian. We summarize this result in the following proposition.

Proposition 6.1 *Suppose (A, B) is controllable. Choose $\beta > 0$, and let Q_β be the positive definite controllability Gramian defined in (6.1). Then $K := -B^TQ^{-1}$ is an asymptotically stabilizing feedback matrix, that is, $A + BK = A - BB^TQ^{-1}$ is Hurwitz.*

Since the controllability Gramian Q may be difficult to compute, this result is primarily of theoretical interest, but it serves as a foundation for further results; see the Notes and References.

An example will illustrate the proposition.

Example 6.1 Consider the controlled harmonic oscillator system

$$\dot{x} = Ax + bu := \begin{bmatrix} 0 & 1 \\ -1 & 0 \end{bmatrix} x + \begin{bmatrix} 0 \\ 1 \end{bmatrix} u.$$

We have

$$e^{tA} = \begin{bmatrix} \cos t & \sin t \\ -\sin t & \cos t \end{bmatrix},$$

and $A^T = -A$. Thus, $e^{-tA^T} = e^{tA}$. Choosing $\beta = 2\pi$, we compute the controllability Gramian in (6.1), and find

$$\begin{aligned} Q_{2\pi} &= \int_0^{2\pi} e^{-tA}BB^Te^{-tA^T}\,dt \\ &= \int_0^{2\pi} \begin{bmatrix} \cos t & -\sin t \\ \sin t & \cos t \end{bmatrix} \begin{bmatrix} 0 \\ 1 \end{bmatrix} \begin{bmatrix} 0 & 1 \end{bmatrix} \begin{bmatrix} \cos t & \sin t \\ -\sin t & \cos t \end{bmatrix} dt \\ &= \int_0^{2\pi} \begin{bmatrix} \sin^2 t & -\sin t\cos t \\ -\sin t \cos t & \cos^2 t \end{bmatrix} dt \\ &= \int_0^{2\pi} \begin{bmatrix} \frac{1}{2} - \frac{1}{2}\cos 2t & -\sin t\cos t \\ -\sin t\cos t & \frac{1}{2} + \frac{1}{2}\cos 2t \end{bmatrix} dt. \end{aligned}$$

Direct integration gives

$$Q_{2\pi} = \begin{bmatrix} \pi & 0 \\ 0 & \pi \end{bmatrix},$$

from which we find the stabilizing feedback matrix given by

$$K = -B^TQ^{-1} = -\begin{bmatrix} 0 & 1 \end{bmatrix} \begin{bmatrix} \frac{1}{\pi} & 0 \\ 0 & \frac{1}{\pi} \end{bmatrix} = \begin{bmatrix} 0 & -\frac{1}{\pi} \end{bmatrix}.$$

This choice does provide the required damping for stabilization; the feedback in this case is velocity feedback given by $u = -\frac{1}{\pi}x_2$. The closed loop coefficient matrix is

$$A + BK = \begin{bmatrix} 0 & 1 \\ -1 & -\frac{1}{\pi} \end{bmatrix},$$

which is Hurwitz. △

More questions about the system in Example 6.1 appear in Exercise 6.2.

Ackermann's Formula for Single-Input Systems

We now consider a single-input controllable system of the form $\dot{x} = Ax + bu$. Our experience with the controller (companion) form for these systems shows that any desired characteristic polynomial with real coefficients may be achieved with a unique choice of linear state feedback. Let $\alpha(s) := \det(sI-(A+bK))$ be the desired characteristic polynomial for the closed loop corresponding to the feedback $u = Kx$. The next proposition gives a formula for the required feedback matrix K which does not require knowledge of the characteristic polynomial of A.

Proposition 6.2 (Ackermann's formula)
Suppose the single-input pair (A, b) is controllable, with controllability matrix $\mathcal{C}(A, b)$. Let $\alpha(s)$ be any monic polynomial with real coefficients. Then there is a unique $1 \times n$ matrix K such that $\alpha(s) := \det(sI - (A + bK))$, and K is given by

$$K = -\begin{bmatrix} 0 & \cdots & 0 & 1 \end{bmatrix} \mathcal{C}(A, b)^{-1} \alpha(A) = -e_n^T \mathcal{C}(A, b)^{-1} \alpha(A), \tag{6.3}$$

where $\alpha(A)$ is the operator polynomial $\alpha(\cdot)$ evaluated at A.

Proof. See Exercise 6.3. □

Equation (6.3) is known as *Ackermann's formula*. This formula says that the feedback matrix required to achieve $\alpha(s)$ as the characteristic polynomial of $A + bK$ is given by the negative of the last row of the inverse of the controllability matrix, multiplied by $\alpha(A)$.

A simple example illustrates the calculations involved in (6.3).

Example 6.2 Consider the system

$$\dot{x} = \begin{bmatrix} 0 & 1 \\ 1 & 0 \end{bmatrix} x + \begin{bmatrix} 0 \\ 1 \end{bmatrix} u,$$

which has an unstable origin when $u = 0$. Suppose we wish to stabilize the system with $u = Kx$, such that the eigenvalues of $A + bK$ are $\lambda_1 = -2, \lambda_2 = -3$. Then the desired characteristic polynomial is $\alpha(s) = (s + 2)(s + 3) = s^2 + 5s + 6$, and we have

$$\alpha(A) = A^2 + 5A + 6I = \begin{bmatrix} 7 & 5 \\ 5 & 7 \end{bmatrix}.$$

The inverse of the controllability matrix is

$$\mathcal{C}(A, b)^{-1} = \begin{bmatrix} 0 & 1 \\ 1 & 0 \end{bmatrix}^{-1} = \begin{bmatrix} 0 & 1 \\ 1 & 0 \end{bmatrix}.$$

Therefore Ackermann's formula (6.3) gives the required feedback as

$$K = -\begin{bmatrix} 0 & 1 \end{bmatrix} \begin{bmatrix} 0 & 1 \\ 1 & 0 \end{bmatrix} \begin{bmatrix} 7 & 5 \\ 5 & 7 \end{bmatrix} = \begin{bmatrix} -7 & -5 \end{bmatrix}.$$

The closed loop matrix is

$$A + bK = \begin{bmatrix} 0 & 1 \\ -6 & -5 \end{bmatrix}.$$

△

The inversion of the controllability matrix required in (6.3) is easily accomplished by a software package for moderate size pairs (A, b).

6.2 LIMITATIONS ON EIGENVALUE PLACEMENT

In this section we discuss the limitations on eigenvalue placement when the pair (A, B) is not controllable. The controllability normal form (Theorem 4.5) is helpful here.

We have seen in Theorem 4.3 that controllability of the single input pair (A, b) provides a strong form of stabilizability for the system $\dot{x} = Ax + bu$, since controllability guarantees arbitrary eigenvalue placement using linear state feedback. We know that controllability implies stabilizability for m-input systems as well.

We begin with some examples to show that a linear system may be stabilizable without being controllable, although there are restrictions on the eigenvalues achievable using feedback. Consider the next example.

Example 6.3 The system

$$\dot{x} = Ax + bu := \begin{bmatrix} -1 & 0 \\ 0 & 2 \end{bmatrix} x + \begin{bmatrix} 0 \\ 1 \end{bmatrix} u$$

is not controllable, since $\operatorname{rank} \mathcal{C}(A, b) = 1$. Matrix A has eigenvalues $\mu_1 = -1$, $\mu_2 = 2$. However, the system is stabilizable. For example, we can take $u = kx := [0 \quad -3]x$ to obtain the closed loop system

$$\dot{x} = (A + bk)x = \begin{bmatrix} -1 & 0 \\ 0 & -1 \end{bmatrix} x$$

having repeated eigenvalue -1. Note that the original eigenvalue $\mu_1 = -1$ cannot be changed by linear feedback, so there are restrictions on the assignment of eigenvalues. △

Example 6.3 shows that a system that is not controllable may still be stabilizable by appropriate linear feedback; however, we should not expect unrestricted eigenvalue placement without controllability.

The next example also has an eigenvalue that remains fixed under any linear feedback. In this example, the A matrix is not diagonal, so the system is not as transparent as the one in Example 6.3. However, the analysis shows a way to understand eigenvalue placement in systems that are not controllable.

Example 6.4 The system

$$\dot{x} = Ax + bu := \begin{bmatrix} -2 & 2 \\ 1 & -1 \end{bmatrix} x + \begin{bmatrix} 1 \\ 1 \end{bmatrix} u$$

is not controllable, and A has eigenvalues $\mu_1 = 0$, $\mu_2 = -3$. The feedback matrix K given by

$$K = \begin{bmatrix} -2 & 0 \end{bmatrix}$$

gives the closed loop matrix

$$A + bK = \begin{bmatrix} -4 & 2 \\ -1 & -1 \end{bmatrix},$$

which has eigenvalues $\lambda_1 = -2$, $\lambda_2 = -3$. Of course, there are many other choices for K that will stabilize this system. Note that the original eigenvalue $\mu_1 = 0$ has a controllable eigenspace; the eigenvector $[1\ 1]^T$ lies in $\mathcal{R}(\mathcal{C}(A, b)) = \text{span}\,\{[1\ 1]^T\}$. Recall now that $\mathcal{N}(A)^{\perp} = \mathcal{R}(A^T)$ and that A and A^T have the same eigenvalues. The eigenvector $[1\ \ -1]^T$ of A^T is orthogonal to the vector $[1\ 1]^T$, and it corresponds to the eigenvalue $\mu_2 = -3$, since

$$A^T \begin{bmatrix} 1 \\ -1 \end{bmatrix} = \begin{bmatrix} -2 & 1 \\ 2 & -1 \end{bmatrix} \begin{bmatrix} 1 \\ -1 \end{bmatrix} = -3 \begin{bmatrix} 1 \\ -1 \end{bmatrix}.$$

The eigenspace of A for $\mu_2 = -3$ does not lie in the range of $\mathcal{C}(A, b)$; one can check that $\mathcal{N}(A + 3I) = \text{span}\,\{[-2\ 1]^T\}$. The controllability normal form helps to explain the situation. We can use the transformation

$$x = Tz = \begin{bmatrix} 1 & 1 \\ 1 & -1 \end{bmatrix} z$$

to produce the controllability normal form

$$\dot{z} = T^{-1}ATz + T^{-1}bu = \begin{bmatrix} 0 & -1 \\ 0 & -3 \end{bmatrix} z + \begin{bmatrix} 1 \\ 0 \end{bmatrix} u.$$

This form clearly displays the eigenvalues $\mu_1 = 0$, $\mu_2 = -3$ on the diagonal. Notice that $\mu_2 = -3$ is unchanged by any linear feedback. On the other hand, the zero eigenvalue may be shifted to a location in the left half plane by appropriate linear feedback. (See Exercise 6.4.) △

The existence of eigenvalues fixed under linear feedback is a general feature of uncontrollable linear systems; it is a fundamental limitation on eigenvalue placement by linear feedback. To understand the general situation, we consider further the controllability normal form given by Theorem 4.5.

Construction of a Controllability Normal Form

Suppose (A, B) is not controllable. Recall the controllability normal form for this pair, given by

$$\dot{z} = \begin{bmatrix} A_{11} & A_{12} \\ 0 & A_{22} \end{bmatrix} z + \begin{bmatrix} B_1 \\ 0 \end{bmatrix} u, \tag{6.4}$$

where A_{11} is $r \times r$ and (A_{11}, B_1) is controllable.

Suppose we partition z as $[z_1^T\ z_2^T]^T$, compatible with the block structure of the normal form coefficient matrices, so that $z_1 \in \mathbf{R}^r$ and $z_2 \in \mathbf{R}^{n-r}$. Theorem 4.5 implies that for any initial condition $z_0 = [z_{10}^T\ 0^T]^T$, and any input $u(t)$, the solution $z(t)$ will evolve within the subspace defined by

$z_2 = 0$, and this trajectory will be determined by the dynamics of the controllable system $\dot{z}_1 = A_{11}z_1 + B_1u$ (since $z_2(t) \equiv 0$).

Transforming a system to controllability normal form requires, in particular, knowledge of a basis for $R(\mathcal{C}(A, B))$. For vector-input systems this requires a choice of linearly independent columns of the controllability matrix, and there is no general recipe for making that choice. For single-input systems the situation is simpler, and we now consider this case.

Suppose $\operatorname{rank} \mathcal{C}(A, b) = r < n$. Which columns of $\mathcal{C}(A, b)$ form a linearly independent set of r vectors? Let m be the smallest integer such that

$$\operatorname{rank} \begin{bmatrix} b & Ab & \cdots & A^m b \end{bmatrix} = \operatorname{rank} \begin{bmatrix} b & Ab & \cdots & A^m b & A^{m+1} b \end{bmatrix}.$$

A smallest such integer must exist since $0 < \operatorname{rank} \mathcal{C}(A, b) = r < n$. Then $A(A^m b)$ is a linear combination of the columns $b, Ab, \ldots, A^m b$. It follows by induction that

$$\operatorname{rank} \begin{bmatrix} b & Ab & \cdots & A^m b \end{bmatrix} = \operatorname{rank} \begin{bmatrix} b & Ab & \cdots & A^m b & \cdots & A^{m+k} b \end{bmatrix}$$

for all k such that $1 \leq k \leq n - m$. Thus, from the assumption that rank $\mathcal{C}(A, b) = r < n$, we conclude that $r = m$ as defined above. We have proved the following lemma.

Lemma 6.1 *If* $\operatorname{rank} \mathcal{C}(A, b) = r < n$, *then*

$$\operatorname{rank} \mathcal{C}(A, b) = \operatorname{rank} \begin{bmatrix} b & Ab & \cdots & A^{r-1} b \end{bmatrix}.$$

A coordinate transformation to the normal form may be defined by $x = Tz$, where T has the form

$$T = \begin{bmatrix} b & Ab & \cdots & A^{r-1} b & v_1 & \cdots & v_{n-r} \end{bmatrix},$$

and the vectors $v_1, \ldots, v_{n-r}$ are any linearly independent vectors in the complement of $R(\mathcal{C}(A, b))$. In particular, we may choose vectors v_j such that $\operatorname{span}\{v_1, \ldots, v_{n-r}\} = R(\mathcal{C}(A, b))^{\perp}$. This was the construction of the transformation T in the simple case of Example 6.4.

For a statement of the general restriction on eigenvalue placement when (A, B) is not controllable, see the paragraphs below headed **Invariance of Stabilizability** and **General Restriction on Eigenvalue Placement**.

Construction of an Observability Normal Form

This is a convenient place to consider the construction of an observability normal form when the pair (C, A) is not observable. We will first discuss the SISO case, and then we offer some comments on the general MIMO case.

We have the following dual version of Lemma 6.1.

Lemma 6.2 *Suppose the matrix pair* (c, A) *has* $\operatorname{rank} \mathcal{O}(c, A) = \rho < n$. *Then*

$$\operatorname{rank} \begin{bmatrix} c \\ cA \\ \vdots \\ cA^{\rho-1} \end{bmatrix} = \operatorname{rank} \mathcal{O}(c, A).$$

Proof. Note that the transpose of the observability matrix for the pair (c, A) is the controllability matrix for the pair (A^T, c^T). The row space of $\mathcal{O}(c, A)$ equals the column space of $\mathcal{C}(A^T, c^T)$. Apply Lemma 6.1 to this controllability matrix, which has rank ρ, and the lemma follows. □

Assume that $\operatorname{rank} \mathcal{O}(c, A) = \rho < n$, and consider the null space of $\mathcal{O}(c, A)$, $N(\mathcal{O}(c, A)) \subset \mathbf{R}^n$. If x_0 is a nonzero vector in $N(\mathcal{O}(c, A))$, and $x(t)$ is the solution of $\dot{x} = Ax$ with $x(0) = x_0$, then we have $x(t) = e^{tA}x_0$ and $y(t) = ce^{tA}x_0 \equiv 0$. Since the output $y(t)$ is also identically zero for the zero solution corresponding to $x_0 = 0$, it follows that $x_0 \in N(\mathcal{O}(c, A))$ cannot be distinguished from the zero vector using only the information from the output and its derivatives. The states in $N(\mathcal{O}(c, A))$ are called *unobservable modes*. Now consider the orthogonal complement $(N(\mathcal{O}(c, A)))^{\perp}$; the vectors in this complement are called *observable modes*. Note that $N(\mathcal{O}(c, A))$ is invariant under A, while $(N(\mathcal{O}(c, A)))^{\perp} = R((\mathcal{O}(c, A))^T) = R(\mathcal{C}(A^T, c^T))$ is invariant under A^T.

The identification of unobservable modes and observable modes can be accomplished, in principle, by a coordinate transformation to an observability normal form (Theorem 5.5). If $\operatorname{rank} \mathcal{O}(c, A) = \rho$, let $w_1, \ldots, w_{n-\rho}$ be linearly independent vectors such that

$$N(\mathcal{O}(c, A)) = \operatorname{span} \{w_1, \ldots, w_{n-\rho}\}.$$

Let $x = Tz$ where

$$T := \begin{bmatrix} c^T & A^T c^T & \cdots & (A^{\rho-1})^T c^T & w_1 & \cdots & w_{n-\rho} \end{bmatrix}.$$

Then the columns of T are linearly independent, so T is nonsingular and therefore a valid coordinate transformation. Moreover, we have a basis of $N(\mathcal{O}(c, A))$ and a basis for its orthogonal complement. Partition z by $[z_1^T \; z_2^T]^T$, where $z_1 \in \mathbf{R}^{\rho}$ and $z_2 \in \mathbf{R}^{n-\rho}$. When we apply the transformation $x = Tz$, we obtain a system of the form (assuming a zero-input matrix for the moment)

$$\dot{z} = \begin{bmatrix} A_{11} & 0 \\ A_{21} & A_{22} \end{bmatrix} z,$$
$$y = \begin{bmatrix} c_1 & 0 \end{bmatrix} z.$$

Entry A_{11} is $\rho \times \rho$, A_{22} is $(n-\rho) \times (n-\rho)$, and c_1 is $1 \times \rho$. It is a recommended exercise (Exercise 6.8) to show directly (under the stated assumption that $\operatorname{rank} \mathcal{O}(c, A) = \rho < n$) that the pair (c_1, A_{11}) is observable.

If the SISO system in the original x coordinates had been supplied with an input matrix b, then the transformed input matrix would have the form

$$\hat{b} = T^{-1}b = \begin{bmatrix} \hat{b}_1 \\ \hat{b}_2 \end{bmatrix},$$

and, in the absence of more specific knowledge about the system, there is no special structure to the $\hat{b}_1, \hat{b}_2$ blocks. A similar comment applies in the case of a multi-input matrix B.

When (C, A) is not observable, the actual construction of an observability normal form by a coordinate transformation requires a choice of linearly independent columns of $R((\mathcal{O}(c, A))^T)$ as basis vectors, and this is not necessarily as straightforward as in the scalar output case. However, a direct argument along the lines given above establishes the *existence* of the observability normal form in Theorem 5.5. (Our earlier argument for Theorem 5.5 was a duality argument using the controllability normal form in Theorem 4.5. See also Exercise 6.8.) It is also worth noting that one can obtain an observability normal form for the pair (C, A) by constructing a controllability normal form for the dual pair (A^T, C^T).

Along with the controllability normal form, the observability normal form is useful in establishing the Kalman decomposition of a linear time invariant system. (See Exercise 6.14.)

Invariance of Stabilizability

Consider once more a controllability normal form

$$\dot{z} = \begin{bmatrix} A_{11} & A_{12} \\ 0 & A_{22} \end{bmatrix} z + \begin{bmatrix} B_1 \\ 0 \end{bmatrix} u,$$

where A_{11} is $r \times r$ and (A_{11}, B_1) is controllable. It is clear that the controllable z_1 component is stabilizable with arbitrary eigenvalue placement for r out of the n system eigenvalues. The other $n - r$ eigenvalues, which are precisely the eigenvalues of matrix A_{22}, are fixed under linear feedback. Thus, linear feedback $u = Kz$ can be chosen such that all solutions with initial conditions of the form $z_0 = [z_{10}^T \; z_{20}^T]^T = [z_{10}^T \; 0^T]^T$ will converge to the origin as $t \to \infty$, and, in principle, the rate of convergence may be chosen arbitrarily. It should also be clear that the normal form itself is a stabilizable linear system if and only if matrix A_{22} is Hurwitz. We want to say that this is a definitive conclusion, but we should first show that stabilizability is invariant under linear coordinate changes.

Lemma 6.3 *If the pair (A, B) is stabilizable and T is nonsingular, then the pair (TAT^{-1}, TB) is also stabilizable.*

Proof. Let K be a feedback matrix such that $A + BK$ is Hurwitz. We must show that there exists $\hat{K}$ such that $TAT^{-1} + TB\hat{K}$ is also Hurwitz. Set

$\hat{K} := KT^{-1}$; then we have

$$TAT^{-1} + TB\hat{K} = TAT^{-1} + TBKT^{-1} = T(A + BK)T^{-1}.$$

Since the eigenvalues of $A + BK$ are a similarity invariant, the transformed pair (TAT^{-1}, TB) is stabilizable. □

Since a controllability normal form is obtained by linear coordinate change, we may conclude (as we wanted to above) that (A, B) is stabilizable if and only if the matrix A_{22} in any controllability normal form is Hurwitz.

General Restriction on Eigenvalue Placement

When the MIMO system pair (A, B) is not controllable, the general restriction on eigenvalue placement by feedback can be stated in a coordinate independent way, without reference to a specific normal form. The next theorem provides such a statement. Recall that $\mathcal{R}(\mathcal{C}(A, B))$ is an invariant subspace for A, and therefore $\mathcal{R}(\mathcal{C}(A, B))^{\perp}$ is an invariant subspace for A^T; moreover, $\mathcal{R}(\mathcal{C}(A, B))^{\perp} = \mathcal{N}(\mathcal{C}(A, B)^T) = \mathcal{N}(\mathcal{O}(B^T, A^T))$.

Theorem 6.1 *Suppose (A, B) is not controllable. Let λ be an eigenvalue of A and w a nonzero vector such that $A^T w = \lambda w$ and $B^T w = 0$. Then λ is also an eigenvalue of $A + BK$ for any $m \times n$ matrix K.*

Proof. See Exercise 6.5. □

6.3 THE PBH STABILIZABILITY TEST

We want to establish the following rather intuitive statement: *The pair (A, B) is stabilizable if and only if every eigenvalue of A associated with uncontrollable modes has negative real part.* This statement is plausible because all other eigenvalues can be moved to the left half-plane by linear feedback, as shown by the normal form.

Theorem 6.2 (PBH Stabilizability Test)
The pair (A, B) is stabilizable if and only if

$$\operatorname{rank}\,[A - \lambda I \;\; B] = n \tag{6.5}$$

for all eigenvalues λ of A with $\operatorname{Re}\lambda \geq 0$.

Proof. The proof is left to Exercise 6.10. □

6.4 EXERCISES

Exercise 6.1 *The Controllability Gramian is Positive Definite*
Show that for any $\beta > 0$ the matrix Q_β in (6.1) must be symmetric positive definite, hence invertible.

Exercise 6.2 *More on the Controllability Gramian*
Refer to the system of Example 6.1.

(a) For an arbitrary $\beta > 0$, find the controllability Gramian Q_β and the corresponding stabilizing feedback matrices $K_\beta := B^T Q_\beta^{-1}$.
(b) For which values of $\beta > 0$ is the feedback $u = -B^T Q_\beta^{-1} x$ velocity feedback only? Is it possible to avoid oscillations in the closed loop system using these particular velocity feedbacks?
(c) Find a velocity feedback that results in an overdamped closed loop system; that is, the closed loop should be asymptotically stable with no oscillations.

Exercise 6.3 Prove Ackermann's formula (6.3). *Hint*: If needed, see the Notes and References.

Exercise 6.4 Use appropriate feedbacks in the normal form in Example 6.4 to shift the eigenvalue $\mu_1 = 0$ to the following positions in the open left half of the complex plane: $(a) - 1$, and $(b) - 5$. Determine the corresponding feedbacks for the original system description.

Exercise 6.5 Prove Theorem 6.1.

Exercise 6.6 *Different Input Channels I*
Consider the system

$$\dot{x} = Ax + bu := \begin{bmatrix} 2 & 1 & 0 & 1 \\ 0 & 2 & 0 & 2 \\ 0 & 0 & -1 & 0 \\ 0 & -1 & 0 & -1 \end{bmatrix} x + \begin{bmatrix} b_1 \\ b_2 \\ b_3 \\ b_4 \end{bmatrix} u.$$

Let e_j be the j-th standard basis vector. Determine a controllability normal form for the pair (A, b) when (i) $b = e_1$, (ii) $b = e_2$, (iii) $b = e_3$, and (iv) $b = e_4$. In each case, determine whether the pair (A, b) is stabilizable based on the normal form obtained.

Exercise 6.7 *Different Input Channels II*
For each of the four pairs (A, b) considered in Exercise 6.6, determine whether the pair is stabilizable by applying the PBH stabilizability test.

Exercise 6.8 Prove Theorem 5.5 without using duality, by showing how to construct a linear coordinate transformation to an observability normal form. Remember to show directly that the pair (C_1, A_{11}) in the observability normal form is observable.

Exercise 6.9 Find an observability normal form for the single-output system

$$\dot{x} = \begin{bmatrix} 15 & -10 & 5 \\ 9 & -2 & 7 \\ 5 & -6 & 15 \end{bmatrix} x,$$
$$y = \begin{bmatrix} 1 & 2 & -1 \end{bmatrix} x.$$

Exercise 6.10 Prove Theorem 6.2.

Exercise 6.11 Consider the system

$$\dot{x} = Ax + bu := \begin{bmatrix} 2 & -2 \\ -1 & 1 \end{bmatrix} x + \begin{bmatrix} 1 \\ 1 \end{bmatrix} u.$$

(a) Show that $\lambda = 3$ is an eigenvalue of $A + bK$ for any 1×2 matrix K. *Hint*: Apply the PBH test.
(b) Show that the system $\dot{x} = Ax + bu$ satisfies rank $[\text{A B}] = 2$. Note: This is Brockett's necessary condition for smooth stabilization; see Theorem 8.3.
(c) Is it possible that feedback by a smooth function $u = \alpha(x)$ can make $x = 0$ a locally asymptotically stable equilibrium for the nonlinear system $\dot{x} = Ax + b\alpha(x)$? *Hint*: Consider the dynamics of the variable $z := x_1 - x_2$.

Exercise 6.12 Given the system

$$\dot{x} = Ax + bu := \begin{bmatrix} -2 & 0 & 3 \\ 0 & -1 & 0 \\ 3 & 0 & -2 \end{bmatrix} x + \begin{bmatrix} 1 \\ 0 \\ 1 \end{bmatrix} u,$$

show that A is not Hurwitz and (A, B) is not controllable. Use the PBH test to show that the system is stabilizable by linear feedback. Determine the eigenvalues fixed under linear feedback. Find a controllability normal form and determine a stabilizing feedback control.

Exercise 6.13 Consider the system

$$\dot{x} = Ax + bu := \begin{bmatrix} -1 & -1 & 0 \\ -1 & 2 & 1 \\ 0 & 0 & -3 \end{bmatrix} x + \begin{bmatrix} 0 \\ 1 \\ 0 \end{bmatrix} u.$$

(a) Show that (A, b) is a stabilizable pair.

(b) Does there exist a feedback matrix K such that the eigenvalues of $A + bK$ all lie to the left of $-2.5 + i0$ in the complex plane? How about to the left of $-5 + i0$?

(c) Can the system be stabilized using a feedback matrix of the form $K = [k_1 \quad 0 \quad 0]$?

(d) What are the restrictions on eigenvalue placement using a feedback matrix of the form $K = [0 \quad k_2 \quad 0]$?

Exercise 6.14 *Kalman Decomposition*

This exercise presents the *Kalman decomposition*. Suppose the matrix triple (C, A, B) is given.

(a) Show that the intersection $R(\mathcal{C}(A, B)) \cap N(\mathcal{O}(C, A))$ is an invariant subspace for A.

(b) Show that there is a nonsingular matrix T such that

$$T^{-1}AT = \begin{bmatrix} A_{11} & 0 & A_{13} & 0 \\ A_{21} & A_{22} & A_{23} & A_{24} \\ 0 & 0 & A_{33} & 0 \\ 0 & 0 & A_{43} & A_{44} \end{bmatrix}, \quad T^{-1}B = \begin{bmatrix} B_1 \\ B_2 \\ 0 \\ 0 \end{bmatrix},$$

$$CT = \begin{bmatrix} C_1 & 0 & C_3 & 0 \end{bmatrix}.$$

(c) Verify that the following matrix pairs are observable:

$$(C_1, A_{11}) \quad \text{and} \quad \left(\begin{bmatrix} C_1 & C_3 \end{bmatrix}, \begin{bmatrix} A_{11} & A_{13} \\ 0 & A_{33} \end{bmatrix} \right).$$

(d) Verify that the following matrix pairs are controllable:

$$(A_{11}, B_1) \quad \text{and} \quad \left(\begin{bmatrix} A_{11} & 0 \\ A_{21} & A_{22} \end{bmatrix}, \begin{bmatrix} B_1 \\ B_2 \end{bmatrix} \right).$$

6.5 NOTES AND REFERENCES

For linear algebra background, see [75]. Helpful text references on systems and control are [6], [9], [16], [82], [91], [104].

The argument leading to Proposition 6.1 follows [26]. The result can be generalized to provide stabilizing linear time varying feedback for linear time varying systems; see [85] (Problem 6.3, page 279) and references given there. Proofs of Ackermann's formula (6.3) may be found in [3], [6], [10], [26], and [53].

Another important setting where stabilizing feedback can be determined by a formulaic prescription, without assuming controllability, is the linear quadratic regulator (LQR) problem, which is discussed in Chapter 7, Section 7.5, as it requires the concept of detectability (Definition 7.1) in addition to stabilizability.

The basic ideas on the controllability normal form, observability normal form, and the PBH test for stabilizability, are standard; see [26], [53].

Chapter Seven

Detectability and Duality

The main goal of this chapter is an understanding of asymptotic state estimation, and the use of such estimation for feedback stabilization. We study real linear systems

$$\dot{x} = Ax + Bu, \tag{7.1}$$
$$y = Cx. \tag{7.2}$$

where A is $n \times n$, C is $p \times n$, and B is $n \times m$. The chapter begins with an example of an observer system used for asymptotic state estimation. We define the detectability property, and establish the PBH detectability test and the duality of detectability and stabilizability. We explain the role of detectability and stabilizability in defining observer systems. We discuss the role of observer systems in observer-based dynamic controllers for feedback stabilization. We also discuss general linear dynamic controllers and stabilization. The final section of the chapter is a brief look at the algebraic Riccati equation, its connection with the linear quadratic regulator problem, and its role in generating stabilizing linear feedback controls. Detectability and stabilizability are the key properties involved in all these results.

7.1 AN EXAMPLE OF AN OBSERVER SYSTEM

Consider an undamped, inverted pendulum with a unit mass at the end of a pendulum arm of unit length, with unit gravitational constant. The scalar second order equation is $\ddot{\theta} + \sin\theta = u$, where θ is the angular displacement from the natural, vertical down, equilibrium defined by $\theta = 0$. The linearization of this equation about its vertical equilibrium position at $\theta = \pi$ is given by $\ddot{y} - y = u$, where $y := \theta - \pi$ is the displacement angle of the pendulum arm. To see this, use the expansion

$$\sin\theta = \theta - \frac{\theta^3}{3!} + \cdots$$

to write

$$\ddot{y} = \ddot{\theta} = -\sin\theta + u = -\sin(y+\pi) + u = -(\sin y \cos\pi + \sin\pi\cos y) + u = \sin y + u.$$

Thus, $\ddot{y} - \sin y = u$, and the linearization at $y = 0$ is $\ddot{y} - y = u$. We take y as the output of this linear system. It is easy to verify that we cannot asymptotically stabilize y and $\dot{y}$ to zero by using linear feedback $u = ky$ of the displacement alone. Note that the characteristic equation of $\ddot{y} - y = ky$ is $\lambda^2 - (1 + k) = 0$, with roots $\lambda = \pm\sqrt{1+k}$, one of which must have $\text{Re}\,\lambda \geq 0$ no matter the value of the real number k. This rules out asymptotic stabilization with $u = ky$.

Writing $x_1 = y$ and $x_2 = \dot{y}$, we have the equivalent system

$$\begin{aligned} \dot{x}_1 &= x_2, \\ \dot{x}_2 &= x_1 + u, \\ y &= x_1. \end{aligned}$$

Is there a way to stabilize the pendulum, using only the information from the output $y = x_1$ and the known system coefficients? Since the full state vector is not directly available, an estimate of the state must be produced from the known data. Since the system is observable, we can uniquely determine the system state using sufficiently many derivatives of the output and input; in principle, no more than $n - 1 = 2 - 1 = 1$ derivative of the output is needed. However, differentiation of the output may not be the most practical approach, since unavoidable and unacceptable errors may be introduced in a numerical computation of $y(t)$, $\dot{y}(t)$, and $u(t)$ in the reconstruction of $x(t)$. Since we wish to stabilize the origin asymptotically, the state estimate only needs to be accurate in an asymptotic sense, as opposed to an accurate instantaneous estimate at each instant of a shorter time interval. If an asymptotic state estimate is available, then we can attempt to use the estimated state in a feedback control for the purpose of stabilization.

An Asymptotic State Estimator for the Inverted Pendulum. The next example constructs an effective asymptotic state estimator for the undamped inverted pendulum with position output.

Example 7.1 (*Asymptotic State Estimation*)
Write the system in the standard matrix format (using B instead of b in order to match general formulas more easily later on)

$$\dot{x} = Ax + Bu := \begin{bmatrix} 0 & 1 \\ 1 & 0 \end{bmatrix} x + \begin{bmatrix} 0 \\ 1 \end{bmatrix} u$$

with output

$$y = Cx := \begin{bmatrix} 1 & 0 \end{bmatrix} x.$$

This is an observable system. The available information consists of the position output $y(t)$ and input $u(t)$, as well as the known matrix triple (C, A, B). Although the full state $x(t)$ is not known to us, we can simulate a dynamical system that uses only the known data in generating an asymptotic estimate of the state. Consider the auxiliary system defined by

$$\dot{\xi} = A\xi + Bu - L(y - C\xi), \tag{7.3}$$

where ξ is an auxiliary state that can be initialized at any vector $\xi(0)$, and the 2×1 matrix L is an *output error gain* to be chosen so that $x - \xi \to 0$ as $t \to \infty$. Note that the inputs to the auxiliary system (7.3) are the original input $u(t)$ and the output $y(t)$. With $y(t)$ known, $\xi(0)$ given, and ξ computed from (7.3), the difference $y(t) - C\xi(t)$ is the only direct measure of the true error $x(t) - \xi(t)$.

If we define the exact error vector by

$$e := x - \xi,$$

then our objective is to choose L such that $e \to 0$ as $t \to \infty$. We obtain the dynamics of the error vector by subtracting the equation for $\dot{\xi}$ from the equation for $\dot{x}$ and substituting $y = Cx$; we have

$$\dot{e} = (A + LC)e.$$

The really interesting fact emerges now. Since the pair (C, A) is observable, we know by duality (or direct computation) that the pair (A^T, C^T) is controllable; thus, there exists a matrix L^T such that $A^T + C^T L^T$ is Hurwitz. The eigenvalues of $A^T + C^T L^T$ are the same as the eigenvalues of its transpose $A + LC$. Hence there exists a matrix L such that $A + LC$ is Hurwitz. With such a choice of L, we have $e \to 0$ as $t \to \infty$. For the pendulum example, we have

$$A + LC = \begin{bmatrix} 0 & 1 \\ 1 & 0 \end{bmatrix} + \begin{bmatrix} l_1 \\ l_2 \end{bmatrix} \begin{bmatrix} 1 & 0 \end{bmatrix} = \begin{bmatrix} l_1 & 1 \\ l_2 + 1 & 0 \end{bmatrix},$$

where $L = [l_1 \quad l_2]^T$. The characteristic equation of $A + LC$ is

$$\lambda^2 - l_1\lambda - (l_2 + 1) = 0.$$

Consequently, $A + LC$ is Hurwitz if $l_1 < 0$ and $l_2 + 1 < 0$. If $l_1 = -3$ and $l_2 = -3$, for example, then the eigenvalues of $A + LC$ are $\lambda_1 = -1$ and $\lambda_2 = -2$. With this choice, the error dynamics are given by

$$\dot{e} = \begin{bmatrix} -3 & 1 \\ -2 & 0 \end{bmatrix} e.$$

This solves the state estimation problem for this example, since we have $x_i(t) - \xi_i(t) \to 0$ as $t \to \infty$ for $i = 1, 2$, as desired. We used only the coefficients C, A, and the observability of the pair (C, A) for the pendulum system. △

Equation (7.3) is an example of a dynamic asymptotic state estimator. The term *dynamic* is used because the estimate ξ is the output of a dynamical system. In practice, an auxiliary system and its integration (its numerical simulation) might be realized in an electronic circuit, for example.

DYNAMIC FEEDBACK STABILIZATION OF THE INVERTED PENDULUM. We continue with the inverted pendulum example, for which we have an asymptotic state estimator, and we now consider its stabilization. The system of Example 7.1 is indeed stabilizable by linear state feedback; in fact, the system is controllable. Thus, there exists a 1×2 matrix K such that $A + BK$ is Hurwitz; however, the problem, as emphasized earlier, is that direct state feedback $u = Kx$ cannot be implemented since the full state $x(t)$ is unknown. We can use the estimated state ξ in place of x in the control $u = K\xi$; the remaining question is whether this feedback of the estimated state will do the job of stabilization of $x(t)$ to the origin. We certainly have $Kx(t) - K\xi(t) \to 0$ as $t \to \infty$, since, by the previous design of L, we have $x - \xi \to 0$ as $t \to \infty$.

We now address the remaining question of stabilization of the inverted pendulum using an asymptotic state estimate in a feedback control.

Example 7.2 (*Observer-Based Dynamic Feedback Stabilization*)
Let $K = [k_1\ k_2]$ and $L = [l_1\ l_2]^T$. If we let $u = K\xi$ and substitute $y = Cx$ in (7.3), then the combined system for x and ξ is

$$\begin{bmatrix} \dot{x} \\ \dot{\xi} \end{bmatrix} = \begin{bmatrix} A & BK \\ -LC & A + BK + LC \end{bmatrix} \begin{bmatrix} x \\ \xi \end{bmatrix}. \tag{7.4}$$

Written out in detail, (7.4) is

$$\begin{bmatrix} \dot{x}_1 \\ \dot{x}_2 \\ \dot{\xi}_1 \\ \dot{\xi}_2 \end{bmatrix} = \begin{bmatrix} 0 & 1 & 0 & 0 \\ 1 & 0 & k_1 & k_2 \\ -l_1 & 0 & l_1 & 1 \\ -l_2 & 0 & k_1 + l_2 + 1 & k_2 \end{bmatrix} \begin{bmatrix} x_1 \\ x_2 \\ \xi_1 \\ \xi_2 \end{bmatrix}.$$

We must be sure that solutions of the combined system remain bounded. (In practical terms, we must avoid an auxiliary component "burning up.") Using the relation $\xi = x - e$, it is not difficult to see that the equivalent system using x and e is

$$\begin{bmatrix} \dot{x} \\ \dot{e} \end{bmatrix} = \begin{bmatrix} A + BK & -BK \\ 0 & A + LC \end{bmatrix} \begin{bmatrix} x \\ e \end{bmatrix}. \tag{7.5}$$

From the block triangular structure of this system, both the boundedness and the convergence we desire will follow from an appropriate and separate design of the matrices K and L. Written out in detail, (7.5) is

$$\begin{bmatrix} \dot{x}_1 \\ \dot{x}_2 \\ \dot{e}_1 \\ \dot{e}_2 \end{bmatrix} = \begin{bmatrix} 0 & 1 & 0 & 0 \\ k_1 + 1 & k_2 & -k_1 & -k_2 \\ 0 & 0 & l_1 & 1 \\ 0 & 0 & l_2 + 1 & 0 \end{bmatrix} \begin{bmatrix} x_1 \\ x_2 \\ e_1 \\ e_2 \end{bmatrix}.$$

Using the determinant property $\det(XY) = \det X \det Y$, the characteristic polynomial of the block triangular matrix in (7.5) is given by

$$\det[\lambda I - (A + BK)] \det[\lambda I - (A + LC)],$$

which is the product of the characteristic polynomials for $A + BK$ and $A + LC$. Thus, we can choose the values k_i separately from the previously chosen values $l_1 = -3$, $l_2 = -3$, which gave -1, -2 as the eigenvalues of $A + LC$. Now choose the k_i so that $A + BK$ has eigenvalues -3, -3; thus, take $k_1 = -10$, $k_2 = -6$. These choices guarantee that $(x(t), e(t)) = (x_1(t), x_2(t), e_1(t), e_2(t))$ approaches the origin in R^4 as $t \to \infty$, independently of the initial conditions $(x(0), e(0)) = (x_1(0), x_2(0), e_1(0), e_2(0))$, and therefore independently of the initial conditions for the auxiliary state, $\xi(0) = (\xi_1(0), \xi_2(0))$.

This solves the stabilization problem for the inverted pendulum system, using a state estimate that draws only on the available information about the system. △

Three further remarks on the example should be noted.

First, if we happen to have $\xi(0) = x(0)$, a fact which is impossible to determine if $x(0)$ is unknown, then we would have $e(0) = 0$ and therefore $e(t) \equiv 0$, giving an exact state estimate.

Second, we were able to choose the l_i parameters independently of the k_i parameters, by virtue of the block triangular form (7.5). This is an important simplification, because it means that the design of the estimator error gain L can be made independently of the design of the feedback gain K.

Third, the assumptions of controllability and observability can be relaxed. In the pendulum example we had a controllable pair (A, B), but we only needed the existence of K such that $A + BK$ was Hurwitz. Moreover, in the example we had an observable pair (C, A) (equivalent to controllability of (A^T, C^T)), but we only needed the existence of L such that $A + LC$ was Hurwitz.

In the next section, we introduce the concept of *detectability* and establish the duality of stabilizability and detectability. This will allow us to discuss the properties just mentioned in a general setting.

7.2 DETECTABILITY, THE PBH TEST, AND DUALITY

Detectability is a property of the state-to-output interaction of a system.

Definition 7.1 *The linear system $\dot{x} = Ax$ with output $y = Cx$ is* detectable *if, for any solution $x(t)$ of $\dot{x} = Ax$,*

$$y(t) = Cx(t) = 0, \;\; (t \geq 0) \implies \lim_{t\to\infty} x(t) = 0.$$

With reference to system (7.1)–(7.2) at the start of this chapter, note that detectability is a property of the unforced system with $u = 0$. When a system is detectable we also say that the matrix pair (C, A) is detectable. It is easy to see that detectability of (C, A) guarantees that, for any input matrix B, for any input function $u(t)$ defined on $[0, \infty)$, and for any pair of trajectories $x(t)$, $\hat{x}(t)$ for $\dot{x} = Ax + Bu$, the following implication holds:

$$Cx(t) \equiv C\hat{x}(t) \ (t \geq 0) \implies \lim_{t \to \infty} (x(t) - \hat{x}(t)) = 0.$$

We say that a linear system defined by the triple (C, A, B) is detectable if and only if the pair (C, A) is detectable. The meaning of detectability is that, in principle, the output can detect a difference in trajectories based on differing asymptotic behavior.

The next lemma is intuitively clear from Definition 7.1.

Lemma 7.1 *Suppose* (C, A) *is detectable. If* $Av = \lambda v$, $v \neq 0$, *and* $Cv = 0$, *then* $\mathrm{Re}\,\lambda < 0$.

Proof. If $Av = \lambda v$, $v \neq 0$, then the function $x(t) = e^{\lambda t} v$ is a solution of $\dot{x} = Ax$. If, in addition, $Cv = 0$, then $Cx(t) = e^{\lambda t} Cv = 0$ for all $t \geq 0$. By detectability, $x(t) \to 0$ as $t \to \infty$, which implies that $\mathrm{Re}\,\lambda < 0$. □

Suppse (C, A) is detectable. Then, from the contrapositive of Lemma 7.1, we conclude immediately that an eigenvector corresponding to an eigenvalue with nonnegative real part (if such exists) cannot lie in $N(\mathcal{O}(C, A))$.

We can now establish the PBH detectability test.

Theorem 7.1 (PBH Detectability Test)
The pair (C, A) *is detectable if and only if*

$$\mathrm{rank} \begin{bmatrix} A - \lambda I \\ C \end{bmatrix} = n \tag{7.6}$$

for all eigenvalues λ *of* A *with* $\mathrm{Re}\,\lambda \geq 0$.

Proof. Suppose (C, A) is detectable. Let λ be an eigenvalue of A with $\mathrm{Re}\,\lambda \geq 0$. Then, by Lemma 7.1, the matrix in (7.6) cannot have a nonzero nullspace. Thus the rank condition (7.6) holds for λ.

Conversely, suppose that (7.6) holds for every eigenvalue λ of A with $\mathrm{Re}\,\lambda \geq 0$. By transposition we also have

$$\mathrm{rank}\,[A^T - \lambda I \quad C^T] = n$$

for every eigenvalue λ of A with $\mathrm{Re}\,\lambda \geq 0$. Thus, by the PBH stabilizability test, the pair (A^T, C^T) is stabilizable. Consequently, there exists a matrix L^T such that $A^T + C^T L^T$ is Hurwitz, and therefore $A + LC$ is Hurwitz. Now suppose that $x(t)$ is a solution of $\dot{x} = Ax$ such that $y(t) = Cx(t) = 0$ for all $t \geq 0$. Then $x(t)$ is also a solution of $\dot{x} = (A + LC)x$, and therefore $x(t) \to 0$ as $t \to \infty$. Hence, (C, A) is detectable. □

Theorem 7.1 is equivalent to the following statement: (C, A) is detectable if and only if the matrix

$$\begin{bmatrix} A - \lambda I \\ C \end{bmatrix} \tag{7.7}$$

has zero nullspace for every eigenvalue of A with nonnegative real part.

Example 7.3 An unforced mass, in rectilinear motion with velocity measurement, moves according to

$$\begin{bmatrix} \dot{x}_1 \\ \dot{x}_2 \end{bmatrix} = \begin{bmatrix} 0 & 1 \\ 0 & 0 \end{bmatrix} \begin{bmatrix} x_1 \\ x_2 \end{bmatrix}$$

with output

$$y = \begin{bmatrix} 0 & 1 \end{bmatrix} x.$$

Note that $A + LC$ has a zero eigenvalue for any 2×1 matrix L. Consider the nullspace of the matrix

$$\begin{bmatrix} A - \lambda I \\ C \end{bmatrix} = \begin{bmatrix} -\lambda & 1 \\ 0 & -\lambda \\ 0 & 1 \end{bmatrix}.$$

If λ is a nonzero eigenvalue, then the nullspace is the zero subspace. However, for eigenvalue $\lambda = 0$, the nullspace is exactly the x_1-axis; therefore the system is not detectable. The position of the mass cannot be observed or asymptotically estimated from the given output. △

The next corollary follows directly from Theorem 7.1 and its proof; it documents the algebraic duality between detectability (Definition 7.1) and stabilizability (Definition 4.5).

Corollary 7.1 *The following statements are true.*

(a) *The matrix pair (C, A) is detectable if and only if the dual pair (A^T, C^T) is stabilizable.*
(b) *The matrix pair (C, A) is detectable if and only if there exists an $n \times p$ matrix L such that $A + LC$ is Hurwitz.*

Here is a simple example to illustrate the application of Corollary 7.1 (b).

Example 7.4 Consider the system matrices

$$A = \begin{bmatrix} -2 & 2 \\ 1 & -1 \end{bmatrix} \quad \text{and} \quad C = \begin{bmatrix} 1 & 0 \end{bmatrix}.$$

Note that

$$L = \begin{bmatrix} 0 \\ -1 \end{bmatrix} \Longrightarrow A + LC = \begin{bmatrix} -2 & 2 \\ 0 & -1 \end{bmatrix}$$

and therefore $A + LC$ is Hurwitz. Hence, (C, A) is detectable. △

Finally, we state a dual version of Theorem 6.1 giving restrictions on eigenvalue placement.

Theorem 7.2 *Suppose (C, A) is not observable, and let λ be an eigenvalue of A corresponding to an eigenvector v with $Cv = 0$. Then λ is also an eigenvalue of $A+LC$ for any $n \times p$ matrix L. Consequently, such eigenvalues are fixed under any output injection operation $v = Ly = LCx$ in the system $\dot{x} = Ax + v(t)$.*

Proof. The proof is recommended as Exercise 7.4. □

7.3 OBSERVER-BASED DYNAMIC STABILIZATION

The goals of this section are

(i) to summarize the construction of an observer system for state estimation when the pair (C, A) is detectable, and
(ii) to summarize the use of dynamic observer-based feedback for stabilization purposes, when (A, B) is stabilizable and (C, A) is detectable.

We also show that detectability is necessary for the observer construction in (i), and that stabilizability is necessary for the observer-based dynamic feedback controller in (ii). In this section, the constructions and the reasoning on which they are based follow the pattern of analysis carried out for (i) in Example 7.1, and for (ii) in Example 7.2.

General Observer Construction

Consider the MIMO linear time invariant system

$$\dot{x} = Ax + Bu, \tag{7.8}$$

$$y = Cx. \tag{7.9}$$

Given (7.8)–(7.9), consider also an auxiliary system

$$\dot{\xi} = A\xi + Bu - L(y - C\xi), \tag{7.10}$$

where ξ is an auxiliary state that can be initialized at any vector $\xi(0)$.

Definition 7.2 *System* (7.10) *is called an* observer system *for* (7.8)–(7.9) *if there exists an $n \times p$ matrix L such that $x(t) - \xi(t) \to 0$ as $t \to \infty$, independently of initial conditions $x(0)$ and $\xi(0)$.*

Matrix L is called an *output error gain*.

The next result is the main theorem on asymptotic estimation by an observer system of the form (7.10).

Corollary 7.2 *There exists an observer system of the form* (7.10) *for* (7.8)–(7.9) *if and only if the pair (C, A) is detectable.*

Proof. Define the error variable

$$e := x - \xi,$$

where x is the state vector in (7.8) and ξ is the auxiliary state vector in (7.10). Subtracting (7.10) from $\dot{x} = Ax + Bu$ and using (7.9), we obtain the error dynamics

$$\dot{e} = (A + LC)e.$$

The result is now immediate from Corollary 7.1 (b): There exists L such that $A + LC$ is Hurwitz if and only if (C, A) is detectable. □

An observer system provides a way of approximating solutions $x(t)$ of (7.8) for larger time values, assuming that the system matrices C and A are known, even when $x(0)$ is not known.

Observer-Based Dynamic Feedback Stabilization

If, in the auxiliary system (7.10), we set

$$u := K\xi, \tag{7.11}$$

then we have

$$\dot{\xi} = (A + BK + LC)\xi - Ly(t), \tag{7.12}$$
$$u = K\xi, \tag{7.13}$$

where $y(t) = Cx(t)$ is known. We can view (7.12)–(7.13) as a linear system for ξ with output given by u; matrix L is the input matrix and matrix K is the output matrix. The output (7.13) is the control u which is to be applied to (7.8) for control of that system.

The interconnection of (7.8)-(7.9) with (7.12)–(7.13) is a system with state (x, ξ) given by the equations

$$\begin{bmatrix} \dot{x} \\ \dot{\xi} \end{bmatrix} = \begin{bmatrix} A & BK \\ -LC & A + LC + BK \end{bmatrix} \begin{bmatrix} x \\ \xi \end{bmatrix}. \tag{7.14}$$

Definition 7.3 *An auxiliary system of the form* (7.12)–(7.13) *is called an* observer-based dynamic controller *for* (7.8)–(7.9). *It is called an (observer-based)* dynamic output stabilizer *if the $n \times p$ matrix L and $m \times n$ matrix K are such that the origin of system* (7.14) *is asymptotically stable.*

The next theorem is the main result on observer-based dynamic feedback stabilization.

Theorem 7.3 *There exists an observer-based dynamic output stabilizer* (7.12)–(7.13) *for system* (7.8)–(7.9) *if and only if (C, A) is detectable and (A, B) is stabilizable.*

Proof. As noted above, the interconnection of (7.8)–(7.9) with (7.12)–(7.13) is given by system (7.14). It is easy to check that the change of variables

$$\begin{bmatrix} x \\ e \end{bmatrix} := \begin{bmatrix} I & 0 \\ I & -I \end{bmatrix} \begin{bmatrix} x \\ \xi \end{bmatrix}$$

produces the block triangular system

$$\begin{bmatrix} \dot{x} \\ \dot{e} \end{bmatrix} = \begin{bmatrix} A+BK & -BK \\ 0 & A+LC \end{bmatrix} \begin{bmatrix} x \\ e \end{bmatrix}. \tag{7.15}$$

The characteristic polynomial of the coefficient matrix in (7.15) is the product of the characteristic polynomials of $A+BK$ and $A+LC$. Therefore this coefficient matrix can be made Hurwitz if and only if (C, A) is detectable and (A, B) is stabilizable. □

As noted in Example 7.2, an important feature of the construction of an observer-based dynamic stabilizing controller is that the feedback matrix K and the controller matrix L can be designed independently while ensuring that the overall interconnected system is asymptotically stable. This independent design feature is called the *separation principle.*

It is not difficult to see that stabilization of (7.8) by static output feedback $u = Ky = KCx$ implies detectability of (C, A); if $A+BKC$ is Hurwitz, then (C, A) is detectable by Corollary 7.1 (b). But detectability is strictly weaker, since not every $n \times p$ matrix L can be written in the factored form $L = BK$ for some K; see also Exercise 7.1.

7.4 LINEAR DYNAMIC CONTROLLERS AND STABILIZERS

Observer-based dynamic controllers (7.12)–(7.13) are a special form of *dynamic output controller*, defined next.

Definition 7.4 *A* linear dynamic output controller *for* (7.1)–(7.2) *is an auxiliary system of the form*

$$\dot{\xi} = A_c\xi + L_c y(t), \tag{7.16}$$
$$u = K_c\xi + D_c y(t), \tag{7.17}$$

where $y(t)$ is given by the output equation (7.2)*, and the matrices A_c, L_c, K_c, and D_c can be chosen freely. It is a* linear dynamic output stabilizer *for* (7.1)–(7.2) *if A_c, L_c, K_c, and D_c are such that the origin of the interconnection of* (7.1)–(7.2) *with* (7.16)–(7.17) *given by*

$$\begin{bmatrix} \dot{x} \\ \dot{\xi} \end{bmatrix} = \begin{bmatrix} A+BD_cC & BK_c \\ L_cC & A_c \end{bmatrix} \begin{bmatrix} x \\ \xi \end{bmatrix} =: \mathcal{A} \begin{bmatrix} x \\ \xi \end{bmatrix} \tag{7.18}$$

is asymptotically stable, that is, matrix $\mathcal{A}$ is Hurwitz.

Note that there is no restriction on the dimension of the state space for the controller (7.16).

Note that the output $y(t)$ of the original system (for any initial condition $x(0)$) is the input to the dynamic output controller (7.16) with state ξ. The output (7.17) of the controller is the input u for the original system. And, as claimed, an observer-based dynamic controller (7.12)–(7.13) matches the form (7.16)–(7.17) with

$$A_c = A + BK + LC, \quad L_c = -L, \quad K_c = K, \quad D_c = 0,$$

where K and L can be chosen freely. A direct feedthrough term $D_c y$ is included in (7.17), although this term is zero for the observer-based dynamic controller.

The class of linear dynamic controllers of the form (7.16)–(7.17) includes the class of state feedback controllers of the form $u = Kx$, since (7.16) may be absent and we can have $\xi = x$ and $D_c = 0$ in (7.17). However, it is a much larger class. It is reasonable to ask if we have enlarged the class of linear systems which can be stabilized, simply by using dynamic feedback rather than state feedback. Our experience with observer-based controllers, and specifically the decomposition in (7.15), shows that dynamic stabilization by such a controller requires both stabilizability of (A, B) and detectability of (C, A). Is it possible that some other linear controller of the form (7.16)–(7.17) might stabilize the origin of $\dot{x} = Ax + Bu$ even if (A, B) is not a stabilizable pair? The answer is given in the next theorem.

Theorem 7.4 *There exists a linear dynamic output stabilizer* (7.16)–(7.17) *for system* (7.1)–(7.2) *if and only if* (A, B) *is stabilizable and* (C, A) *is detectable.*

Proof. First, the sufficiency (the *if* part) follows from Theorem 7.3, because (A, B) stabilizable and (C, A) detectable imply the existence of an observer-based dynamic output stabilizer.

It remains to show that the existence of a linear dynamic output stabilizer (7.16)–(7.17) implies stabilizability and detectability of (7.1)–(7.2). The closed loop system formed by the interconnection of (7.1)–(7.2) and (7.16)–(7.17) is given by (7.18), repeated here for convenience:

$$\begin{bmatrix} \dot{x} \\ \dot{\xi} \end{bmatrix} = \begin{bmatrix} A + BD_cC & BK_c \\ L_cC & A_c \end{bmatrix} \begin{bmatrix} x \\ \xi \end{bmatrix} =: \mathcal{A} \begin{bmatrix} x \\ \xi \end{bmatrix}.$$

The proof hinges on showing that

(i) every eigenvalue of A corresponding to uncontrollable modes in (7.1) is also an eigenvalue of $\mathcal{A}$;

(ii) every eigenvalue of A corresponding to unobservable modes in (7.1)–(7.2) is also an eigenvalue of $\mathcal{A}$.

After (i) and (ii) are established, the assumption of asymptotic stability for the closed loop system implies that any eigenvalue of A associated with either uncontrollable modes or unobservable modes must have negative real part. Therefore (A, B) must be stabilizable and (C, A) must be detectable. Thus, we focus now on establishing (i) and (ii).

For (i), let λ be an eigenvalue of A corresponding to uncontrollable modes. Thus, there exists $v \neq 0$ such that

$$A^T v = \lambda v \quad \text{and} \quad B^T v = 0.$$

We want to show that λ is an eigenvalue of $\mathcal{A}$. By forming the transpose $\mathcal{A}^T$, an easy calculation shows that

$$\mathcal{A}^T \begin{bmatrix} v \\ 0 \end{bmatrix} = \lambda \begin{bmatrix} v \\ 0 \end{bmatrix},$$

and since $v \neq 0$, this proves that λ is an eigenvalue of $\mathcal{A}^T$; hence, λ is an eigenvalue of $\mathcal{A}$. This establishes statement (i) above; thus, all eigenvalues of A corresponding to uncontrollable modes are also eigenvalues of $\mathcal{A}$.

To show (ii), let λ be an eigenvalue of A corresponding to unobservable modes. There exists $v \neq 0$ such that

$$Av = \lambda v \quad \text{and} \quad Cv = 0.$$

We want to show that λ is an eigenvalue of $\mathcal{A}$. Another easy calculation shows that

$$\mathcal{A} \begin{bmatrix} v \\ 0 \end{bmatrix} = \lambda \begin{bmatrix} v \\ 0 \end{bmatrix},$$

so that λ is an eigenvalue of $\mathcal{A}$. This establishes statement (ii) above; thus, all eigenvalues of A corresponding to unobservable modes are also eigenvalues of $\mathcal{A}$. This completes the proof. □

A simple example will illustrate Theorem 7.4.

Example 7.5 Consider the system

$$\dot{x} = Ax + Bu := \begin{bmatrix} 2 & -2 \\ -1 & 1 \end{bmatrix} x + \begin{bmatrix} 1 \\ 1 \end{bmatrix} u$$

with output given by

$$y = Cx := \begin{bmatrix} 1 & -1 \end{bmatrix} x.$$

Can this system be stabilized by a linear dynamic output stabilizer? The eigenvalues of A are $\lambda_1 = 0$ and $\lambda_2 = 3$, and $\lambda_2 = 3$ corresponds to uncontrollable modes, as can be seen by a PBH controllability test:

$$\text{rank} \begin{bmatrix} A - 3I & B \end{bmatrix} = \text{rank} \begin{bmatrix} -1 & -2 & 1 \\ -1 & -2 & 1 \end{bmatrix} = 1;$$

thus, (A, B) is not stabilizable. By Theorem 7.4, this system cannot be stabilized by a linear dynamic output controller. We would draw the same conclusion if we first carried out a PBH detectability test:

$$\operatorname{rank}\begin{bmatrix} A \\ C \end{bmatrix} = \operatorname{rank}\begin{bmatrix} 2 & -2 \\ -1 & 1 \\ 1 & -1 \end{bmatrix} = 1;$$

thus, (C, A) is not detectable, and this fact alone would be sufficient to conclude from Theorem 7.4 that there is no linear dynamic output stabilizer. (We note that eigenvalue $\lambda_2 = 3$ does not correspond to unobservable modes because

$$\operatorname{rank}\begin{bmatrix} A - 3I \\ C \end{bmatrix} = \operatorname{rank}\begin{bmatrix} -1 & -2 \\ -1 & -2 \\ 1 & -1 \end{bmatrix} = 2.$$

And eigenvalue $\lambda_1 = 0$ does not correspond to uncontrollable modes since $\operatorname{rank}[A \quad B] = 2$). △

One final comment is in order. Note that in Example 7.5, for any matrix K, the matrix $A+BK$ also has eigenvalue $\lambda = 3$ (see Proposition 6.1). And, for any L, the matrix $A+LC$ also has eigenvalue $\lambda = 0$ (see Proposition 7.2). Thus, an *observer-based* dynamic feedback controller (7.10)–(7.11) cannot shift these two eigenvalues to the open left half-plane. But note that the proof of Theorem 7.4 says more than this; it says that $\lambda = 3$ and $\lambda = 0$ must be eigenvalues of the coefficient matrix $\mathcal{A}$ in (7.18), for any proposed linear dynamic controller (7.16)–(7.17).

Theorem 7.4 makes it clear that linear dynamic output controllers do not enlarge the class of systems $\dot{x} = Ax + Bu$ that can be stabilized by linear controllers; if a linear system can be stabilized by a dynamic output controller, then it must be stabilizable by linear state feedback. However, as shown by simple examples and exercises of this chapter, dynamic controllers can provide more performance options. We end this section with an example and discussion of dynamic control and constant disturbances.

Dynamic Control and Constant Disturbances

Real physical systems are often subject to disturbances which may be modeled but are of unknown magnitude. Thus, an effective controller system must compensate for the disturbance, by removing the effects of the disturbance from the operation of the physical system. In the next example, we consider constant force disturbances in a simple model of a pendulum arm.

Example 7.6 Consider again the motion of the inverted pendulum arm, which is frictionless and massless, with an attached unit mass at the end of the arm, subject to constant unit gravity. The second order equation is

$$\ddot{\theta} - \theta = u(t),$$

where $u(t)$ is an external force and θ is the angular displacement of the unit mass from the vertical upright equilibrium defined by $\theta = 0$ and $\dot{\theta} = 0$. In the presence of a constant disturbing force w, the model is

$$\ddot{\theta} - \theta = u(t) + w.$$

The equivalent system, taking account of the disturbance, is

$$\begin{bmatrix} \dot{x}_1 \\ \dot{x}_2 \end{bmatrix} = \begin{bmatrix} 0 & 1 \\ 1 & 0 \end{bmatrix} \begin{bmatrix} x_1 \\ x_2 \end{bmatrix} + \begin{bmatrix} 0 \\ 1 \end{bmatrix} u + \begin{bmatrix} 0 \\ w \end{bmatrix}. \tag{7.19}$$

where $x_1 = \theta$ is the angular position, and $x_2 = \dot{\theta}$ the angular velocity, of the unit mass. Due to the disturbing force w, the uncontrolled system with $u = 0$ has a new equilibrium state given by $x^w := (x_1^w, x_2^w) = (-w, 0)$. With $u = 0$, this new equilibrium is unstable. Suppose that the full state x is available for feedback; then full-state feedback can be used to stabilize the *nominal* system (where $w = 0$) at the origin, for example, by taking $u = Kx = [-2 \quad -2]x$. In the presence of the disturbance, however, the closed loop resulting from this feedback is

$$\begin{bmatrix} \dot{x}_1 \\ \dot{x}_2 \end{bmatrix} = \begin{bmatrix} 0 & 1 \\ -1 & -2 \end{bmatrix} \begin{bmatrix} x_1 \\ x_2 \end{bmatrix} + \begin{bmatrix} 0 \\ w \end{bmatrix},$$

and, due to the disturbance $w \neq 0$, all solutions approach a different equilibrium at $\hat{x}^w := (w, 0)$ as $t \to \infty$. This is not acceptable, even for arbitrarily small values of w, if the pendulum arm must reach the vertical upright position to accomplish its purpose.

A dynamic controller provides a solution to the problem of stabilizing the system to the origin. The disturbance w affects the position of the new equilibrium. We can use the integral of position to define a one-dimensional dynamic controller, which augments the original system to give the combined system

$$\begin{bmatrix} \dot{\xi} \\ \dot{x}_1 \\ \dot{x}_2 \end{bmatrix} = \begin{bmatrix} 0 & 1 & 0 \\ 0 & 0 & 1 \\ 0 & 1 & 0 \end{bmatrix} \begin{bmatrix} \xi \\ x_1 \\ x_2 \end{bmatrix} + \begin{bmatrix} 0 \\ 0 \\ 1 \end{bmatrix} u + \begin{bmatrix} 0 \\ 0 \\ w \end{bmatrix}. \tag{7.20}$$

The equations for x_1 and x_2 are as before, but they can be modified by dynamic feedback from ξ as follows. By setting $u = Kx := k_0\xi + k_1x_1 + k_2x_2$, we have an example of a proportional-integral-derivative (or PID) control. The resulting closed loop system is given by

$$\begin{bmatrix} \dot{\xi} \\ \dot{x}_1 \\ \dot{x}_2 \end{bmatrix} = \begin{bmatrix} 0 & 1 & 0 \\ 0 & 0 & 1 \\ k_0 & 1 + k_1 & k_2 \end{bmatrix} \begin{bmatrix} \xi \\ x_1 \\ x_2 \end{bmatrix} + \begin{bmatrix} 0 \\ 0 \\ w \end{bmatrix}.$$

Note that, with an appropriate choice of the k_i, the origin will be an asymptotically stable equilibrium of this system when $w = 0$. But more importantly, when $w \neq 0$, and for the same choice of the k_i, the closed loop system has a unique, asymptotically stable equilibrium at the point

$x_K^w := (-\frac{w}{k_0}, 0, 0)$. Thus, both the position and the velocity of the pendulum arm approach zero, as desired. (See Exercise 7.6.) In other words, dynamic feedback with the PID controller effectively shifts the steady-state error from the physical position to the auxiliary state ξ, independently of the unknown disturbance magnitude. The auxiliary controller component absorbs the steady state error so that the physical system does not have to. △

There are indications in the literature that integral control is a relevant model for certain control mechanisms in biological systems. See the Notes and References for this chapter.

7.5 LQR AND THE ALGEBRAIC RICCATI EQUATION

There is an established theory for the design of stabilizing linear feedback controls for linear systems $\dot{x} = Ax + Bu$, which involves an integral performance index which is to be minimized, called Linear Quadratic Regulator (LQR) theory. The solution of the LQR problem involves the matrix *algebraic Riccati equation (ARE)* given by

$$PA + A^T P + Q - PBR^{-1}B^T P = 0, \quad \text{where} \quad Q := C^T C. \tag{7.21}$$

A solution P of (7.21) is called a *stabilizing solution* if $P^T = P$ and the matrix $A - BR^{-1}B^T P$ is Hurwitz, corresponding to feedback by $u = -R^{-1}B^T Px$. If a stabilizing solution exists, then it is unique ([26], page 191). The main purpose of this section is to emphasize the important role that stabilizability and detectability play in generating stabilizing linear feedback controls by means of positive semidefinite solutions of (7.21).

We begin by setting up some notation that allows us to state the linear quadratic regulator problem and a few essential facts about it. Consider the linear system

$$\dot{x} = Ax + Bu, \tag{7.22}$$

$$y = Cx, \tag{7.23}$$

where the pair (A, B) is stabilizable. Given an initial condition x_0, a control is to be chosen to minimize the integral performance index

$$V(x_0, u(\cdot)) := \int_0^\infty (x^T(s)C^T Cx(s) + u^T(s)Ru(s))\, ds, \quad u(s) \in \mathbf{R}^m, \tag{7.24}$$

where R is positive definite and the set of admissible controls may be taken to be the space $L^2([0, \infty), \mathbf{R}^m)$ of square integrable functions from $[0, \infty)$ to $\mathbf{R}^m$. The function $V : \mathcal{U} \to \mathbf{R} \cup \{\infty, -\infty\}$ is called the *performance value function*. The value $V(x_0, u(\cdot))$ is the cost of carrying out the control $u(t)$, $0 \le t < \infty$, when the system is started in the initial state $x(0) = x_0$. Since $R > 0$ and $C^T C \ge 0$, the integrand in (7.24) penalizes large control magnitudes and large excursions of the state x away from $x = 0$. Note also that any positive semidefinite matrix Q can be factored as $Q := C^T C$ for some matrix C, so the state-dependent term in the integrand could be written $x^T Qx$ with no loss in generality.

We can now state the LQR problem: For each initial condition x_0, find the optimal (minimal) performance value

$$V(x_0) := \inf \left\{ \int_0^\infty (\|y(s)\|_2^2 + u^T(s)Ru(s))\, ds : u \in L^2([0,\infty), \mathbf{R}^m) \right\}$$

subject to

$$\dot{x} = Ax + Bu, \quad y = Cx, \quad x(0) = x_0,$$

and determine the optimal control u.

We now indicate the importance of stabilizability and detectability for the LQR problem.

If (A, B) is not stabilizable, then the infimum defining $V(x_0)$ need not exist. For example, the system

$$\dot{x} = \begin{bmatrix} 1 & 0 \\ 0 & 0 \end{bmatrix} x + \begin{bmatrix} 0 \\ 1 \end{bmatrix} u, \quad y = \begin{bmatrix} 1 & 0 \end{bmatrix} x$$

is not stabilizable. If the initial state component $x_{10} \neq 0$, then

$$\int_0^\infty (\|y(s)\|_2^2 + u^T(s)Ru(s))\, ds \geq \int_0^\infty e^{2s}|x_{10}|^2\, ds = +\infty.$$

Therefore the infimum over u does not exist for any $x_{10} \neq 0$. However, if (A, B) is stabilizable, then there exists K such that $A+BK$ is Hurwitz, and there exist constants $M > 0$ and $\alpha > 0$ such that $\|x(t)\| \leq M\|x_0\|e^{-\alpha t}$ for all $t \geq 0$. With $y = Cx$ and $u = Kx$, it is not difficult to show that

$$\int_0^\infty (\|y(s)\|_2^2 + u^T(s)Ru(s))\, ds \leq \beta\|x_0\|^2$$

for some finite $\beta > 0$. Thus, when (A, B) is stabilizable, the infimum defining $V(x_0)$ exists for each x_0. Moreover, it is plausible that a stabilizing feedback should produce a smaller value in (7.24) than a nonstabilizing control.

In addition to (A, B) stabilizable, we assume that the pair (C, A) is detectable. The intuition behind this assumption is that detectability guarantees that any unstable modes are revealed through the term $x^T Qx = x^T C^T Cx = y^T y$ in the performance index. (It is not difficult to see that if $Q = C^T C$, then (C, A) is detectable if and only if (Q, A) is detectable.) Simple examples show that detectability is not necessary for the existence of a stabilizing feedback via (7.21); however, without detectability, there may be no stabilizing control. See Exercise 7.9 for examples illustrating these facts.

The LQR problem is a constrained minimization problem known as a Lagrange problem, in which the integral of a functional is to be minimized subject to differential constraints. The Euler-Lagrange equations for this constrained minimization problem provide necessary conditions for a solution, and they imply the existence of a positive semidefinite matrix P satisfying the algebraic Riccati equation (7.21). In a more complete development of LQR theory, it is shown that such a solution P is indeed a minimizer for the performance index (7.24), and that the minimum is given by $V(x_0) := x_0^T P x_0$. The minimizing control is in feedback form, given by $u = -R^{-1}B^T Px$. Thus, the solution of the LQR problem for any initial state x_0 is achieved using the same P and the same optimal feedback control. (For a proof of the minimizing property of the positive semidefinite solution P of the ARE, see B. N. Datta, *Numerical Methods for Linear Control Systems*, Elsevier Academic Press, 2004, where guidance on numerical solution routines may also be found.)

For our purposes, the next theorem is the most important result of the section, as it shows the importance of stabilizability, detectability, and the algebraic Riccati equation in generating stabilizing feedback controls.

Theorem 7.5 (Asymptotic Stabilization via the ARE)
Assume R is positive definite. If (A, B) is stabilizable and (C, A) is detectable, then the algebraic Riccati equation (7.21) has a unique positive semidefinite solution P which is also the unique stabilizing solution, that is, the unique solution such that $A - BR^{-1}B^T P$ is Hurwitz.

Proof. See [104] for a proof of the existence of a unique positive semidefinite solution P of (7.21). For the uniqueness of stabilizing solutions, see [26] (page 191). Assuming the existence and uniqueness of $P \geq 0$, we proceed to show that P is stabilizing. Start by rewriting the algebraic Riccati equation as

$$(A - BR^{-1}B^T P)^T P + P(A - BR^{-1}B^T P) + C^T C + PBR^{-1}B^T P = 0. \quad (7.25)$$

We must show that $A - BR^{-1}B^T P$ is Hurwitz.

Suppose that $(A - BR^{-1}B^T P)w = \lambda w$ with $w \neq 0$. Pre-multiply and post-multiply (7.25) by w^* and w, respectively, to obtain

$$(\bar{\lambda} + \lambda) w^* P w + w^* C^T C w + w^* PBR^{-1}B^T P w = 0.$$

Suppose $\bar{\lambda} + \lambda = 0$, that is, $\operatorname{Re}\lambda = 0$. Positive semidefiniteness of the remaining terms, and the fact that $R^{-1} > 0$, imply that $Cw = 0$ and $B^T P w = 0$. Thus, λ is an eigenvalue of A with eigenvector w such that $Cw = 0$. This contradicts detectability. Suppose that $\bar{\lambda} + \lambda > 0$, that is, $\operatorname{Re}\lambda > 0$. Since each term on the left side of (7.25) is nonnegative, we must have $w^* P w \leq 0$. Since $P \geq 0$, we have $w^* P w \geq 0$, and therefore $Pw = 0$. Then, as before, $Cw = 0$ and $B^T P w = 0$. Therefore λ is an eigenvalue of A with eigenvector w such that $Cw = 0$. This again contradicts detectability. Consequently, if $(A - BR^{-1}B^T P)w = \lambda w$ with $w \neq 0$, then the real part of λ must be negative, which shows that $A - BR^{-1}B^T P$ is Hurwitz. □

If the hypothesis of detectability of (C, A) in Theorem 7.5 is strengthened to observability, then the unique solution P of (7.21) can be shown to be positive definite.

We end the section with two examples to illustrate Theorem 7.5.

Example 7.7 The system

$$\dot{x} = \begin{bmatrix} 1 & 0 \\ 0 & -1 \end{bmatrix} x + \begin{bmatrix} 1 \\ 0 \end{bmatrix} u,$$
$$y = \begin{bmatrix} 1 & 1 \end{bmatrix} x$$

has (A, B) stabilizable and (C, A) observable. Define the performance index (7.24) by $Q = C^T C$ and $R = 1$. Write

$$P = \begin{bmatrix} p_1 & p_2 \\ p_2 & p_3 \end{bmatrix}.$$

Finding a symmetric solution of the algebraic Riccati equation (7.21) is equivalent to finding a solution of the algebraic system

$$\begin{aligned} 2p_1 - p_1^2 + 1 &= 0, \\ -p_1 p_2 + 1 &= 0, \\ -2p_3 - p_2^2 + 1 &= 0. \end{aligned}$$

Since we are looking for a solution $P \geq 0$, we take $p_1 = 1 + \sqrt{2}$ as the solution of the first equation. Then, by the second equation, $p_2 = \frac{1}{1+\sqrt{2}}$. Finally, by the third equation, we get $p_3 = \frac{1}{1+\sqrt{2}}$. Thus,

$$P = \begin{bmatrix} 1+\sqrt{2} & \dfrac{1}{1+\sqrt{2}} \\ \dfrac{1}{1+\sqrt{2}} & \dfrac{1}{1+\sqrt{2}} \end{bmatrix},$$

which is positive definite, and

$$A - BB^T P = \begin{bmatrix} 1 & 0 \\ 0 & -1 \end{bmatrix} - \begin{bmatrix} 1 & 0 \\ 0 & 0 \end{bmatrix} \begin{bmatrix} 1+\sqrt{2} & \dfrac{1}{1+\sqrt{2}} \\ \dfrac{1}{1+\sqrt{2}} & \dfrac{1}{1+\sqrt{2}} \end{bmatrix} = \begin{bmatrix} -\sqrt{2} & -\dfrac{1}{1+\sqrt{2}} \\ 0 & -1 \end{bmatrix}$$

is Hurwitz. △

The next example shows that if (C, A) is merely detectable, then the positive semidefinite P may not be positive definite, but $A - BR^{-1}B^T P$ is indeed Hurwitz.

Example 7.8 Consider the system

$$\dot{x} = \begin{bmatrix} 1 & 0 \\ 0 & -1 \end{bmatrix} x + \begin{bmatrix} 1 \\ 0 \end{bmatrix} u,$$
$$y = \begin{bmatrix} 1 & 0 \end{bmatrix} x,$$

which has (A, B) stabilizable and (C, A) detectable (but not observable). If we write

$$P = \begin{bmatrix} p_1 & p_2 \\ p_2 & p_3 \end{bmatrix},$$

then the algebraic Riccati equation is equivalent to the system

$$2p_1 - p_1^2 + 1 = 0,$$
$$-p_1 p_2 = 0,$$
$$-2p_3 - p_2^2 = 0.$$

If $p_1 = 0$ we contradict the first equation, so we take $p_2 = 0$, and $p_3 = 0$ then follows from the third equation. We take $p_1 = 1 + \sqrt{2}$, and therefore

$$P = \begin{bmatrix} 1+\sqrt{2} & 0 \\ 0 & 0 \end{bmatrix},$$

which is positive semidefinite, and

$$A - BB^T P = \begin{bmatrix} 1 & 0 \\ 0 & -1 \end{bmatrix} - \begin{bmatrix} 1 \\ 0 \end{bmatrix} \begin{bmatrix} 1 & 0 \end{bmatrix} \begin{bmatrix} 1+\sqrt{2} & 0 \\ 0 & 0 \end{bmatrix} = \begin{bmatrix} -\sqrt{2} & 0 \\ 0 & -1 \end{bmatrix}$$

is indeed Hurwitz. △

7.6 EXERCISES

Exercise 7.1 *Dynamic vs. Static Output Feedback I*

The straight line motion of an unattached mass, with position measurement, is modeled by

$$\dot{x} = Ax + Bu := \begin{bmatrix} 0 & 1 \\ 0 & 0 \end{bmatrix} x + \begin{bmatrix} 0 \\ 1 \end{bmatrix} u, \quad y = Cx := \begin{bmatrix} 1 & 0 \end{bmatrix} x.$$

(a) Show that this system is stabilizable and detectable, but that it is not stabilizable by static output feedback $u = ky$.

(b) Design an observer-based dynamic stabilizer, using $u = K\xi$, by finding K and L such that the eigenvalues of $A + BK$ are $\lambda_1 = -1$, $\lambda_2 = -1$, and the eigenvalues of $A + LC$ are $\lambda_3 = -2$, $\lambda_4 = -2$.

Exercise 7.2 *Dynamic vs. Static Output Feedback II*

Consider a pendulum system with unit damping, modeled by the equations

$$\dot{x}_1 = x_2,$$
$$\dot{x}_2 = -x_1 - x_2 + u,$$
$$y = x_1.$$

Denote the system in the usual way by the matrix triple (C, A, B).

(a) Set $u = Ky = KCx$, find the eigenvalues of the closed loop $\dot{x} = (A+BKC)x$ as a function of the scalar K, and describe the dynamic behavior of solutions for all real K, including an exponential form norm estimate.

(b) Is it possible to design an observer-based dynamic stabilizer such that $A+BK$ has eigenvalues $\lambda_1 = -3$, $\lambda_2 = -3$, and $A+LC$ has eigenvalues $\lambda_3 = -2$, $\lambda_4 = -2$?

(c) Is it possible to achieve arbitrary eigenvalue placement of the four eigenvalues of system (7.14)? Explain.

Exercise 7.3 *Invariance of Detectability*

(a) Show that detectability is invariant under linear coordinate transformations.

(b) Prove the converse of Lemma 7.1: If, for every eigenvalue λ of A, we have

$$[Av = \lambda v, \quad v \neq 0, \quad \text{and} \quad Cv = 0] \Longrightarrow \operatorname{Re}\lambda < 0,$$

then the pair (C, A) is detectable. *Hint*: Use Theorem 5.5.

(c) Show that if (C, A) is detectable then $(C, A + LC)$ is detectable for any $n \times p$ matrix L. In other words, detectability is preserved by the output injection operation $v = Ly = LCx$ in $\dot{x} = Ax + v(t)$.

Exercise 7.4 Prove Theorem 7.2.

Exercise 7.5 *Existence of an Observer System*
Consider the system

$$\begin{aligned} \dot{x} &= Ax + Bu, \quad x(0) = x_0, \\ y &= Cx, \end{aligned}$$

with

$$A = \begin{bmatrix} -10 & 10 & 0 \\ 28 & -1 & 0 \\ 0 & 0 & -\frac{8}{3} \end{bmatrix}, \quad B = \begin{bmatrix} b_1 \\ b_2 \\ b_3 \end{bmatrix},$$
$$C = \begin{bmatrix} c_1 & c_2 & c_3 \end{bmatrix}.$$

Determine conditions on the entries b_i, c_i, $i = 1, 2, 3$, that are necessary and sufficient for the existence of an observer system (7.10).

Exercise 7.6 *On Example 7.6*

(a) Find a feedback matrix $K = [k_1\ k_2\ k_3]$ such that the three-dimensional augmented system (7.20) in Example 7.6 has an asymptotically stable origin when $w = 0$. Then verify that the same feedback works when

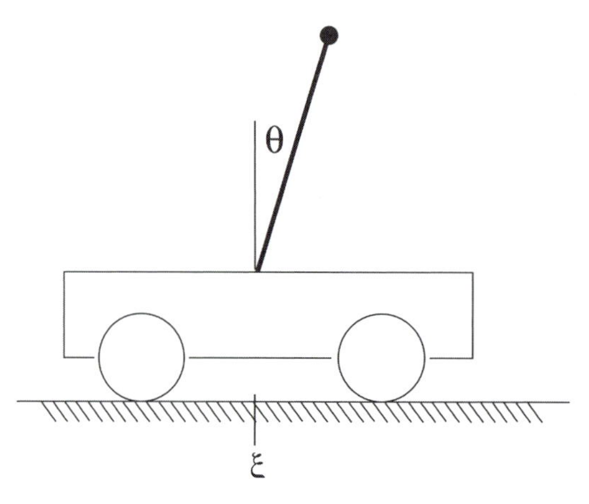

Figure 7.1 The inverted pendulum on a cart—Exercise 7.7.

$w \neq 0$ to guarantee that for all initial conditions, $\xi \to -\frac{w}{k_0}$, $x_1 \to 0$ and $x_2 \to 0$, as $t \to \infty$.

(b) Suppose that, in system (7.19), a full-state feedback matrix $K = [k_1 \quad k_2]$ is used to stabilize the system when $w = 0$. Show that, when $w \neq 0$, it takes arbitrarily large feedback gains in $K = [k_1 \quad k_2]$ to place the new stable equilibrium arbitrarily close to the desired location at the origin.

Exercise 7.7 *Inverted Pendulum on a Cart*
An inverted pendulum is attached to a rolling cart. The cart is free to move along a horizontal line, and the pendulum arm is free to move in a plane that passes through the line of motion of the cart. (See Figure 7.1.) The system may be controlled through a control force u applied to the cart. Variable ξ gives the horizontal distance to a point located directly below the pendulum pivot point, and θ is the angle of the pendulum arm measured from the vertical (unstable) equilibrium position. The equations of motion are given by

$$\begin{aligned}(M+m)\ddot{\xi} + ml(\ddot{\theta}\cos\theta - \dot{\theta}^2\sin\theta) + b\dot{\xi} &= u,\\ ml\ddot{\xi}\cos\theta + ml^2\ddot{\theta} - mgl\sin\theta) &= 0,\end{aligned}$$

where M is the mass of the cart, m is the mass at the end of the pendulum arm, l is the length of the arm, g is the acceleration due to gravity, and b is a friction coefficient for the motion of the cart.

(a) Using the state variables $x_1 = \xi$, $x_2 = \dot{\xi}$, $x_3 = \theta$, $x_4 = \dot{\theta}$, write the state equations in the form $\dot{x} = f(x) + g(x)u$ where $x = (x_1, x_2, x_3, x_4)$.

(b) Write the linearization $\dot{x} = Ax + Bu := Df(0) + g(0)u$ of the system in (a) about the equilibrium at the origin.

(c) Let the output be $y = x_3 = \theta$, the angle of the pendulum arm from the vertical. For simplicity, take the values of the constants to be $M = 10$, $m = 1$, $l = 1$, $g = 1$, and $b = 1$. Show that the linearization is controllable, but not observable. (Note: These conclusions are independent of the numerical values chosen for the constants.)

(d) Show that the linearization cannot be asymptotically stabilized by linear output feedback alone, using $u = ky = kx_3$.

(e) Show that the linearization has an observer-based dynamic stabilizer of dimension four.

(f) Can you find a dynamic output stabilizer with state dimension less than four?

Exercise 7.8 Consider the double integrator system $\dot{x}_1 = x_2$, $\dot{x}_2 = u$, with output $y = x_1$, and performance index (7.24) defined by $Q = C^T C$ and $R = 1$. Verify that Theorem 7.5 applies to this system. Find the positive definite solution P of the algebraic Riccati equation (7.21), and verify that $A - BB^T P$ is Hurwitz.

Exercise 7.9 Consider the LQR problem defined by the data

$$A = \begin{bmatrix} 0 & 1 \\ 0 & 1 \end{bmatrix}, \quad B = \begin{bmatrix} 0 \\ 1 \end{bmatrix}, \quad C = \begin{bmatrix} \beta & 1 \end{bmatrix}, \quad R = 1.$$

(a) Verify that Theorem 7.5 applies if $\beta \neq 0, -1$.

(b) Show that if $\beta = 0$, there are two symmetric solutions of the algebraic Riccati equation (7.21), $P_1 \geq 0$ and $P_2 \leq 0$, and neither is stabilizing.

(c) Show that if $\beta = -1$, there are three symmetric solutions of the algebraic Riccati equation (7.21): $P_1 > 0$, which is stabilizing, $P_2 \geq 0$, not stabilizing, and $P_3 \leq 0$.

7.7 NOTES AND REFERENCES

For linear algebra background, see [75]. Helpful text references on systems and control are [9], [16], [82], [91], [104].

The statement of Theorem 7.5 is from [104] (Theorem 12.2, page 282) together with the statement on uniqueness of stabilizing solutions of the ARE from [26] (page 191). More on the linear quadratic regulator problem can be found in specific chapters in the books [5], [26], [91], and [104].

In this chapter we considered outputs in the form $y = Cx$, rather than $y = Cx + Du$, where D is the *direct feedthrough* matrix. The development of controllability, observability, stabilizability, and detectability can be carried through with direct feedthrough terms; see [26]. We discussed a full order observer only, that is, an observer system for which the auxiliary state had the same dimension as the system state. When $\operatorname{rank} C = r$, there are r linearly independent equations relating the state components, and an observer of order $n - r$ can be obtained; see [70], [71] for more on observers.

Example 7.6 is from the discussion of integral control in [91] (Chapter 1). A classical problem involving dynamic feedback control is the *absolute stability* problem, which involves nonlinear feedback in the linearization of a system; an exposition of this problem appears in the differential equations text [15] (pages 261–273). More recently, the importance of integral control in biological system modeling has been noted in the text by U. Alon, *An Introduction to Systems Biology: Design Principles of Biological Circuits*, Chapman and Hall/CRC, 2007, and in the paper by B. P. Ingalls, T-M. Yi, and P. A. Iglesias, Using control theory to study biology, in *System Modeling in Cellular Biology*, edited by Z. Szallasi, J. Stelling, and V. Periwal, MIT Press, Cambridge, MA, 2006 pages 243–267.

The equations for the inverted pendulum on a cart in Exercise 7.7 are from [45] (page 69).

Stabilizability and detectability of the linearization of a nonlinear system are necessary for the solvability of certain output tracking and output regulation problems; see [20], [45], and [47].

Chapter Eight

Stability Theory

In this chapter we present the basic Lyapunov theorems on stability and asymptotic stability based on the analysis of Lyapunov functions. We explain which problems of stability and (smooth) stabilization for equilibria can be resolved based solely on the Jacobian linearization at the equilibrium. The other cases are called *critical problems* of stability and stabilization and they must be investigated by methods that take into account the nonlinear terms of the system. The general Lyapunov theorems and the important invariance theorem, also included here, can address critical problems as well as noncritical ones. We illustrate these theorems with simple examples. We also indicate how Lyapunov functions can help to estimate the basin of attraction of an asymptotically stable equilibrium.

8.1 LYAPUNOV THEOREMS AND LINEARIZATION

As we turn to the stability of nonlinear systems, it may be helpful to revisit, as needed, our initial nonlinear examples in the phase plane as well as Definition 3.3.

The condition that a system is Hamiltonian is too restrictive as a general condition for Lyapunov stability; nevertheless, Hamiltonian systems provide instructive and important examples of Lyapunov stable equilibria. The first general result on Lyapunov stability of equilibria involves an energy-like function $V(x)$ which is merely nonincreasing along solutions of (8.1), as in Example 2.16. Such functions provide a general way of deducing stability of equilibria for nonlinear systems. A second general result, on asymptotic stability, applies to systems where an energy-like function is known to be *strictly* decreasing along solution curves near equilibrium, as in Example 2.14. Such functions provide a general way of deducing local asymptotic stability of equilibria for nonlinear systems. The two general results just mentioned are both included in Theorem 8.1.

For noncritical problems of asymptotic stability, those in which the Jacobian linearization is sufficient to determine asymptotic stability of an equilibrium, we will see that a local energy function may be found as a quadratic form, based on the Jacobian linearization at the equilibrium.

8.1.1 Lyapunov Theorems

We continue to study a system of the form

$$\dot{x} = f(x). \tag{8.1}$$

The unique solution of (8.1) with initial condition x at time $t = 0$ will be denoted by $\phi_t(x)$ or by $\phi(t, x)$. Recall that we are assuming an equilibrium for (8.1) at the origin in $\mathbf{R}^n$, that is, $f(0) = 0$. We often use the computation and notation given by

$$\frac{d}{dt} V(\phi(t,x))\Big|_{t=0} = \nabla V(x) \cdot f(x) =: L_f V(x),$$

where $\phi(t, x)$ is the solution of (8.1) that satisfies $\phi(0, x) = x$. Also note that the inner product, or dot product, in the definition above is equivalent to the matrix product $(\nabla V)^T(x) f(x)$, which can also be written as $\frac{\partial V(x)}{\partial x} f(x)$. The dot product notation or one of the matrix product notations may be preferable when both terms resulting from a chain rule computation should be displayed. Another commonly used notation is $\dot{V}_A(x) := \nabla V(x) \cdot f(x)$. (These notational options are also given in Appendix A on notation.)

First we define some standard concepts which help in analyzing asymptotic behavior. We will need the definition of *ω-limit point* and *ω-limit set* of a solution $\phi(t, x)$ of (8.1) when this solution is defined for all $t \geq 0$.

Definition 8.1 *A point q is an ω-*limit point *of the solution $\phi(t, x)$ if there exists a sequence $t_k \to \infty$ as $k \to \infty$ such that* $\lim_{k\to\infty} \phi(t_k, x) = q$. *The set of all ω-limit points of $\phi(t, x)$ is denoted $\omega(\phi(t, x))$, or just $\omega(x)$, and it is called the* ω-limit set *of x. Finally, for an arbitrary function $\psi : [0, \infty) \to \mathbf{R}$ the ω-limit set of ψ is defined similarly.*

Thus, the ω-limit set of a solution consists of all points that are asymptotically approached by a subsequence $\phi(t_k, x)$ of points on the trajectory, where the subsequence is indexed by an unbounded sequence of time values t_k.

Example 8.1 Consider the planar system given in polar coordinates by

$$\begin{aligned}\dot{\theta} &= 1,\\ \dot{r} &= r(1 - r).\end{aligned}$$

The general solution of this system is

$$\begin{aligned}\theta(t) &= \theta_0 + t,\\ r(t) &= \frac{r_0}{r_0 + (1 - r_0)e^{-t}},\end{aligned}$$

with initial conditions given by $(r(0), \theta(0)) = (r_0, \theta_0)$. For example, the unique solution with initial condition $(r_0, \theta_0) = (2, 0)$ is

$$\theta(t) = t,$$
$$r(t) = \frac{2}{2 - e^{-t}}.$$

Every point on the unit circle is an ω-limit point of this solution. (A phase portrait strongly suggests this fact.) To show it, fix θ_0 and consider the point $(1, \theta_0)$ on the unit circle. Define $t_k = \theta_0 + 2k\pi$, and set $(r(t_k), \theta(t_k)) = (\frac{2}{2-e^{-t_k}}, t_k)$, for positive integer k. This sequence of points converges to $(1, \theta_0)$ as $k \to \infty$. On the other hand, it is not possible for a point off the unit circle to be an ω-limit point of this solution. Therefore the ω-limit set for this solution is exactly the unit circle. △

It is also convenient to have the concept of an invariant set for a differential equation.

Definition 8.2 (Invariant Set)
A set M is an invariant set *for* (8.1) *if*

$$x \in M \Longrightarrow \phi(t, x) \in M, \quad -\infty < t < \infty.$$

Similarly, M is forward (respectively, backward) invariant *if*

$$x \in M \Longrightarrow \phi(t, x) \in M, \quad 0 \le t \le \infty \text{ (respectively, } -\infty < t \le 0).$$

An invariant set must consist of whole solution trajectories. In Example 8.1, the unit circle is an invariant set; moreover, the closed unit disk and the (open) complement of the closed unit disk are also invariant sets in that example.

When the initial condition is understood, we may denote a solution by $x(t)$. The next result gives two direct Lyapunov theorems.

Theorem 8.1 *Let D be a nonempty open set in $\mathbf{R}^n$ containing the origin. Suppose that $f : D \to \mathbf{R}^n$ is a continuously differentiable mapping and $x_0 = 0$ is an equilibrium point for* (8.1). *Let $V : D \to \mathbf{R}^n$ be a continuously differentiable function such that $V(0) = 0$. Then the following statements are true:*

(a) *If $V(x) > 0$ for all nonzero x in D and $L_f V(x) \le 0$ for all x in D, then $x_0 = 0$ is (Lyapunov) stable.*

(b) *If $V(x) > 0$ and $L_f V(x) < 0$ for all nonzero x in D, then $x_0 = 0$ is asymptotically stable.*

Proof. (a): Let $\epsilon > 0$. Since D is open, we can choose r with $0 < r < \epsilon$ such that $B_r(0) \subset D$. Since V is continuous and positive on the boundary sphere where $\|x\| = r$, we have

$$\alpha := \min_{\|x\|=r} V(x) > 0.$$

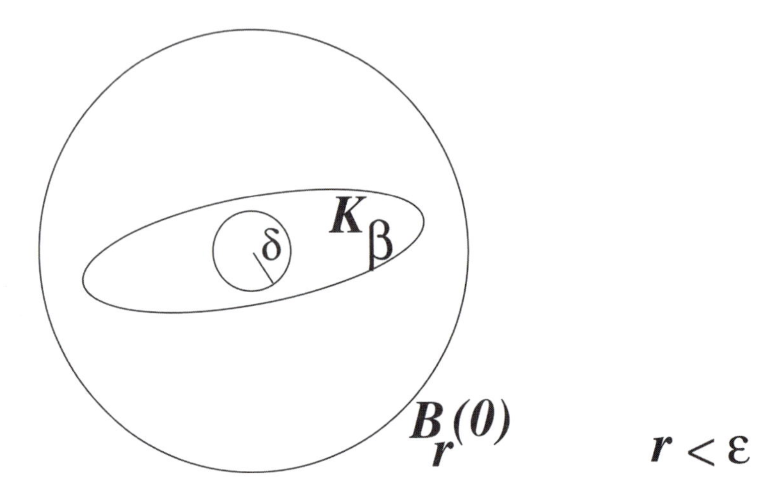

Figure 8.1 The proof of stability in Theorem 8.1.

Let $0 < \beta < \alpha$, and define the set

$$K_\beta := \{x \in B_r(0) \,:\, V(x) \leq \beta\}.$$

Since $\beta < \alpha$, K_β is contained in the interior of $B_r(0)$ and K_β is compact (closed by continuity of V, and bounded by containment in $B_r(0)$). Moreover, K_β is forward invariant for the system. To see this, let $x(t)$ be a solution with initial condition $x_0 \in K_\beta$ and defined on the maximal interval J. Note that

$$L_f V(x(t)) \leq 0 \Longrightarrow V(x(t)) \leq V(x(0)) \leq \beta, \quad \text{for all } t \in J \cap [0, \infty);$$

thus, $x(t)$, being continuous, must remain within $B_r(0)$ for all $t \geq 0$ for which it is defined. Therefore the forward orbit remains in the compact set K_β. By Theorem C.3 of Appendix C, the solution $x(t)$ exists for all $t \geq 0$ when $x_0 \in K_\beta$. Given $\beta > 0$ as above, the continuity of V at the origin implies that there exists $\delta > 0$ such that $\delta < r$ and

$$\|x\| < \delta \Longrightarrow V(x) < \beta.$$

Therefore we have

$$B_\delta(0) \subseteq K_\beta \subseteq B_r(0) \subseteq B_\epsilon(0),$$

and if $x_0 \in B_\delta(0)$, then the solution $x(t)$ with $x(0) = x_0$ satisfies $x(t) \in B_r(0)$, hence $\|x(t)\| < \epsilon$, for all $t \geq 0$. This shows that $x = 0$ is Lyapunov stable according to Definition 3.3 (a). (See Figure 8.1.)

(b): Clearly, the argument just given for part (a) applies here as well, so $x = 0$ is Lyapunov stable. To show asymptotic stability, then, it suffices to show that the origin is attractive for all initial conditions in some ball centered at the origin. Let $B_r(0)$ be a ball as in the proof of (a), and let $0 < \beta < r$, with K_β defined as before, so that $K_\beta \subset B_r(0)$. Again let $\delta > 0$

such that $\delta < r$ and $\|x\| < \delta$ implies $V(x) < \beta$. Our conclusion is that if $x_0 \in B_\delta(0)$ then $x(t) \in B_r(0)$ for all $t \geq 0$. We now show that if $x_0 \in B_\delta(0)$, then $V(x(t)) \to 0$ as $t \to \infty$. Since $V(x(t))$ is decreasing and bounded below by zero, we have

$$V(x(t)) \to c \geq 0 \quad \text{as}\, t \to \infty.$$

Suppose now that $c > 0$; we will show this leads to a contradiction. Consider the set

$$K_c = \{x \in B_r(0) \,:\, V(x) \leq c\}.$$

By continuity of V, we can find a $d > 0$ such that $d < r$ and $B_d(0) \subset K_c$. Since $V(x(t))$ is monotone decreasing to c, we have that $x(t)$ lies outside the ball $B_d(0)$ for all $t \geq 0$. Now consider the compact set $K = \{x \,:\, d \leq \|x\| \leq r\}$. Since the function $x \mapsto L_f V(x)$ is continuous on K, it has a maximum there:

$$\max_{x \in K} L_f V(x) =: -\gamma, \quad \text{where } \gamma > 0.$$

Thus, a solution $x(t)$ with $x(0) = x_0 \in B_d(0)$ satisfies

$$V(x(t)) = V(x(0)) + \int_0^t L_f V(x(s))\, ds \leq V(x(0)) - \gamma t \quad \text{for all } t \geq 0.$$

But the right-hand side is negative for sufficiently large $t > 0$, contradicting the fact that V is nonnegative. Thus, $c = 0$ and $V(x(t)) \to 0$ as $t \to \infty$. We now argue that $x(t) \to 0$ using the concept of ω-limit point. Let q be an ω-limit point of $x(t)$, with $x(t_k) \to q$ as $t \to \infty$; at least one such point exists since $x(t)$ is bounded for $t \geq 0$. Then we have a monotone decreasing sequence $V(x(t_k))$, and by continuity of V, $V(x(t_k)) \to V(q) = 0$. Since V is positive definite, we have $q = 0$, as we wanted to show. If $x(t)$ does not converge to zero as $t \to \infty$, we can find a nonzero ω-limit point for the forward solution, which is a contradiction of the fact just shown. We may now conclude that if $x_0 \in B_\delta(0)$, then $x(t) \to 0$ as $t \to \infty$; hence (b) is proved. □

Definition 8.3 (Lyapunov Function)
A smooth, real-valued, positive definite function V as in Theorem 8.1 (part (a) or (b)) is called a Lyapunov function*; a function V as in part (b) is called a* strict Lyapunov function.

This is a convenient place to recommend Exercise 8.5 and Exercise 8.6.

Suppose we wish to determine the stability properties of an equilibrium of a nonlinear system $\dot{x} = f(x)$, where f is continuously differentiable. Under what conditions can we make this determination based only on a study of the Jacobian linearization at the equilibrium? The conditions that allow a determination of stability properties based on the linearization are set out completely in the hypotheses for parts (a) and (b) of the next theorem.

Theorem 8.2 *Suppose $f : D \to \mathbf{R}^n$ is C^1 on a neighborhood D of $x = 0$ and $f(0) = 0$. Let*

$$A := Df(0) = \frac{\partial f}{\partial x}(x)\Big|_{x=0} .$$

(a) *If Re $\lambda < 0$ for each eigenvalue λ of A, then the equilibrium $x = 0$ of (8.1) is asymptotically stable.*
(b) *If Re $\lambda > 0$ for at least one eigenvalue λ of A, then the equilibrium $x = 0$ of (8.1) is unstable.*

Proof. For use in both parts (a) and (b), we may write

$$\dot{x} = f(x) = Ax + g(x),$$

where g is C^1 in D and $Dg(0) = 0$. By the mean value theorem (Theorem B.5), for any $\epsilon > 0$ there is a $\delta > 0$ such that

$$\|g(x)\| \le \epsilon \|x\| \quad \text{for } \|x\| < \delta. \tag{8.2}$$

(See Exercise 8.2.) We may use any vector norm in (8.2).

Proof of (a): We will find a quadratic Lyapunov function. Since A is assumed to be Hurwitz, Theorem 3.7 implies that there is a unique real positive definite matrix P which satisfies the Lyapunov equation

$$A^T P + PA = -I.$$

In fact, by Proposition 3.2 we have

$$P = \int_0^\infty e^{tA^T} e^{tA}\, dt.$$

Define $V : B_\delta(0) \to \mathbf{R}^n$ by

$$V(x) = x^T P x .$$

We want to show that the restriction of V to some neighborhood of the origin is a strict Lyapunov function. Thus, we compute

$$\begin{aligned} L_f V(x) &= \dot{x}^T P x + x^T P \dot{x} \\ &= (Ax + g(x))^T P x + x^T P(Ax + g(x)) \\ &= x^T A^T P x + x^T P A x + g(x)^T P x + x^T P g(x) \\ &= x^T (A^T P + PA) x + 2 g(x)^T P x \quad (\text{since } P^T = P) \\ &= -x^T x + 2 g(x)^T P x. \end{aligned}$$

Since $Dg(0) = 0$ and $Dg(x)$ is continuous at $x = 0$, there is a $\delta > 0$ such that, if $\|x\|_2 < \delta$ then $\|g(x)\|_2 \le \frac{1}{4\|P\|_2} \|x\|_2$, where $\|P\|_2$ is the matrix norm induced by the Euclidean vector norm. Now use the Cauchy-Schwartz inequality to get

$$\begin{aligned} 2g(x)^T Px &\le 2\,|g(x)^T Px| \\ &\le 2\,\|g(x)\|_2\;\|P\|_2\;\|x\|_2 \\ &\le \frac{1}{2}\|x\|_2^2, \end{aligned}$$

and thus

$$L_f V(x) \le -\frac{1}{2}\|x\|_2^2 \quad \text{for } 0 < \|x\|_2 < \delta.$$

Therefore V is a strict Lyapunov function on $B_\delta(0)$ and the proof of (a) is complete.

(b) The idea of the proof is that if there is an eigenvalue with positive real part, then, for small $\|x\|$, there are dominant linear terms that allow for exponential growth of some solution near the origin, and this precludes stability of the origin. The method of proof is by contradiction.

First, we set up some notation. Define new coordinates by $x = Sz$, which gives the equation

$$\dot{z} = Jz + S^{-1}g(Sz), \tag{8.3}$$

where $J := S^{-1}AS$ is the complex Jordan form of A. The equilibrium $z = 0$ of (8.3) is stable if and only if the equilibrium $x = 0$ of $\dot{x} = f(x)$ is stable. Observe that z may be complex for real x; however, Sz must be real, so $g(Sz)$ is well defined. By appropriate choice of the transformation S, we can arrange that J takes the form

$$J = \begin{bmatrix} B_1 & 0 \\ 0 & B_2 \end{bmatrix}, \tag{8.4}$$

where B_1 is a $k \times k$ diagonal matrix with the full list of eigenvalues having positive real part. (We assume there are k such eigenvalues.) Matrix B_2 has eigenvalues with nonpositive real part. Since J is the complex Jordan form, the entries on the main diagonal are the eigenvalues. Any and all nonzero entries on the superdiagonal of J may be taken to equal $\gamma > 0$, where γ may be chosen as small as we wish by appropriate choice of S. (See Exercise 3.10.) Write a given solution of (8.3) as $z(t) = [z_1(t) \;\cdots\; z_n(t)]^T$, and let

$$R^2(t) := \sum_{i=1}^{k} |z_i(t)|^2 \quad \text{and} \quad \rho^2(t) := \sum_{i=k+1}^{n} |z_i(t)|^2.$$

Choose $\sigma > 0$ such that every eigenvalue λ of B_1 has $\operatorname{Re}\lambda > \sigma$. Write $G(z) := S^{-1}g(Sz)$, and note that, given $\epsilon > 0$ there exists a $\delta > 0$ such that

$$\|G(z)\| \le \epsilon\|z\| \quad \text{for } \|z\| < \delta. \tag{8.5}$$

In order to proceed with the proof by contradiction, we suppose that $z = 0$ is a stable equilibrium of (8.3). Then, given $\delta > 0$ as in (8.5), there exists $\delta_1 > 0$ such that any solution of (8.3) with initial condition satisfying the norm bound

$$R(0) + \rho(0) < \delta_1$$

must also satisfy

$$R(t) + \rho(t) < \delta \quad \text{for all } t \geq 0.$$

Let $z(t)$ denote such a solution, keeping in mind that ϵ can be chosen freely. The i-th equation of (8.3) may be written

$$\dot{z}_i = \lambda_i z_i + [\gamma] z_{i+1} + G_i(z),$$

where $[\gamma] := \gamma$ if γ is present in the equation and $[\gamma] := 0$ otherwise. Note that $R^2(t) = \sum_{i=1}^{k} z_i^* z_i$ implies that (suppressing the t variable, $t \geq 0$)

$$\begin{aligned} 2R\dot{R} &= \sum_{i=1}^{k} \dot{z}_i^* z_i + z_i^* \dot{z}_i \\ &= \sum_{i=1}^{k} \left[(\bar{\lambda}_i z_i^* + [\gamma] z_{i+1}^* + G_i^*(z)) z_i + z_i^* (\lambda_i z_i + [\gamma] z_{i+1} + G_i(z)) \right]. \end{aligned}$$

Using (8.3), (8.4), (8.5), $\lambda_i + \bar{\lambda}_i = 2\mathrm{Re}\,\lambda_i$, and straightforward absolute value estimates, one arrives at the estimate

$$2R\dot{R} \geq 2\sigma R^2 - 2\gamma R^2 - 2\epsilon R(R + \rho).$$

Now choose $\gamma > 0$ and $\epsilon > 0$ so that

$$\dot{R} \geq (\sigma - \gamma - \epsilon) R - \epsilon \rho \geq \frac{\sigma}{2} R - \frac{\sigma}{10} \rho. \tag{8.6}$$

Observe that this can be accomplished by choosing $\epsilon < \frac{\sigma}{10}$ and $\gamma < \frac{\sigma}{10}$. Similarly, $\rho^2(t) = \sum_{i=k+1}^{n} z_i^* z_i$ implies that (again suppressing the t variable, $t \geq 0$)

$$\begin{aligned} 2\rho\dot{\rho} &= \sum_{i=k+1}^{n} \dot{z}_i^* z_i + z_i^* \dot{z}_i \\ &= \sum_{i=k+1}^{n} \left[(\bar{\lambda}_i z_i^* + [\gamma] z_{i+1}^* + G_i^*(z)) z_i + z_i^* (\lambda_i z_i + [\gamma] z_{i+1} + G_i(z)) \right], \end{aligned}$$

but now $2\mathrm{Re}\,\lambda_i \leq 0$. Thus, by (8.3), (8.4), (8.5), and standard absolute value estimates, one obtains

$$2\rho\dot{\rho} \leq 2\gamma\rho^2 + 2\epsilon\rho(R + \rho),$$

from which follows

$$\dot{\rho} \leq \gamma\rho + \epsilon(R + \rho).$$

Observe now that, by taking $\epsilon < \frac{\sigma}{10}$ and $\gamma < \frac{\sigma}{10}$, we have

$$\dot{\rho} \leq \frac{\sigma}{10} \rho + \frac{\sigma}{10} (R + \rho). \tag{8.7}$$

Thus, by the choices $\epsilon < \frac{\sigma}{10}$, $\gamma < \frac{\sigma}{20}$, and (8.6), (8.7), we have

$$\begin{aligned}\dot{R} - \dot{\rho} &\geq \frac{\sigma}{2}R - \frac{\sigma}{10}\rho - \frac{\sigma}{10}\rho - \frac{\sigma}{10}(R+\rho)\\ &= \frac{4}{10}\sigma R - \frac{3}{10}\sigma\rho\\ &\geq \frac{3}{10}\sigma(R-\rho).\end{aligned}$$

Hence, we have

$$R(t) - \rho(t) \geq [R(0) - \rho(0)]\, e^{\frac{3}{10}\sigma t} \quad \text{for all } t \geq 0.$$

Now choose a particular solution $z(t)$ as above, satisfying $R(t) + \rho(t) < \delta$ for all $t \geq 0$ and having initial condition $z(0)$ such that $R(0) = 2\rho(0) \neq 0$ and $R(0) + \rho(0) < \delta_1$. For this solution, we have

$$R(t) \geq R(t) - \rho(t) \geq \rho(0)\, e^{\frac{3}{10}\sigma t} \quad \text{for all } t \geq 0.$$

But this contradicts the fact that $R(t) \leq R(t) + \rho(t) < \delta$ for all $t \geq 0$. Hence $z = 0$ is unstable. Therefore $x = 0$ is unstable, completing the proof of (b). □

The cases covered by Theorem 8.2 are called *noncritical problems* of stability. All other cases are called *critical problems* of stability; they are the problems for which the linearization is inconclusive. Thus, a critical problem is one for which each eigenvalue of the Jacobian linearization at the equilibrium has real part less than or equal to zero, and some eigenvalue has real part equal to zero. For critical problems, the nonlinear terms in the vector field $f(x)$ must be taken into account in order to determine stability properties of equilibria.

It is important to realize that Theorem 8.2 (a) implies only local asymptotic stability of the equilibrium of the nonlinear system. Recall that Definition 3.3 (b) describes a local property.

Example 8.2 Consider the system $\dot{x} = Ax + g(x)$, where

$$A = \begin{bmatrix} -1 & 3 & 0 \\ -2 & -1 & 0 \\ 0 & 0 & -2 \end{bmatrix}, \quad g(x_1, x_2, x_3) = \begin{bmatrix} x_1^2 + x_2^2 + x_3 x_2^2 \\ x_3^2 + x_1 x_2 x_3 \\ x_1 x_3 \end{bmatrix}.$$

It is straightforward to verify that $Dg(0) = 0$, and the eigenvalues of A are $\lambda_1 = -3$ and $\lambda_{2,3} = -1 \pm i\sqrt{6}$. Therefore the origin is (at least locally) asymptotically stable by Theorem 8.2 part (a). △

We now define the *basin of attraction* of an asymptotically stable equilibrium point.

Definition 8.4 (Basin of Attraction)
Suppose that $x = 0$ is an asymptotically stable equilibrium of (8.1). *The* basin of attraction *of $x = 0$ is the set of all points x_0 such that*

$$\phi(t, x_0) \to 0 \quad \text{as } t \to \infty.$$

The next example emphasizes the local nature of asymptotic stability as determined from the linearization.

Example 8.3 Let $b > 0$ be a fixed constant, and consider the scalar equation

$$\dot{x} = -bx + x^3 = -x(b - x^2).$$

The equilibrium solutions are $x = 0$ and $x = \pm\sqrt{b}$. The linearization at the origin is $\dot{z} = -bz$; thus the eigenvalue of the linearization is $-b$. By Theorem 8.2 (a), the origin is asymptotically stable. Since solutions with small $|x(0)|$ remain bounded in forward time, they exist for all $t \geq 0$. Further, it is straightforward to check that the basin of attraction of $x = 0$ is the open interval $(-\sqrt{b}, \sqrt{b})$. This is so because, as noted, any initial condition with $|x(0)| < \sqrt{b}$ yields a solution that remains bounded and is thus defined for all forward time. Then one can argue as in Theorem 8.1 (b), using $V(x) = x^2$, to conclude that the solution must converge to the origin as $t \to \infty$. Note that the basin of attraction therefore depends on the nonlinear term. Finally, note that the equilibria $x = \pm\sqrt{b}$ are unstable. (See Exercise 8.3.) △

This is a convenient place to characterize the stability properties of scalar differential equations $\dot{x} = f(x)$. Let $f(0) = 0$, and suppose that the lowest order derivative of f which is nonzero at the origin is the k-th derivative, so that $f^{(k)}(0) \neq 0$ and $f^{(j)}(0) = 0$ for $1 \leq j < k$. If $f^{(k+1)}$ is defined throughout some open set about $x = 0$, then by Taylor's theorem (Theorem B.6) we may write

$$f(x) = ax^k + \frac{f^{(k+1)}(\xi)}{(k+1)!}x^{k+1}, \tag{8.8}$$

where $a := f^{(k)}(0)/k! \neq 0$ and, for each x near zero, the point $\xi = \xi(x)$ satisfies $0 < |\xi| < |x|$. If $f^{(k+1)}$ is bounded on a neighborhood of $x = 0$, then there is an $M > 0$ such that $|f^{(k+1)}(\xi)| \leq M$ for ξ near zero. We have the following proposition.

Proposition 8.1 *Consider the scalar equation $\dot{x} = f(x)$ with $f(0) = 0$. Assume that f satisfies* (8.8) *with $k \geq 1$ and that $f^{(k+1)}$ is bounded on a neighborhood of $x = 0$. The following statements are true:*

(a) *If k is odd and $a < 0$, then the equilibrium $x = 0$ is asymptotically stable.*
(b) *If k is odd and $a > 0$, then $x = 0$ is unstable.*
(c) *If k is even, then $x = 0$ is unstable.*

Proof. See Exercise 8.4. □

We note that there are differential equations with solutions that exhibit very slow approach to the origin, even slower than those of $\dot{x} = -x^k$, with $k > 0$, k odd. An example, taken from [32] (page 284), is the scalar equation $\dot{x} = f(x)$, where

$$f(x) = -x^3 \exp(-x^{-2}) \quad \text{for } x \neq 0, \quad \text{and} \quad f(0) = 0;$$

the general solution is

$$x(t, x_0) = \left[\ln(\exp(x_0^{-2}) + 2t)\right]^{-\frac{1}{2}}, \quad \text{for } x_0 \neq 0.$$

8.1.2 Stabilization from the Jacobian Linearization

The next result is a very important corollary of Theorem 8.2, and it may justify the use of a linear system model in many applications.

Corollary 8.1 (Stabilization from the Jacobian Linearization)
Consider the system $\dot{x} = f(x) + g(x)u$, where $f(0) = 0$ and f and g are C^1 on a neighborhood of the origin.

(a) *If the matrix pair $(A, B) := (Df(0), g(0))$ is stabilizable, and K is such that $A+BK$ is Hurwitz, then the linear feedback $u = Kx$ makes the origin an asymptotically stable equilibrium for the closed loop system $\dot{x} = f(x) + g(x)Kx$.*

(b) *If the pair $(A, B) := (Df(0), g(0))$ is not stabilizable, then the origin of the nonlinear system is not stabilizable by any C^1 feedback $u = \alpha(x)$ satisfying $\alpha(0) = 0$.*

Proof. See Exercise 8.7. □

To emphasize that the linear feedback of Corollary 8.1 is only locally stabilizing for the nonlinear system, we can keep a simple scalar example in mind (as in Example 8.3).

Example 8.4 Consider the scalar equation

$$\dot{x} = x + x^3 + u.$$

With $u = 0$, there is an unstable equilibrium at $x = 0$. The linearization at the origin is $\dot{x} = x + u$, and the linearized system is asymptotically stabilized by setting $u = -2x$. This same feedback applied to the nonlinear system gives $\dot{x} = -x + x^3$, and the origin has basin of attraction equal to the open interval $(-1, 1)$. △

In many applications, it is important to estimate the size of the basin of attraction of the equilibrium, since the basin tells us how far the initial condition can be from equilibrium and still correspond to a solution that approaches the equilibrium asymptotically. The basin may thus represent a region of safe or effective operation for the dynamic process being modeled.

Obtaining an estimate of the basin of attraction often requires ingenuity and experimentation with different Lyapunov functions as well as numerical exploration of the phase portrait. In general, if A is Hurwitz, then one might estimate the basin of attraction for $x = 0$ in a system of the form $\dot{x} = Ax + g(x)$, $Dg(0) = 0$, by solving the Lyapunov equation

$$A^T P + PA = -Q,$$

where Q is positive definite, yielding a quadratic Lyapunov function $V(x) = x^T Px$. Different choices of Q give different solutions for P, and this may result in additional information on the region of attraction for the nonlinear system, by taking the set union of the different estimates for the basin.

The negative result in Corollary 8.1 (b) deals with feedback $u = \alpha(x)$ having $\alpha(0) = 0$. This is a good place to mention the following consideration: If $f(0) = 0$ and $g(0) = B = 0$, then the linearization pair $(A, B) := (Df(0), g(0))$ is not stabilizable; however, this situation does not rule out the possibility of stabilizing the origin by a continuous feedback $u = \alpha(x)$ with $\alpha(0) \neq 0$.

Example 8.5 For the planar single-input system $\dot{x} = Ax + g(x)u$ defined by

$$\dot{x}_1 = (2x_1 + x_2)u,$$
$$\dot{x}_2 = x_2 - x_1 u,$$

the linearization pair (A, b) at the origin is not stabilizable, since $b := g(0) = 0$. By Corollary 8.1 (b), the origin cannot be stabilized by a C^1 feedback $u = \alpha(x_1, x_2)$ with $\alpha(0, 0) = 0$. However, a constant control $u \equiv u_0$ gives a closed loop linear system having eigenvalues

$$\lambda = \frac{1 + 2u_0}{2} \pm \frac{1}{2}\sqrt{1 - 4u_0}.$$

Therefore the eigenvalues are negative real for any number $u_0 < -2$. A similar phenomenon occurs with any system

$$\dot{x} = Ax + f_2(x) + [Gx + G_2(x)]u$$

that satisfies $Df_2(0) = 0$ and $DG_2(0) = 0$, and for which there exists a constant $u_0 \neq 0$ such that $A + Gu_0$ is Hurwitz. △

8.1.3 Brockett's Necessary Condition

We note that part (a) of Corollary 8.1 is an important result which provides a *sufficient* condition for local stabilization of an equilibrium of a nonlinear system. The main goal of this subsection is to state an important *necessary* condition for local asymptotic stabilization by a smooth feedback control. This condition is often relatively easy to check.

Consider a control system of the form

$$\dot{x} = f(x, u), \tag{8.9}$$

where $f : U \times V \subseteq \mathbf{R}^n \times \mathbf{R}^m \to \mathbf{R}^n$ is C^1 and U, V are open sets in $\mathbf{R}^n$, $\mathbf{R}^m$, respectively. The system is locally asymptotically stabilized to x_0 by the feedback $u = k(x)$, where $k : U \to \mathbf{R}^m$ is a C^1 mapping of an open neighborhood U of $x = x_0$, if the system

$$\dot{x} = f(x, k(x)) \tag{8.10}$$

has an asymptotically stable equilibrium at x_0.

For an n-dimensional linear system, $\dot{x} = Ax + Bu$, the PBH stabilizability test implies, in particular, that stabilizability of (A, B) entails the condition rank $[A\ B] = n$. This rank condition says that the linear mapping $(x, u) \mapsto Ax + Bu$ must be onto $\mathbf{R}^n$, and for a linear mapping, being onto the range space is equivalent to the condition that the range contains an open neighborhood of the origin. The latter formulation is Brockett's Necessary Condition for smooth (C^1) stabilization of the nonlinear system (8.9).

Theorem 8.3 (R. W. Brockett's Necessary Condition)
Suppose that the C^1 system (8.9) *is locally asymptotically stabilizable to an equilibrium x_0 by C^1 state feedback. Then the image of the mapping $f : U \times V \to \mathbf{R}^n$ contains an open neighborhood of the point x_0.*

Proof. A complete proof appears in [91] (pp. 253–255). □

We note that the system of Exercise 6.11 satisfies Brockett's necessary condition, but it is not stabilizable by any smooth feedback. See also Exercise 8.7.

8.1.4 Examples of Critical Problems

What about the cases not covered by Theorem 8.2? The remaining cases are exactly those cases for which Re $v\lambda \le 0$ for all eigenvalues λ of $Df(0)$, while at least one eigenvalue has a zero real part. As noted earlier, these cases are called *critical problems* of stability.

For linear systems, we have seen that eigenvalues with zero real part preclude asymptotic stability, but may allow Lyapunov stability of the equilibrium. There is a case we should single out as rather special. Recall that an eigenvalue of a matrix A is called simple if the geometric multiplicity of the eigenvalue equals the algebraic multiplicity of that eigenvalue. Thus, for simple eigenvalues the maximum number of linearly independent eigenvectors is the same as the algebraic multiplicity. Consider a linear system $\dot{x} = Ax$, with Re $(\lambda) \le 0$ for every eigenvalue of A. If every eigenvalue with zero real part is simple, then the origin $x = 0$ is Lyapunov stable for the linear system. (See Exercise 3.13.) On the other hand, the existence of an eigenvalue of A that has zero real part and is not simple implies that

$x=0$ is an unstable equilibrium of $\dot{x}=Ax$. The simplest illustration of this fact is provided by the system $\dot{x}_1 = x_2$, $\dot{x}_2 = 0$. Even in a case of simple eigenvalues with zero real part, however, a nonlinear perturbation $g(x)$ (with $g(0) = 0$) may render the origin of the system $\dot{x} = Ax + g(x)$ stable, unstable, or asymptotically stable depending on the perturbation $g(x)$. (For examples, see Exercise 2.10.) For nonlinear systems, the stability properties of the equilibrium $x = 0$, in a case where some eigenvalue(s) of $A = Df(0)$ have zero real part, cannot be determined without considering the nonlinear terms of the vector field. In critical problems for nonlinear systems, the linearization simply fails to decide stability issues for equilibria.

The next two examples deal with the nonlinear spring equation $m\ddot{x}+b\dot{x}+kx^3=0$, where the mass m, damping constant b, and spring coefficient k are nonnegative constants. The restoring force of the spring, when $k > 0$, is a cubic nonlinear term which models a *soft* spring having a slower response to small displacements from equilibrium than would occur with a linear Hooke's law restoring force.

Example 8.6 We set the damping constant $b = 0$, and consider the nonlinear spring without damping, $m\ddot{x} + kx^3 = 0$. We take $m = k = 1$ for this example. With $x_1 = x$ and $x_2 = \dot{x}$, the system is

$$\begin{aligned}\dot{x}_1 &= x_2,\\ \dot{x}_2 &= -x_1^3.\end{aligned}$$

The Jacobian linearization at $(0,0)$ is

$$\dot{x} = \begin{bmatrix} 0 & 1 \\ 0 & 0 \end{bmatrix} x,$$

and $A = Df(0)$ has only one linearly independent eigenvector. The zero eigenvalue is not simple, and, as we noted above, the origin is actually unstable for the linearization. Theorem 8.2 does not apply to the nonlinear spring without damping. However, we can determine Lyapunov stability for the origin by using a conservation of energy argument. The spring force is conservative, that is, $\ddot{x} = -x^3 = -\frac{d\phi}{dx}(x)$, where $\phi(x) = \phi(x_1) = \frac{1}{4}x_1^4$. Our Lyapunov function candidate is the total energy given by

$$V(x_1, x_2) = \frac{1}{4}x_1^4 + \frac{1}{2}x_2^2 = \text{potential energy} + \text{kinetic energy},$$

which is a positive definite function. A calculation using the chain rule shows that energy is conserved along any solution curve $(x_1(t), x_2(t))$:

$$\frac{d}{dt}V(x_1(t), x_2(t)) = x_1^3 x_2 + x_2(-x_1^3) = 0.$$

Therefore stability of the origin follows from Theorem 8.1 (a). However, the origin is not asymptotically stable, because trajectories are trapped within closed level curves of the energy V; see Exercise 8.8. △

Example 8.6 can be extended as follows. Suppose $p, q \in \mathbf{R}^n$. If the origin is an equilibrium for a Hamiltonian system,

$$\begin{aligned}\dot{q} &= \frac{\partial H}{\partial p}(q,p),\\ \dot{p} &= -\frac{\partial H}{\partial q}(q,p),\end{aligned}$$

where $H(q,p)$ is a positive definite function, then the conservation of the Hamiltonian energy $H(q,p)$ along trajectories shows that $H(q,p)$ is a Lyapunov function guaranteeing the stability of the origin, by Theorem 8.1 (a). (We have written $\frac{\partial H}{\partial p}(q,p)$ for the Jacobian of $H(q,p)$ with respect to the p coordinates, and similarly $\frac{\partial H}{\partial q}(q,p)$ is the Jacobian with respect to the q coordinates.)

The nonlinear spring with damping provides another illustration of Theorem 8.1 (a), as the next example shows.

Example 8.7 Again we take $m = 1$, $b = 1$, and $k = 1$ for the mass, friction constant, and spring coefficient, respectively. The system for the nonlinear spring with damping is then

$$\begin{aligned}\dot{x}_1 &= x_2,\\ \dot{x}_2 &= -x_1^3 - x_2.\end{aligned}$$

Again let V be the total energy given by $V(x_1, x_2) = \frac{1}{4}x_1^4 + \frac{1}{2}x_2^2$, and compute the rate of change of V along a trajectory to get

$$\dot{V} = \frac{d}{dt}V(x_1(t), x_2(t)) = x_1^3 x_2 + x_2(-x_1^3 - x_2) = -x_2^2 \leq 0.$$

By Theorem 8.1 (a), we conclude (merely) stability of the origin. Note that $\dot{V} = 0$ at all points of the x_1 axis. Experience with real springs leads us to believe that the damping should result in asymptotic stability of the origin. It is important to realize that the hypothesis of Theorem 8.1 (b) does not hold with the V we have chosen above. And Theorem 8.2 does not apply. We still want to show that this particular model actually reflects real experience with damped springs; hence, we want to establish rigorously that the origin is an asymptotically stable equilibrium of the model.

In order to show that the origin is asymptotically stable, we require either (i) a special argument for this example (an ad hoc argument), or (ii) another theorem. We give the special argument here, and save the important additional theorem (the invariance theorem) for the next section.

Suppose that $(x_1(t), x_2(t))$ is a nonzero solution. If we could show that $V(x_1(t), x_2(t))$ is a *strictly* decreasing function of t for $t > 0$, then, since V is C^1, we would actually have the situation dealt with in the *proof* of Theorem 8.1 (b); this would be sufficient to establish asymptotic stability. In order to show that $V(x_1(t), x_2(t))$ is strictly decreasing, we have to show that a trajectory that hits the set $\{x : \dot{V}(x) = 0\}$ (the x_1-axis), say at time $t = \hat{t}$, must cross the axis transversely and therefore cannot remain on that axis throughout a time interval $[\hat{t}, \hat{t} + \delta)$ with $\delta > 0$. Here is a geometric argument. The fact that a trajectory must cross the x_1-axis transversely follows from the fact that the velocity vector has the form $[\dot{x}_1 \quad \dot{x}_2]^T = [0 \quad -x_1^3]^T$ at the intersection point with the axis. On the one hand, if $x_1 > 0$ at the intersection point, then the tangent vector to the trajectory points vertically down into the fourth quadrant. On the other hand, if $x_1 < 0$ at intersection, then the tangent vector points vertically upward into the second quadrant. Therefore the solution curve crosses the axis transversely at a single point, and hence spends no positive time on the x_1-axis. In other words, $V(x_1(t), x_2(t))$ is a strictly decreasing function of t for $t > 0$, as we wished to show. Alternatively, here is an analytic argument, by contradiction. Suppose a nonzero solution intersects and remains on the x_1-axis during a time interval $[\hat{t}, \hat{t} + \delta)$ where $\delta > 0$. Then, during that time period, $x_2(t) \equiv 0$. Now argue from the system equations that this leads to a contradiction: Note that x_1, and thus $\dot{x}_2$, must be nonzero at intersection time $\hat{t}$, and by continuity, both are also nonzero on some positive time interval $[\hat{t}, \hat{t} + \tilde{\delta})$ where $\tilde{\delta} \leq \delta$; but this contradicts $x_2(t) \equiv 0$ for $t \in [\hat{t}, \hat{t} + \delta)$. Again, we conclude that $V(x_1(t), x_2(t))$ is a strictly decreasing function of t for $t > 0$. △

With the discussion of Example 8.7 in mind, we can imagine more complicated systems $\dot{x} = f(x)$ for which a candidate Lyapunov function V satisfies $L_f V(x) \leq 0$, but the set $\{\, x : L_f V(x) = 0\}$ is complicated to the point where it is too difficult to make special arguments for the strict decrease of $V(x(t))$. In some cases, it may be possible to use a different Lyapunov function V for which we can determine that $L_f V(x)$ is negative definite near the origin. (See Exercise 8.9 for one example, and Exercise 8.11 for another look at the nonlinear spring of Example 8.7.) However, a general result is needed, one which allows for wide application, ideally based on a simple principle. The answer to this need has proven to be the invariance theorem of the next section.

8.2 THE INVARIANCE THEOREM

Since it can be difficult to find Lyapunov functions that are strictly decreasing, a general result is needed to address such cases. The desired general result is usually called the invariance theorem and is stated below as Theorem 8.4. We will need some additional definitions and a few preliminary results to establish the invariance theorem.

Recall the Lie derivative notation $L_f V(x) = \nabla V(x) \cdot f(x)$ for the directional derivative. We will study cases involving a possibly nontrivial set $\{x : \dot{V}(x) = 0\} = (L_f V)^{-1}(0) = \{x : \nabla V(x) \cdot f(x) = 0\}$. Stated in simplest terms, the invariance theorem deals with the situation of a nontrivial set $(L_f V)^{-1}(0)$ for a Lyapunov function V by guaranteeing that trajectories with initial condition close to the equilibrium must asymptotically approach a subset of $(L_f V)^{-1}(0)$ consisting of complete solution curves of the system. For the nonlinear spring of Example 8.7, the only such subset of the x_1-axis is $\{0\}$, the set consisting of the origin itself; thus, in this case, the invariance theorem says that trajectories must approach the origin. Note the shift in emphasis from Example 8.7: Instead of trying to show a strict decrease of $V(x_1(t), x_2(t))$, the invariance theorem guarantees a conclusion about asymptotic behavior of trajectories if we can identify the largest set of complete trajectories contained in the set $\{x : \dot{V}(x) = 0\} = (L_f V)^{-1}(0)$, and this is often an easier task.

We now develop some important properties of ω-limit sets needed for the invariance theorem.

Lemma 8.1 *Suppose $\phi(\cdot, x_0) : [0, \infty) \to \mathbf{R}^n$ is a forward solution of $\dot{x} = f(x)$, and let $\omega(\phi) = \omega(x_0)$ be its ω-limit set. Then the following statements are true:*

(a) *$\omega(x_0)$ is forward and backward invariant under the flow of $\dot{x} = f(x)$.*
(b) *$\omega(x_0)$ is a closed set.*
(c) *$\omega(x_0) = \emptyset$ (the empty set) if and only if $\|\phi(\cdot, x_0)\| \to \infty$ as $t \to \infty$.*

Proof. The proof of (b) and (c) is left to Exercise 8.10. We prove (a), which is the most important property of $\omega(x_0)$ used later on. Thus, suppose that $q \in \omega(x_0)$. Then there exists $t_k \to \infty$ as $k \to \infty$ such that $\phi(t_k, x_0) \to q$ as $k \to \infty$. For any real t for which the solution $\phi(t, q)$ is defined, we have $\phi(t+t_k, x_0) = \phi(t, \phi(t_k, x_0)) \to \phi(t, q)$ as $k \to \infty$, by continuity of the time t map in the second (space) variable. It follows that $\phi(t, q) \in \omega(x_0)$, for each real t for which $\phi(t, q)$ is defined. Therefore $\omega(x_0)$ is invariant. □

We note that statements (b) and (c) of Lemma 8.1 are true for any function $\phi : [0, \infty) \to \mathbf{R}^n$, not just solutions of differential equations.

The next lemma is important in the development of the invariance theorem. Recall that a function (or curve) $x : [0, \infty) \to \mathbf{R}^n$ *approaches* the set $M \subset \mathbf{R}^n$ if

$$\lim_{t \to \infty} \operatorname{dist}(x(t), M) = 0.$$

Lemma 8.2 *Let $\phi : [0, \infty) \to \mathbf{R}^n$ be a forward solution of $\dot{x} = f(x)$ and $\omega(\phi)$ its ω-limit set. If ϕ is bounded, then $\omega(\phi)$ is nonempty and compact, and ϕ approaches $\omega(\phi)$.*

Proof. If ϕ is constant, then the conclusions of statement (a) clearly hold. If ϕ is not constant, then, since ϕ is continuous and bounded, the range of ϕ must be a bounded interval; hence, the Bolzano-Weierstrass theorem implies that $\omega(\phi)$ is nonempty. It is easy to see that $\omega(\phi)$ must be bounded.

Since $\omega(\phi)$ is closed by Lemma 8.1 (b), $\omega(\phi)$ is compact in $\mathbf{R}^n$. From the definition of $\omega(\phi)$, it is easy to see that ϕ approaches $\omega(\phi)$. □

We can now state and prove the invariance theorem. Note that the V function in this theorem is not assumed to be positive definite near the equilibrium.

Theorem 8.4 (Invariance Theorem)
Let $V(x)$ be a C^1, real-valued function, and assume that $f(0) = 0$ in (8.1). Suppose that there exists a value of c such that the set

$$O_c := \{x : V(x) < c\}$$

is a bounded subset of the domain D of f. Suppose further that

$$L_f V(x) \le 0 \quad \text{for } x \in O_c.$$

Then every solution of $\dot{x} = f(x)$ with initial condition $x(0) = x_0 \in O_c$ remains in O_c for $t \ge 0$ and, as $t \to \infty$, approaches the largest invariant set M contained in the set

$$E := \{x \in O_c \,:\, L_f V(x) = 0\,\}.$$

If $M = \{0\}$ then the origin is asymptotically stable.

Proof. Assume c is given as in the hypothesis. Choose in O_c an initial condition $x_0 \neq 0$, and let $\phi(t) := \phi(t, x_0)$ be the corresponding forward solution. By the hypotheses on V, the function $V(\phi(t))$ is nonincreasing for as long as the solution is defined. In fact, $\phi(t)$ must be defined for all $t \ge 0$ by the assumption that O_c is a bounded subset of the domain D of f. By hypothesis, O_c is a forward invariant set for the system. Since V is continuous and O_c is contained in a compact subset of D,

$$\lim_{t\to\infty} V(x(t)) =: m$$

exists and $m < c$. If q is a point in the limit set $\omega(\phi)$, then $V(q) = m$ as well. It follows that $\omega(\phi) \subset O_c$. Since $\omega(\phi)$ is invariant, $\frac{\partial V}{\partial x}(q)f(q) = 0$ for q in $\omega(\phi)$. Hence, $\omega(\phi) \subset E$, and since $\omega(\phi)$ is an invariant set, it must be contained in M. Since $\phi(t)$ is bounded for $t \ge 0$, we may conclude that $\phi \to M$ as $t \to \infty$. This proves the first conclusion of the theorem statement.

If $M = \{0\}$, then, by the previous arguments, the origin is attractive for all trajectories starting in O_c. The stability of the origin follows from the boundedness of O_c and the fact that $L_f V(x) \le 0$ for x in O_c. (Since V is continuous, V attains a minimum on the closure of O_c; then the argument of Theorem 8.1 (a) can be applied to $V + k$ for a constant k such that $V(x)+k$ is positive definite near equilibrium.) Hence, $x = 0$ is asymptotically stable. □

Several remarks on Theorem 8.4 may be useful.

As noted, the function V in Theorem 8.4 need not be positive definite. Part of the conclusion of the theorem is that the set O_c is contained in the basin of attraction of the origin. In this way, the invariance theorem helps in estimating a basin of attraction.

The condition $M = \{0\}$ is equivalent to the statement that no solution other than $x = 0$ remains in set E for all $t \geq 0$. If it happens that V is positive definite and $L_f V(x)$ is negative definite throughout O_c, then $E = \{0\} = M$, and we conclude that the origin is asymptotically stable via Theorem 8.1 (b). But Theorem 8.4 applies in many situations where the statement of Theorem 8.1 (b) does not apply. Simple examples include the cubic nonlinear spring model (see Example 8.8 below) and the nonlinear pendulum with friction.

Theorem 8.4 says that bounded solutions $\phi(t)$ approach M, in the sense that

$$\text{dist}\,(\phi(t), M) \to 0 \quad \text{as } t \to \infty.$$

Recall this means that for every $\epsilon > 0$ there is an $N > 0$ such that, if $t \geq N$, then $\text{dist}\,(\phi(t), M) < \epsilon$. The main point of Theorem 8.4 is that we may not know anything specific about the limit set $\omega(\phi)$, but with a judicious choice of the function V, we may at least find information on invariant sets that contain $\omega(\phi)$.

Example 8.8 The system for the nonlinear spring with damping from Example 8.7 is

$$\begin{aligned} \dot{x}_1 &= x_2, \\ \dot{x}_2 &= -x_1^3 - x_2. \end{aligned}$$

With the total mechanical energy given by $V(x_1, x_2) = \frac{1}{4}x_1^4 + \frac{1}{2}x_2^2$, we have $\dot{V}(x) = -x_2^2 \leq 0$, and the set where $\dot{V}(x) = 0$ is the x_1-axis, that is, $\{x : x_2 = 0\}$. It is straightforward to check that the level sets $V = c > 0$ are closed curves, and that the sublevel sets defined by $V(x) \leq c$ are bounded. By the invariance theorem, every solution $x(t)$ approaches the largest invariant set M contained in the x_1-axis. However, a trajectory which starts and remains on the x_1-axis must satisfy $\dot{x}_1 = 0$ for all $t \geq 0$, so $x_1(t) = k$, a constant. But then the equation for x_2 says that $0 = \dot{x}_2 = -x_1^3$, so $x_1 = 0$. Therefore $M = \{0\}$, and it follows that the equilibrium at the origin is asymptotically stable. (See also Exercise 8.11.) $\triangle$

We now address the important issue of boundededness for the sublevel sets defined by

$$\Omega_c = \{x : V(x) \leq c\}. \tag{8.11}$$

Definition 8.5 (Proper Mapping)
Let $V : D \subset \mathbf{R}^n \to \mathbf{R}^m$. V is called proper *(or a* proper mapping*) if the preimage $V^{-1}(K) = \{x : V(x) \in K\}$ is compact for each compact set K.*

Note that if V is continuous, then for compact K the preimage $V^{-1}(K)$ is closed, so a hypothesis that a continuous V is proper means that $V^{-1}(K)$ is bounded for each compact K.

If V is positive definite, then the set Ω_c in (8.11) is bounded for sufficiently small $c > 0$. (See Proposition E.1 in Appendix E.) However, if V is not positive definite, then Ω_c may not be bounded; simple examples are given by the functions $V(x_1, x_2) = x_1 x_2$ and $V(x_1, x_2) = (x_2 - x_1)^2$. If $V : D \to \mathbf{R}$ is positive definite and proper, then the preimage $V^{-1}([0, a]) = \{x : V(x) \le a\}$ is compact for each $a \ge 0$. If $\frac{\partial V}{\partial x}(x) f(x) \le 0$, then the sets $V^{-1}([0, a])$ are compact forward invariant sets for the system $\dot{x} = f(x)$.

The next definition is convenient.

Definition 8.6 (Radially Unbounded Function)
Let $V : \mathbf{R}^n \to \mathbf{R}$. V *is called* radially unbounded *if* $V(x) \to \infty$ *as* $\|x\| \to \infty$.

If $V : \mathbf{R}^n \to [0, \infty)$ is continuous and radially unbounded, then, for each c, the closed set $\Omega_c = \{x : V(x) \le c\}$ is also bounded; hence, V is proper. Conversely, if $V : \mathbf{R}^n \to [0, \infty)$ is continuous and proper, then V must be radially unbounded. (See Exercise 8.12.)

It is worth noting that the behavior of a positive definite function along the coordinate axes alone does not reveal whether the function is radially unbounded. A good example to keep in mind is the positive definite function $V : \mathbf{R}^2 \to \mathbf{R}$ given by

$$V(x_1, x_2) = \frac{x_1^2 + x_2^2}{1 + x_1^2 + x_2^2} + (x_2 - x_1)^2.$$

We have $V(x_1, 0) \to \infty$ as $|x_1| \to \infty$, and $V(0, x_2) \to \infty$ as $|x_2| \to \infty$, but V is not radially unbounded. Along the line $x_2 = x_1$, as $\|x\| \to \infty$ we have $V(x) \le 1$.

The next result is often useful in deducing global asymptotic stability of an equilibrium.

Theorem 8.5 *Suppose that the system* $\dot{x} = f(x)$ *is defined on* $D = \mathbf{R}^n$, $f(0) = 0$, *and the* C^1 *function* $V(x)$ *is defined on* $D = \mathbf{R}^n$, *is radially unbounded and satisfies*

$$\begin{aligned} V(x) &> 0 \quad \text{for } x \neq 0, \\ L_f V(x) &\le 0 \quad \text{for } x \text{ in } \mathbf{R}^n. \end{aligned}$$

Suppose that the largest invariant set M *contained in the set* $\{x : L_f V(x) = 0\}$ *is* $\{0\}$. *Then the origin is globally asymptotically stable.*

Proof. By the hypotheses on V, for each $c > 0$, the set $\Omega_c = \{x : V(x) \le c\}$ is bounded and forward invariant. Thus, we may take any of the sets Ω_c as a bounded domain, and apply Theorem 8.4. Hence, the origin is asymptotically stable; moreover, each of the sets Ω_c, for $c > 0$, is contained in the basin of attraction. Finally, since V is radially unbounded, every initial condition is in the interior of a sublevel set Ω_c for some $c > 0$. □

Returning to Example 8.8 on the cubic nonlinear spring with damping, we see that the V function there is radially unbounded, and the example is covered by the global result in Theorem 8.5. Thus, for the cubic nonlinear spring with damping, the origin is globally asymptotically stable.

Consider again the assumptions of Theorem 8.5. Dropping only the hypothesis that V is radially unbounded, it is possible to have $V(x) > 0$ and $L_f V(x) \leq 0$ for all $x \neq 0$, and $M = \{0\}$, yet the origin is not globally asymptotically stable; see Exercise 8.20.

8.3 BASIN OF ATTRACTION

This section presents some general facts about invariant sets, including the basin of attraction of an asymptotically stable equilibrium point (Definition 8.4).

Lemma 8.3 *The closure of any invariant set (forward, backward, or both) for* (8.1) *is an invariant set of the same type.*

Proof. Let M be an invariant set. Choose the time interval $\mathcal{I}$ according to the invariance type of M : $[0, \infty)$, $(-\infty, 0]$, or $(-\infty, \infty)$. We must show that the closure of M is invariant of the same type. If $x \in M$, then by hypothesis the trajectory $\phi(t, x)$ is in M for all $t \in \mathcal{I}$. For a boundary point x of M, with $x \notin M$, let x_n be a sequence of points in M that converges to x, and consider the solutions $\phi(t, x_n)$ with $\phi(0, x_n) = x_n$. Fix $t \in \mathcal{I}$. By continuous dependence on initial conditions, we have

$$\lim_{n \to \infty} \phi(t, x_n) = \phi(t, x).$$

Hence, $\phi(t, x)$ is in the closure of M for each $t \in \mathcal{I}$. □

Observe that an open invariant set has an invariant closure, by Lemma 8.3. Therefore the boundary of an open invariant set must itself consist of whole solution trajectories. This simple observation leads to the next result.

Theorem 8.6 *The basin of attraction of an asymptotically stable equilibrium point of* (8.1) *is an open set. Hence, the boundary of the basin of attraction consists of whole solution trajectories.*

Proof. Suppose x_0 is an asymptotically stable equilibrium of $\dot{x} = f(x)$. The basin of attraction M is clearly forward invariant; thus, by the remark just before the theorem statement, we only need to show that the basin is an open set. By the definition of asymptotic stability, there exists $\delta > 0$ such that the open ball $B_\delta(x_0)$ is contained in M. Suppose $x \in M$, and consider the forward trajectory of x, $\phi(t, x)$. There exists $T > 0$ sufficiently large that $\|\phi(t,x)-x_0\|<\frac{\delta}{2}$ for $t\geq T$. Invoking the continuous dependence of solutions on initial conditions, choose $\delta_1>0$ such that $\|y-x\|<\delta_1$ implies $\|\phi(T,y)-\phi(T,x)\|<\frac{\delta}{2}$. By the triangle inequality, it is immediate that $\phi(T,y)\in B_\delta(x_0)\subset M$. Hence, the forward trajectory $\phi(t,\phi(T,y))=\phi(t+T,y)$, $t\geq 0$, converges to x_0 and therefore y is in M. This shows that $B_{\delta_1}(x)\subset M$. Since x was an arbitrary point in M, we conclude that M is open. □

We consider a simple example with an estimate of the basin.

Example 8.9 One form of the time-reversed Van der Pol system is given by

$$\dot{x}_1 = x_2 + \left(\frac{x_1^3}{3} - x_1\right),$$
$$\dot{x}_2 = -x_1.$$

There is only one equilibrium point, at the origin, and it is asymptotically stable; this can be deduced from the linearization. Take

$$V(x_1,x_2)=\frac{1}{2}(x_1^2+x_2^2).$$

Then

$$\dot{V}(x_1,x_2)=x_1\dot{x}_1+x_2\dot{x}_2=x_1\left(\frac{x_1^3}{3}-x_1\right)=x_1^2\left(\frac{x_1^2}{3}-1\right).$$

Therefore $\dot{V}(x_1,x_2)\leq 0$ if $x_1^2\leq 3$. Hence, within the disk $x_1^2+x_2^2<3$, we have $\dot{V}(x_1,x_2)\leq 0$, and we conclude that this disk lies within the basin of attraction. For this time-reversed Van der Pol system, there is known to be a periodic solution that encloses the origin. Solutions starting near this periodic orbit move away from it in forward time. The periodic orbit is therefore called an *unstable limit cycle*. By the calculation above, this periodic solution must lie outside the disk $x_1^2+x_2^2<3$. △

Further examples of basins of attraction appear later on within the text and in exercises.

Another important property of the basin of attraction is that it is contractible to a point. Writing M for the basin, contractibility of M to the point z means that there exists a continuous mapping

$$H:[0,1]\times M\to M$$

such that $H(0,x)=x$ for all x and $H(1,x)=z$ for all x. In fact we may take z to be the attracting equilibrium point, $z=x_0$. For details on contractibility of the basin, see [91] (page 251).

8.4 CONVERSE LYAPUNOV THEOREMS

Theorem 8.1 presented two direct Lyapunov theorems, one for stability and one for asymptotic stability; those results assert that certain stability properties are implied by the existence of Lyapunov functions with specific properties. A *converse Lyapunov theorem* is a result which asserts that some stability property implies the existence of a Lyapunov function V with specific properties. There are many versions of converse Lyapunov theorems; see [32] (Chapter VI).

Consider the system

$$\dot{x} = f(x), \tag{8.12}$$

where x is in $\mathbf{R}^n$, $f : W \subseteq \mathbf{R}^n \to \mathbf{R}^n$ is locally Lipschitz, and $f(x_0) = 0$ for some x_0 in W.

We first state a converse theorem for asymptotic stability, including both a local and a global result.

Theorem 8.7 (J. L. Massera) (Converse Theorem for Asymptotic Stability)

(a) *If the equilibrium x_0 of* (8.12) *is asymptotically stable, then there exist a neighborhood U of x_0 and a C^1 Lyapunov function $V : U \to \mathbf{R}$ such that $L_f V(x) < 0$ for all $x \neq x_0$ in U.*

(b) *If x_0 is globally asymptotically stable, then there exists a C^1, radially unbounded Lyapunov function $V : \mathbf{R}^n \to \mathbf{R}$.*

We also have a converse theorem for Lyapunov stability.

Theorem 8.8 (K. P. Persidskii) (Converse Theorem for Stability)
If the equilibrium x_0 of (8.12) *is Lyapunov stable, then there exist a neighborhood U of x_0 and a function $V : U \times [0, \infty) \to \mathbf{R}$, continuously differentiable in (x, t), such that $\dot{V}(x,t) := \frac{\partial V}{\partial x}(x,t)f(x) + \frac{\partial V}{\partial t}(x,t) \leq 0$ for all $x \neq x_0$ in U.*

The converse theorem for Lyapunov stability guarantees only a time-dependent Lyapunov function $V(x,t)$ such that $\dot{V}(x,t) \leq 0$ along solutions. In general, a Lyapunov function independent of t need not exist; see [32] (page 228). In attempting to verify Lyapunov stability for an equilibrium x_0, it is *sufficient* to find a function $V(x)$ such that $\frac{\partial V}{\partial x}(x)f(x) \leq 0$ in a neighborhood of x_0 (Theorem 8.1 (a)).

8.5 EXERCISES

Exercise 8.1 *Lagrange's Principle*
A scalar second order equation is called *conservative* if it can be written in the form $\ddot{x} = -\frac{d\phi}{dx}(x)$, where $\phi : \mathbf{R} \to \mathbf{R}$. The function ϕ is called the

potential function. Assume that the potential function ϕ is C^2, $\frac{d\phi}{dx}(0) = 0$, and $\frac{d^2\phi}{dx^2}(0) \neq 0$.

(a) With the stated assumptions on ϕ, show that if ϕ has a relative minimum at $x = 0$, then $(x, \dot{x}) = (0, 0)$ is a stable equilibrium of the system.
(b) Show that part (a) applies to the pendulum system without friction, $\ddot{x} + \sin x = 0$, and to the cubic nonlinear spring system, $\ddot{x} + x^3 = 0$.
(c) State conditions on the function $g(x)$ such that part (a) applies to the equation $\ddot{x} + g(x) = 0$.
(d) With the stated assumptions on ϕ, show that if ϕ has a relative maximum at $x = 0$, then $(x, \dot{x}) = (0, 0)$ is an unstable equilibrium of the system.

Exercise 8.2 Suppose g is C^1 in a neighborhood of the origin, $g(0) = 0$, and $Dg(0) = 0$. Use the mean value theorem (Theorem B.5) to show that for any $\epsilon > 0$ there is a $\delta > 0$ such that $\|x\| < \delta$ implies $\|g(x)\| \leq \epsilon \|x\|$.

Exercise 8.3 Consider the equation $\dot{x} = -bx + x^3$ in Example 8.3.

(a) Verify that the equilibria $x = \pm\sqrt{b}$ are unstable. Do this by a phase line analysis of the sign of $\dot{x}$, and also as an application of Theorem 8.2.
(b) Show that this differential equation is not forward complete.

Exercise 8.4 Prove Proposition 8.1.

Exercise 8.5 *Limit of a Function and ω-limit Points*
Let $\psi : [0, \infty) \to \mathbf{R}$ be a bounded function. Show that $\lim_{t\to\infty} \psi(t) = 0$ if and only if the ω-limit set of ψ equals $\{0\}$.

Exercise 8.6 *Stability Properties and Local Coordinate Transformations*
Consider a system $\dot{x} = f(x)$ with $f(0) = 0$. Let $z = T(x)$ be a coordinate transformation where $T(0) = 0$, $T(x)$ is C^1 and locally invertible in a neighborhood of $x = 0$, and the local inverse T^{-1} is also C^1. The transformed system is

$$\dot{z} = \frac{\partial T}{\partial x}(T^{-1}(z)) f(T^{-1}(z)) =: F(z).$$

(a) Show that $x = 0$ is an isolated equilibrium of $\dot{x} = f(x)$ if and only if $z = 0$ is an isolated equilibrium of $\dot{z} = F(z)$.
(b) Show that $x = 0$ is a stable (asymptotically stable, unstable) equilibrium of $\dot{x} = f(x)$ if and only if $z = 0$ is a stable (asymptotically stable, unstable) equilibrium of $\dot{z} = F(z)$.

Exercise 8.7 *On Stabilization from the Linearization*

(a) Prove parts (a) and (b) of Corollary 8.1.

(b) Use Corollary 8.1 (b) to show that the origin of the single-input planar system

$$\begin{aligned}\dot{x}_1 &= 4x_1 + x_2^2 u,\\ \dot{x}_2 &= x_2 + u\end{aligned}$$

cannot be asymptotically stabilized by a C^1 state feedback $\alpha(x)$ with $\alpha(0) = 0$. Does this system satisfy Brockett's necessary condition for stabilization at the origin?

Exercise 8.8 *On Example 8.6*

Consider the nonlinear spring without damping in Example 8.6.

(a) Show that the origin is not asymptotically stable, by showing that the level curves of positive total energy are closed curves surrounding the origin.

(b) Answer the following questions: How do we know that solutions of this system are defined for all $t \geq 0$? How do we know that nonzero solutions are periodic?

Exercise 8.9 *Different Lyapunov Functions*

Consider the linear spring equation with damping, $\ddot{x} + \dot{x} + kx = 0$, and write the corresponding system with $x_1 = x$ and $x_2 = \dot{x}$.

(a) Show that the total mechanical energy

$$V(x_1, x_2) = \frac{1}{2}kx_1^2 + \frac{1}{2}x_2^2$$

is a Lyapunov function with $\dot{V}(x_1, x_2) \leq 0$ and $\dot{V}(x_1, x_2) = 0$ on the line $x_2 = 0$.

(b) Let $k = 2$. Find a positive definite function $V(x_1, x_2) = \alpha x_1^2 + \beta x_1 x_2 + \gamma x_2^2$ such that $\dot{V}(x_1, x_2) < 0$ for $(x_1, x_2) \neq (0, 0)$.

Exercise 8.10 Complete the details of the proofs for Lemma 8.1 (b), (c).

Exercise 8.11 *Nonlinear Spring Again*

Consider the nonlinear spring equation with damping, $\ddot{x} + \dot{x} + x^3 = 0$, and the corresponding system with $x_1 = x$ and $x_2 = \dot{x}$ (Example 8.8). Let $V(x_1, x_2) = ax_1^4 + bx_1^2 + cx_1x_2 + dx_2^2$, and show that a, b, c, d may be chosen such that

$$\dot{V}(x_1, x_2) = -x_1^4 - x_2^2 < 0 \quad \text{for } (x_1, x_2) \neq (0, 0).$$

Verify that the resulting V is positive definite.

Exercise 8.12 Suppose $V : \mathbf{R}^n \to [0, \infty)$ is continuous. Show that V is proper if and only if V is radially unbounded.

Exercise 8.13 Consider the nonlinear system

$$\begin{aligned}\dot{x}_1 &= x_2,\\ \dot{x}_2 &= x_1 - x_1^3.\end{aligned}$$

(a) Show that this system is Hamiltonian.
(b) Show that the origin is unstable.
(c) Find all the other equilibria and determine their stability properties.
(d) Sketch the global phase portrait of the system, showing the local behavior near each equilibrium. *Hint on global behavior*: There are two complete solution curves γ_1, γ_2, each corresponding to a solution that approaches the origin as $t \to \infty$ and as $t \to -\infty$. Therefore the set union $\gamma_1 \cup \gamma_2 \cup \{0\}$ is an invariant set; it is both the global stable manifold and the global unstable manifold for the origin.

Exercise 8.14 Let $k > 0$. Show that the origin of the planar system

$$\begin{aligned}\dot{x} &= -x + xz,\\ \dot{z} &= -kz\end{aligned}$$

is globally asymptotically stable.

Exercise 8.15 *Estimating a Basin*
The origin of the planar system

$$\begin{aligned}\dot{x} &= -x + x^2 z,\\ \dot{z} &= -z\end{aligned}$$

is asymptotically stable. Obtain an estimate of the basin of attraction of the origin using only a quadratic Lyapunov function of your choice.

Exercise 8.16 Show that the origin of the planar system

$$\begin{aligned}\dot{x} &= z - x^3,\\ \dot{z} &= -x - z^3\end{aligned}$$

is globally asymptotically stable.

Exercise 8.17 *Estimating Another Basin*
Suppose $a > 0$ and $b > 0$. For the equation

$$\ddot{x} + a\dot{x} + 2bx + 3x^2 = 0,$$

estimate the basin of attraction of the equilibrium $(x, \dot{x}) = (0, 0)$ using only the total mechanical energy of the system as a Lyapunov function.

Exercise 8.18 *Asymptotic Stabilization via the ARE: (C, A) Observable* Assume the pair (A, B) is stabilizable. Use the invariance theorem to show that if (C, A) is observable and P is the unique positive semidefinite solution of the algebraic Riccati equation (7.21), then the closed loop coefficient matrix $A - BR^{-1}B^T P$ is Hurwitz.

Exercise 8.19 Consider the system

$$\dot{x}_1 = x_2^3,$$
$$\dot{x}_2 = -x_2.$$

(a) Show that the origin presents a critical problem of stability.
(b) Show that the function $V(x_1, x_2) = \frac{1}{2}(x_1^2 + x_2^2)$ is a Lyapunov function for $(0, 0)$ in a neighborhood of the origin. Describe the set where $\dot{V}(x_1, x_2) = 0$. Can you rule out asymptotic stability of $(0, 0)$, and if so, how?
(c) Replace the first equation with $\dot{x}_1 = -x_2^3$, and repeat (a) and (b); are the results the same or different?

Exercise 8.20 Let $(x_1, x_2) \in \mathbf{R}^2$, and let $u := 1 + x_1^2$. Consider the system

$$\dot{x}_1 = -\frac{6x_1}{u^2} + 2x_2,$$
$$\dot{x}_2 = -\frac{2x_1 + 2x_2}{u^2}.$$

(a) Show that $V(x_1, x_2) := \frac{x_1^2}{u} + x_2^2 = \frac{x_1^2}{1+x_1^2} + x_2^2$ is not radially unbounded and the level curves $V(x_1, x_2) = c$ are not bounded for $c > 1$.
(b) Show that the largest invariant set M in the set $\{(x_1, x_2) : \dot{V}(x_1, x_2) = 0\}$ is $M = \{0\}$.
(c) Show that, for points on the hyperbola defined by $x_2 = 2/(x_1 - \sqrt{2})$, the tangent slope of solutions given by $\dot{x}_2/\dot{x}_1$ is larger negative than the slope of the tangent line to the hyperbola, provided $x_1 > \sqrt{2}$. Conclude that the origin is not globally asymptotically stable.

8.6 NOTES AND REFERENCES

For more on phase plane analysis of two-dimensional autonomous systems, see [21], [41]. For analysis, see [7].

An English translation of A. M. Lyapunov's original memoir on stability is available in [73]. The proof of the instability result in Theorem 8.2 (b) follows E. A. Coddington and N. Levinson, *Theory of Ordinary Differential Equations*, McGraw-Hill, New York 1955 (pages 317–318). The relation between the full nonlinear system $\dot{x} = f(x)$ and its linearization is not limited to the stability results of Theorem 8.2. See also the Hartman-Grobman theorem [25] (page 311), which says that, for hyperbolic equilibria, the phase portrait of the linearized system is homeomorphic (topologically equivalent) to the local phase portrait of the nonlinear system in a neighborhood of the equilibrium. A *hyperbolic equilibrium* is one for which each eigenvalue of the linearization at equilibrium has nonzero real part.

Two instability theorems which do not depend on linearization appear in [15](Theorems 5.3–5.4, page 195, with the Theorem 5.3 proof on pages 207–208).

The invariance theorem (Theorem 8.4) is due to [8]; the presentation here is influenced in part by the statement of the result in [69]. The basic facts about ω-limit sets are from [33].

The general facts on invariant sets and the basin of attraction are from [32]. The discussion of Van der Pol's equation in Example 8.9 is adapted from [33] (pages 317–318); see also [50]. As mentioned in the text, for contractibility of the basin, see [91] (page 251).

The converse theorem for asymptotic stability (Theorem 8.7) is from [74]; the converse theorem for stability (Theorem 8.8) is taken from [32] (page 226) where it is attributed to K. P. Persidskii, Über die Stabilität nach der ersten Näherung, *Mat. Sbornik*, 40 : 284–293, 1933 (in Russian, German summary).

Exercise 8.17 is taken from [33] (page 318). The system in Exercise 8.20 is from [32] (page 109).

The books [25], [33], and [81] all have detailed material on the implicit function theorem, the inverse function theorem, the contraction mapping theorem, and their application in deducing properties of solutions of differential equations.

Lyapunov functions have played a role in the study of long-term memory in neuronal networks; see [103] (Chapter 14: Lyapunov Functions and Memory).

Chapter Nine

Cascade Systems

In this chapter we consider interconnected systems in *cascade* form

$$\dot{x} = f(x, z), \tag{9.1}$$

$$\dot{z} = g(z), \tag{9.2}$$

where $x \in \mathbf{R}^n$, $z \in \mathbf{R}^m$, $f(0,0) = 0$, $g(0) = 0$, and $f : \mathbf{R}^n \times \mathbf{R}^m \to \mathbf{R}^n$ and $g : \mathbf{R}^m \to \mathbf{R}^m$ are locally Lipschitz. Equation (9.2) defines the *driving system* of the cascade, while (9.1) is the *driven system* of the cascade. We will study the stability properties of the equilibrium at $(x, z) = (0, 0)$ for (9.1)–(9.2), under some natural assumptions on the stability properties of the *decoupled subsystems* defined by $\dot{x} = f(x, 0)$ and $\dot{z} = g(z)$. For local properties, the functions f and g are defined on some open subsets of $\mathbf{R}^n \times \mathbf{R}^m$ and $\mathbf{R}^m$ containing the respective origins.

We will sometimes use the abbreviation LAS in reference to the local asymptotic stability property, and GAS in reference to global asymptotic stability. In broad overview, here are the main results of the chapter:

- If the equilibrium $x = 0$ of the subsystem $\dot{x} = f(x, 0)$ is LAS and the equilibrium $z = 0$ of $\dot{z} = g(z)$ is LAS, then the equilibrium $(x, z) = (0, 0)$ of the cascade connection (9.1)–(9.2) is also LAS.
- If the equilibrium $x = 0$ of $\dot{x} = f(x, 0)$ and GAS and the equilibrium $z = 0$ of $\dot{z} = g(z)$ is GAS, then it need not be the case that the equilibrium $(x, z) = (0, 0)$ of the cascade is GAS. Global asymptotic stability of $(x, z) = (0, 0)$ for (9.1)–(9.2) generally requires additional conditions beyond the GAS assumption for the subsystem $\dot{x} = f(x, 0)$.

9.1 THE THEOREM ON TOTAL STABILITY

Consider a system of the form

$$\dot{x} = f(x), \tag{9.3}$$

where $x \in \mathbf{R}^n$, $f(0) = 0$, and f is locally Lipschitz. The idealization of a dynamic process represented by the model (9.3) is likely to ignore the effects of some small disturbances, such as those caused by friction in mechanical joints, heat build-up in circuits, or inexact measurements in the modeling process leading to $f(x)$. We will assume that the perturbed system has the form

$$\dot{x} = f(x) + g(x,t), \tag{9.4}$$

where $g(x,t)$ is the perturbation term. We will consider perturbations $g(x,t)$ that are piecewise continuous in t and Lipschitz in x in a neighborhood of $x=0$. Under these conditions, system (9.4) has solutions that depend uniquely on initial conditions. The Lipschitz condition is that there exists a constant $L > 0$ such that

$$\|g(x,t) - g(\hat{x},t)\| \leq L\|x - \hat{x}\| \tag{9.5}$$

for all $t \geq 0$ and all x, $\hat{x}$ in a neighborhood U of $x = 0$. We write $\phi(t, t_0, x_0)$ for the general solution of (9.4) with $\phi(t_0, t_0, x_0) = x_0$.

The perturbations modeled by $g(x,t)$ may involve some uncertainty, and may even prevent the persistence of an equilibrium at the point $x = 0$, so it may not be reasonable to assume that $g(0,t) = 0$ for all t. We still want to express the idea that solutions of the perturbed equation (9.4) with small $\|x(0)\|$ have essentially "stable equilibrium-like" behavior, provided the perturbation remains sufficiently small in norm for small $\|x\|$.

Example 9.1 Consider the scalar equation

$$\dot{x} = -x + \beta \sin(x+t),$$

defined for $(t,x) \in \mathbf{R}^2$, where $\beta \neq 0$ may be a function of (t,x) which is known to be bounded, say $|\beta| \leq \bar{\beta}$. The nominal, or unperturbed, equation, $\dot{x} = -x$, has an asymptotically stable equilibrium at $x = 0$. Note that $x = 0$ is not an equilibrium for the perturbed equation. Since $x = 0$ is asymptotically stable when $\beta = 0$, we might expect that if the amplitude $|\beta|$ is sufficiently small then the solution $x(t)$ will remain close to the origin provided $|x(0)|$ is sufficiently small. A direct calculation using the variation of parameters formula shows that any solution $x(t)$ must remain bounded in forward time, and therefore each solution $x(t)$ is defined for all $t \geq 0$. In fact, using simple estimates based on variation of parameters, it can be shown that

$$|x(t)| \leq |x(0)| + \bar{\beta} \quad \text{for all } t \geq 0.$$

Given $\epsilon > 0$, suppose there exists $\delta_1(\epsilon) > 0$ and $\delta_2(\epsilon) > 0$ such that

$$|x_0| \leq \delta_1(\epsilon) < \frac{\epsilon}{2},$$

and

$$|g(x,t)| \leq |\beta| \leq \delta_2(\epsilon) < \frac{\epsilon}{2} \quad \text{for all } |x| \leq \epsilon \text{ and all } t \geq 0.$$

Then, once again by estimates based on variation of parameters, one can show that

$$|x(t)| \leq |x(0)| + |\beta| \leq \frac{\epsilon}{2} + \frac{\epsilon}{2} = \epsilon.$$

(See Exercise 9.1). Thus, solutions remain small in norm provided the initial condition is small in norm and the perturbation term remains small for small $|x|$. This is the type of persistence of stability-like behavior that we want to express by a general concept. △

See also Exercise 9.2 on a perturbed scalar system and solution bounds.

The general concept we need is expressed in the definition of the *total stability* concept. There are variations on the total stability concept [32], but the following definition is appropriate for our use.

Definition 9.1 (Total Stability)
The equilibrium $x = 0$ of (9.3) is called totally stable *if for each $\epsilon > 0$, there exist numbers $\delta_1(\epsilon) > 0$ and $\delta_2(\epsilon) > 0$ such that, if*

$$\|x_0\| \leq \delta_1(\epsilon)$$

and

$$\|g(x,t)\| \leq \delta_2(\epsilon) \quad \textit{for all } \|x\| < \epsilon \textit{ and all } t \geq 0,$$

then the solution $\phi(t, 0, x_0)$ of (9.4) with $\phi(0, 0, x_0) = x_0$ satisfies

$$\|\phi(t, 0, x_0)\| \leq \epsilon \quad \textit{for all} \quad t \geq 0.$$

In [32] (page 275) the class of allowable perturbations is taken to be all perturbations $g(x,t)$ such that (9.4) has a solution for each initial condition x_0 near $x = 0$, and the solution exists for $t \geq 0$; time varying nominal systems $\dot{x} = f(x,t)$ are considered as well.

It is not difficult to see that if the origin is merely Lyapunov stable for the nominal system (9.3), then we cannot expect it to be totally stable. If the (origin of the) driven system of a cascade is not totally stable, then we cannot guarantee Lyapunov stability of the (origin of the) cascade; see Exercise 9.3. However, a general result on total stability can be proved under the assumption that the nominal system has an asymptotically stable equilibrium. The next theorem is called the *theorem on total stability*.

Theorem 9.1 (Total Stability)
If the equilibrium $x = 0$ of (9.3) is asymptotically stable, then it is totally stable.

Proof. We invoke the converse Lyapunov Theorem 8.7, and let $\bar{U}$ be an open neighborhood of $x = 0$ and $V : \bar{U} \to \mathbf{R}$ a C^1 positive definite function such that

$$L_f V(x) = \frac{\partial V}{\partial x}(x) f(x) < 0 \quad \text{for } x \in \bar{U}.$$

Since V is C^1, there exist a neighborhood U of $x = 0$ and a number $M > 0$ such that

$$\left\| \frac{\partial V}{\partial x}(x) \right\|_2 \leq M \quad \text{for } x \in U \cap \bar{U}.$$

Let $\epsilon > 0$ be fixed and sufficiently small that $B_\epsilon(0) \subset U \cap \bar{U}$. Writing

$$\Omega_c := \{x : V(x) \leq c\},$$

we may choose $c > 0$ such that, if $x \in \Omega_c$ then $x \in B_\epsilon(0) \subset U \cap \bar{U}$. By shrinking the neighborhood $U \cap \bar{U}$ further, if necessary, we may assume that $g(x,t)$ satisfies (9.5) on $U \cap \bar{U}$; thus, we may choose $\delta_2 = \delta_2(\epsilon) > 0$ such that $\|x\| < \epsilon$ and $t \geq 0$ imply $\|g(x,t)\|_2 \leq \delta_2(\epsilon)$; moreover, if x is a point on the boundary of Ω_c (that is, $V(x) = c$), then

$$\frac{\partial V}{\partial x}(x)[f(x) + g(x,t)] = L_f V(x) + \frac{\partial V}{\partial x}(x) g(x,t) \leq L_f V(x) + M\,\delta_2(\epsilon) < 0.$$

Now let $x_0 = \phi(0,0,x_0)$ be an initial condition in the interior of Ω_c. Then the corresponding solution $x(t) := \phi(t,0,x_0)$ of (9.4) is defined for all $t \geq 0$ and remains in Ω_c for all $t \geq 0$; consequently, $\|x(t)\| \leq \epsilon$ for all $t \geq 0$. Now choose $\delta_1 = \delta_1(\epsilon)$ sufficiently small that $\|x_0\| \leq \delta_1(\epsilon)$ implies x_0 is in the interior of Ω_c. This completes the proof. □

9.1.1 Lyapunov Stability in Cascade Systems

As an application of Theorem 9.1, we now address Lyapunov stability for an equilibrium of a cascade system. It is worth noting that simple examples show that "stability plus stability (for the decoupled subsystems) does not imply stability for the cascade"; see Exercise 9.3. By requiring asymptotic stability of $x = 0$ for $\dot{x} = f(x,0)$, we obtain the following result on stability of the origin of the combined cascade system.

Corollary 9.1 (Stability in a Cascade)
Suppose the equilibrium $x = 0$ of $\dot{x} = f(x,0)$ is asymptotically stable and the equilibrium $z = 0$ of $\dot{z} = g(z)$ is stable. Then the equilibrium $(x,z) = (0,0)$ of the cascade (9.1)–(9.2) is stable.

Proof. Given $z_0 = z(0)$, the solution $z(t,z_0)$ of the driving system is completely determined. By hypothesis, for any $\epsilon > 0$ there exists a $\delta(\epsilon) > 0$ such that, if $z(0) = z_0$ satisfies $\|z_0\| < \delta$, then the solution $z(t,z_0)$ satisfies $\|z(t,z_0)\| < \epsilon$ for all $t \geq 0$. Express the differential equation for x as

$$\dot{x} = f(x, z(t)) = f(x,0) + g(x,t,z_0), \tag{9.6}$$

where

$$g(x,t,z_0) := f(x, z(t,z_0)) - f(x,0).$$

The perturbation term depends on z_0 as well as on (x,t), but since f is C^1, there exist constants $\eta > 0$ and $M > 0$ such that

$$\|g(x,t,z_0)\| \leq M\|z(t,z_0)\| \leq M\epsilon \quad \text{for all} \quad \|x\| < \eta,\ \|z_0\| < \delta,\ t \geq 0. \tag{9.7}$$

Note that the bound on $\|g(x,t,z_0)\|$ in (9.7) can be made arbitrarily small by choosing a sufficiently small $\delta > 0$ and $\|z_0\| < \delta$. Thus, the stability of $(x, z) = (0, 0)$ now follows by an argument as in the proof of Theorem 9.1. □

It is good to keep in mind a simple example.

Example 9.2 Consider the planar system

$$\dot{x} = -x^3 - xz,$$
$$\dot{z} = 0.$$

The hypotheses of Corollary 9.1 hold for this cascade. It is not difficult to see that the origin $(x, z) = (0, 0)$ is stable but not asymptotically stable, and it is instructive to sketch a phase portrait for the system. See Exercise 9.4. △

9.2 ASYMPTOTIC STABILITY IN CASCADES

This section addresses asymptotic stability in cascade systems. A few moments of thought will show that the following two conditions are necessary for asymptotic stability of the origin of the cascade (9.1)–(9.2):

(LASx) $x = 0$ is an asymptotically stable equilibrium for $\dot{x} = f(x, 0)$.

(LASz) $z = 0$ is an asymptotically stable equilibrium for $\dot{z} = g(z)$.

Thus, in this section we strengthen the hypothesis on the driving system (as compared with Corollary 9.1) from Lyapunov stability to asymptotic stability.

9.2.1 Examples of Planar Systems

We begin with a simple planar system.

Example 9.3 Consider the planar system

$$\dot{x} = -x + x^2 z,$$
$$\dot{z} = -z.$$

We have $f(x, z) = -x + x^2 z$ and $g(z) = -z$. The origin $x = 0$ is globally asymptotically stable for $\dot{x} = f(x, 0) = -x$ and $z = 0$ is globally asymptotically stable for $\dot{z} = -z$. We will show that $(x, z) = (0, 0)$ is locally (but not globally) asymptotically stable for the cascade. The local property follows from Jacobian linearization, of course. However, we want to consider the global property as well, and to do so in this first example it will be useful to consider the exact solutions.

Note that the z-axis is invariant, and for solutions off that axis we may use the change of variable $\eta = 1/x$, replacing the x equation by $\dot{\eta} = \eta - z(t)$, which can be solved explicitly. Since we have $z(t) = e^{-t}z_0$, we may use variation of parameters to write

$$\begin{aligned}
\eta(t) &= e^t \left(\eta_0 - \int_0^t e^{-s} z(s)\, ds \right) \\
&= e^t \left(\eta_0 - \int_0^t e^{-2s} z_0\, ds \right) \\
&= e^t \left(\eta_0 + \frac{1}{2} e^{-2t} z_0 - \frac{1}{2} z_0 \right) \\
&= e^t \eta_0 + \frac{1}{2} e^{-t} z_0 - \frac{1}{2} e^t z_0 .
\end{aligned}$$

From this expression for $\eta(t)$, it is straightforward to deduce that

$$x(t) = \frac{x_0}{e^t - \frac{x_0 z_0}{2}(e^t - e^{-t})} \tag{9.8}$$

for $x_0 \neq 0$, z_0 arbitrary, and $t \geq 0$ for as long as $x(t)$ is defined. In fact (9.8) is also valid when $x_0 = 0$. The behavior of $x(t)$ for all initial conditions not on the coordinate axes may be deduced from (9.8). We summarize as follows (see also Figure 9.1):

- For initial conditions on the hyperbola $x_0 z_0 = 2$, we have $x(t) = e^t x_0 \to \infty$ as $t \to \infty$. And since $z(t) = e^{-t} z_0$, it follows that each branch of this hyperbola is a complete trajectory of the system.
- Initial conditions in the region $x_0 z_0 < 2$, between the branches of the hyperbola $x_0 z_0 = 2$, guarantee that $x(t) \to 0$ as $t \to \infty$. In the second and fourth quadrants, where $x_0 z_0 < 0$, this is clear from (9.8). In the first and third quadrants, where $0 < x_0 z_0 < 2$, the denominator in (9.8) remains positive for $t \geq 0$ and approaches ∞ as $t \to \infty$.
- Initial conditions in the region $x_0 z_0 > 2$ correspond to solutions $x(t)$ that have a *finite positive escape time*; by definition, this means that there exists some $\hat{t} > 0$ such that $x(t)$ is defined on $[0, \hat{t})]$ and has the property that $|x(t)| \to \infty$ as $t \to \hat{t}$. (See Exercise 9.5.)

Note that for any $x_0 \neq 0$, even for arbitrarily small $|x_0|$, there exists $|z_0|$ large enough to cause unboundedness in $x(t)$, due to the interconnection term $x^2 z$. It is important to note that the decoupled subsystems are globally asymptotically stable to their origins, but the origin of the cascade is only locally asymptotically stable. △

In Example 9.3, we could have deduced the asymptotic stability of the origin from the fact that the linearization is Hurwitz. In the next example we cannot rely on linearization.

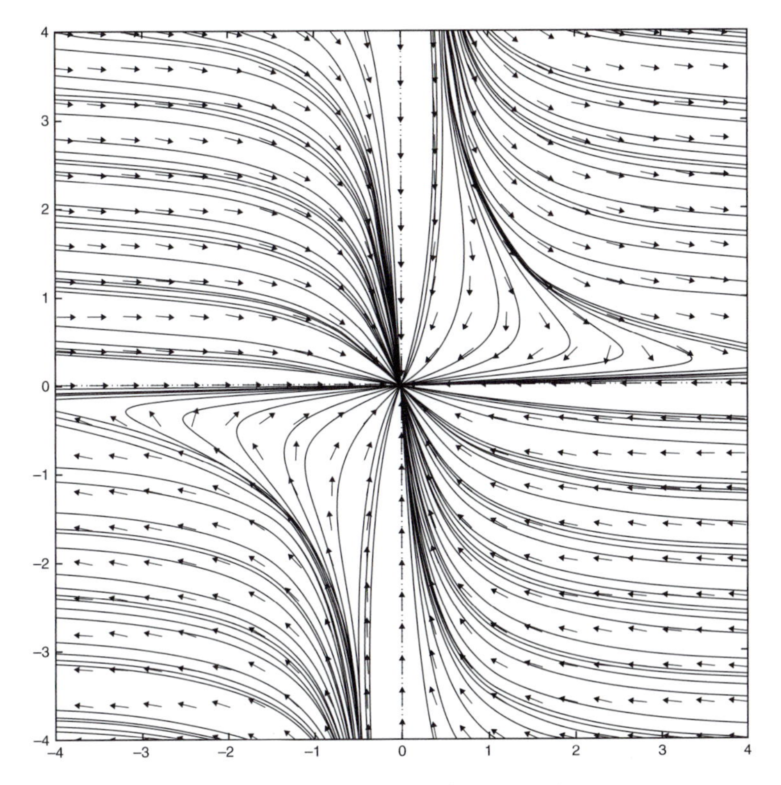

Figure 9.1 Phase portrait for Example 9.3.

Example 9.4 Consider the planar system

$$\dot{x} = -x^3 + x^2 z,$$
$$\dot{z} = -z + z^3.$$

There are five equilibria, at $(0,0)$, $(0,1)$, $(0,-1)$, $(1,1)$, and $(-1,-1)$. So the, origin cannot be globally asymptotically stable. It is not difficult to show that there is a forward invariant region of the form

$$\{(x,z) \, : \, |x| \le \epsilon, \; |z| \le \delta\}$$

for some $0 < \epsilon < 1$ and $0 < \delta < 1$. In fact, for a given solution $z(t)$ of the second equation, we have $\dot{x} = -x^3 + x^2 z(t)$, which presents us with the situation dealt with in the theorem on total stability. Given an ϵ in the range $0 < \epsilon < 1$, it is straightforward to verify that we may choose $\delta_1 = \epsilon$ and $\delta_2 = \epsilon^3/2$ and obtain

$$|x^2 z(t)| \le \frac{\epsilon^3}{2}$$

simply by choosing $|x_0| \le \epsilon$ and $|z_0| \le \epsilon/2$. For this example, it is not difficult to see that if $0 < \epsilon < 1$ is given, then the rectangle

$$S = \{(x,z) \, : \, |x| \le \epsilon, \; |z| \le \epsilon/2\}$$

is forward invariant; this can be verified by examining the slope field on the boundary segments of the rectangle. Thus, every forward trajectory in this rectangle has a nonempty, compact ω-limit set. Since the coordinate axes are invariant, if there were a periodic orbit in the rectangle it would have to be contained entirely within one of the four open quadrants. However, a careful look at the slope field rules out periodic orbits, because both $x(t)$ and $z(t)$ are monotonic in the indicated rectangle. Thus, for this planar system, we may invoke the Poincaré-Bendixson theorem (Theorem 9.3 in the Appendix to this chapter) to conclude that the ω-limit set of the forward trajectory of any initial condition in the rectangle consists of the origin alone. This proves that the origin is attractive; the stability of the origin follows from Corollary 9.1; thus the origin is asymptotically stable. We also have an estimate for the basin of attraction. $\triangle$

It must be noted that in dimensions greater than $n = 2$, there is no analogue of the Poincaré-Bendixson theorem. So we need a different, more general argument to deduce attractivity of the origin $(x, z) = (0, 0)$ when the decoupled subsystems are asymptotically stable to their respective origins.

9.2.2 Boundedness of Driven Trajectories

The next lemma is a consequence of total stability. It is useful because it establishes a forward invariant region enclosing $(x, z) = (0, 0)$, as we had in Example 9.4. First, some notation: If the decoupled subsystems are locally asymptotically stable to their origins, we write $\mathcal{A}_x$ for the basin of attraction of $x = 0$ for $\dot{x} = f(x, 0)$, and $\mathcal{A}_z$ for the basin of attraction of $z = 0$ for $\dot{z} = g(z)$.

Lemma 9.1 *Suppose that the decoupled subsystems are asymptotically stable to their respective origins, with basins of attraction $\mathcal{A}_x$ and $\mathcal{A}_z$. Then there exist numbers $\epsilon > 0$ and $\delta > 0$ such that the set*

$$S := \{(x, z) : \|x\| < \epsilon, \|z\| < \delta\} \tag{9.9}$$

has the following property: For any initial condition $(x_0, z_0) \in S$, $z(t) \to 0$ as $t \to \infty$, and the forward solution of $\dot{x} = f(x, z(t))$ with $x(0) = x_0$ is bounded and remains in $\mathcal{A}_x$ for all $t \geq 0$.

Proof. We may use the splitting in (9.6); let $\eta > 0$, $M > 0$, and $\delta > 0$ be such that the estimate (9.7) holds, where ϵ satisfies $0 < \epsilon < \eta$ and is such that $\{x : \|x\| < \epsilon\} \subset \mathcal{A}_x$. As noted, the bound on $\|g(x, t, z_0)\|$ in (9.7) can be made arbitrarily small by choosing a sufficiently small $\delta > 0$ and $\|z_0\| < \delta$. Thus we can keep $\|x(t)\| \leq \epsilon$ for all $t \geq 0$ by choosing δ sufficiently small and such that $\{z : \|z\| < \delta\} \subset \mathcal{A}_z$. $\square$

The next theorem shows that, if $z(t)$ is a solution satisfying $z(t) \to 0$ as $t \to \infty$, and if the forward solution $x(t)$ of $\dot{x} = f(x, z(t))$ is bounded and remains in the basin $\mathcal{A}_x$ of the origin for $\dot{x} = f(x, 0)$, then $x(t)$ must approach zero. This result applies even if $z = 0$ is not an asymptotically stable equilibrium for $\dot{z} = g(z)$.

Theorem 9.2 (E. D. Sontag)
Consider system (9.1)–(9.2) *and suppose that the equilibrium $x = 0$ of $\dot{x} = f(x, 0)$ is asymptotically stable. Let (x_0, z_0) be such that the solution $z(t)$ of $\dot{z} = g(z)$ satisfying $z(0) = z_0$ is defined for all $t \geq 0$ and $z(t) \to 0$ as $t \to \infty$, and the solution $x(t)$ of $\dot{x} = f(x, z(t))$ satisfying $x(0) = x_0$ is defined for all $t \geq 0$, is bounded on the interval $t \geq 0$, and is such that $x(t) \in \mathcal{A}_x$ for all $t \geq 0$. Then*

$$\lim_{t \to \infty} x(t) = 0.$$

Proof. As in the theorem on total stability, since the equilibrium $x = 0$ of $\dot{x} = f(x, 0)$ is locally asymptotically stable, we may split $f(x, z(t))$ as in (9.6) and use arguments similar to those used in the proof of Corollary 9.1 in order to conclude the following: Given any $\rho > 0$, we may choose $\delta_1(\rho) > 0$ and $\delta(\rho) > 0$ such that, if $\|x_0\| \leq \delta_1$ and $\|z(t)\| \leq \delta$ for all $t \geq 0$, then the solution $x(t)$ of the initial value problem

$$\dot{x} = f(x, z(t)), \quad x(0) = x_0$$

satisfies $\|x(t)\| \leq \rho$ for all $t \geq 0$.

Let $(x(t), z(t))$ be as in the statement of the theorem, with initial condition $(x(0), z(0)) = (x_0, z_0)$. In order to show that $x(t) \to 0$ as $t \to \infty$, we must show that for every $\rho > 0$, there exists $T = T(\rho) > 0$ such that

$$\|x(t)\| \leq \rho \quad \text{for all } t \geq T. \tag{9.10}$$

Thus, given ρ, we must show that it is possible to find a time $T > 0$ such that $\|x(T)\| \leq \delta_1$, and $\|z(t)\| \leq \delta$ for all $t \geq T$. Then the estimate from the preceding paragraph applies and (9.10) holds; and since ρ is arbitrary, the theorem will be proved.

To proceed with the proof, let $T_1 > 0$, set

$$z_{T_1}(t) = \begin{cases} z(t) & \text{if } t \leq T_1, \\ 0 & \text{if } t > T_1, \end{cases}$$

and let $x_{T_1}(t)$ denote the integral curve of

$$\dot{x} = f(x, z_{T_1}(t))$$

satisfying $x_{T_1}(0) = x_0$ (the same initial condition as $x(t)$). Indeed, $x_{T_1}(t) = x(t)$ for all $0 \leq t \leq T_1$. Moreover, for $t > T_1$, $x_{T_1}(t)$ is a solution of

$$\dot{x} = f(x, 0)$$

and hence tends to 0 as $t \to \infty$, because $x_{T_1}(T_1) = x(T_1)$ and, by hypothesis, $x(T_1)$ is in the basin $\mathcal{A}_x$. In particular, there exists $T_2 > 0$ such that

$$\|x_{T_1}(T_1 + T_2)\| \leq \frac{\delta_1}{2}. \tag{9.11}$$

The time T_2 may depend on $x_{T_1}(T_1)$ and hence on T_1; however, by hypothesis $x(t)$ is bounded for $t \geq 0$; hence, there exists a T_2, which depends only on the initial condition x_0, for which the inequality (9.11) holds for any T_1.

We proceed with T_2 having been chosen, and we set

$$T = T_1 + T_2.$$

Now the goal is to show that, if T_1 is large enough, then

(a) $\|z(t)\| \leq \delta$ for all $t \geq T$, and
(b) $\|x(T) - x_{T_1}(T)\| \leq \frac{\delta_1}{2}$.

This will conclude the proof, since the triangle inequality then implies that $\|x(T)\| \leq \delta_1$ and consequently (9.10) holds.

Property (a) is an immediate consequence of the fact that $z(t)$ converges to 0 as $t \to \infty$.

To prove (b), we proceed as follows. Observe that $x(t)$ and $x_{T_1}(t)$ are integral curves of

$$\dot{x} = f(x, z(t))$$

and

$$\dot{x} = f(x, 0),$$

respectively, satisfying $x(T_1) = x_{T_1}(T_1)$. Thus, for $t \geq T_1$,

$$x(t) = x(T_1) + \int_{T_1}^{t} f(x(s), z(s))\, ds$$

and

$$x_{T_1}(t) = x(T_1) + \int_{T_1}^{t} f(x_{T_1}(s), 0)\, ds\,.$$

Since $x(s)$, $x_{T_1}(s)$, and $z(s)$ are defined and bounded for $s \geq T_1$, and $f(x, z)$ is locally Lipschitz, there exist $L > 0$ and $M > 0$ such that

$$\|\, f(x(s), z(s)) - f(x_{T_1}(s), 0)\, \| \leq L\|x(s) - x_{T_1}(s)\| + M\|z(s)\|$$

for all $s \geq T_1$. Thus, by standard estimates,

$$\|x(t) - x_{T_1}(t)\| \leq L \int_{T_1}^{t} \|x(s) - x_{T_1}(s)\|\, ds + M \int_{T_1}^{t} \|z(s)\|\, ds\,.$$

Since $z(s) \to 0$ as $t \to \infty$, given any $\gamma > 0$, there is a $T_1 > 0$ such that $\|z(s)\| \leq \gamma$ for all $t \geq T_1$. Thus, for all $t \geq T_1$ we can write

$$\|x(t) - x_{T_1}(t)\| \leq L \int_{T_1}^{t} \|x(s) - x_{T_1}(s)\|\, ds + M\gamma(t - T_1)\,. \tag{9.12}$$

On the interval $T_1 \le t \le T = T_1 + T_2$, the last term on the right side of (9.12) may be replaced by a constant upper bound; thus,

$$\|x(t)-x_{T_1}(t)\| \le L \int_{T_1}^{t} \|x(s)-x_{T_1}(s)\|\, ds+\gamma M T_2 \quad \text{for } T_1 \le t \le T. \tag{9.13}$$

By Gronwall's inequality (Lemma C.2), we have

$$\|x(t) - x_{T_1}(t)\| \le \gamma M T_2 e^{L(t-T_1)} \quad \text{for } T_1 \le t \le T = T_1 + T_2.$$

Set $t = T$ to get

$$\|x(T) - x_{T_1}(T)\| \le \gamma M T_2 e^{L T_2}.$$

Note that M, L, and T_2 are fixed numbers; however, γ can be made arbitrarily small by choosing an appropriately large T_1, using only the convergence of $z(t)$ to zero. This proves statement (b) above, and completes the proof of the theorem. □

The next example emphasizes the generality of Theorem 9.2, in that the theorem does not require a hypothesis of asymptotic stability for the driving system of the cascade.

Example 9.5 Consider the planar system

$$\begin{aligned}\dot{x} &= -x^3 + xz,\\ \dot{z} &= z^2.\end{aligned}$$

Clearly, $z = 0$ is attracting only for $z_0 \le 0$. After an explicit solution of $\dot{z} = z^2$ for $z_0 < 0$, we have that $x(t)$ satisfies

$$\dot{x} = -x^3 + \frac{z_0}{1 - z_0 t}\, x.$$

By setting $V(x) = \frac{1}{2}x^2$ and computing that

$$\dot{V}(x) = x\dot{x} = -x^4 + \frac{z_0}{1 - z_0 t}\, x^2 \le 0 \quad \text{for } z_0 < 0 \text{ and } t \ge 0,$$

we conclude that $x(t)$ must be bounded for $t \ge 0$. Therefore Theorem 9.2 applies to any solution of this cascade with initial condition (x_0, z_0) for which $z_0 \le 0$. Note that the origin $(x, z) = (0, 0)$ is not Lyapunov stable. △

In the next section we apply Theorem 9.2 to get results on asymptotic stability.

9.2.3 Local Asymptotic Stability

As noted earlier, the conditions (LASx) and (LASz) are necessary for asymptotic stability of the origin of the cascade (9.1)–(9.2). The next result is a corollary of Theorem 9.2; it states that, taken in conjunction, (LASx) and (LASz) are sufficient for asymptotic stability of the origin of the cascade.

Corollary 9.2 (LAS in Cascades)
If the equilibrium $x = 0$ *of* $\dot{x} = f(x,0)$ *and the equilibrium* $z = 0$ *of* $\dot{z} = g(z)$ *are both asymptotically stable, then the equilibrium* $(x,z) = (0,0)$ *of the cascade* (9.1)–(9.2) *is asymptotically stable.*

Proof. By Corollary 9.1, the equilibrium $(x,z) = (0,0)$ is stable. Hence we only need to show attractivity of the origin. Consider a set S as in (9.9) of Lemma 9.1. Theorem 9.2 applies to the trajectory $(x(t), z(t))$ determined by any initial condition $(x_0, z_0) \in S$. Consequently, the origin is attractive, hence asymptotically stable, and S is contained in its basin of attraction. This completes the proof. □

We now consider some examples to illustrate the previous results.

Example 9.6 Consider the planar system

$$\begin{aligned} \dot{x} &= -x^3 + xz, \\ \dot{z} &= -z^3. \end{aligned}$$

The asymptotic stability of the origin follows from Corollary 9.2. △

Example 9.3 showed that even for converging $z(t)$, there were initial conditions in the basin of attraction of $\dot{x} = f(x,0)$ that corresponded to unbounded trajectories of the cascade. In that example, $\dot{x} = f(x,0)$ was globally asymptotically stable. In the next example, we see that initial conditions in the basin of attraction of a locally asymptotically stable origin for $\dot{x} = f(x,0)$ can be driven out of that basin by converging trajectories $z(t) \to 0$, and, in fact, the corresponding $x(t)$ (and hence $(x(t), z(t))$) can be unbounded.

Example 9.7 Consider the planar system

$$\begin{aligned} \dot{x} &= -x^3 + x^5 + xz^2, \\ \dot{z} &= -z - z^3. \end{aligned}$$

Note that the subsystem $\dot{x} = f(x,0) = -x^3 + x^5$ has an equilibrium at $x = 0$ which is LAS (but not GAS). Consequently, the origin $(0,0)$ cannot be GAS for the cascade. There are other ways to see that GAS of $(0,0)$ is ruled out; for example, $\dot{x} > 0$ for $x \geq 1$ and arbitrary z; alternatively, the existence of multiple equilibria, which occur at $(0,0)$ and at $(\pm 1, 0)$, also precludes GAS for the origin. The equilibria at $(\pm 1, 0)$ are unstable hyperbolic points; we note that the Jacobian matrix of the vector field at each of these points is

$$J_{(\pm 1,0)} = \begin{bmatrix} 2 & 0 \\ 0 & -1 \end{bmatrix}.$$

Among other facts, we show here that there are solutions $(x(t), z(t))$, having initial conditions (x_0, z_0) with $x_0 \in \mathcal{A}_x$, for which $x(t)$ can be driven out of that basin by converging trajectories $z(t) \to 0$.

First, since the basin of attraction of the origin is an open set, it certainly contains an open disk $x^2+z^2 < d^2$. We will obtain below a better description of the basin by calling on the Poincaré-Bendixson theorem (Theorem 9.3 in the Appendix to this chapter) and the stable manifold theory for hyperbolic equilibria.

The basin of attraction for $\dot{x} = f(x,0) = -x^3 + x^5$ is the interval $|x| < 1$. Now consider any initial condition of the form $(x(0), z(0)) = (1, z_0)$ with $z_0 \neq 0$. There is a disk about the point $(1, z_0)$ such that every trajectory starting in the disk crosses the line $x = 1$ and remains in the region $x > 1$ thereafter. In fact, for such initial conditions the forward trajectory $(x(t), z(t))$ must be unbounded. (If it remains bounded in forward time, then the ω-limit set of the trajectory is a nonempty, closed, bounded invariant set in the region $x > 1$. But $\dot{x} > 0$ for $x > 1$, and therefore the ω-limit set contains no equilibrium points or nontrivial periodic orbits in the region $x > 1$. But then the Poincaré-Bendixson theorem implies that the ω-limit set cannot be bounded, a contradiction. Therefore the forward trajectory $(x(t), z(t))$ must be unbounded for the stated initial conditions.) A similar analysis applies to trajectories that start in a neighborhood of any point $(-1, z_0)$ with $z_0 \neq 0$. From these arguments, we conclude that the basin must lie within the infinite strip $-1 \leq x \leq 1$.

Further insight, and a better estimate of the basin of attraction of the origin, is available by considering the global stable invariant manifold for each of the unstable hyperbolic equilibria at the points $(\pm 1, 0)$. These invariant manifolds must lie within the infinite strip $-1 \leq x \leq 1$, according to our argument above, and they are asymptotic to the z-axis for increasing $|z|$. In the present example, the basin of the origin must be bounded by the union of the global stable manifold of $(1,0)$ and the global stable manifold of $(-1,0)$. A phase portrait should help in visualizing the basin boundary, which consists of a total of six whole trajectories; three trajectories comprise each of these global stable manifolds; and the entire basin lies within the infinite strip $-1 \leq x \leq 1$. (See Figure 9.2.)

Thus, by stable manifold theory, the basin of attraction of the origin contains the entire z-axis, although the basin is very narrow along the extremes of the z-axis. We conclude that there are arbitrarily small initial conditions $|x_0|$ associated with unbounded trajectories of the cascade, because such x_0 may be paired with z_0 such that $|x(t)| \to \infty$ as $t \to \infty$. $\triangle$

See the Notes and References for more on stable manifold theory. The next example indicates the use of Corollary 9.2 for larger cascades as well.

Example 9.8 Consider the system

$$\begin{aligned}
\dot{x} &= -x^3 + x^2 z + w, \\
\dot{z} &= -z - z^3 + xzw, \\
\dot{w} &= -w^3 + w^5,
\end{aligned}$$

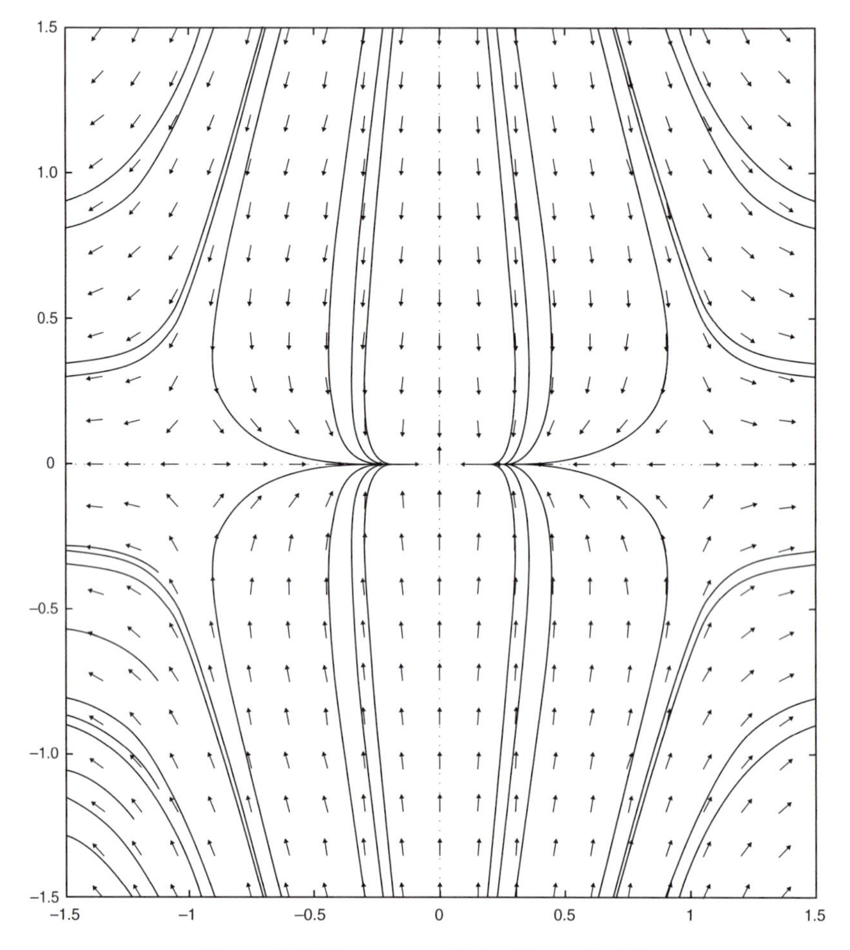

Figure 9.2 Phase portrait for Example 9.7.

with x, z, and w real. Note that when $w = 0$ the first two equations form the undriven subsystem

$$\begin{aligned}\dot{x} &= -x^3 + x^2 z,\\ \dot{z} &= -z - z^3.\end{aligned}$$

By Corollary 9.2, this planar subsystem has asymptotically stable origin $(x, z) = (0, 0)$. Since $w = 0$ is asymptotically stable for $\dot{w} = -w^3 + w^5$, we apply Corollary 9.2 once more to the full three-dimensional system, and conclude that the origin $(0, 0, 0)$ is asymptotically stable. It is not globally asymptotically stable because there are multiple equilibria. △

9.2.4 Boundedness and Global Asymptotic Stability

We now consider global asymptotic stability (GAS) of the cascade (9.1)–(9.2). Clearly, we must assume that the GAS property holds for the origin in each of the decoupled subsystems, $\dot{x} = f(x,0)$ and $\dot{z} = g(z)$. The next result emphasizes the advantage of the boundedness of driven trajectories, as a consequence of Theorem 9.2.

Corollary 9.3 *Suppose that the equilibrium $x = 0$ of $\dot{x} = f(x,0)$ is globally asymptotically stable and the equilibrium $z = 0$ of $\dot{z} = g(z)$ is globally asymptotically stable. Then every bounded forward solution of the cascade system (9.1)–(9.2) converges to $(x,z) = (0,0)$ as $t \to \infty$.*

Proof. First, local asymptotic stability of $(x,z) = (0,0)$ follows from Corollary 9.2. By the GAS hypothesis on the decoupled subsystems, for each initial condition (x_0, z_0), we have $z(t) \to 0$ as $t \to \infty$, and $x(t) \in \mathcal{A}_x = \mathbf{R}^n$ for all $t \geq 0$. Therefore Theorem 9.2 applies to every solution $(x(t), z(t))$ for which $x(t)$ remains bounded for $t \geq 0$, that is, every bounded forward solution $(x(t), z(t))$. Thus, every bounded forward solution converges to the origin. $\square$

The next example applies Corollary 9.3 to deduce global asymptotic stability in a cascade.

Example 9.9 Consider the system

$$\begin{aligned}
\dot{x} &= -x^3 + xz^2 w, \\
\dot{z} &= -z - zw^2 - zx^2 w, \\
\dot{w} &= -w^3.
\end{aligned}$$

Note that $w = 0$ is GAS for the last equation. When $w = 0$, the first two equations form the subsystem

$$\begin{aligned}
\dot{x} &= -x^3, \\
\dot{z} &= -z,
\end{aligned}$$

which has GAS origin $(x,z) = (0,0)$. By Corollary 9.2, we know that $(0,0,0)$ is LAS. In order to show that $(0,0,0)$ is GAS, it is sufficient, according to Corollary 9.3, to show that all solutions of the driven system given by

$$\begin{aligned}
\dot{x} &= -x^3 + xz^2 w(t), \\
\dot{z} &= -z - z[w(t)]^2 - zx^2 w(t),
\end{aligned}$$

remain bounded. The boundedness of these solutions is relatively easy to show. Let $V(x,z) = x^2 + z^2$ and compute that

$$\frac{d}{dt}V(x(t), z(t)) = 2x(t)\dot{x}(t) + 2z(t)\dot{z}(t) = -2x^4(t) - 2z^2(t) - 2z^2(t)[w(t)]^2 < 0$$

for $(x(t), z(t)) \neq (0,0)$ and all $w(t)$. Therefore every closed disk $x^2 + z^2 \leq C$ is forward invariant, and every solution $(x(t), z(t))$ of the driven system is bounded for $t \geq 0$. Hence, by Corollary 9.3, the origin $(0,0,0)$ is globally asymptotically stable. $\triangle$

Some additional examples of cascades appear in the exercises.

9.3 CASCADES BY AGGREGATION

In this section we show that an initial regrouping of equations can yield a cascade form to which the previous results apply. We begin with an example; afterward, we discuss the generality of the approach taken in the example.

Example 9.10 [99] Consider the system

$$\begin{aligned}\dot{z}_1 &= -z_1^3 + z_4,\\ \dot{z}_2 &= (1 - z_1)z_3 + z_1 z_2,\\ \dot{z}_3 &= -z_2 + (z_4 - 1)z_3,\\ \dot{z}_4 &= -z_1,\end{aligned}$$

where $z = (z_1, z_2, z_3, z_4) \in \mathbf{R}^4$. The Jacobian matrix of the vector field at the equilibrium $z = 0$ is easily verified to be

$$A = \begin{bmatrix} 0 & 0 & 0 & 1 \\ 0 & 0 & 1 & 0 \\ 0 & -1 & -1 & 0 \\ -1 & 0 & 0 & 0 \end{bmatrix}.$$

The eigenvalues of A are i, $-i$, and $-\frac{1}{2} \pm \frac{i\sqrt{3}}{2}$. Thus, we have a critical problem of stability for $z = 0$. The first and fourth equations define a decoupled subsystem, and the variables z_1 and z_4 act as inputs to a subsystem for z_2 and z_3 defined by the second and third equations. To make this more precise, and to conform with notation later in this section, we define

$$x_1 = [x_{11} \quad x_{12}]^T := [z_1 \quad z_4]^T \quad \text{and} \quad x_2 = [x_{21} \quad x_{22}]^T := [z_2 \quad z_3]^T;$$

then we rewrite the system as

$$\dot{x}_1 := \begin{bmatrix} \dot{x}_{11} \\ \dot{x}_{12} \end{bmatrix} = \begin{bmatrix} -x_{11}^3 + x_{12} \\ -x_{11} \end{bmatrix} =: f_1(x_1),$$

$$\dot{x}_2 := \begin{bmatrix} \dot{x}_{21} \\ \dot{x}_{22} \end{bmatrix} = \begin{bmatrix} (1 - x_{11})x_{22} + x_{11}x_{21} \\ -x_{21} + (x_{12} - 1)x_{22} \end{bmatrix} =: f_2(x_1, x_2).$$

When we set the driving variable x_1 equal to zero in the equations for x_2, we have the following decoupled subsystems:

$$\dot{x}_1 = \begin{bmatrix} \dot{x}_{11} \\ \dot{x}_{12} \end{bmatrix} = \begin{bmatrix} -x_{11}^3 + x_{12} \\ -x_{11} \end{bmatrix}, \tag{9.14}$$

$$\dot{x}_2 = \begin{bmatrix} \dot{x}_{21} \\ \dot{x}_{22} \end{bmatrix} = \begin{bmatrix} x_{22} \\ -x_{21} - x_{22} \end{bmatrix}. \tag{9.15}$$

Suppose we wish to investigate asymptotic stability of $z = 0$ via Corollary 9.2, for example. Note that (9.15) is a linear system with Hurwitz coefficient matrix, so (9.15) has asymptotically stable origin. For the driving system (9.14), the function $V_1(x_1) = x_{11}^2 + x_{12}^2$ is a Lyapunov function, since $\dot{V}_1(x_1) = -2x_{11}^4$. An application of the invariance theorem shows that (9.14) has asymptotically stable origin. Thus, Corollary 9.2 applies and the origin $z = 0$ is asymptotically stable. See also Exercise 9.10. △

From Example 9.10, it is clear that some regrouping, or aggregation, of equations can be helpful even for critical problems of stability. The decomposition used in Example 9.10 is an example of a general approach to the decomposition of a system of differential equations.

Given a system of differential equations

$$\begin{aligned} \dot{z}_1 &= g_1(z_1, \ldots, z_n), \\ \vdots\ &\quad \vdots \\ \dot{z}_n &= g_n(z_1, \ldots, z_n), \end{aligned} \tag{9.16}$$

where $z = (z_1, \ldots, z_n) \in \mathbf{R}^n$, we present below a general approach to rewriting the system by aggregating equations to obtain a hierarchical cascade of autonomous systems,

$$\begin{aligned} \dot{x}_1 &= f_1(x_1), \\ \dot{x}_2 &= f_2(x_1, x_2), \\ \vdots\ &\quad \vdots \\ \dot{x}_i &= f_i(x_1, \ldots, x_i), \\ \vdots\ &\quad \vdots \\ \dot{x}_m &= f_m(x_1, \ldots, x_m). \end{aligned} \tag{9.17}$$

We note first that it is straightforward to extend Corollary 9.2 to the hierarchical cascade (9.17).

Corollary 9.4 *Consider a system in hierarchical form* (9.17), *denoted*

$$\mathcal{S}: \quad \dot{x}_i = f_i(x_1, \ldots, x_i) \quad \textit{for } i = 1, \ldots, m,$$

and the associated isolated subsystems, denoted

$$\mathcal{S}_i: \quad \dot{x}_i = f_i(0, \ldots, 0, x_i) \quad \textit{for } i = 1, \ldots, m.$$

Suppose that for each $i = 1, \ldots, m$, f_i is a C^1 function of its vector variable $(x_1, \ldots, x_i)$ in a neighborhood of its origin, and $f_i(0, \ldots, 0) = 0$. Then, $x = 0$ is an asymptotically stable equilibrium of $\mathcal{S}$ if and only if $x_i = 0$ is an asymptotically stable equilibrium of $\mathcal{S}_i$ for each $i = 1, \ldots, m$.

Proof. The "if" part is proved by induction on the number of isolated subsystems; Corollary 9.2 starts the induction and completes the induction step. The "only if" part is left as an exercise, with the reminder that it is necessary to show stability plus attractiveness for each $\mathcal{S}_i$. See Exercise 9.12. □

The general approach we alluded to for obtaining the cascade (9.17) from (9.16) does not involve general coordinate transformations; instead, only the renumbering and aggregation of components is required, as in Example 9.10. The remainder of this section addresses this general aggregation technique and shows that the decomposition (9.17) is generally obtainable.

Suppose (9.16) is given. We will define a directed graph $\mathcal{D}$ on n vertices, $v_1, \ldots, v_n$, which are labeled in one-to-one order-preserving correspondence with the components of the vector variable z; this directed graph is intended to model the dynamic interaction pattern of system (9.16). The constructive definition of $\mathcal{D}$ is as follows: Draw a directed edge from v_j to v_i if the function g_i depends explicitly on the variable z_j, for $j \neq i$; that is, if $\frac{\partial g_i}{\partial z_j}(z_1, \ldots, z_n)$ is not identically zero. When all such edges are drawn for all functions g_i, the resulting directed graph is $\mathcal{D}$. Note that $\mathcal{D}$ contains a directed edge from v_j to v_i if and only if z_j directly affects the dynamics of z_i by an explicit appearance on the right side of the equation for $\dot{z}_i$.

When the directed graph $\mathcal{D}$ has been determined, it can be used to find the desired aggregation of components to achieve (9.17). A pair (v_i, v_j) of vertices is called *strongly connected* if $i = j$, or, if there is a directed edge from v_i to v_j and a directed edge from v_j to v_i in $\mathcal{D}$. The relation of *strong connectedness* is an equivalence relation on the set $V := \{v_1, \ldots, v_n\}$. Denote the equivalence classes by $V_1, \ldots, V_m$. Renumber these equivalence classes, if necessary, such that, if $v_a \in V_i$ and $v_b \in V_j$ with $i < j$, then there is no directed edge from v_b to v_a in $\mathcal{D}$. In Example 9.10, we worked with the equivalence classes $V_1 = \{v_1, v_4\}$ and $V_2 = \{v_2, v_3\}$. (Such a numbering is always possible; for some systems it might be achieved in more than one way.) The result of this numbering is that the collection of vertices in the equivalence class V_i indicates precisely the set of components of vector z which form the variable x_i in (9.17). Thus, for each $i = 1, \ldots, m$, we have

$$(\text{the set of components of } x_i) := \{z_j \, : \, v_j \in V_i\},$$

and, in accordance with that choice, the functions f_i of (9.17) are determined; thus, for each $i = 1, \ldots, m$, we have

$$(\text{the set of components of } f_i) := \{g_j \, : \, v_j \in V_i\}.$$

In summary, these definitions allow system (9.16) to be rewritten in the form (9.17) by renumbering and aggregation of the components of the vector $z = [z_1 \; \cdots \; z_n]^T$ and the vector field $g = [g_1 \; \cdots \; g_n]^T$.

9.4 APPENDIX: THE POINCARÉ-BENDIXSON THEOREM

The Poincaré-Bendixson theorem is one of the most important results concerning the asymptotic behavior of trajectories of planar autonomous systems. For convenience we recall here (from Definition 8.1) the definitions of ω-limit point and the ω-limit set of a trajectory of $\dot{x} = f(x)$. A point q is an ω-limit point of the solution $\phi(t, x)$ if there exists a sequence $t_k \to \infty$ as $k \to \infty$ such that $\lim_{k\to\infty} \phi(t_k, x) = q$. The set of all ω-limit points of $\phi(t, x)$ is denoted $\omega(\phi(t, x))$, or just $\omega(x)$, and it is called the ω-limit set for that solution.

We can now state the Poincaré-Bendixson theorem.

Theorem 9.3 (H. Poincaré — I. Bendixson)
Suppose $f : D \subset \mathbf{R}^2 \to \mathbf{R}^2$ is C^1. A nonempty compact ω-limit set of a trajectory of a planar system $\dot{x} = f(x)$, which contains no equilibrium point, must be a closed periodic trajectory.

A complete proof of Theorem 9.3 appears in [41]; see also [40] or [46].

9.5 EXERCISES

Exercise 9.1 *On Example 9.1*
Consider the nominal scalar equation $\dot{x} = -x^3$ and the perturbed equation $\dot{x} = -x^3 + \beta \sin(x + t)$ with the function $\beta \neq 0$ and bounded by $|\beta| \leq \bar{\beta}$, $\bar{\beta}$ constant. Under the assumptions stated in Example 9.1, verify the bounds given there on a solution $x(t)$ of the perturbed equation.

Exercise 9.2 *A Perturbed Equation*
Consider the scalar equation $\dot{x} = -x^3 + \beta x$, where $\beta > 0$, as a perturbation of the equation $\dot{x} = -x^3$.

(a) Show that, if $\beta > 0$ is constant, then the perturbed equation has three equilibria; the origin is unstable, and the other two equilibria are asymptotically stable.

(b) Suppose $\beta = ae^{-2t}$, with $a > 0$ constant; thus, the perturbation is $g(x, t) = ae^{-2t}x$. Given ϵ with $0 < \epsilon < \sqrt{a}$, let $\delta_1(\epsilon) = \epsilon$ and $\delta_2(\epsilon) = a\epsilon$. Show that, by choosing $|x_0| = |x(0)| \leq \delta_1(\epsilon)$, we have

$$|x| < \epsilon \quad \text{and} \quad t \geq 0 \quad \Longrightarrow \quad |g(x, t)| \leq \delta_2(\epsilon)$$

and $|x(t)| \leq \epsilon$ for all $t \geq 0$.

Exercise 9.3 *Complement to Corollary 9.1*

(a) Give an example of a cascade system (9.1)–(9.2) to show that stability of $z = 0$ for $\dot{z} = g(z)$ plus stability of $x = 0$ for $\dot{x} = f(x, 0)$ does not imply stability of $(x, z) = (0, 0)$ for the cascade.

(b) Give an example to show that asymptotic stability of $z = 0$ for $\dot{z} = g(z)$ plus stability of $x = 0$ for $\dot{x} = f(x,0)$ does not imply stability of the origin for the cascade (9.1)–(9.2).

Exercise 9.4 Sketch a phase portrait for the system of Example 9.2.

(a) Show directly from the definition that the origin is stable: For every $\epsilon > 0$, find a $\delta(\epsilon) > 0$ such that $\|(x(0), z(0))\|_2 = \|(x_0, z_0)\|_2 < \delta(\epsilon)$ implies $\|(x(t), z(t))\|_2 < \epsilon$ for all $t \geq 0$.

(b) Identify all the equilibria of the system and their stability properties.

Exercise 9.5 *On Example 9.3*

(a) For the system in Example 9.3, show that initial conditions in the region $x_0 z_0 > 2$ correspond to solutions $x(t)$ with finite positive escape time.

(b) Show the invariance of the branches of the hyperbola $xz = 2$ by differentiating the hyperbola equation with respect to time.

Exercise 9.6 *Varying the Convergence Rate*
Let $k > 0$ be constant, and consider the system

$$\begin{aligned}\dot{x} &= -x + x^2 z,\\ \dot{z} &= -kz.\end{aligned}$$

(a) Determine the basin of attraction of the origin, and, if possible, the positive escape time of unbounded trajectories.

(b) For this example, is it possible to include any given compact subset of the plane in the basin of attraction by choosing k sufficiently large?

Exercise 9.7 Consider the system

$$\begin{aligned}\dot{x} &= -x + xz^2,\\ \dot{z} &= -z.\end{aligned}$$

Use the Lyapunov function $V(x, z) = x^2 e^{z^2} + z^2$ to show that the equilibrium $(x, z) = (0, 0)$ is globally asymptotically stable.

Exercise 9.8 Consider the system

$$\begin{aligned}\dot{x} &= -x^3 + x^3 z,\\ \dot{z} &= -z.\end{aligned}$$

Show that for any initial condition $x(0) \neq 0$, there are initial conditions $z(0)$ such that $x(t)$ has finite escape time. *Hint*: Use separation of variables on the x equation together with the known solution for $z(t)$.

Exercise 9.9 *Does Converging $z(t)$ Imply Bounded $x(t)$?*
The systems in this exercise group might be considered as motivators for the *input-to-state stability* concept considered later in the book.

(a) Show that the origin of the planar system

$$\dot{x} = -x^3 + xz,$$
$$\dot{z} = -z^3$$

is globally asymptotically stable. *Hint*: Use the function $V(x) = x^2$ to show that every forward solution of $\dot{x} = -x^3 + xz(t)$ is bounded; write $\dot{V}(x,t) := x\dot{x} = -x^4 + x^2 z(t) = -(1-\delta)x^4 - \delta x^4 + x^2 z(t)$, where $0 < \delta < 1$, and argue that $\dot{V}(x,t) < 0$ for sufficiently large $|x|$.

(b) Consider the cascade

$$\dot{x}_1 = x_1(x_2 + x_3) - x_1^3,$$
$$\dot{x}_2 = x_2 x_3 - x_2^3,$$
$$\dot{x}_3 = -x_3^3.$$

Show that the origin is asymptotically stable. Is the origin globally asymptotically stable? *Hint*: Use the result of (a).

(c) Show that the origin of the planar system

$$\dot{x} = -x^3 + x^2 z,$$
$$\dot{z} = -z^3$$

is asymptotically stable. Can you prove that the origin is globally asymptotically stable? *Hint*: Follow an approach similar to the hint for part (a) of this exercise group.

Exercise 9.10 *On Example 9.10*

(a) For the system in Example 9.10, show that the origin is stable for the linearization, $\dot{z} = Az$. Does it follow, on the basis of linearization alone, that the origin is stable for the full nonlinear system?

(b) Show that $z = 0$ is globally asymptotically stable. (Note that this is not a deduction from Corollary 9.2 or Corollary 9.3.)

Exercise 9.11 *What if $A := \frac{\partial f}{\partial x}(0,0)$ is Hurwitz?*
Consider the cascade

$$\dot{x} = f(x, z),$$
$$\dot{z} = g(z),$$

where $z = 0$ is an asymptotically stable equilibrium of $\dot{z} = g(z)$ and the matrix $A := \frac{\partial f}{\partial x}(0,0)$ is Hurwitz. Let $z(0)$ be such that $z(t) \to 0$ as $t \to \infty$. For $\|x(0)\| = \|x_0\|$ sufficiently small, use a variation of parameters formula for the solution $x(t)$ of $\dot{x} = f(x, z(t))$, together with standard norm estimates, to show that $(x, z) = (0, 0)$ is an asymptotically stable equilibrium of the cascade. *Hint*: Write $\dot{x}(t) = Ax(t) + [f(x(t), z(t)) - Ax(t)]$.

Exercise 9.12 Write a detailed proof of Corollary 9.4.

Exercise 9.13 *LAS in a Cascade*
Sketch the phase portrait of the planar system

$$\begin{aligned}\dot{x} &= -x(1 - x^2)(3 - x) + x(x - 1)z^2,\\ \dot{z} &= -z.\end{aligned}$$

Show all equilibria as well as a qualitative sketch of the invariant manifolds associated with each equilibrium point. Indicate on the sketch an estimate of the basin of attraction of the origin.

Exercise 9.14 *More LAS in a Cascade*
Use qualitative analysis to sketch the phase portrait of the planar system

$$\begin{aligned}\dot{x} &= -x(1 - x^2)(3 - x) + x(x - 1)z^2,\\ \dot{z} &= -z + z^3.\end{aligned}$$

Show all equilibria as well as a qualitative sketch of the invariant manifolds associated with each equilibrium point. Indicate on the sketch an estimate of the basin of attraction of the origin.

Exercise 9.15 *Again LAS in a Cascade*
Sketch the phase portrait of the planar system

$$\begin{aligned}\dot{x} &= -x^3 + x^5 + xz^2,\\ \dot{z} &= -z + z^3.\end{aligned}$$

Include detail as in Exercises 9.13–9.14.

Exercise 9.16 *Cascade by Aggregation*
Aggregate components in two different ways to recognize the following system as a cascade:

$$\begin{aligned}\dot{x}_1 &= -x_1 + x_4 x_1^2 + x_2 x_3 x_5,\\ \dot{x}_2 &= -2x_2 + x_5,\\ \dot{x}_3 &= -3x_3 + x_2 x_5,\\ \dot{x}_4 &= -x_4^3 + x_2 + x_3 + x_5^2,\\ \dot{x}_5 &= -5x_5 + x_5^2 x_2.\end{aligned}$$

Show that the equilibrium at the origin presents a critical problem and the origin is asymptotically stable.

9.6 NOTES AND REFERENCES

For phase plane analysis of two-dimensional autonomous systems, see [21], [41]. For a proof of the stable manifold theorem for planar autonomous systems with saddle equilibria (such as the points $(\pm 1, 0)$ in Example 9.7), see [41] or [34]; see [41] (pages 168–174) for the local stable curve result, and [34] (pages 292–298) for the local result as well as a discussion of the global stable and unstable manifolds. As noted earlier, a complete proof of the Poincaré-Bendixson theorem (Theorem 9.3) appears in [41]; see also [40] or [46]. The Poincaré-Bendixson theorem is helpful in the search for periodic orbits of planar autonomous systems; for this aspect, see [58].

The material on total stability of an equilibrium of $\dot{x} = f(x)$ and Theorem 9.1 is guided by [32] (pages 275–276) and [48] (pages 11–12). The result in Theorem 9.2 is due to E. D. Sontag, Further facts about input to state stabilization, *IEEE Trans. Aut. Contr.*, AC-35:473–476, 1990. Our presentation of Theorem 9.2 and Corollaries 9.2–9.3 follows [48] (pages 15–17).

Example 9.10 is from [99]. The general procedure for obtaining the hierarchical form (9.17) from (9.16) appears in [99] and [100]. We considered only time invariant systems; however, reference [100] has results on asymptotic stability for time varying cascade systems.

The system in Exercise 9.7 is taken from [86] (Example 4.9, pages 131–132).

As noted in Exercise 9.9, the concept of *input-to-state stability* (ISS), discussed later in the text, provides an approach to the question of boundedness of driven trajectories in cascades. Those who wish to look ahead to the definition of ISS should see, in addition, Exercises 16.2, 16.7, 16.8, and revisit the systems in Exercise 9.9.

Cascade decompositions arise not only by aggregation of the original state vector components but also by coordinate transformations, as in the study of zero dynamics in Chapter 11.

Chapter Ten

Center Manifold Theory

In this chapter we show how the basic ideas of center manifold theory are used in addressing critical problems of stability. Recall that a critical problem of stability is one for which all eigenvalues of the linearization at equilibrium have nonpositive real part and there is at least one eigenvalue with zero real part. Critical problems of stability for nonlinear systems are exactly those for which the linearization cannot decide the issue; they are the local problems not covered by Theorem 8.2.

We state the three main theorems of center manifold theory. These theorems cover (i) the existence of a center manifold, (ii) the reduction of a critical problem to a lower-dimensional system, and (iii) the approximation of a center manifold which enables the reduction in (ii). We illustrate these results with simple examples where the computations are fairly easy. We include a Lyapunov-theoretic proof of asymptotic stability via center manifold reduction, and we leave other proofs to the references.

Appendix D, Section D.1 on "Manifolds and the Preimage Theorem" provides a discussion of submanifolds of $\mathbf{R}^n$ that can be read along with this chapter.

10.1 INTRODUCTION

In this section, we begin with a motivating example for the development of center manifold theory; then we define invariant sets and invariant manifolds for a nonlinear system; and last, we discuss special coordinates for the study of critical problems.

10.1.1 An Example

We begin with a nonlinear system in the plane having stability properties we can determine easily, and we compare it with a second system whose behavior, it appears, should be quite similar to that of the first system. This example illustrates a critical problem of stability and helps to motivate the definition of a center manifold.

Consider first the system

$$\dot{x} = ax^3, \tag{10.1}$$

$$\dot{y} = -y + y^2, \tag{10.2}$$

where the parameter a is constant. These scalar equations are uncoupled, hence, they can be considered separately for stability analysis. Indeed, for the uncoupled system (10.1)–(10.2), it is not difficult to see that the origin $(x, y) = (0, 0)$ is (i) asymptotically stable if $a < 0$, (ii) unstable if $a > 0$, and (iii) stable, but not asymptotically stable, if $a = 0$.

Now consider the system given by

$$\dot{x} = ax^3 + x^2 y, \tag{10.3}$$

$$\dot{y} = -y + y^2 - x^3 . \tag{10.4}$$

System (10.3)–(10.4) is a coupled system in which each of the right-hand sides depends on both x and y; it is not a cascade.

It is straightforward to check that each of these planar systems has a linearization at the origin having eigenvalues 0 and -1. The stability property of the origin in (10.3)–(10.4) is not immediately apparent, although we might make a reasonable guess by a comparison with the results described above for the uncoupled system (10.1)–(10.2). For example, we might expect that if $a < 0$ then the origin in (10.3)–(10.4) is asymptotically stable. It is best to consider this a guess at this point, rather than a confident conjecture, due to the presence of the additional terms $x^2 y$ and $-x^3$, each having the same order as the stabilizing term in the first equation.

The relatively simple analysis of (10.1)–(10.2) may be viewed as follows. The x-axis, described by the curve $y = 0$, is an invariant curve for (10.1). That is, for any initial condition $(x_0, 0)$ on that curve, the corresponding solution remains on that curve (on the x-axis). Note that this invariant curve is tangent, at the origin, to the eigenspace for eigenvalue $\lambda = 0$ of the Jacobian matrix at the origin. Moreover, it is the dynamics on this invariant curve that determines the stability property of the origin. With this observation, attention is shifted a bit from the parameter a to the idea of the invariant curve and its dynamics.

In principle, the reduction of a stability problem to a lower-dimensional manifold (such as an invariant curve, which is a one-dimensional invariant manifold) can always be accomplished for a critical problem by means of center manifold analysis. We now introduce precise definitions that enable the discussion of this reduction.

10.1.2 Invariant Manifolds

Consider the system

$$\dot{x} = f(x, y), \tag{10.5}$$

$$\dot{y} = g(x, y), \tag{10.6}$$

where $f(x, y)$ and $g(x, y)$ are real-valued C^1 functions on $\mathbf{R}^2$ with $f(0, 0) = 0$ and $g(0, 0) = 0$. A curve $y = h(x)$ defined for small $|x|$, that is, near equilibrium, is called a (local) *invariant curve* for (10.5)–(10.6) if each solution

$(x(t), y(t))$ having initial condition $(x(0), y(0)) = (x_0, h(x_0))$ remains on the curve $y = h(x)$ for sufficiently small t, that is, $y(t) = h(x(t))$ for $|t|$ small. In system (10.1)–(10.2), it is the behavior on the invariant curve $y = 0$ that determines the stability property of the origin.

Now consider the system

$$\dot{z} = F(z), \tag{10.7}$$

where $F : D \to \mathbf{R}^N$, $D \subseteq \mathbf{R}^N$ is an open set containing the origin, and $F(0) = 0$. We have seen the meaning and importance of the idea of a local invariant curve for a planar system. We now want to define more general invariant manifolds.

Definition 10.1 (Local Invariant Sets, Local Invariant Manifolds)
A set $S \subset \mathbf{R}^N$ is a local invariant set *for* (10.7) *if, for each $z_0 \in S$, there exists $T > 0$ such that the solution $z(t)$ with $z(0) = z_0$ lies in S for all $|t| < T$. If we can always take $T = \infty$ for $z_0 \in S$, then S is an* invariant set *(see Definition 8.2). If a local invariant set S for* (10.7) *is also a k-dimensional submanifold of $\mathbf{R}^N$, then S is a* local invariant manifold *of* (10.7).

A local invariant curve for (10.5)–(10.6) is a local invariant set for that system; it is also a local invariant manifold for the system.

Note that Definition 10.1 merely requires a local invariant set to be a union of solution trajectories of $\dot{z} = F(z)$; a local invariant set need not be a submanifold of $\mathbf{R}^N$. For example, consider the system

$$\begin{aligned} \dot{x}_1 &= x_3, \\ \dot{x}_2 &= -x_2, \\ \dot{x}_3 &= -x_1. \end{aligned}$$

The x_1,x_3-plane is a two-dimensional invariant manifold of this system. On the other hand, the union of the x_1,x_3-plane and the positive x_2-axis is an invariant set, but not a k-dimensional submanifold of $\mathbf{R}^3$.

10.1.3 Special Coordinates for Critical Problems

Let $D \subseteq \mathbf{R}^N$, $F : D \to \mathbf{R}^N$ with $F(0) = 0$, and suppose that the stability problem for the origin is a critical problem. Assume that $DF(0)$ has n eigenvalues with zero real part and m eigenvalues with negative real part; thus $N = n + m$. Then there exists an invertible real matrix Q such that

$$Q^{-1}[DF(0)]Q =: \begin{bmatrix} A & 0 \\ 0 & B \end{bmatrix}, \tag{10.8}$$

where all eigenvalues of the $n \times n$ matrix A have zero real part, and all eigenvalues of the $m \times m$ matrix B have negative real part. Writing x for column vectors in $\mathbf{R}^n$ and y for column vectors in $\mathbf{R}^m$, the linear coordinate change defined by $z = Q[x^T \; y^T]^T$ transforms $\dot{z} = F(z)$ into the form

$$\dot{x} = Ax + f(x, y), \tag{10.9}$$

$$\dot{y} = By + g(x, y). \tag{10.10}$$

We assume that F is C^2 in a neighborhood of the origin; then the functions f and g are also C^2 near the origin, that is, all second order partial derivatives of f and g are continuous in a neighborhood of $(x, y) = (0, 0)$. From the fact that $F(0) = 0$ and (10.8), we have $f(0, 0) = 0$, $g(0, 0) = 0$, and $Df(0, 0) = 0$, $Dg(0, 0) = 0$. We also assume that equation (10.10) is actually present, that is, $m \geq 1$.

For the remainder of this chapter, we assume that we have a decomposition (10.9)–(10.10) based on a block diagonal form (10.8) for $DF(0)$.

10.2 THE MAIN THEOREMS

This section gives the definition of a center manifold and presents the three main theorems of center manifold theory. These theorems deal with (i) the existence of a center manifold for a critical equilibrium, (ii) the reduction of the stability problem to a lower-dimensional system, and (iii) the approximation of a center manifold required to carry out the reduction in (ii).

10.2.1 Definition and Existence of Center Manifolds

Suppose for the moment that the system (10.7) is *linear*. Then we have $f = g = 0$ in (10.9)–(10.10). The subset defined by $y = 0$ is an invariant manifold (invariant subspace); in fact, it is the sum of the generalized eigenspaces for the eigenvalues having zero real part. The subset defined by $x = 0$ is also an invariant manifold. Solutions that start at $t = 0$ within the set where $x = 0$ must maintain $x = 0$ for all $t \geq 0$ and approach the origin as $t \to \infty$ since B is Hurwitz. The invariant manifold defined by $x = 0$ is called the stable manifold of the linear system. Similarly, solutions that start at $t = 0$ within the set where $y = 0$ must maintain $y = 0$ for all $t \geq 0$, and the invariant manifold defined by $y = 0$ is called the center manifold of the linear system. Note, in addition, that each solution $(x(t), y(t))$ asymptotically tracks the solution $x(t)$ that evolves within the center manifold itself, because $(x(t), y(t)) - (x(t), 0) = (0, y(t)) \to 0$ as $t \to \infty$.

If (10.7) is nonlinear, the functions f and g are nonzero. Instead of the invariant manifold (for the linearization) defined by $y = 0$, we will define a local invariant manifold for (10.7) in a neighborhood of the origin $(x, y) = (0, 0)$, described by an equation $y = h(x)$, where $h(x)$ is a smooth function defined near $x = 0$ in $\mathbf{R}^n$.

Definition 10.2 (Center Manifold)
A local invariant manifold for (10.9)–(10.10) *is a* (*local*) center manifold *for the equilibrium at the origin if there exists an open set* $W \subseteq \mathbf{R}^n$ *about* $x = 0$

and a smooth mapping $h : W \to \mathbf{R}^m$ *such that*

$$h(0) = 0 \quad \textit{and} \quad Dh(0) = 0.$$

According to Definition 10.2, the equation $y = h(x)$ defines a local center manifold for (10.9)–(10.10) if the graph of h is a local invariant set for (10.9)–(10.10), and, at the origin, the graph of h is tangent to the center subspace for the linearization.

It is important to realize that center manifolds are not uniquely determined; the next two examples illustrate the nonuniqueness.

Example 10.1 Consider the planar system

$$\dot{x}_1 = x_1^2,$$
$$\dot{x}_2 = -x_2.$$

The center subspace of the linearization at $(0,0)$ is defined by $x_2 = 0$. A center manifold for the nonlinear system is given by the line $x_2 = 0$, because this line (the x_1-axis) is a local invariant manifold for the system and we may take $h(x_1) \equiv 0$ in Definition 10.2. However, there are many other center manifolds, which may be found by integrating the separable equation

$$\frac{dx_2}{dx_1} = -\frac{x_2}{x_1^2}$$

and solving for x_2 in terms of x_1 near the origin. The general solution, for $x_1 \neq 0$, is

$$x_2 = C \exp\left(\frac{1}{x_1}\right), \quad x_1 \neq 0,$$

where C is an arbitrary real constant. From the general solution, we may obtain many center manifolds by splicing together at the origin the graph of a solution defined for $x_1 > 0$ with the graph of a solution defined for $x_1 < 0$, provided we respect the tangency condition of Definition 10.2. Suppose we define the function h_C, for constant C, by

$$\begin{cases} h_C(x_1) = C \exp\left(\dfrac{1}{x_1}\right) & \text{for} \quad x_1 < 0, \\ h_C(x_1) = 0 \quad \text{for} \quad x_1 \geq 0. \end{cases}$$

Then it is easy to check that the graph of each function h_C is a center manifold for the equilibrium at the origin. Thus, there are infinitely many center manifolds for this equilibrium. △

Another very simple example may be helpful.

Example 10.2 Consider the system

$$\dot{x}_1 = -x_1^3,$$
$$\dot{x}_2 = -x_2.$$

The center subspace of the linearization at $(0,0)$ is defined by $x_2 = 0$ (the x_1 axis). Again, we can integrate a separable differential equation to find center manifolds for the origin. Define $h_C(x_1)$ by

$$\begin{cases} h_C(x_1) = C \exp\left(-\frac{1}{2}x_1^{-2}\right) & \text{if } x_1 < 0, \\ h_C(x_1) = 0 \quad \text{if } x_1 \geq 0. \end{cases}$$

As domain of h_C, we may take $W = (-\delta, \delta)$ for some $\delta > 0$, or even $\delta = \infty$. For each real number C, we have $h_C(0) = 0$ and $Dh_c(0) = h'_C(0) = 0$. Thus, the graph of each h_C is a center manifold for the equilibrium at the origin. In particular, the entire x_1-axis is a center manifold, corresponding to $C = 0$ in the family of center manifolds just defined. The family of center manifolds defined by the functions h_C above does not exhaust all possible center manifolds for $(0,0)$; see Exercise 10.1. △

It is of fundamental importance that a local center manifold exists, under the assumptions on (10.7) which led us to (10.9)–(10.10).

Theorem 10.1 (Existence of a Center Manifold)
Suppose that, in (10.9)–(10.10), *f and g are C^2 in a neighborhood of the origin, every eigenvalue of A has zero real part, and B is Hurwitz. Then there exist a $\delta > 0$ and a smooth mapping $h : B_\delta(0) \subset \mathbf{R}^n \to \mathbf{R}^m$ such that the equation $y = h(x)$ defines a local center manifold of* (10.9)–(10.10).

Proof. Proofs are available in [23] (pages 16–19) and [58]. □

According to Definition 10.2, in order for the graph of a function $h(x)$ to be a center manifold for the origin of (10.9)–(10.10), it must be a local invariant manifold for (10.9)–(10.10), in addition to satisfying the tangency conditions

$$h(0) = 0 \quad \text{and} \quad Dh(0) = 0. \tag{10.11}$$

Any point on the graph of h with coordinates (x, y) must satisfy $y = h(x)$, of course, and by invariance, the vector field at (x, y) must satisfy $\dot{y} = Dh(x)\dot{x}$. Therefore we substitute the expressions for $\dot{x}$ and $\dot{y}$ from (10.9)–(10.10), and $y = h(x)$, into $\dot{y} = Dh(x)\dot{x}$ to obtain

$$Dh(x)\,[\,Ax + f(x, h(x))\,] = Bh(x) + g(x, h(x)). \tag{10.12}$$

Equation (10.12) must be satisfied by each solution $(x(t), h(x(t))$ on the center manifold.

The next subsection addresses the question of how an invariant center manifold enables a determination of the stability property of the origin in (10.9)–(10.10).

10.2.2 The Reduced Dynamics

Let (x_0, y_0) be a point near equilibrium and lying in a center manifold for (10.9)–(10.10) defined by $h : B_\delta(0) \subset \mathbf{R}^n \to \mathbf{R}^m$. By invariance, the solution $(x(t), y(t))$ with initial condition $(x(0), y(0)) = (x_0, y_0) = (x_0, h(x_0))$ satisfies $y(t) = h(x(t))$ for all t near $t_0 = 0$. Thus, solutions $(x(t), y(t))$ on the center manifold itself are uniquely determined by initial conditions for the equation

$$\dot{u} = Au + f(u, h(u)), \quad u \in \mathbf{R}^n, \quad \|u\| < \delta, \tag{10.13}$$

because, by invariance, the function $y(t) = h(u(t))$ then necessarily satisfies (10.10). Consequently, we call system (10.13) the *reduced dynamics* associated with the center manifold $y = h(x)$. We usually refer to (10.13) simply as the reduced dynamics, when the center manifold is understood.

If we had $f \equiv 0$ in (10.9), then (10.13) would be exactly the dynamics $\dot{u} = Au$ on the center *subspace* for the linearization. However, note that even in the decoupled systems in Examples 10.1–10.2, we had to take account of the term $f(u, h(u))$.

As the earlier examples suggest, the reduced dynamics (10.13) embodies all the information required to determine the stability property of the origin in (10.9)–(10.10). The intuition is that, near the origin, the behavior of y is dominated by the Hurwitz matrix B, hence the stability of the origin $(x, y) = (0, 0)$ should be determined by the stability property of $x = 0$ relative to the invariant center manifold $y = h(x)$. The next theorem confirms this intuition.

Theorem 10.2 (Reduced Dynamics)
Under the assumptions of Theorem 10.1, let $y = h(x)$ define a center manifold near the origin of system (10.9)–(10.10) *with reduced dynamics* (10.13). *Then the following statements are true:*

(a) *If the origin of* (10.13) *is unstable, then the origin of* (10.9)–(10.10) *is unstable.*

(b) *Suppose the origin of* (10.13) *is stable. Then there exists a neighborhood U of $(x, y) = (0, 0)$ such that, for every initial condition $(x(0), y(0)) = (x_0, y_0)$ in U, there is a solution $u(t)$ of* (10.13) *such that*

$$x(t) = u(t) + O(e^{-\gamma t}),$$
$$y(t) = h(u(t)) + O(e^{-\gamma t})$$

as $t \to \infty$, where γ is a constant that depends only on the matrix B in (10.10).

(c) *If the origin of* (10.13) *is stable, then the origin of* (10.9)–(10.10) *is stable.*

(d) *If the origin of* (10.13) *is asymptotically stable, then the origin of* (10.9)–(10.10) *is asymptotically stable.*

Proof. Statement (a) is immediate from the invariance of the given center manifold. It is not difficult to see that (b) implies (c), and (b) implies (d). A proof of (b) is in [23] (pages 21–25).

We now give a direct proof of the asymptotic stability result in part (d), which will illustrate the use of direct and converse Lyapunov theorems.

Proof of (d): Start with (10.9)–(10.10), let $y = h(x)$ be a center manifold, and make the simple change of coordinates $(x, w) := (x, y - h(x))$, which defines w as the deviation from the given center manifold. This coordinate change transforms (10.9)–(10.10) into the system

$$\dot{x} = Ax + f(x, w + h(x)) \tag{10.14}$$

$$\dot{w} = B[w + h(x)] + g(x, w + h(x)) - \frac{\partial h}{\partial x}(x)[\,Ax + f(x, w + h(x))\,]. \tag{10.15}$$

The center manifold is given by $w = 0$. We have seen that, by invariance of the center manifold, the function $h(x)$ must satisfy the partial differential equation (10.12) in addition to the tangency conditions (10.11). Rewrite (10.12) as

$$0 = Bh(x) + g(x, h(x)) - \frac{\partial h}{\partial x}(x)[\,Ax + f(x, h(x))\,]. \tag{10.16}$$

Add and subtract $f(x, h(x))$ on the right-hand side of (10.14), and then subtract (10.16) from (10.15), to get

$$\dot{x} = Ax + f(x, h(x)) + \psi_1(x, w), \tag{10.17}$$

$$\dot{w} = Bw + \psi_2(x, w). \tag{10.18}$$

where

$$\psi_1(x, w) := f(x, w + h(x)) - f(x, h(x))$$

and

$$\psi_2(x, w) := g(x, w + h(x)) - g(x, h(x)) - \frac{\partial h}{\partial x}(x)\psi_1(x, w).$$

It is straightforward to check that

$$\psi_j(x, 0) \equiv 0 \quad \text{for } j = 1, 2, \qquad \frac{\partial \psi_j}{\partial w}(0, 0) \equiv 0 \quad \text{for } j = 1, 2.$$

Thus, in a ball defined by

$$\|(x, w)\|_2 < \rho,$$

we have estimates of the form

$$\|\psi_1(x, w)\|_2 \le k_1 \|w\|_2 \quad \text{and} \quad \|\psi_2(x, w)\|_2 \le k_2 \|w\|_2,$$

where the constants k_1 and k_2 depend on ρ and may be taken arbitrarily small by choosing $\rho > 0$ sufficiently small. If the reduced dynamics (10.13) has asymptotically stable origin, then by the converse Lyapunov theorem

(Theorem 8.7) there exists a C^1 strict Lyapunov function $V(x)$ for $\dot{x} = Ax + f(x, h(x))$, such that

$$\frac{\partial V}{\partial x}(x)[\,Ax + f(x, h(x))\,] \leq -\alpha(\|x\|_2) \quad \text{for } x \neq 0,$$

where $\alpha(\|x\|_2)$ is increasing in $\|x\|_2$ and $\alpha(0) = 0$. (The function $\alpha(\cdot)$ is a class $\mathcal{K}$ function; see Lemma E.2 of Appendix E.) Write

$$\nu(x, w) := V(x) + \sqrt{w^T P w},$$

where P is a symmetric positive definite matrix such that $B^T P + PB = -I$. We note that $\nu(x, w)$ and $\dot{\nu}(x, w)$ are continuous everywhere except for $w=0$, but since the solution behavior on $w=0$ is assumed known, the estimates given below will show that the statement of the basic Lyapunov theorem will still hold for (10.17)–(10.18). Along solutions of (10.17)–(10.18) with w never equal to zero, we have

$$\begin{aligned}
\dot{\nu}(x, w) &= \frac{\partial V}{\partial x}(x)[\,Ax + f(x, h(x)) + \psi_1(x, w)\,] \\
&\quad + \frac{1}{2\sqrt{w^T P w}}[\,w^T(B^T P + PB)w + 2w^T P \psi_2(x, w)\,] \\
&\leq -\alpha(\|x\|_2) + \frac{\partial V}{\partial x}(x)\psi_1(x, w) + \frac{1}{2\sqrt{w^T P w}}[-w^T w + 2w^T P \psi_2(x, w)] \\
&\leq -\alpha(\|x\|_2) + k_1 m \|w\|_2 + \frac{1}{2\sqrt{w^T P w}}[-w^T w + 2w^T P \psi_2(x, w)],
\end{aligned}$$

where $\|\frac{\partial V}{\partial x}(x)\|_2 \leq m$ for $\|x\|$ near zero, using the fact that V is C^1 with a minimum at $x = 0$. Using standard norm estimates, we find that

$$\dot{\nu}(x, w) \leq -\alpha(\|x\|_2) + k_1 m \|w\|_2 - \frac{1}{2\sqrt{\|P\|_2}}\|w\|_2 + k_2 \frac{\|P\|_2}{\lambda_{\min}(P)}\|w\|_2,$$

where $\lambda_{\min}(P)$ is the minimum eigenvalue of P, which must be positive. We can rearrange the right hand side of the last inequality, and write

$$\begin{aligned}
\dot{\nu}(x, w) &\leq -\alpha(\|x\|_2) - \frac{1}{4\sqrt{\|P\|_2}}\|w\|_2 \\
&\quad - \left[\frac{1}{4\sqrt{\|P\|_2}} - k_1 m - k_2 \frac{\|P\|_2}{\lambda_{\min}(P)}\right]\|w\|_2.
\end{aligned}$$

Since k_1 and k_2 can be made arbitrarily small by making ρ smaller, we have that, for $\|(x, w)\|_2 < \rho$ sufficiently small,

$$\frac{1}{4\sqrt{\|P\|_2}} - k_1 m - k_2 \frac{\|P\|_2}{\lambda_{\min}(P)} > 0 \quad \text{for } (x, w) \neq (0, 0).$$

Hence, for nonzero (x, w) near zero, we have

$$\dot{\nu}(x,w) < -\alpha(\|x\|_2) - \frac{1}{4\sqrt{\|P\|_2}}\|w\|_2 < 0.$$

Thus the origin of (10.17)–(10.18) is asymptotically stable, and therefore so is the origin of (10.9)–(10.10). This completes the proof of (d). □

Note that, in attempting to verify the hypothesis of Lyapunov stability for (10.13), it is sufficient to find a function $V(u)$ such that $\frac{\partial V}{\partial u}(u)(Au + f(u,h(u))) \le 0$ in a neighborhood of $u = 0$. However, recall that the converse theorem for Lyapunov stability guarantees only a Lyapunov function $V(u,t)$ such that $\dot{V}(u,t) := \frac{\partial V}{\partial u}(u,t)(Au + f(u,h(u))) + \frac{\partial V}{\partial t}(u,t) \le 0$ along solutions of (10.13), and there need not exist a Lyapunov function independent of t; see [32] (page 228).

We note that the reduced dynamics (10.13) cannot be replaced by simply setting $h(u) = 0$ there, or equivalently, setting $y = 0$ in (10.9); such a substitution is equivalent to reducing the full system dynamics to the dynamics on the tangent approximation of the center manifold given by the center subspace of the linearization at $(0,0)$. In general, this substitution does not yield correct predictions about (10.9)–(10.10), as shown in the next example.

Example 10.3 Consider the system

$$\begin{aligned}\dot{x} &= -x^3 + xy,\\ \dot{y} &= -y + x^2.\end{aligned}$$

The linearization has center subspace given by the x-axis ($y = 0$). Note that the parabola $y = h(x) = x^2$ consists of the full set of equilibria for this system; thus it is an invariant manifold, and clearly it satisfies the tangency condition for a center manifold. The reduced dynamics (10.13) are given by $\dot{u} = -u^3 + uh(u) = 0$, for which $u = 0$ is stable with Lyapunov function $V(u) = u^2$. Therefore, by Theorem 10.2 (c), the origin $(x,y) = (0,0)$ is stable. However, if we set $y = 0$ in the first differential equation, we have $\dot{x} = -x^3$, but this cannot predict the correct result.

One can also consider the slight variation given by

$$\begin{aligned}\dot{x} &= x^3 - xy,\\ \dot{y} &= -y + x^2.\end{aligned}$$

The linearization at $(0,0)$ is the same as before, and the parabola $y = h(x) = x^2$ is a center manifold of equilibria. As before, we deduce from Theorem 10.2 (c) that the origin $(x,y) = (0,0)$ is stable. Setting $y = 0$ in the first differential equation, we have $\dot{x} = x^3$, which cannot predict the correct result. △

In order to apply Theorem 10.2, we need a method of computing center manifolds, at least to an order of accuracy sufficient to enable a determination of the stability property of the origin of the reduced dynamics (10.13). And that is the next subject.

10.2.3 Approximation of a Center Manifold

We have seen that conditions (10.11) and equation (10.12) must be satisfied exactly by $h(x)$ on a neighborhood of $x = 0$ in order to solve for an exact center manifold. However, it is usually too much to ask for an exact solution for $h(x)$, since that would be equivalent to solving the original system (10.9)–(10.10). What is really needed is an approximation of $h(x)$ that is sufficiently accurate to enable stability determinations for the reduced dynamics (10.13). For this purpose, we need to know that a center manifold can be approximated, in principle, to any desired degree of accuracy, as measured by the order term $O(\|x\|^k)$. Theorem 10.3 below states the desired result.

The following notation is helpful. If $\phi : W \subset \mathbf{R}^n \to \mathbf{R}^m$ is C^1 in a neighborhood W of the origin, then define

$$(M\phi)(x) := D\phi(x)\,[\,Ax + f(x, \phi(x))\,] - B\phi(x) - g(x, \phi(x)). \qquad (10.19)$$

If $y = h(x)$ defines an exact center manifold as a graph of h over an open set W about $x = 0$, then $(Mh)(x) = 0$ for all $x \in W$.

Theorem 10.3 (Approximation of a Center Manifold)
Let $h(x)$ be an exact solution of (10.12) *for x in an open set W about $x = 0$ in $\mathbf{R}^n$. Suppose that $\phi : W \to \mathbf{R}^m$ is C^1, $\phi(0) = 0$, and $D\phi(0) = 0$, and there is an integer $q > 1$ such that $(M\phi)(x) = O(\|x\|^q)$ as $x \to 0$. Then*

$$\|\, h(x) - \phi(x)\,\| = O(\|x\|^q) \quad \text{as } x \to 0.$$

Proof. See [23] (pages 25–28) or [58]. □

Note that both of the relevant error terms in Theorem 10.3 have the same order; that is, if $(M\phi)(x) = O(\|x\|^q)$, then $\|h(x) - \phi(x)\| = O(\|x\|^q)$.

Theorem 10.3 says that, if $h(x)$ describes an exact center manifold, then we can get an approximation $\phi(x)$ which agrees with $h(x)$ through terms of order $O(\|x\|^n)$ for small $\|x\|$ by finding a function ϕ that satisfies (10.12) through terms of order $O(\|x\|^n)$; equivalently, the requirement is that $(M\phi)(x) = O(\|x\|^{n+1})$. Such an approximate solution $\phi(x)$ for the exact equation $(Mh)(x) = 0$ of a center manifold will be called an *order-n approximation* of a center manifold.

Taylor expansions for all the terms in (10.19) are a convenient tool for obtaining approximate center manifolds with sufficient accuracy for stability determinations. We illustrate the procedure on the system (10.3)–(10.4).

Example 10.4 Equations (10.3)–(10.4) are

$$\dot{x} = ax^3 + x^2 y, \tag{10.20}$$

$$\dot{y} = -y + y^2 - x^3\,. \tag{10.21}$$

These equations already have the special coordinate form discussed above. We identify the relevant terms as $A = [0]$, $B = [-1]$, $f(x, y) = ax^3 + x^2 y$ and $g(x, y) = y^2 - x^3$. Let $y = h(x)$ define a center manifold. In order to approximate $h(x)$ by $\phi(x)$, we set

$$(M\phi)(x) := D\phi(x)\,[\,ax^3 + x^2\phi(x)\,] + \phi(x) - \phi^2(x) + x^3 = 0 \tag{10.22}$$

where $D\phi(x) = \frac{d\phi}{dx}(x)$. We wish to satisfy equation (10.22) to order $O(\|x\|^n)$, where n is sufficiently large to allow a stability determination for the origin of the reduced dynamics. If $\phi(x) = O(\|x\|^3)$ (as $\|x\| \to 0$), then, for $a \neq 0$,

$$(M\phi)(x) = O(\|x\|^5) + \phi(x) - \phi^2(x) + x^3,$$

and for $a = 0$, we have

$$(M\phi)(x) = O(\|x\|^7) + \phi(x) - \phi^2(x) + x^3.$$

Suppose $a \neq 0$. Then $\phi(x) = -x^3$ implies

$$(M\phi)(x) = O(\|x\|^5) - \phi^2(x) = O(\|x\|^5).$$

By Theorem 10.3, we have $h(x) = -x^3 + O(\|x\|^5)$. By Theorem 10.2, the stability of the origin $(0, 0)$ is determined by the stability property of $u = 0$ in the reduced dynamics given by

$$\begin{aligned}\dot{u} = Au + f(u, h(u)) &= au^3 + u^2 h(u)\\ &= au^3 + u^2(-u^3 + O(\|u\|^5))\\ &= au^3 - u^5 + O(\|u\|^7),\end{aligned}$$

if indeed we can make the determination for $u = 0$. The origin of this scalar equation is asymptotically stable if $a < 0$ and unstable if $a > 0$. By Theorem 10.2, the origin of (10.20)–(10.21) is asymptotically stable if $a < 0$ and unstable if $a > 0$. Now suppose $a = 0$. With $\phi(x) = -x^3$, we have

$$(M\phi)(x) = O(\|x\|^7) + \phi(x) - \phi^2(x) + x^3 = O(\|x\|^7) - \phi^2(x) = O(\|x\|^6).$$

Thus, we have $h(x) = -x^3 + O(\|x\|^6)$ and the reduced dynamics (for $a = 0$)

$$\dot{u} = Au + f(u, h(u)) = u^2(-u^3 + O(\|u\|^6)) = -u^5 + O(\|u\|^8).$$

Hence, for $a = 0$, the origin in (10.20)–(10.21) is asymptotically stable. △

The next example includes several approximations to a center manifold.

Example 10.5 Let β be a nonzero real constant. Then the system

$$\dot{x} = -x + \beta z^2,$$

$$\dot{z} = xz$$

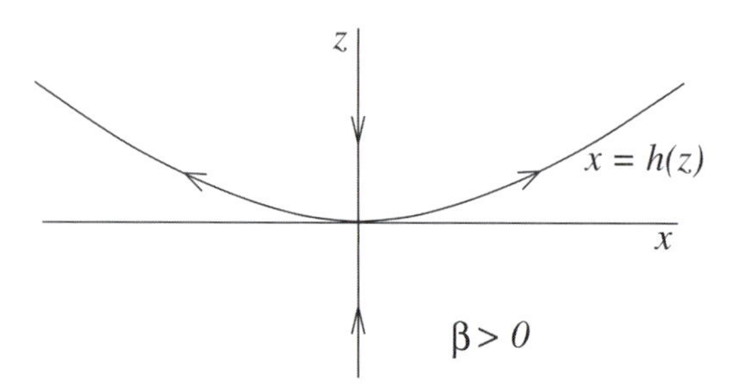

Figure 10.1 An unstable origin in Example 10.5.

has a single equilibrium at $(0, 0)$, and the linearization at $(0, 0)$ has matrix

$$A = \begin{bmatrix} -1 & 0 \\ 0 & 0 \end{bmatrix}.$$

The center subspace of A lies along the z-axis. By Theorem 10.1, there exists a center manifold defined by the graph of $x = h(z)$, where $h(0) = 0$ and $Dh(0) = \frac{dh}{dz}(0) = 0$. The function h must satisfy the exact equation (10.12) for a center manifold; after some rearrangement, that equation is

$$\frac{dh}{dz}(z)[h(z)z] + h(z) - \beta z^2 = 0. \tag{10.23}$$

The reduced dynamics is given by

$$\dot{u} = h(u)u.$$

Express $h(z)$ by a Taylor expansion

$$h(z) = az^2 + bz^3 + cz^4 + dz^5 + \cdots$$

and substitute this series into (10.23). Using the expansion for $h(z)$, we have

$$(2az + 3bz^2 + 4cz^3 + \cdots)(az^3 + bz^4 + cz^5 + \cdots) + az^2 + bz^3 + cz^4 + \cdots - \beta z^2 = 0.$$

Suppose we want an order-2 approximation for h. Then we equate to zero all coefficients of powers z^0, z, z^2, and ignore terms of order z^3 and higher (Theorem 10.3). This requires only that

$$a = \beta,$$

so that the approximation has the form $h(z) = \beta z^2 + O(|z|^3)$ for small $|z|$. The reduced dynamics have the form

$$\dot{z} = h(z)z = \beta z^3 + O(|z|^4) \quad \text{for small } |z|.$$

This is accurate enough to make a determination of stability properties. If $\beta > 0$, then $z = 0$ is unstable; hence, the origin $(x, z) = (0, 0)$ is unstable. (See Figure 10.1 for this unstable case.) If $\beta < 0$, then $z = 0$ is asymptotically stable; hence, the origin $(x, z) = (0, 0)$ is asymptotically stable.

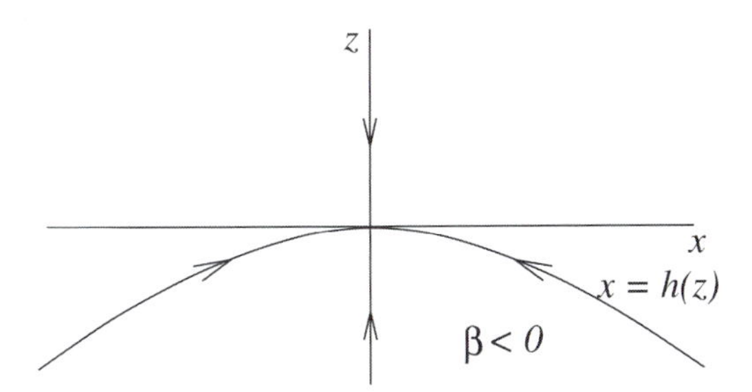

Figure 10.2 An asymptotically stable origin in Example 10.5.

(See Figure 10.2 for the asymptotically stable case.) Both conclusions follow from Theorem 10.2.

If we had asked for an order-3 approximation of h, we would require in addition that $b = 0$, so that $h(z) = \beta z^2 + O(|z|^4)$. The reduced dynamics is given by $\dot{z} = h(z)z = \beta z^3 + O(|z|^5)$ and the conclusions for $\beta \neq 0$ are, of course, the same.

To compute one further approximation, suppose we ask for an order-4 approximation of h. Then, in addition to the previous coefficients, we also set $2a^2 + c = 0$. Since $a = \beta$, this gives $h(z) = \beta z^2 - 2\beta^2 z^4 + O(|z|^5)$, with reduced dynamics given by $\dot{z} = \beta z^3 - 2\beta^2 z^5 + O(|z|^6)$. This approximation adds only more information on the shape of the invariant curve $x = h(z)$, as the stability properties were determined already (for $\beta \neq 0$) by the order-2 approximation.

Finally, it is instructive to see what happens if we do have $\beta = 0$. Then the system is $\dot{x} = -x$, $\dot{z} = xz$. The z-axis consists of equilibria, and therefore $x = h(z) = 0$ defines a center manifold near the origin. Thus, if $\beta = 0$, then the origin $(x, z) = (0, 0)$ is only Lyapunov stable. We note that the approximation approach yields $h(z) = O(|z|^n)$ for every order-n approximation. However, without prior identification of the center manifold $x = h(z) = 0$, this approximation fact would be inconclusive in itself, since there exist polynomial systems with no analytic center manifold (Exercise 10.8); however, the fact is consistent with $x = h(z) = 0$ being a center manifold near the origin. Finally, we note that it is inconclusive to consider the dynamics resulting from the calculation $\frac{dz}{dx} = -z$ for $x \neq 0$, since that construction ignores the equilibria and the center manifold $x = h(z) = 0$. $\triangle$

10.3 TWO APPLICATIONS

This section applies center manifold theory (i) to show how stabilizability is preserved by adding an integrator to a system, and (ii) to deduce a very special case of the asymptotic stability result for cascades.

10.3.1 Adding an Integrator for Stabilization

We consider here planar systems of the form

$$\dot{x} = f(x, y), \tag{10.24}$$
$$\dot{y} = u, \tag{10.25}$$

with state $(x, y) \in \mathbf{R}^2$, $f(0,0) = 0$, and scalar control u. We ask the following question: Under what conditions is the origin $(x, y) = (0, 0)$ asymptotically stabilizable by smooth state feedback? Several cases can be settled easily based on the linearization at the origin. These cases are given in the next lemma; the proof is left to Exercise 10.3.

Lemma 10.1 *Consider system* (10.24)–(10.25). *The following statements are true:*

(a) *If* $\frac{\partial f}{\partial y}(0,0) \neq 0$, *then the origin can be asymptotically stabilized by linear state feedback.*
(b) *If* $\frac{\partial f}{\partial y}(0,0) = 0$ *and* $\frac{\partial f}{\partial x}(0,0) < 0$, *then the origin can be asymptotically stabilized by linear feedback* $u = ky$ *alone.*
(c) *If* $\frac{\partial f}{\partial y}(0,0) = 0$ *and* $\frac{\partial f}{\partial x}(0,0) > 0$, *then the origin cannot be stabilized by any* C^1 *state feedback* $u = \alpha(x, y)$.

The cases not covered by Lemma 10.1 are those for which $\nabla f(0,0) = 0$. Using center manifold theory, sufficient conditions for stabilization of the origin of (10.24)–(10.25) can be given for these critical cases. The next proposition says that if (10.24) is a stabilizable system with input variable y, then augmenting the system with the integrator $\dot{y} = u$ gives a composite system that is stabilizable as well.

Proposition 10.1 *Consider equation* (10.24) *as a control system with input* y, *and assume that* $\nabla f(0,0) = 0$. *If there exists a* C^2 *feedback* $y = k(x)$, *with* $k(0) = 0$, *such that* $\dot{x} = f(x, k(x))$ *has asymptotically stable origin* $x = 0$, *then there exists a* C^1 *state feedback* $u = \alpha(x, y)$ *such that the system*

$$\dot{x} = f(x, y),$$
$$\dot{y} = \alpha(x, y)$$

has asymptotically stable origin $(x, y) = (0, 0)$.

Proof. Define new coordinates by $v = x$ and $w = y - k(x)$. In the new coordinates, (10.24)–(10.25) take the form

$$\dot{v} = f(v, w + k(v)), \tag{10.26}$$
$$\dot{w} = u - \frac{\partial k}{\partial v}(v)\, f(v, w + k(v)). \tag{10.27}$$

Define the feedback

$$u = \alpha(v, w) := -w + \frac{\partial k}{\partial v}(v) f(v, w + k(v)) \tag{10.28}$$

and substitute it into (10.27) to get the closed loop system

$$\dot{v} = f(v, w + k(v)) =: F(v, w), \tag{10.29}$$

$$\dot{w} = -w. \tag{10.30}$$

Since $\nabla f(0,0) = 0$, the linearization of (10.29)–(10.30) at $(v, w) = (0, 0)$ has matrix

$$A = \begin{bmatrix} \frac{\partial F}{\partial v}(0,0) & \frac{\partial F}{\partial w}(0,0) \\ 0 & -1 \end{bmatrix} = \begin{bmatrix} 0 & 0 \\ 0 & -1 \end{bmatrix}.$$

Thus, asymptotic stabilization will be due to the nonlinear terms in the feedback u. The manifold defined by $w = 0$ (the v-axis) is invariant for (10.29)–(10.30); thus, $w = 0$ qualifies as a center manifold for the origin. The reduced dynamics is given by

$$\dot{v} = f(v, k(v)),$$

and, by hypothesis, $v = 0$ is asymptotically stable for this equation. By Theorem 10.2, (10.29)–(10.30) has asymptotically stable origin, hence the feedback (10.28) asymptotically stabilizes (10.24)-(10.25). □

The next example illustrates the feedback construction in the proof of Proposition 10.1.

Example 10.6 Consider the system

$$\dot{x} = x^3 + xy,$$

$$\dot{y} = u.$$

It is straightforward to check that the hypotheses of Proposition 10.1 are satisfied. The feedback $y = k(x) = -2x^2$ stabilizes the origin $x = 0$ of the first equation: $\dot{x} = x^3 + x(-2x^2) = -x^3$. Setting $v = x$ and $w = y - k(x)$, the feedback in (10.28) gives

$$\begin{aligned} u = \alpha(v, w) &:= -w + \frac{\partial k}{\partial v}(v) f(v, w + k(v)), \\ &= -(y + 2x^2) - 4x(x^3 + xy), \end{aligned}$$

which asymptotically stabilizes $(x, y) = (0, 0)$. That is, the origin $(0, 0)$ of the closed loop system

$$\dot{x} = x^3 + xy,$$

$$\dot{y} = -y - 2x^2 - 4x^2 y - 4x^4$$

is asymptotically stable. △

10.3.2 LAS in Special Cascades: Center Manifold Argument

Here we will use center manifold arguments to prove a special case of Corollary 9.2 on asymptotic stability in cascades. We consider the special case of Corollary 9.2 in which the driving system is linear with a Hurwitz coefficient matrix.

Proposition 10.2 *Consider the cascade*

$$\dot{x} = f(x, z), \tag{10.31}$$

$$\dot{z} = Az\,, \tag{10.32}$$

where $f(0,0) = 0$. *If the equilibrium* $x = 0$ *of* $\dot{x} = f(x, 0)$ *is asymptotically stable and the matrix* A *is Hurwitz, then the equilibrium* $(x, z) = (0, 0)$ *of* (10.31)–(10.32) *is asymptotically stable.*

Proof. Begin by expanding $f(x, z)$ as

$$f(x, z) = Fx + Gz + S(x, z),$$

where $F := \frac{\partial f}{\partial x}(0,0)$, $G := \frac{\partial f}{\partial z}(0,0)$, and $DS(0,0) = 0$. We can apply a linear coordinate transformation

$$\begin{bmatrix} w \\ z \end{bmatrix} = \begin{bmatrix} T & 0 \\ 0 & I \end{bmatrix} \begin{bmatrix} x \\ z \end{bmatrix},$$

with T nonsingular, such that

$$TFT^{-1} = \begin{bmatrix} F_1 & 0 \\ 0 & F_2 \end{bmatrix}, \tag{10.33}$$

where all eigenvalues of F_1 have zero real part and all eigenvalues of F_2 have negative real part. Writing $w = [w_1^T \;\; w_2^T]^T$, we have the following system in (w_1, w_2, z):

$$\dot{w}_1 = F_1 w_1 + S_1(w_1, w_2, z), \tag{10.34}$$

$$\dot{w}_2 = F_2 w_2 + S_2(w_1, w_2, z), \tag{10.35}$$

$$\dot{z} = Az. \tag{10.36}$$

In addition, the functions S_1 and S_2 satisfy

$$S_1(0,0,0) = 0 \quad \text{and} \quad S_2(0,0,0) = 0,$$

as well as

$$\frac{\partial S_1}{\partial(w_1, w_2, z)}(0,0,0) = 0 \quad \text{and} \quad \frac{\partial S_2}{\partial(w_1, w_2, z)}(0,0,0) = 0.$$

With $z = 0$ in (10.34)–(10.35), we have the subsystem

$$\dot{w}_1 = F_1 w_1 + S_1(w_1, w_2, 0), \tag{10.37}$$

$$\dot{w}_2 = F_2 w_2 + S_2(w_1, w_2, 0). \tag{10.38}$$

By hypothesis, the equilibrium $(w_1, w_2) = (0, 0)$ of this subsystem is asymptotically stable. (However, we are not assuming that the Jacobian matrix (10.33) is Hurwitz.) Suppose the graph of $w_2 = h(w_1)$ is a center manifold for the origin of (10.37)–(10.38). Thus, $h(w_1)$ satisfies

$$\frac{\partial h}{\partial w_1}(w_1)(F_1 w_1 + S_1(w_1, h(w_1), 0) = F_2\, h(w_1) + S_2(w_1, h(w_1), 0).$$

Moreover, the reduced dynamics

$$\dot{u} = F_1 u + S_1(u, h(u), 0)$$

necessarily has an asymptotically stable equilibrium at $u = 0$. (Otherwise, the hypothesis of the proposition is contradicted, due to Theorem 10.2.)

Now, the cascade (10.34)–(10.36) for (w_1, w_2, z), taken as a whole, must have a center manifold at the origin described by equations of the form

$$z = k_1(w_1) \quad \text{and} \quad w_2 = k_2(w_1),$$

and such that

$$\frac{\partial k_2}{\partial w_1}(F_1 w_1 + S_1(w_1, k_2(w_1), k_1(w_1))\,) = F_2 k_2(w_1) + S_2(w_1, k_2(w_1), k_1(w_1)),$$
$$\frac{\partial k_1}{\partial w_1}(F_1 w_1 + S_1(w_1, k_2(w_1), k_1(w_1))\,) = A k_1(w_1). \tag{10.39}$$

It is straightforward to verify that these last two equations have the solution

$$k_1(w_1) = 0 \quad \text{and} \quad k_2(w_1) = h(w_1).$$

Thus, the reduced dynamics associated with this center manifold is given by

$$\dot{u} = F_1 u + S_1(u, h(u), 0);$$

but this is exactly the reduced dynamics for (10.37)–(10.38) above, which is asymptotically stable. Theorem 10.2 implies asymptotic stability for $(w_1, w_2, z) = (0, 0, 0)$, hence for $(x, z) = (0, 0)$ in (10.31)–(10.32). □

See Exercise 10.4 to extend the center manifold argument of Proposition 10.2 when the linear system $\dot{z} = Az$ in (10.32) is replaced by certain systems of the form $\dot{z} = Az + p(x, z)$, with A still assumed Hurwitz. (Note that the resulting system is in general, no longer a cascade.)

10.4 EXERCISES

Exercise 10.1 *Non-unique Center Manifolds*
Construct the functions $h_C(x_1)$ in Example 10.2 by integrating the separable equation

$$\frac{dx_2}{dx_1} = \frac{x_2}{x_1^3}$$

and splicing solutions at the origin to obtain invariant center manifolds. Find an infinite family of center manifolds through the origin not included in the family defined by the functions h_C in Example 10.2.

Exercise 10.2 *Center Manifold Reduction*
Consider the system

$$\begin{aligned}\dot{x} &= ax^3 + x^2 y,\\ \dot{y} &= -y + y^2 + xy - x^3 .\end{aligned}$$

Use center manifold reduction to investigate the stability property of the origin for any value of the constant a.

Exercise 10.3 Prove Lemma 10.1. For each statement of the lemma, give an example system that illustrates the statement.

Exercise 10.4 Consider the system

$$\begin{aligned}\dot{x} &= f(x, z),\\ \dot{z} &= Az + p(x, z) ,\end{aligned}$$

where the equilibrium $x = 0$ of $\dot{x} = f(x, 0)$ is asymptotically stable, A is Hurwitz, and $p(x, z)$ is such that

$$p(x, 0) = 0 \quad \text{for } x \text{ near } 0, \quad \text{and} \quad \frac{\partial p}{\partial z}(0, 0) = 0.$$

Modify the center manifold argument of Proposition 10.2 to show that the origin $(x, z) = (0, 0)$ is asymptotically stable in this case. *Hint*: Verify that only (10.36) and (10.39) need to be modified; write the appropriate modifications, and then check that the statements after (10.39) in the proof remain valid.

Exercise 10.5 Consider the system

$$\begin{aligned}\dot{x} &= xy,\\ \dot{y} &= u.\end{aligned}$$

(a) Use center manifold theory to show that no linear feedback $u = \alpha x + \beta y$ can make $(x, y) = (0, 0)$ asymptotically stable. *Hint*: Be sure that the linearization at the origin is a block diagonal form.

(b) Use center manifold theory to show that feedback of the form $u = \alpha x^2 - y$, $\alpha < 0$, makes the origin asymptotically stable.

Exercise 10.6 Show that the origin of the system

$$\begin{aligned}\dot{x}_1 &= -x_2^3,\\ \dot{x}_2 &= -x_2 + z^2,\\ \dot{z} &= -z + zx_1^2\end{aligned}$$

is asymptotically stable. Is it globally asymptotically stable? Explain.

Exercise 10.7 Consider the system

$$\begin{aligned}\dot{y} &= z^2 + u,\\ \dot{z} &= y^2 - z^5,\end{aligned}$$

which has a unique equilibrium at the origin when $u = 0$.

(a) Show that the origin is unstable when $u = 0$.
(b) Show that the system satisfies Brockett's necessary condition for stabilization.
(c) Thinking of y as the output of the system, show that there is no output feedback $u = g(y)$ with $g(0) = 0$ and $\frac{dg}{dy}(0) \neq 0$ which can asymptotically stabilize the origin.

Exercise 10.8 *No Real Analytic Center Manifold*
Show that the system

$$\begin{aligned}\dot{x}_1 &= -x_1^3,\\ \dot{x}_2 &= -x_2 + x_1^2\end{aligned}$$

does not have a real analytic center manifold. That is, there is no function $x_2 = h(x_1) = \sum_{k=2}^{\infty} a_k x_1^k$ given by an infinite, convergent Taylor series in an open neighborhood of $x_1 = 0$, which can define a center manifold.

Exercise 10.9 *A Global Phase Portrait*
For the system

$$\begin{aligned}\dot{x} &= x^2 y - x^5,\\ \dot{y} &= -y + x^2,\end{aligned}$$

find all the equilibria, determine their stability properties, and sketch a global phase portrait.

10.5 NOTES AND REFERENCES

In [23], the proof of existence of a center manifold (Theorem 10.1), the proof of the tracking estimate (Theorem 10.2 (b)), and the approximation result in Theorem 10.3 all depend on contraction mapping arguments.

The Lyapunov-theoretic proof of the asymptotic stability result in Theorem 10.2(d) follows an argument in [58].

The application of center manifold theory in Proposition 10.1 is taken from [49]. Proposition 10.2 and its extension in Exercise 10.4 follow [47].

The system of Exercise 10.2 is considered in [23]. The system of Exercise 10.8 is from [23] (page 29); another example in [23] (page 29) gives a polynomial (hence, real analytic) vector field for which there is no C^∞ center manifold.

Exercise 10.5 was suggested by the analysis of this planar system in [4] (pages 292–293).

The system of Exercise 10.7 is from the paper by C. I. Byrnes and A. Isidori, New results and examples in nonlinear feedback stabilization, *Syst. Contr. Lett.*, 12:437–442, 1989; among other things, this paper shows that there is no smooth output feedback $u = g(y)$ that can asymptotically stabilize the origin of this system; however, there does exist a dynamic output feedback stabilizing control.

Center manifold theory is an essential tool in the study of bifurcations of equilibria and the associated stability analysis. The examples and exercises of this chapter provide only an initial experience; they cannot indicate the full complexity of center manifold analysis. For more on center manifolds and connections with bifurcation theory, see [30], [66]. Reference [77] presents ideas and techniques of normal form theory for dynamical systems.

Chapter Eleven

Zero Dynamics

In this chapter we define the relative degree at a point for single-input single-output nonlinear systems. When the relative degree at an equilibrium point is defined and less than the dimension of the state space, there is a coordinate transformation that allows the system to be represented in a normal form which is a cascade system with a linear driving system. Using the cascade normal form, we define the zero dynamics manifold near equilibrium and the zero dynamics subsystem. The stability analysis of the zero dynamics system is important in the study of asymptotic stabilization, including critical problems. This chapter provides some foundations for the discussion of feedback linearization discussed later on; but, most important, the ideas of this chapter are important in many other problems of control and systems theory.

11.1 THE RELATIVE DEGREE AND NORMAL FORM

We study the single-input single-output system

$$\dot{x} = f(x) + g(x)u, \qquad (11.1)$$
$$y = h(x), \qquad (11.2)$$

where f and g are smooth vector fields and h is a smooth real-valued function, all defined in some open set about an equilibrium point for the unforced equation. We generally take this equilibrium to be $x_0 = 0$; thus, $f(0) = 0$.

Definition 11.1 below is motivated by the form of the terms that appear when the output function $y = h(x)$ is differentiated repeatedly along solutions of (11.1). Recall that the Lie derivative of a real-valued function h along a vector field g is the real-valued function $L_g h$ defined by

$$L_g h(x) = dh(x) \cdot g(x),$$

where $dh(x)$ is the differential (row gradient) of h at x. For iterated derivatives of this type, we write $L_g^0 h = h$, and $L_g^k h = L_g\,(L_g^{k-1} h)$.

Definition 11.1 (Relative Degree: SISO System)
The system (11.1)–(11.2) *has* relative degree $r = 1$ *at the point* x_0 *if* $L_g h(x_0) \neq 0$*; and it has* relative degree $r \geq 2$ *at* x_0 *if the following two*

conditions hold:

(i) *for each* $0 \le k \le r-2$, $L_g L_f^k h(x) = 0$ *for all* x *in a neighborhood of* x_0*;*
(ii) $L_g L_f^{r-1} h(x_0) \neq 0$.

When f and g are understood, we also refer to the relative degree of the function h at a given point, meaning the relative degree at that point of the system defined by the triple (h, f, g).

Let $u(t)$ be given, and let $x(t)$ be the solution of (11.1) with $x(0) = x_0$. Then the relative degree of h at x_0 is the number of differentiations of $h(x(t))$ needed to make the input u appear explicitly (Exercise 11.1). There may be points x_0 where a relative degree for h is not defined. Note that if the function $L_g L_f^{r-1} h$ is continuous at x_0, then Definition 11.1 enables us to speak of a function h having relative degree r *in an open set* U containing x_0. Thus, with this continuity assumption, the set of points at which the relative degree of a function is defined is an open subset of the domain of the function.

Example 11.1 Consider the system

$$\dot{x}_1 = \sin x_2,$$
$$\dot{x}_2 = -x_1^2 + u.$$

The function $y = h(x) = x_2$ has relative degree one at all points since $L_g h(x) = 1 \neq 0$. On the other hand, the output $y = \hat{h}(x) = x_1$ satisfies $L_g \hat{h}(x) = 0$ for all x, and $L_g L_f \hat{h}(x) = L_g(\sin x_2) = \cos x_2$, so $\hat{h}$ has relative degree 2 at the origin. The output $y = \hat{h}(x) = x_1$ does not have a well-defined relative degree at any of the points (x_1, x_2) such that $x_2 = (2n+1)\frac{\pi}{2}$, where n is an integer. $\triangle$

The relative degree of a function h is undefined at a point x_0 if the first function of the sequence

$$L_g h(x),\ L_g L_f h(x),\ \ldots\ ,\ L_g L_f^k h(x),\ \ldots$$

which is *not* indentically zero in a neighborhood of x_0 nevertheless has a zero at the point x_0 itself. This is illustrated by the output $y = \hat{h} = x_1$ in Example 11.1.

The next example addresses the calculation of the relative degree for a SISO linear system.

Example 11.2 Consider the linear system

$$\dot{x} = Ax + bu,$$
$$y = cx,$$

where b is $n \times 1$ and c is $1 \times n$. We write $f(x) = Ax$, $g(x) = b$, and $h(x) = cx$, and compute that

$$L_f^k h(x) = cA^k x.$$

Therefore

$$L_g L_f^k h(x) = cA^k b,$$

and the defining conditions for relative degree r are as follows:

$$\begin{aligned} cA^k b &= 0 \quad \text{for all } 0 \le k \le r-2, \\ cA^{r-1} b &\ne 0. \end{aligned}$$

For linear systems defined by a triple (c, A, b) such that $cA^q b \ne 0$ for some nonnegative integer q, the relative degree exists and has the same value r at all points x_0. The example

$$c = \begin{bmatrix} 1 & 0 \end{bmatrix}, \quad A = \begin{bmatrix} 0 & 0 \\ 0 & 1 \end{bmatrix}, \quad b = \begin{bmatrix} 0 \\ 1 \end{bmatrix}$$

shows that a triple (c, A, b) need not have a well-defined relative degree. △

A *local coordinate change* is defined by a mapping $z = T(x)$, where $DT(x_0)$ is nonsingular for some point x_0 of interest. A *regular feedback transformation* is defined by $u = \alpha(x) + \beta(x)v$, where α and β are smooth functions defined in a neighborhood of x_0 and taking values in the same space as u, such that $\beta(x_0)$ is nonsingular (for m-input systems), or $\beta(x_0) \ne 0$ (for single-input systems). The variable v is called the new *reference input*. It is important that the relative degree at x_0 of the output h of a SISO system is invariant under local coordinate transformations and regular feedback transformations defined near x_0. (See Exercise 11.2.)

We return to Example 11.1 to illustrate an important consideration.

Example 11.3 Consider again the system

$$\begin{aligned} \dot{x}_1 &= \sin x_2, \\ \dot{x}_2 &= -x_1^2 + u, \end{aligned}$$

with output $y = x_2$. It is easy to linearize the input-output behavior, using a feedback transformation. Let $u = x_1^2 + v$, which produces the linear input-output relation $\dot{y} = v$. However, by considering only $\dot{y} = v$, some of the original dynamics remains hidden. An important question concerns the stability of the dynamics hidden from the input-output behavior. For example, suppose we wanted to use the feedback $u = x_1^2 + v$ to hold the output $y = x_2$ at a constant value $x_2 = \sigma$; the input $v = 0$ will do this. However, the x_1 solution is then $x_1(t) = x_1(0) + t \sin \sigma$, so that $x_1 \to \infty$ as $t \to \infty$, unless we happen to have $\sin \sigma = 0$. △

The remainder of this section is devoted to the development of a normal form for a single-input single-output system having relative degree $r < n$

at x_0. Using the normal form, the issue of unobservable dynamics can be addressed in a systematic way. In order to obtain the normal form, we need to know that, under the assumption that the relative degree exists at a point x_0 of interest, the functions $h(x), L_f h(x), \ldots, L_f^{r-1}(x)$ can be used as part of a coordinate change. That is, we need to know that the differentials $dh(x), dL_f h(x), \ldots, dL_f^{r-1}h(x)$ are linearly independent near the point x_0. Proposition 11.1 below proves a bit more. But first we need some standard notation.

Definition 11.2 (Lie Bracket of Vector Fields)
The Lie bracket $[g_1, g_2]$ *of two smooth vector fields g_1, g_2 is the vector field*

$$\begin{aligned}[g_1, g_2](x) &:= \frac{\partial g_2}{\partial x}(x) g_1(x) - \frac{\partial g_1}{\partial x}(x) g_2(x) \\ &= Dg_2(x) g_1(x) - Dg_1(x) g_2(x). \end{aligned} \tag{11.3}$$

There is a standard notation that helps with iterated Lie brackets. Write $ad^0_{g_1} g_2 := g_2$, $ad_{g_1} g_2 := [g_1, g_2]$, and, in general, $ad^k_{g_1} g_2 := [g_1, ad^{k-1}_{g_1} g_2]$ for positive integers k. This is called the *ad-notation* for brackets.

The next example shows that Lie brackets are indeed important, even for linear systems.

Example 11.4 If $f(x) = Ax$ and $g(x) = b$, then $[f, g](x) = -Ab$. Also, $ad^2_f g(x) = [f, [f, g]](x) = A^2 b$, and in general, $ad^k_f g(x) = (-1)^k A^k b$. △

Some additional standard terminology is worth mentioning: The Lie bracket $[f, g]$ of vector fields f and g is also called the *Lie derivative of g along f*. Note that $[g, f] = -[f, g]$.

The algebraic property that shows the connection between the Lie bracket operation on vector fields and the Lie derivative operation on functions is the *Jacobi identity*. The Jacobi identity states that if v, w are smooth vector fields and λ is a smooth function, then

$$(L_{[v,w]}\lambda)(x) = L_v L_w \lambda(x) - L_w L_v \lambda(x).$$

(See Exercise 11.4.) By suppressing the argument x, and using the *ad*-notation, we have the identity

$$L_{ad_v w}\lambda = L_{[v,w]}\lambda = L_v L_w \lambda - L_w L_v \lambda.$$

Proposition 11.1 *If the system* (11.1)–(11.2) *has relative degree r in an open set U, then for all $x \in U$,*

(a) *the row vectors $dh(x), dL_f h(x), \ldots, dL_f^{r-1}h(x)$ are linearly independent;*
(b) *the vectors $g(x)$, $ad_f g(x)$, $\ldots$, $ad_f^{r-1} g(x)$ are linearly independent.*

Proof. The proof hinges on showing that relative degree r implies that the matrix product given by

$$\begin{bmatrix} dh(x) \\ dL_f h(x) \\ \cdots \\ dL_f^{r-1} h(x) \end{bmatrix} \begin{bmatrix} g(x) & ad_f g(x) & \cdots & ad_f^{r-1} g(x) \end{bmatrix}$$

$$= \begin{bmatrix} L_g h(x) & L_{ad_f g} h(x) & \cdots & \cdots & L_{ad_f^{r-1} g} h(x) \\ L_g L_f h(x) & & & L_{ad_f^{r-2} g} L_f h(x) & \cdots \\ \cdots & & & & \cdots \\ L_g L_f^{r-1} h(x) & \cdots & \cdots & \cdots & \star \end{bmatrix} \tag{11.4}$$

yields a matrix which is lower right triangular with nonzero entries on the skew-diagonal. The proposition then follows.

We suppress the x arguments of all functions in what follows. By relative degree r, the first column has the required form. Proceed by induction on the columns, using the Jacobi identity. The k, l entry in the matrix is $(d\,L_f^k h) \cdot (ad_f^l g)$, for $0 \le k, l \le r-1$. The skew-diagonal entries in question are those for which $k+l = r-1$. Assume the desired property for column l: thus, assume that $L_{ad_f^l g}\, L_f^k h = 0$ for $k+l \le r-2$, and $L_{ad_f^l g}\, L_f^{r-1-l} h \neq 0$. For column $l+1$ we need $L_{ad_f^{l+1} g}\, L_f^k h = 0$ for $k \le r-3-l$, and $L_{ad_f^{l+1} g}\, L_f^{r-2-l} h \neq 0$. Using the Jacobi identity, the entries in column $l+1$ are given by

$$(k, l+1) \text{ entry} = (d\,L_f^k h) \cdot (ad_f^{l+1} g) = L_f L_{ad_f^l g}\, L_f^k h - L_{ad_f^l g}\, L_f^{k+1} h. \tag{11.5}$$

Apply the induction hypothesis to (11.5) for $k = 0, \ldots, r-3-l$ to get zero. For $k = r-2-l$, only the last term in (11.5) contributes to the skew-diagonal entry, which is

$$-L_{ad_f^l g}\, L_f^{r-1-l} h \neq 0.$$

This completes the induction step. Note also that this is the *negative* of the skew-diagonal entry in column l. Since the first column has last entry $L_g L_f^{r-1} h(x)$, the diagonal entry in column l must be $(-1)^l L_g L_f^{r-1} h(x) \neq 0$ for $l = 0, \ldots, r-1$. Thus, matrix (11.4) is lower right triangular with nonzero skew-diagonal entries, for all $x \in U$. This completes the proof. □

Proposition 11.1 (a) says that when h has relative degree r, then necessarily $r \le n$ and the r functions $h, L_f h, \ldots, L_f^{r-1} h$ may be used as part of a coordinate transformation defined in the set U. Proposition 11.2, below, explains how the remaining coordinate functions may be chosen to simplify the system in the new coordinates. Proposition 11.2 will invoke the Frobenius theorem (Theorem D.4). (This is a good time to see Appendix D for more background on the Frobenius theorem; see "Distributions and the Frobenius Theorem.")

The Frobenius theorem states that a nonsingular (that is, constant dimension) distribution is completely integrable if and only if it is involutive. The case of the Frobenius theorem needed in Proposition 11.2, stated next, involves only a one-dimensional distribution. Observe that any nonsingular one-dimensional distribution $\mathcal{D}$ is the span of a smooth vector field $g(x)$, $\mathcal{D}(x) = \text{span}\,\{g(x)\}$, and must be involutive since $[g,g] = 0$. Thus, if $\mathcal{D}(x) = \text{span}\,\{g(x)\}$ is nonsingular, then it is completely integrable. And $\text{span}\,\{g(x)\}$ is nonsingular on a domain U if and only if $g(x) \neq 0$ on U.

Proposition 11.2 (Transformation to Normal Form)
Suppose the system (11.1)–(11.2) *has relative degree* r *at* x_0. *Let*

$$\begin{aligned}
\phi_1(x) &:= h(x),\\
\phi_2(x) &:= L_f h(x),\\
\cdots &:= \cdots,\\
\phi_r(x) &:= L_f^{r-1} h(x).
\end{aligned}$$

If $r < n$, *then it is possible to find* $n - r$ *additional functions* $\phi_{r+1}(x)$, $\ldots, \phi_n(x)$ *such that the mapping*

$$\Phi(x) = \begin{bmatrix} \phi_1(x) \\ \cdots \\ \phi_n(x) \end{bmatrix}$$

defines a coordinate transformation in a neighborhood of x_0. *The value of the additional functions* $\phi_{r+1}(x), \ldots, \phi_n(x)$ *at* x_0 *may be chosen arbitrarily. Finally, it is possible to choose* $\phi_{r+1}(x), \ldots, \phi_n(x)$ *such that*

$$L_g \phi_j(x) = 0$$

for all $r + 1 \leq j \leq n$ *and all* x *near* x_0.

Proof. By the relative degree assumption, we must have $g(x_0) \neq 0$, so there is an open set containing x_0 on which the distribution $G(x) := \text{span}\,\{g(x)\}$ is nonsingular, which in this case means constant dimension equal to one at each x in that open set. Since G is nonsingular, by the Frobenius theorem (Theorem D.4) there exist $n - 1$ real-valued smooth functions $\psi_1(x), \ldots, \psi_{n-1}(x)$, defined on a neighborhood of x_0, such that, at each point x of that neighborhood, the differentials of these $n - 1$ functions span the orthogonal complement of $G(x)$. That is,

$$G^{\perp}(x) = \text{span}\,\{d\psi_1(x), \ldots, d\psi_{n-1}(x)\}. \tag{11.6}$$

The subspace

$$G^{\perp}(x_0) + \text{span}\,\{dh(x_0), d(L_f h)(x_0), \ldots, d(L_f^{r-1} h)(x_0)\}$$

must have dimension n. To see this, suppose to the contrary that there is a nonzero vector in

$$G(x_0) \cap (\text{span}\,\{dh(x_0), d(L_f h)(x_0), \ldots, d(L_f^{r-1}h)(x_0)\})^{\perp}.$$

This nonzero vector must be a scalar multiple of $g(x_0)$; thus we have

$$d(L_f^k h)(x_0) \cdot g(x_0) = 0, \quad 0 \le k \le r-1.$$

But this contradicts the fact that $d(L_f^{r-1}h)(x_0) \cdot g(x_0) \ne 0$. Therefore

$$\dim\,(G^{\perp}(x) + \text{span}\,\{dh(x), d(L_f h)(x), \ldots, d(L_f^{r-1}h)(x)\}) = n \qquad (11.7)$$

for all x in a neighborhood of x_0. By (11.6) and (11.7), we may choose, within the set of functions $\{\psi_1, \ldots, \psi_{n-1}\}$, $n-r$ functions, which we may assume (without relabeling) are the functions $\psi_1, \ldots, \psi_{n-r}$, such that the n differentials $dh, d(L_f h), \ldots, d(L_f^{r-1}h), d\psi_1, \ldots, d\psi_{n-r}$ are linearly independent at x_0. Letting $\phi_{r+j} = \psi_j$ for $1 \le j \le n-r$, it follows by construction that

$$d\phi_{r+j}(x) \cdot g(x) = 0, \quad \text{for } 1 \le j \le n-r \text{ and all } x \text{ near } x_0.$$

These last conditions are also satisfied by any functions of the form $\phi_{r+j}(x) + \sigma_j$, $1 \le j \le n-r$, where the σ_j are constants. So the values of $\phi_{r+1}, \ldots, \phi_n$ at x_0 can be chosen arbitrarily. □

Given the functions ϕ_j, $1 \le j \le n$, in the proposition, the condition that $L_g\phi_j(x) = 0$ for $r+1 \le j \le n$ and x near x_0 really does provide a simplification of the system description in the new coordinates, because it says that each of the last $n-r$ new coordinate functions satisfies a differential equation whose right-hand side is independent of the input u. (This might be a good exercise here, before reading on, but the straightforward calculation appears momentarily below.)

Given the transformation $z = \Phi(x)$, we now describe the system in the new coordinates defined by $z_i = \phi_i(x)$, $1 \le i \le n$. Let $x(t)$ be any solution corresponding to input $u(t)$; then, following on the relative degree r condition, we have

$$\begin{aligned}
\frac{dz_1}{dt}(t) &= L_f h(x(t)) = \phi_2(x(t)) = z_2(t),\\
\frac{dz_2}{dt}(t) &= L_f^2 h(x(t)) = \phi_2(x(t)) = z_2(t),\\
&\cdots \quad \cdots,\\
\frac{dz_{r-1}}{dt}(t) &= L_f^{r-1}h(x(t)) = \phi_r(x(t)) = z_r(t),
\end{aligned}$$

and finally, the input u appears in

$$\frac{dz_r}{dt}(t) = L_f^r h(x(t)) + L_g L_f^{r-1} h(x(t))u(t).$$

The first $r-1$ of these equations are already expressed fully in terms of the new z coordinates. To express the differential equation for z_r in terms of z, we may write $x(t) = \Phi^{-1}(z(t))$, and then set

$$a(z) = L_g L_f^{r-1} h(\Phi^{-1}(z)),$$
$$b(z) = L_f^r h(\Phi^{-1}(z)).$$

This gives

$$\frac{dz_r}{dt} = b(z(t)) + a(z(t))u(t).$$

By the definition of relative degree r at $x_0 = \Phi^{-1}(z_0)$, we have $a(z_0) \neq 0$, and by continuity the function $a(z)$ is nonzero throughout a neighborhood of z_0.

If there are no conditions on the functions $\phi_{r+1}(x), \ldots, \phi_n(x)$ other than the linear independence of their differentials at x_0, then we cannot expect any special structure in the differential equations for $z_{r+1}, \ldots, z_n$. In general, those differential equations will depend on the input u as well as z. However, by imposing the conditions $L_g\phi_j(x) = 0$ for $r+1 \leq j \leq n$, these last $n-r$ differential equations will have right-hand sides that are independent of the input u. To see this, we compute that, for $r+1 \leq i \leq n$ (and suppressing the t variable),

$$\frac{dz_i}{dt} = L_f\phi_i(x) + L_g\phi_i(x)u = L_f\phi_i(x) = L_f\phi_i(\Phi^{-1}(z)).$$

By writing

$$q_i(z) := L_f\phi_i(\Phi^{-1}(z)) \quad \text{for } r+1 \leq i \leq n,$$

the last $n-r$ differential equations take the form

$$\frac{dz_i}{dt} = q_i(z(t)) \quad \text{for } r+1 \leq i \leq n.$$

The full system in z coordinates is

$$\begin{aligned}
\dot{z}_1 &= z_2, \\
\dot{z}_2 &= z_3, \\
&\cdots \cdots, \\
\dot{z}_{r-1} &= z_r \\
\dot{z}_r &= b(z) + a(z)u, \\
\dot{z}_{r+1} &= q_{r+1}(z), \\
&\cdots \cdots, \\
\dot{z}_n &= q_n(z)
\end{aligned} \tag{11.8}$$

with output given by

$$y = z_1.$$

Note that the output is $y = h(\Phi^{-1}(z)) = \phi_1(\Phi^{-1}(z)) = z_1$, and the functions $a(z)$, $b(z)$, $q_{r+1}(z), \dots, q_n(z)$ are as defined above.

One of the most important features of the form (11.8) can be seen now. By defining the feedback transformation

$$u = -\frac{b(z)}{a(z)} + \frac{1}{a(z)}v,$$

where v is a new input variable, the first r equations from (11.8) take the form of a chain of r integrations of the input v, which is a controllable linear system in the state $\xi := [z_1, \dots, z_r]^T$, with input v. Using v, an additional linear feedback depending only on ξ, say $v = \sum_{i=1}^{r} k_i z_i$, results in a closed loop system that has a cascade form, as follows:

$$\begin{aligned} \dot{\xi} &= A\xi, \\ \dot{z}_{r+1} &= q_{r+1}(\xi, z_{r+1}, \dots, z_n), \\ &\cdots \cdots \\ \dot{z}_n &= q_n(\xi, z_{r+1}, \dots, z_n), \end{aligned} \tag{11.9}$$

where A is a companion matrix with last row equal to $[k_1 \ \cdots \ k_r]$.

The equations $L_g\phi_j(x) = 0$, for $r+1 \le j \le n$, constitute a system of $n-r$ partial differential equations, which may be difficult to solve for the ϕ_i. If we simply choose $\phi_{r+1}(x), \dots, \phi_n(x)$ to complete a coordinate transformation $z = \Phi(x)$, without requiring $L_g\phi_j(x) = 0$ for $r+1 \le j \le n$, then we obtain a system in z coordinates where the first r equations (and the output) are the same as in the previous form (11.8), but now the last $n-r$ differential equations take the following form:

$$\begin{aligned} \dot{z}_{r+1} &= q_{r+1}(z) + p_{r+1}(z)u, \\ &\cdots \quad \cdots \\ \dot{z}_n &= q_n(z) + p_n(z)u, \end{aligned}$$

where the functions $q_i(z)$ and $p_i(z)$ are defined for $r+1 \le i \le n$ by

$$q_i(z) = L_f\phi_i(\Phi^{-1}(z))$$

and

$$p_i(z) = L_g\phi_i(\Phi^{-1}(z)).$$

For this system, however, the feedback constructions described in the previous paragraph do not yield a cascade system.

Consider the next example, which illustrates the normal form (11.8) and its value for stabilization problems.

Example 11.5 Consider the system

$$\begin{aligned}\dot{x}_1 &= e^{-x_2} + u,\\ \dot{x}_2 &= x_1 e^{-x_1} - x_2^3 + x_2 x_3,\\ \dot{x}_3 &= x_1 x_2 - x_3^3,\\ y &= x_1\end{aligned}$$

with real scalar input u. It is not difficult to see that the origin is an unstable equilibrium when $u = 0$. We want to use the output to achieve the normal form and reveal useful structure to help in stabilizing the origin. Since $y = x_1$, we have $\dot{y} = \dot{x}_1 = e^{-x_2} + u$, so the relative degree at the origin is one. Set $\xi = x_1$ and $u = -e^{-x_2} + v$, where v is a new reference input. The resulting equations are

$$\begin{aligned}\dot{\xi} &= v,\\ \dot{x}_2 &= \xi e^{-\xi} - x_2^3 + x_2 x_3,\\ \dot{x}_3 &= \xi x_2 - x_3^3.\end{aligned}$$

This is a useful normal form; note that a linear feedback to stabilize ξ will yield the form (11.9) above. Clearly, we can stabilize ξ to zero by setting $v = -\xi$. Observe that ξ is the driving variable for the x_2, x_3 subsystem; note the cascade structure. Most important, when $\xi = 0$, the decoupled x_2, x_3 system, the driven system of the cascade, is given by

$$\begin{aligned}\dot{x}_2 &= -x_2^3 + x_2 x_3,\\ \dot{x}_3 &= -x_3^3;\end{aligned}$$

and this system (which is a cascade itself) has an asymptotically stable origin $(x_2, x_3) = (0, 0)$. In summary, using the combined feedback defined by

$$u = -e^{-x_2} + v = -e^{-x_2} - \xi = -e^{-x_2} - x_1,$$

we have stabilized the origin of the original three-dimensional system. △

It is interesting to note that when we set $\xi = 0$, the x_2, x_3 subsystem in Example 11.5 gives the dynamics compatible with identically zero output $y = x_1 = \xi \equiv 0$; this was accomplished by means of the initial feedback, denoted here by $u_{zd} := -e^{-x_2}$. This subsystem played a valuable role in the stabilization problem.

We consider another example of the normal form (11.8), which illustrates several important points.

Example 11.6 Consider the system

$$\begin{aligned}\dot{x}_1 &= x_2,\\ \dot{x}_2 &= x_3 + (\sin x_1)u,\\ \dot{x}_3 &= (\cos x_1)u.\end{aligned}$$

This is not in normal form already. Suppose we wish to study the system near $x_0 = 0$. The output function $y = h(x) = x_3$ has relative degree one at the origin, as is easily verified. We will define a coordinate transformation

$$z = \Phi(x) = \begin{bmatrix} h(x) \\ \phi_2(x) \\ \phi_3(x) \end{bmatrix} = \begin{bmatrix} x_3 \\ \phi_2(x) \\ \phi_3(x) \end{bmatrix}$$

to the normal form (11.8). The condition $L_g\phi(x) = 0$ for x near zero, reads

$$\begin{bmatrix} \frac{\partial \phi}{\partial x_1}(x) & \frac{\partial \phi}{\partial x_2}(x) & \frac{\partial \phi}{\partial x_3}(x) \end{bmatrix} \begin{bmatrix} 0 \\ \sin x_1 \\ \cos x_1 \end{bmatrix} = \sin x_1 \frac{\partial \phi}{\partial x_2}(x) + \cos x_1 \frac{\partial \phi}{\partial x_3}(x) = 0$$

for x near zero, and we require that both ϕ_2 and ϕ_3 satisfy this first order partial differential equation. Essentially by inspection, note that by taking $\frac{\partial \phi}{\partial x_2} = -\cos x_1$ and $\frac{\partial \phi}{\partial x_3} = \sin x_1$, we obtain the solution

$$\phi_2(x) = -x_2 \cos x_1 + x_3 \sin x_1.$$

And by taking $\frac{\partial \phi}{\partial x_2} = 0$ and $\frac{\partial \phi}{\partial x_3} = 0$, we obtain the solution

$$\phi_3(x) = x_1,$$

for example. Then the transformation is

$$\begin{bmatrix} z_1 \\ z_2 \\ z_3 \end{bmatrix} = \begin{bmatrix} x_3 \\ -x_2 \cos x_1 + x_3 \sin x_1 \\ x_1 \end{bmatrix}$$

and its Jacobian matrix at the origin,

$$D\Phi(0) = \begin{bmatrix} 0 & 0 & 1 \\ 0 & -1 & 0 \\ 1 & 0 & 0 \end{bmatrix},$$

is nonsingular. Direct substitutions now yield the following differential equations in the z coordinates:

$$\begin{aligned} \dot{z}_1 &= (\cos z_3)u, \\ \dot{z}_2 &= x_2^2 \sin z_3 + (x_2 - 1)z_1 \cos z_3, \\ \dot{z}_3 &= x_2, \end{aligned}$$

where

$$x_2 = \frac{-z_2 + z_1 \sin z_3}{\cos z_3}.$$

Note that the control variable u is absent from the last two equations, as guaranteed by the normal form.

The z_1 component can be stabilized to zero by using the feedback transformation $u = (1/\cos z_3)v$ with $v = -z_1$, which yields the well-defined feedback $u = -(1/\cos z_3)z_1$ near $z = 0$. The closed loop system is then in cascade form with driving system $\dot{z}_1 = -z_1$. If we set $z_1 = 0$ and note that $x_2 = -z_2/\cos z_3$ when $z_1 = 0$, then we obtain the decoupled subsystem

$$\dot{z}_2 = \frac{z_2^2}{\cos^2 z_3} \sin z_3, \tag{11.10}$$

$$\dot{z}_3 = -\frac{z_2}{\cos z_3}. \tag{11.11}$$

We have a critical problem of stability for the origin of (11.10)–(11.11); it can be examined more closely in Exercise 11.6. However, observe that by Corollary 9.2, if the origin $(z_2, z_3) = (0, 0)$ is asymptotically stable, then the origin of the full, closed loop z system is also asymptotically stable. However, if $(z_2, z_3) = (0, 0)$ is merely stable, or if $(z_2, z_3) = (0, 0)$ is unstable, then we cannot draw a conclusion about the stability of $z = 0$ in the closed loop without further analysis of the full closed loop system. △

11.2 THE ZERO DYNAMICS SUBSYSTEM

The main goal of this section is to define the *zero dynamics* (or *zero dynamics subsystem*) of a single-input single-output nonlinear system. These dynamics are the internal state dynamics of the system, in a neighborhood of a point x_0, which are compatible with the output function being identically zero on some nontrivial time interval about the initial time $t = 0$. This was exactly the case with the x_2, x_3 subsystem we considered in Example 11.5. In addition to the definition of zero dynamics, this section includes an important observation about certain eigenvalues of the linearization of the zero dynamics subsystem.

Consider the normal form of a system as given in (11.8), before any feedback transformations. Write

$$\xi = \begin{bmatrix} z_1 \\ \vdots \\ z_r \end{bmatrix}$$

and

$$\eta = \begin{bmatrix} z_{r+1} \\ \vdots \\ z_n \end{bmatrix}.$$

Then the normal form for a single-input single-output system with relative degree $r < n$, at a point $z_0 = \Phi(x_0)$, may be written

$$\begin{aligned} \dot{z}_1 &= z_2, \\ \dot{z}_2 &= z_3, \\ &\cdots\cdots, \\ \dot{z}_{r-1} &= z_r \\ \dot{z}_r &= b(\xi,\eta) + a(\xi,\eta)u, \\ \dot{\eta} &= q(\xi,\eta), \\ y &= z_1. \end{aligned} \tag{11.12}$$

Recall that we may write

$$\begin{aligned} a(\xi,\eta) &= L_g L_f^{r-1} h(\Phi^{-1}(\xi,\eta)), \\ b(\xi,\eta) &= L_f^r h(\Phi^{-1}(\xi,\eta)). \end{aligned}$$

The normal form is obtained via coordinate transformation, so we should be explicit about how our standard assumptions are transformed to the normal form. If we assume, for the original system (11.1)–(11.2), that $f(x_0) = 0$ and $h(x_0) = 0$, then clearly we have $\xi = 0$ at x_0. We have seen that the value of η may be chosen arbitrarily at x_0; therefore it is possible to choose $\eta = 0$ at x_0 (Proposition 11.2). Thus, if x_0 is an equilibrium for the unforced equation (11.1), then, by the choices just made, the point $(\xi,\eta) = (0,0)$ is an equilibrium for the unforced system (11.12). Therefore we have

$$\begin{aligned} b(\xi,\eta) &= 0 \quad \text{at } (\xi,\eta) = (0,0), \\ q(\xi,\eta) &= 0 \quad \text{at } (\xi,\eta) = (0,0)\ . \end{aligned}$$

Suppose we want to describe the internal dynamics of the system, in a neighborhood of $(x_0, u_0) = (0,0)$, compatible with the constraint that the output is identically zero on some time interval about $t = 0$. That is, we want to find all initial conditions x such that, for some input $u(t)$ defined on an interval about $t = 0$, the corresponding output function $y(t) \equiv 0$ for all t in a neighborhood of $t = 0$. To identify these dynamics, we analyze the normal form. The output in the normal form is $y(t) = z_1(t)$, so the constraint that $y(t) = 0$ for all t near zero implies the following additional constraints:

$$z_2(t) = z_3(t) = \cdots = z_r(t) = 0,$$

and therefore $\xi(t) = 0$ for all t near zero. Note that the same constraints in the original x coordinates are as follows:

$$h(x) = 0, \quad L_f h(x) = 0, \ \ldots\ , \quad L_f^{r-1} h(x) = 0.$$

By the linear independence of the differentials of these functions at x_0, these equations define a manifold $Z \subset \mathbf{R}^n$ of dimension $n - r$. In addition, since we must have $\dot{z}_r(t) = 0$ for all t, the scalar input $u(t)$ is uniquely determined by the equation

$$0 = b(0, \eta(t)) + a(0, \eta(t))u(t);$$

the uniqueness of $u(t)$ follows from $a(0, \eta(t)) \neq 0$ for t near zero. Finally, $\eta(t)$ must evolve according to the differential equation

$$\dot{\eta}(t) = q(0, \eta(t)). \tag{11.13}$$

In summary, in order to zero the output $y(t)$, the initial condition must have the form $(\xi(0), \eta(0)) = (0, \eta_0)$, where η_0 is arbitrary; given η_0, the unique input $u(t)$ that maintains the constraint of zero output can be expressed by

$$u_{zd}(t) = -\frac{b(0, \eta(t))}{a(0, \eta(t))} = -\frac{L_f^r h(\Phi^{-1}(0, \eta(t)))}{L_g L_f^{r-1} h(\Phi^{-1}(0, \eta(t)))}, \tag{11.14}$$

where $\eta(t)$ is the solution of the differential equation (11.13) with $\eta(0) = \eta_0$. For each choice of initial condition $(0, \eta(0))$, the input (11.14) is the unique input that maintains zero output. Note that this input can be expressed in feedback form

$$u_{zd} := -\frac{b(0, \eta)}{a(0, \eta)} = -\frac{L_f^r h(\Phi^{-1}(0, \eta))}{L_g L_f^{r-1} h(\Phi^{-1}(0, \eta))}. \tag{11.15}$$

Definition 11.3 *The feedback* (11.15) *is called the* zero dynamics feedback.

The zero dynamics feedback is the unique feedback that renders the manifold

$$Z := \{x : h(x) = 0, \; L_f h(x) = 0, \; \ldots \; , \; L_f^{r-1} h(x) = 0\} \tag{11.16}$$

invariant for the resulting closed loop system in x.

Definition 11.4 *The manifold Z in* (11.16) *is called the* zero dynamics manifold.

Note that the zero dynamics feedback can be expressed easily in terms of the original x coordinates for x in Z; we have

$$u_{zd} := -\frac{L_f^r h(x)}{L_g L_f^{r-1} h(x)} \quad \text{for } x \in Z. \tag{11.17}$$

Definition 11.5 *The subsystem in* (11.13), *that is,*

$$\dot{\eta} = q(0, \eta) \tag{11.18}$$

is called the zero dynamics subsystem, *or simply the* zero dynamics.

The zero dynamics subsystem is important in the study of asymptotic stabilization and many other control problems.

We end the section with some useful facts about the linearization of the zero dynamics subsystem. First, the linear approximation at $\eta = 0$ of the zero dynamics of a nonlinear system coincides with the zero dynamics of the linear approximation of the entire system at $x = 0$. (See Exercise 11.7.)

In the remainder of this section, we prove the following important lemma.

Lemma 11.1 *Eigenvalues associated with uncontrollable modes of the linearization at the origin of* (11.1)–(11.2), *if any exist, must be eigenvalues of the linearization at the origin of the zero dynamics subsystem* (11.18).

Proof. Begin with the normal form (11.12) obtained by a transformation $[\xi^T \ \eta^T]^T = \Phi(x)$, where $\xi = [z_1 \ \cdots \ z_r]^T$, and observe that its linearization at $(\xi, \eta) = (0, 0)$ is given by

$$\begin{aligned} \dot{z}_1 &= z_2, \\ \dot{z}_2 &= z_3, \\ &\cdots \cdots, \\ \dot{z}_{r-1} &= z_r, \\ \dot{z}_r &= R\xi + S\eta + Ku, \\ \dot{\eta} &= P\xi + Q\eta, \end{aligned} \tag{11.19}$$

where

$$R := \frac{\partial b}{\partial \xi}(0,0), \qquad S := \frac{\partial b}{\partial \eta}(0,0),$$

$$P := \frac{\partial q}{\partial \xi}(0,0), \qquad Q := \frac{\partial q}{\partial \eta}(0,0),$$

and

$$K := a(0,0).$$

Suppose the linearization is uncontrollable, and let λ be an eigenvalue associated with uncontrollable modes. It will be convenient to write $R =: [r_1 \ r_2 \ \cdots \ r_r]$. By the PBH controllability test, the $n \times (n+1)$ matrix

$$\begin{bmatrix} \lambda & -1 & 0 & \cdots & 0 & 0 & 0 \\ 0 & \lambda & -1 & \cdots & 0 & 0 & 0 \\ \vdots & \vdots & \ddots & \vdots & \vdots & \vdots & \vdots \\ 0 & 0 & 0 & \cdots & -1 & 0 & 0 \\ -r_1 & -r_2 & -r_3 & \cdots & \lambda - r_r & -S & K \\ & & -P & & & \lambda I - Q & 0 \end{bmatrix} \tag{11.20}$$

has rank less than n. Note that the next-to-last block column has dimension $n \times (n - r)$. We wish to conclude that λ is an eigenvalue of Q. To see it, observe that $K = L_g L_f^{r-1}(0) \neq 0$, so that the first r rows are clearly linearly independent. If $\lambda I - Q$ were nonsingular, then the matrix (11.20) would have full row rank n, contrary to the hypothesis that λ is associated with uncontrollable modes. Therefore $\lambda I - Q$ is singular and λ is an eigenvalue of Q, as we wished to show. □

11.3 ZERO DYNAMICS AND STABILIZATION

We begin this section with some terminology used frequently in the control literature. Then we proceed to discuss stabilization by zero dynamics analysis.

Some Terminology for Stability Properties

We assume as before that the transformation $z = \Phi(x)$ of (11.1)–(11.2) to normal form maps the origin to the origin. If the zero dynamics subsystem (11.18) has an asymptotically stable equilibrium at $\eta = 0$, then (11.1)–(11.2) is called a *minimum phase* system. If the zero dynamics has merely a Lyapunov stable equilibrium at $\eta = 0$, then (11.1)–(11.2) is called a *weakly minimum phase* system. (There are sometimes technical conventions in the literature. In [86], the *minimum phase* and *weak minimum phase* conditions presuppose a C^2 Lyapunov function for the zero dynamics. The terms *minimum phase* and *weak minimum phase* originate with the frequency domain (transfer function) analysis of linear time invariant systems; for more background, see M. Schetzen, *Linear Time-Invariant Systems*, IEEE Press, 2003 (Chapter 9)).

Local Asymptotic Stabilization

Consider again the normal form (11.12). Suppose that we apply feedback of the form

$$u = \frac{1}{a(\xi,\eta)}(-b(\xi,\eta) - k_0 z_1 - k_1 z_2 - \cdots - k_{r-1} z_r), \tag{11.21}$$

where $k_0, \ldots, k_{r-1}$ are real numbers; the result is a closed loop system of the form

$$\dot{\xi} = A\xi, \tag{11.22}$$

$$\dot{\eta} = q(\xi,\eta), \tag{11.23}$$

where

$$A = \begin{bmatrix} 0 & 1 & 0 & \cdots & 0 \\ 0 & 0 & 1 & \cdots & 0 \\ \cdot & \cdot & \cdot & \cdots & 1 \\ -k_0 & -k_1 & -k_2 & \cdots & -k_{r-1} \end{bmatrix}$$

is a companion matrix. The characteristic polynomial of A is

$$p(\lambda) = \lambda^r + k_{r-1}\lambda^{r-1} + \cdots + k_1\lambda + k_0. \tag{11.24}$$

Due to the cascade form of (11.22)–(11.23), we have the following proposition.

Proposition 11.3 (LAS via Zero Dynamics)
Consider the normal form (11.12) *of system* (11.1)–(11.2). *Suppose that the equilibrium* $\eta = 0$ *of the zero dynamics is asymptotically stable and that the coefficients* $k_0, k_1, \ldots, k_{r-1}$ *in* (11.24) *are chosen such that all the roots of the polynomial* $p(\lambda)$ *have negative real part. Then the feedback* (11.21) *asymptotically stabilizes the equilibrium* $(\xi, \eta) = (0, 0)$.

Proposition 11.3 follows directly from the general result on local asymptotic stability for cascades (Corollary 9.2) applied to the closed loop (11.22)–(11.23) defined by the feedback (11.21). The new feature here is that the output function is essentially free to choose, based on specific knowledge of the system or in a more experimental sense.

Of course, with an appropriate choice of the k_i, expressing (11.21) directly in terms of the original x coordinates gives a stabilizing feedback $u = \alpha(x)$ for the origin of (11.1)–(11.2).

Example 11.7 Consider the system

$$\begin{aligned}
\dot{x}_1 &= x_2 - x_1^3, \\
\dot{x}_2 &= -x_1 + x_2 + 2x_3 + u, \\
\dot{x}_3 &= x_2 - 3x_3, \\
y &= x_2,
\end{aligned}$$

which has relative degree one at the origin. The zero dynamics feedback is $u_{zd} = x_1 - x_2 - 2x_3$. We may use the coordinate transformation $z_1 = x_2$, $z_2 = x_1$, $z_3 = x_3$, and apply the feedback transformation $u = u_{zd} + v$ to write the system as

$$\begin{aligned}
\dot{z}_1 &= v, \\
\dot{z}_2 &= z_1 - z_2^3, \\
\dot{z}_3 &= z_1 - 3z_3.
\end{aligned}$$

Note that the zero dynamics subsystem is given by

$$\begin{aligned}
\dot{z}_2 &= -z_2^3, \\
\dot{z}_3 &= -3z_3.
\end{aligned}$$

Thus, the zero dynamics has an asymptotically stable origin. However, the linearization of the zero dynamics subsystem is not asymptotically stable. By Proposition 11.3, the full system has an asymptotically stable origin after the application of the additional feedback $v = -z_1$. In summary, since $z_1 = x_2$, the feedback

$$u = u_{zd} + v = (x_1 - x_2 - 2x_3) - x_2 = x_1 - 2x_2 - 2x_3$$

makes the origin of the original system asymptotically stable. △

In Example 11.7, the possibility of stabilizing the nonlinear system by linear state feedback might have been deduced from the linearization, which is stabilizable. This fact may not be immediately obvious, since the linearization is not in the controllability normal form. However, the eigenvalues might have been computed, and the PBH stabilizability test could have been carried out.

In contrast to Example 11.7, if the linear approximation of the system has uncontrollable modes associated only with eigenvalues on the imaginary axis, then the zero dynamics of the nonlinear system may still be asymptotically stable due to the nonlinear terms. Thus, Proposition 11.3 covers many critical cases of asymptotic stabilization by smooth feedback when the linearization is not stabilizable by linear feedback, and this is the real contribution of the proposition. In the next example, the linearization is not stabilizable by linear feedback, so this is truly a critical problem of stability, solved with the help of Proposition 11.3.

Example 11.8 Consider the system

$$\begin{aligned} \dot{x}_1 &= x_1^2 x_2^5, \\ \dot{x}_2 &= x_2 + u. \end{aligned}$$

If we can maintain the relation $x_2 = \alpha(x_1) := -x_1$ by feedback, then the x_1 component is stabilized, and we can study the zero dynamics of the system defined by the output $y = x_2 - \alpha(x_1) = x_2 + x_1$. This output has relative degree one at the origin. To get a complete set of coordinates for a normal form we may use $(\xi, \eta) := (x_2 + x_1, x_1)$. To see this in detail, note that the zero dynamics feedback, using the output just described, is

$$u_{zd} = -x_2 - x_1^2 x_2^5.$$

After applying the feedback $u = u_{zd} + v$, we have the following equations in the (ξ, η) coordinates:

$$\begin{aligned} \dot{\xi} &= v, \\ \dot{\eta} &= \eta^2 (\xi - \eta)^5. \end{aligned}$$

The zero dynamics subsystem is $\dot{\eta} = -\eta^7$, which has asymptotically stable origin $\eta = 0$. Proposition 11.3 says that the (ξ, η) system has asymptotically stable origin if we set $v = -\xi$, for example. Thus we may write

$$u = u_{zd} + v = -x_2 - x_1^2 x_2^5 - (x_2 + x_1) = -2x_2 - x_1 - x_1^2 x_2^5$$

to get a stabilizing feedback in the original coordinates; the asymptotic stability of the origin in the closed loop system given by

$$\begin{aligned} \dot{x}_1 &= x_1^2 x_2^5, \\ \dot{x}_2 &= -x_2 - x_1 - x_1^2 x_2^5 \end{aligned}$$

is thus assured. △

Is the origin of the closed loop in Example 11.8 globally asymptotically stable? That is a different question, and we make the following general observation about it. In the normal form, the components of the state ξ of the linearized subsystem are the output $h(x)$ and its Lie derivatives $L_f^k h(x)$ up to order $r-1$; these components act as inputs for the nonlinear zero dynamics subsystem according to the equation $\dot{\eta} = q(\xi, \eta)$. The injection of these inputs into the zero dynamics may prevent global asymptotic stabilization of minimum phase systems, although local asymptotic stabilization of minimum phase systems is always possible by Proposition 11.3.

11.4 VECTOR RELATIVE DEGREE OF MIMO SYSTEMS

The concept of relative degree can be extended in a very useful and natural way to systems with multivariable input and output. The concept of vector relative degree leads to a normal form and a concept of zero dynamics for multi-input multi-output nonlinear systems. The full extension can be pursued through a reading of [47] (Chapter 5). In this section, we will present the definition of *vector relative degree* and discuss appropriate notations, since a special case of vector relative degree is used in the chapter on passivity.

Consider a system described by equations of the form

$$\begin{aligned} \dot{x} &= f(x) + \sum_{i=1}^{m} g_i(x) u_i, \\ y_1 &= h_1(x), \\ &\cdots, \\ y_m &= h_m(x), \end{aligned}$$

where the $h_j(x)$, $1 \le j \le m$, are smooth real-valued output functions and the $g_j(x)$, $1 \le j \le m$, are smooth vector fields, all defined on some open set in $\mathbf{R}^n$. It is often convenient, when this context is understood, to write such a system as

$$\dot{x} = f(x) + g(x)u, \tag{11.25}$$

$$y = h(x), \tag{11.26}$$

with the understanding that $u = [u_1 \cdots u_m]^T$, $y = [y_1 \cdots y_m]^T$, and

$$\begin{aligned} g(x) &= [\, g_1(x) \;\; \cdots \;\; g_m(x) \,]_{n \times m}, \\ h(x) &= [\, h_1(x) \;\; \cdots \;\; h_m(x) \,]^T. \end{aligned}$$

For each output component $y_i = h_i(x)$, we might consider its relative degree with respect to each of the inputs u_j. To do so, we can consider the single-input single-output system where the only nonzero input is u_j. Fix an index i. If each of these SISO relative degrees (of y_i with respect to the u_j) is defined at the point x_0, there is a smallest such relative degree, which occurs for some input u_{j_i}, and which we label r_i. Thus, the row vector function

$$[L_{g_1} L_f^{r_i-1} h(x) \quad \cdots \quad L_{g_m} L_f^{r_i-1} h(x)]$$

is nonzero at x_0, and, in particular, $L_{g_{j_i}} L_f^{r_i-1} h(x_0) \neq 0$. The input u_{j_i} is the first input to appear explicitly when the output y_i is differentiated repeatedly along system solutions; moreover, u_{j_i} appears after exactly $r_i - 1$ differentiations.

Suppose we place these row vector functions (one for each output component) as rows of an $m \times m$ matrix, as follows:

$$A(x) := \begin{bmatrix} L_{g_1} L_f^{r_1-1} h(x) & \cdots & L_{g_m} L_f^{r_1-1} h(x) \\ L_{g_1} L_f^{r_2-1} h(x) & \cdots & L_{g_m} L_f^{r_2-1} h(x) \\ \cdots & \cdots & \cdots \\ L_{g_1} L_f^{r_m-1} h(x) & \cdots & L_{g_m} L_f^{r_m-1} h(x) \end{bmatrix}. \tag{11.27}$$

For example, consider the system

$$\begin{aligned} \dot{x}_1 &= f_1(x) + (\sin x_1) u_1 + (\sin x_2) u_2, \\ \dot{x}_2 &= f_2(x) + u_2, \\ y_1 &= x_1, \\ y_2 &= x_2. \end{aligned}$$

For this system, with $x_0 = 0$, we have

$$A(0) = \begin{bmatrix} 0 & 0 \\ 0 & 1 \end{bmatrix}.$$

As this example shows, there is no guarantee that (11.27) is nonsingular in a neighborhood of x_0. We can now state the definition of vector relative degree.

Definition 11.6 (Vector Relative Degree)
System (11.25)–(11.26) *has* vector relative degree $(r_1, r_2, \ldots, r_m)$ *at the point* x_0, *if*

(i) $L_{g_j} L_f^k h_i(x) = 0$ *for all* $1 \leq j \leq m$, *for all* $k \leq r_i - 2$, *for all* $1 \leq i \leq m$, *and for all* x *in a neighborhood of* x_0, *and*
(ii) *the* $m \times m$ *matrix* $A(x)$ *in* (11.27) *is nonsingular at* $x = x_0$.

Note that, as defined earlier for the SISO case, we set $r_i = 1$ *if* $L_{g_j} h_i(x_0) \neq 0$ *for some* j, *and (ii) holds.*

We note briefly that the vector relative degree at a point can be defined in a similar way for systems having a different number of input and output components, but such cases will not be needed in this book.

The notion of a system having *uniform relative degree one* is useful in the chapter on passivity.

Definition 11.7 (Uniform Relative Degree One)
System (11.25)–(11.26) *has* uniform relative degree one *at* x_0 *if the vector relative degree exists at* x_0 *and equals* $(1, 1, \ldots, 1)$.

Example 11.9 Consider the following planar system with the full state available as output:

$$\begin{aligned}
\dot{x}_1 &= a x_1 - \alpha x_1 x_2 + u_1, \\
\dot{x}_2 &= -c x_2 + \gamma x_1 x_2 + u_2, \\
y_1 &= h_1(x) = x_1, \\
y_2 &= h_2(x) = x_2.
\end{aligned}$$

The input vector fields are $g_1(x) = [1 \quad 0]^T$ and $g_2(x) = [0 \quad 1]^T$. It is straightforward to check that this system has uniform relative degree one. From the definition of $A(x)$ in (11.27), we have

$$A(x) = \begin{bmatrix} L_{g_1} h_1(x) & L_{g_2} h_1(x) \\ L_{g_1} h_2(x) & L_{g_2} h_2(x) \end{bmatrix} = \begin{bmatrix} 1 & 0 \\ 0 & 1 \end{bmatrix}.$$

Since $A(x)$ is nonsingular at each x, this system has uniform relative degree one at each point in the plane. △

The following notation is natural and convenient. For system (11.25)–(11.26), write

$$L_g h(x) := \begin{bmatrix} L_{g_1} h_1(x) & \cdots & L_{g_m} h_1(x) \\ \cdots & \cdots & \cdots \\ L_{g_1} h_m(x) & \cdots & L_{g_m} h_m(x) \end{bmatrix}.$$

If system (11.25)–(11.26) has uniform relative degree one at x_0, then

$$A(x) = L_g h(x)$$

is nonsingular in a neighborhood of x_0. Note that this is another way of expressing the fact that when we differentiate the output (11.26) once along system solutions, we obtain

$$\dot{y} = L_f h(x) + L_g h(x) u$$

where $L_g h(x)$ is nonsingular at x_0.

The special case of zero dynamics feedback for a system with uniform relative degree one is helpful in the chapter on passivity. If system (11.25)–(11.26) has uniform relative degree one at x_0, then the feedback

$$u_{zd} := -[L_g h(x)]^{-1} L_f h(x) \tag{11.28}$$

is defined in some neighborhood of x_0. For each initial condition in that neighborhood, the zero dynamics feedback (11.28) makes the closed loop system output (11.26) identically zero on some time interval about the initial time $t_0 = 0$.

11.5 TWO APPLICATIONS

This section includes applications of zero dynamics ideas to (i) the design of a center manifold and reduced dynamics for stabilization, and (ii) a description of the zero dynamics for some linear SISO systems that will be helpful in the study of partially linear cascade systems.

11.5.1 Designing a Center Manifold

We consider a system having critically stable zero dynamics, so that Proposition 11.3 does not apply. In order to show how a center manifold might be designed by feedback, such that the reduced dynamics on the center manifold has asymptotically stable origin, we consider a simple model of airplane flight. In this model, the motion is restricted to a plane perpendicular to the horizontal ground, and the only control action is due to the elevator control surface at the tail of the airplane. We can picture the motion as taking place in a flight path above a runway.

We first consider a simple linear model for landing the airplane.

Example 11.10 *Landing an Airplane*
Suppose the altitude of an airplane in meters is h. The body of the airplane is slanted at an angle of ϕ radians from the horizontal, and the actual flight path of the airplane body forms an angle of α radians with the horizontal. We analyze the situation for α small. The equations of motion are

$$\begin{aligned}\dot{\alpha} &= a(\phi - \alpha),\\ \ddot{\phi} &= -\omega^2(\phi - \alpha - bu),\\ \dot{h} &= c(h)\alpha,\end{aligned}$$

where $\omega > 0$ is a constant representing a natural oscillation frequency, and a and b are positive constants. From the definition of angle α, the term $c(h)$ is identified as the ground speed of the airplane. The system is linear if and only if the ground speed is assumed to be constant, $c(h) \equiv c_0$. For this example, we assume constant ground speed $c_0 > 0$.

Define $x_1 = \alpha$, $x_2 = \phi$, $x_3 = \dot{\phi}$, and $x_4 = h$. This yields the linear system

$$\begin{aligned}\dot{x}_1 &= -ax_1 + ax_2,\\ \dot{x}_2 &= x_3,\\ \dot{x}_3 &= \omega^2 x_1 - \omega^2 x_2 + \omega^2 bu,\\ \dot{x}_4 &= c_0 x_1.\end{aligned}$$

Define the output function $y = x_1 = \alpha$. Straightforward calculations show that this output has relative degree three at the origin, and that the zero dynamics subsystem is given by

$$\dot{x}_4 = 0.$$

The origin in the normal form cannot be asymptotically stabilized by feedback using only the coordinates $z_1 = x_1$, $z_2 := \dot{z}_1$, and $z_3 := \dot{z}_2$ in that form, and these three coordinates depend only on x_1, x_2, and x_3. However, even from the system in the original x coordinates, it is not difficult to see that the origin cannot be made asymptotically stable by linear feedback involving only x_1, x_2, and x_3, since there is a zero eigenvalue which is unchanged by such feedback; it is this zero eigenvalue which is responsible for the Lyapunov stable zero dynamics subsystem. Note that Proposition 11.3 does not apply. This linear system is controllable, but linear feedback using only a measurement of $x_1 = \alpha$ cannot make the origin asymptotically stable. △

We continue with the airplane motion model of the previous example, but now take as a goal the asymptotic tracking of a target altitude. Let z denote the difference between the airplane altitude in meters and the desired target altitude. As an initial modeling assumption, suppose that the altitude dynamics are given by

$$\dot{z} = q(z)\alpha,$$

where $q(0) = 0$; otherwise the model is the same as in Example 11.10.

Example 11.11 *Tracking a Target Altitude*
We again take the measured output as the angle $y = \alpha := x_1$. In addition, we now define the variables

$$\begin{aligned} x_2 &= \dot{x}_1 = a(\phi - \alpha), \\ x_3 &= \dot{x}_2 = a(\dot{\phi} - \dot{\alpha}) = a\dot{\phi} - a^2(\phi - \alpha). \end{aligned}$$

Thus, we have

$$\dot{x}_3 = a\ddot{\phi} - ax_3 = -a\omega^2(\phi - \alpha - bu) - ax_3 = -\omega^2 x_2 - ax_3 + ab\omega^2 u.$$

We have the normal form equations

$$\begin{aligned} \dot{x}_1 &= x_2, \\ \dot{x}_2 &= x_3, \\ \dot{x}_3 &= -\omega^2 x_2 - ax_3 + ab\omega^2 u, \\ \dot{z} &= q(z)x_1 \end{aligned} \tag{11.29}$$

corresponding to the output function $y = x_1 = \alpha$. As in the previous example, this output has relative degree three at the origin and the zero dynamics subsystem is given by

$$\dot{z} = 0.$$

The origin of the zero dynamics is not asymptotically stable, and Proposition 11.3 does not apply. We now consider a nonlinear feedback control. Note that the subsystem for the components x_1, x_2, x_3 is a controllable linear system. The idea is to apply linear feedback of x_1, x_2, x_3, in order to stabilize these variables to the origin, together with nonlinear feedback dependent on z, in order to stabilize the variable z of a center manifold. The resulting indirect dependence of x_1 on z will be the stabilizing influence in the last equation of system (11.29). The problem is to show that the stabilization of all variables may be accomplished with an appropriate scalar feedback. Try a feedback of the form

$$u = -\frac{(-\omega^2 x_2 - a x_3)}{ab\omega^2} + \frac{1}{ab\omega^2}(k_1 x_1 + k_2 x_2 + k_3 x_3 + \beta z^2), \tag{11.30}$$

and consider the closed loop system

$$\begin{aligned} \dot{x}_1 &= x_2, \\ \dot{x}_2 &= x_3, \\ \dot{x}_3 &= k_1 x_1 + k_2 x_2 + k_3 x_3 + \beta z^2, \\ \dot{z} &= q(z) x_1, \end{aligned} \tag{11.31}$$

where k_1, k_2, k_3, and β are to be determined. There are many choices for the k_j. For example, choose $k_1 = -6$, $k_2 = -11$, and $k_3 = -6$ to give eigenvalues $\lambda = -1, -2, -3$ at the origin for the x subsystem. We shall use these values for the k_j in the arguments below. (Before proceeding, it may be helpful to note that the resulting 3×3 companion matrix now corresponds to the B matrix in (10.10).)

The parameter β will be used to design an approximate center manifold for which the reduced dynamics has asymptotically stable origin $z = 0$. Suppose a center manifold is given by

$$\begin{bmatrix} x_1 \\ x_2 \\ x_3 \end{bmatrix} = H(z) = \begin{bmatrix} H_1(z) \\ H_2(z) \\ H_3(z) \end{bmatrix} = O(\|z\|^2).$$

Translating the decomposition (10.9)–(10.10) to the present situation, and using the operator M from (10.19), we have

$$\begin{aligned} (M\,H)(z) &= DH(z)\,[q(z)H_1(z)] - \begin{bmatrix} H_2(z) \\ H_3(z) \\ k_1 H_1(z) + k_2 H_2(z) + k_3 H_3(z) + \beta z^2 \end{bmatrix} \\ &= O(\|z\|^4) - \begin{bmatrix} H_2(z) \\ H_3(z) \\ k_1 H_1(z) + k_2 H_2(z) + k_3 H_3(z) + \beta z^2 \end{bmatrix}, \end{aligned}$$

where we have used the fact that $H(z) = O(\|z\|^2)$ and $q(z) = O(\|z\|)$. Now, to make $(M\,H)(z) = O(\|z\|^4)$, we can choose

$$H_2(z) = 0, \quad H_3(z) = 0, \quad \text{and} \quad H_1(z) = -\frac{\beta}{k_1} z^2.$$

In fact, by the theorem on the approximation of a center manifold (Theorem 10.3), a center manifold may also be written as

$$
\begin{aligned}
x_1 &= H_1(z) = -\frac{\beta}{k_1} z^2 + O(\|z\|^4), \\
x_2 &= H_2(z) = O(\|z\|^4), \\
x_3 &= H_3(z) = O(\|z\|^4).
\end{aligned}
$$

The reduced dynamics, in this case, is given by the scalar equation

$$
\begin{aligned}
\dot{z} &= q(z) H_1(z), \\
&= \left[\frac{\partial q}{\partial z}(0) z + O(\|z\|^2) \right] \left[-\frac{\beta}{k_1} z^2 + O(\|z\|^4) \right], \\
&= -\frac{\beta}{k_1} \frac{\partial q}{\partial z}(0) z^3 + O(\|z\|^4), \qquad (11.32)
\end{aligned}
$$

where $z \in \mathbf{R}$ and $|z| < \delta$ for some $\delta > 0$. In order to achieve asymptotic stability of (11.32) we require $-\frac{\beta}{k_1} \frac{\partial q}{\partial z}(0) < 0$. Since $k_1 = -6 < 0$, we may choose $\beta = -\frac{\partial q}{\partial z}(0) \neq 0$ to achieve asymptotic stability of (11.32). With this choice of β, Theorem 10.2 implies that the origin of the closed loop system (11.31) is asymptotically stable. $\triangle$

It is worth noting that the analysis in Example 11.11 worked for the following reasons:

(i) The nonlinear feedback in z did not destroy the asymptotic stability of the origin $x = 0$ achieved by the choice of the k_j.

(ii) The nonlinear feedback in z was able to affect the structure of the center manifold, and the reduced dynamics, by injecting the variable $x_1 = H_1(z)$ into the center manifold dynamics, while maintaining an adequate order of approximation for the center manifold in order to allow asymptotic stability to be determined and achieved.

It is also interesting to note that the stabilization in Example 11.11 is achieved without full knowledge of the altitude error dynamics. The only assumption was that the error dynamics were autonomous and could be modeled by $\dot{z} = q(z)x_1$, with $\frac{\partial q}{\partial z}(0) \neq 0$; in fact, if they are autonomous but take the form

$$
\dot{z} = q(z)x_1 + O(\|x\|^2), \qquad (11.33)
$$

then a similar analysis shows that the same feedback control will asymptotically stabilize the origin. (See Exercise 11.13.)

11.5.2 Zero Dynamics for Linear SISO Systems

We have seen that zero dynamics analysis is a way to use the system output to reveal system structure. The resulting structure, or normal form, can be helpful in thinking about stabilization as well as other questions mentioned in the Notes and References. In a general sense, the normal form for a *linear* system cannot really improve on the controllability normal form in (6.4). The ideas here have more general interest when applied to nonlinear systems, but certain examples discussed here are helpful in the study of partially linear cascade systems later in the book. An outline of the construction of the normal form written specifically for a linear SISO system appears in a summary/exercise form at the end of this section. This outline can also serve as a useful review and example of the general procedure discussed in this chapter.

We proceed here with illustrations of normal form calculations for specific linear SISO systems. Consider the single-input single-output system

$$\begin{aligned} \dot{x} &= Ax + bu, \\ y &= cx. \end{aligned}$$

Suppose that $cA^k b \neq 0$ for some positive integer k, and let r be the least positive integer for which $cA^{r-1}b \neq 0$. Then r is the relative degree of the single-input single-output linear system. (It is easy to give examples where there is no positive integer k such that $cA^k b \neq 0$. Note that, with direct feedthrough models having output of the form $y = cx + du$, relative degree zero is possible, but we do not consider this here.)

Example 11.12 Consider the system

$$\dot{x} = Ax + bu := \begin{bmatrix} 0 & 1 & 0 \\ -1 & 1 & 2 \\ 0 & 1 & -3 \end{bmatrix} x + \begin{bmatrix} 0 \\ 1 \\ 0 \end{bmatrix} u\,.$$

We note that with $u = 0$, the system has eigenvalues -3.42408, $.71204 \pm .6076i$; hence, the origin is unstable. Suppose we choose $y = x_1$; then the relative degree is two. The zero dynamics feedback is $u_{zd} = x_1 - x_2 - 2x_3$. With $u = u_{zd} + v$, we obtain

$$\begin{aligned} \dot{x}_1 &= x_2, \\ \dot{x}_2 &= v, \\ \dot{x}_3 &= x_2 - 3x_3. \end{aligned}$$

Note that the resulting system has a zero eigenvalue with eigenvector $[1\ 0\ 0]^T$ which can be compensated for by linear feedback of the form $v = k_1 x_1 + k_2 x_2$. The zero dynamics subsystem is $\dot{x}_3 = -3x_3$, which is asymptotically stable. The feedback $v = -x_1 - x_2$ will asymptotically stabilize the x_1 and x_2 components to zero. Thus the feedback

$$u = u_{zd} + v = x_1 - x_2 - 2x_3 - x_1 - x_2 = -2x_2 - 2x_3$$

asymptotically stabilizes the origin of the full system.

Now consider the same dynamical system, but with the new scalar output $y = x_2$; then the relative degree is one. The zero dynamics feedback is the same as before, as is easily checked. By setting $z_1 = x_2$, $z_2 = x_1$, and $z_3 = x_3$, and applying the feedback $u = u_{zd} + v$, we may write the system as

$$\begin{aligned} \dot{z}_1 &= v, \\ \dot{z}_2 &= z_1, \\ \dot{z}_3 &= z_1 - 3z_3. \end{aligned}$$

This system has a zero eigenvalue with eigenvector $[0\ 1\ 0]^T$ which cannot be changed by linear feedback of the form $v = kz_1$. The zero dynamics subsystem is given by

$$\begin{aligned} \dot{z}_2 &= 0, \\ \dot{z}_3 &= -3z_3. \end{aligned}$$

Therefore the zero dynamics has Lyapunov stable, but not asymptotically stable, origin. One may conclude that the full system has Lyapunov stable origin after the application of the feedback $v = -z_1$, for example. But more can be said, because the original pair (A, b) is controllable; hence, the system can be asymptotically stabilized by linear state feedback. The limitation in the present conclusion of Lyapunov stability is due to the limitation of the output $y = x_2$ and the restriction to feedback of the form $v = kz_1$, which did not reveal the stabilizability through the cascade normal form. (Note that $v = -z_1 - z_2 = -x_2 - x_1$ will stabilize the origin.) $\triangle$

Example 11.12 raises a natural question as to which outputs might be helpful in the sense of allowing an application of Proposition 11.3, especially in cases where the pair (A, b) is not controllable. Clearly, in order for Proposition 11.3 to apply, any uncontrollable and unstable modes must be compensated by the feedback $u = u_{zd} + v$. The modes defined by vector ξ in the form (11.42)–(11.43) below are controllable. And, by Lemma 11.1, any eigenvalues associated with uncontrollable modes of the original pair (A, b) must also be eigenvalues of the zero dynamics subsystem. Thus, for linear systems, the normal form via zero dynamics in this chapter cannot be more revealing than the controllability normal form.

The zero dynamics approach is especially valuable when considering nonlinear feedback in nonlinear systems. In particular, in certain cascade systems with a linear driving system, the stability of the zero dynamics of the driving system can play an important role in the feedback stabilization of the cascade, for example in the technique of feedback passivation. Three-dimensional linear systems in controller form provide some interesting examples, and we consider them here.

Example 11.13 (Three-Dimensional Linear SISO Controller Forms) Consider the three-dimensional controller form SISO system,

$$\begin{aligned}\dot{x}_1 &= x_2,\\ \dot{x}_2 &= x_3,\\ \dot{x}_3 &= u,\\ y &= c_1x_1 + c_2x_2 + c_3x_3.\end{aligned}$$

We label this system for later reference as $\dot{x} = Ax + bu$, $y = cx$. (We have assumed that the last row of A is a zero row, since this can always be achieved by a preliminary feedback, if necessary.) We consider the case of an output having relative degree one and we describe the normal form and zero dynamics equations. (The case of relative degree one is important in discussions of *feedback passivity* later in the book.) It is easy to see that the relative degree one condition is equivalent to $c_3 \neq 0$. Thus, with $c_3 \neq 0$, we consider the following two cases.

(i) Suppose that $c_1 = c_2 = 0$; then the relative degree one output is $y = c_3x_3$ and we may use the coordinates $\xi = c_3x_3$, $\eta_1 = x_1$, and $\eta_2 = x_2$ to write the normal form. After applying the feedback $u = u_{zd} + v$ to the original system, we have the normal form

$$\begin{aligned}\dot{\xi} &= v,\\ \dot{\eta}_1 &= \eta_2,\\ \dot{\eta}_2 &= \frac{1}{c_3}\xi.\end{aligned}$$

We set the output equal to zero ($\xi = 0$) to see that the zero dynamics susbsystem is

$$\begin{aligned}\dot{\eta}_1 &= \eta_2,\\ \dot{\eta}_2 &= 0.\end{aligned}$$

Thus the origin $(\eta_1, \eta_2) = (0, 0)$ of the zero dynamics is unstable.

(ii) Now suppose that $c_1^2 + c_2^2 \neq 0$, that is, at least one of c_1, c_2 is nonzero. In this case, the three row vectors c, $(c^T \times b)^T$, and $(b \times (c^T \times b))^T$ are linearly independent, and these vectors give the rows of a transformation to normal form:

$$T = \begin{bmatrix} c_1 & c_2 & c_3 \\ c_2 & -c_1 & 0 \\ c_1 & c_2 & 0 \end{bmatrix}, \qquad \begin{bmatrix} \xi \\ \eta_1 \\ \eta_2 \end{bmatrix} := T \begin{bmatrix} x_1 \\ x_2 \\ x_3 \end{bmatrix}.$$

Note that the last two rows have zero dot product with $b = [0\ 0\ 1]^T$. The zero dynamics subsystem is defined by the equations

$$\dot{\eta}_1 = \begin{bmatrix} c_2 & -c_1 & 0 \end{bmatrix} AT^{-1} \begin{bmatrix} 0 \\ \eta_1 \\ \eta_2 \end{bmatrix},$$

$$\dot{\eta}_2 = \begin{bmatrix} c_1 & c_2 & 0 \end{bmatrix} AT^{-1} \begin{bmatrix} 0 \\ \eta_1 \\ \eta_2 \end{bmatrix}.$$

The inverse T^{-1} may be determined by some algebraic manipulation as

$$T^{-1} = \begin{bmatrix} 0 & \frac{c_2}{w} & \frac{c_1}{w} \\ 0 & -\frac{c_1}{w} & \frac{c_2}{w} \\ \frac{1}{c_3} & 0 & -\frac{1}{c_3} \end{bmatrix}, \quad \text{where } w := c_1^2 + c_2^2.$$

Thus, when $c_1^2 + c_2^2 \neq 0$, the zero dynamics subsystem is

$$\begin{bmatrix} \dot{\eta}_1 \\ \dot{\eta}_2 \end{bmatrix} = \begin{bmatrix} -\frac{c_1 c_2}{w} & \frac{c_2^2}{w} + \frac{c_1}{c_3} \\ -\frac{c_1^2}{w} & \frac{c_1 c_2}{w} - \frac{c_2}{c_3} \end{bmatrix} \begin{bmatrix} \eta_1 \\ \eta_2 \end{bmatrix} =: Q \begin{bmatrix} \eta_1 \\ \eta_2 \end{bmatrix}. \tag{11.34}$$

A straightforward calculation shows that the characteristic equation of matrix Q is

$$\lambda^2 + \frac{c_2}{c_3}\lambda + \frac{c_1}{c_3} = 0. \tag{11.35}$$

Based on (11.35), one can deduce the stability properties of the origin of the zero dynamics for all cases of the output coefficients c_1, c_2, c_3 with $w = c_1^2 + c_2^2 \neq 0$. In order to proceed, observe that $\operatorname{tr} Q = -\frac{c_2}{c_3}$ and $\det Q = \frac{c_1}{c_3}$, then use the algebraic catalog of linear planar systems (Example 2.15); see Exercise 11.15. △

We end the section with a summary/exercise of the construction of the normal form and zero dynamics equations for linear SISO systems.

Normal Form and Zero Dynamics for Linear SISO Systems. Suppose given a linear system triple (c, A, b) with relative degree $r < n$, so that $cA^j b = 0$ for $j \leq r - 2$, while $cA^{r-1}b \neq 0$.

(a) If r is the relative degree, then the row vectors c, cA, $\ldots$, cA^{r-1} are linearly independent.

(b) Define variables z_j by $z_j = cA^{j-1}x$ for $1 \leq j \leq r$. These components satisfy the differential equations

$$\begin{aligned} \dot{z}_1 &= z_2, \\ \dot{z}_2 &= z_3, \\ \cdots &= \cdots, \\ \dot{z}_{r-1} &= z_r, \\ \dot{z}_r &= cA^r x + (cA^{r-1}b)u. \end{aligned} \tag{11.36}$$

Since $cA^{r-1}b \neq 0$, there is a unique state feedback that keeps $y(t) = z_1(t) = 0$ for all t, and this feedback is given by

$$u_{zd} = \alpha(x) = -\frac{cA^r x}{cA^{r-1}b}. \tag{11.37}$$

The feedback in (11.37) is called the *zero dynamics feedback.*

(c) Complete a linear coordinate transformation, incorporating the components z_j of part (b), by choosing linearly independent row vectors $w_1, \ldots, w_{n-r}$ such that the matrix

$$T := \begin{bmatrix} c \\ cA \\ \cdots \\ cA^{r-1} \\ w_1 \\ \cdots \\ w_{n-r} \end{bmatrix}$$

is nonsingular, and let $z_{r+j} = w_j x$ for $1 \leq j \leq n-r$. It is possible to choose the w_i such that $w_i^T b = 0$ for $1 \leq i \leq n-r$. The zero dynamics feedback, expressed in (11.37) in terms of x, can be expressed as a function of the state $z = Tx$ by

$$u_{zd} = \hat{\alpha}(z) := \alpha \circ T^{-1} z = -\frac{cA^r T^{-1} z}{cA^{r-1}b}. \tag{11.38}$$

(d) Using (11.38), apply the feedback transformation

$$u = u_{zd} + v = \hat{\alpha}(z) + v \tag{11.39}$$

to (11.36). Writing $\xi = [z_1 \cdots z_r]^T$, the first r differential equations take the standard controllable form

$$\dot{\xi} = N\xi + dv, \tag{11.40}$$

where N is nilpotent with ones on the superdiagonal and zeros elsewhere, and $d = [0 \cdots 0\, 1]^T$. For the remaining $n-r$ differential equations, write

$$W = \begin{bmatrix} w_1 \\ \cdots \\ w_{n-r} \end{bmatrix}$$

and set $\eta := [z_{r+1} \;\cdots\; z_n]^T = Wx$. The $n-r$ differential equations for η are given by

$$\dot{\eta} = W\dot{x} = W(AT^{-1}z + bu) = WAT^{-1}z + Wbu = WAT^{-1}z,$$

since $Wb = 0$. Write $WAT^{-1} =: [P\; Q]$, where P is $(n-r) \times r$ and Q is $(n-r) \times (n-r)$, to get

$$\dot{\eta} = P\xi + Q\eta. \tag{11.41}$$

(e) Combine (11.40) and (11.41) to obtain the system

$$\dot{\xi} = N\xi + dv, \tag{11.42}$$
$$\dot{\eta} = P\xi + Q\eta, \tag{11.43}$$
$$y = z_1,$$

with input v and output y. When the new input v is a linear feedback $v = K\xi$, a zero initial condition for ξ implies that $\xi(t) = 0$ for all t. With this constraint, (11.43) reduces to

$$\dot{\eta} = Q\eta. \tag{11.44}$$

System (11.44) is called the *zero dynamics* subsystem of the original single-input single-output system. It is the decoupled (or unforced) driven system of a cascade.

(f) Since the matrix pair (N, d) is controllable, there exists linear feedback $v = K\xi = [k_1 \ \cdots \ k_r]\xi$ such that $N + dK$ is Hurwitz. This observation leads to the next proposition, which is Proposition 11.3 specialized to the linear SISO case.

Proposition 11.4 (Linear Zero Dynamics)
If the zero dynamics system (11.44) *is asymptotically stable, that is, Q is Hurwitz, then the origin $(\xi, \eta) = (0, 0)$ of* (11.42)–(11.43) *can be asymptotically stabilized by linear feedback $v = K\xi$.*

If the zero dynamics subsystem is merely Lyapunov stable, then we can conclude that the origin of the original system is also Lyapunov stable, but this may not be the strongest conclusion possible. Any eigenvalues associated with uncontrollable modes determined by the original pair (A, b) must also be eigenvalues of the zero dynamics subsystem, that is, they must be eigenvalues of matrix Q in (11.44). (See Example 11.12.)

11.6 EXERCISES

Exercise 11.1 Let $x(t)$ be the solution of (11.1) with $x(0) = x_0$. Show that the relative degree of h at x_0 is the number of differentiations of $h(x(t))$ needed to make the input u appear explicitly.

Exercise 11.2 Show that the relative degree at x_0 of a SISO system is invariant under coordinate transformations $z = \Phi(x)$ defined near x_0 and regular feedback transformations $u = \alpha(x) + \beta(x)v$, where $\beta(x_0) \neq 0$.

Exercise 11.3 *Relative Degree and Transfer Function*
This exercise considers the relative degree of a SISO linear system (c, A, b) in

terms of the Laplace-transformed solution of the system. *Notation*: Capital letters denote the Laplace transform of the function denoted by the same letter in lower case; thus $\mathcal{L}(x(t)) =: X(s)$.

(a) Take the Laplace transform of the equation $\dot{x} = Ax + bu$, assuming initial condition $x_0 = 0$, and obtain $(sI - A)X(s) = bU(s)$. Show that $sI - A$ is invertible for $|s| > \|A\|$; verify that it does not matter which norm is used. Thus, deduce that

$$X(s) = (sI - A)^{-1}U(s) \quad \text{for } |s| > \|A\|.$$

Hint: Use a geometric series argument.

(b) Take the ratio of transformed output to transformed input to obtain the system *transfer function*

$$H(s) := \frac{Y(s)}{U(s)} = c(sI - A)^{-1}b.$$

(c) Conclude that $H(s)$ has a Laurent series expansion, centered at $s = 0$, which converges for $|s| > \|A\|$.

(d) Show that the relative degree equals r if and only if the coefficient of the largest negative power of s that appears nontrivially in the Laurent expansion of $H(s)$ centered at $s = 0$ is $cA^{r-1}b$, the term in question being

$$\frac{cA^{r-1}b}{s^r}.$$

Exercise 11.4 *The Jacobi Identity*

(a) Prove the Jacobi identity: If v, w are smooth vector fields and λ is a smooth function, then

$$(L_{[v,w]}\lambda)(x) = (L_v L_w \lambda - L_w L_v \lambda)(x).$$

(b) Assume that the SISO system (λ, f, g) has relative degree r at x_0. Use the Jacobi identity to show directly that, for each x in a neighborhood U of x_0, the vectors $ad_f^k g(x)$, for $k = 0, \ldots, r-2$, are in the nullspace of the linear functional $d\lambda(x)$; hence, these vectors form a basis for the nullspace at each x near x_0. *Hint*: Use an induction argument.

Exercise 11.5 Given an example of a linear system triple (c, A, b), with $A \in \mathbf{R}^{3\times 3}$, for which the relative degree equals $r = 2$, and write explicitly the calculation of the matrix in (11.4).

Exercise 11.6 Consider the system of Example 11.6.

(a) Show that neither of the output functions $y = x_2$, $y = x_1$ has a well-defined relative degree at $x_0 = 0$.

(b) With $y = x_3$ as in the example, consider the decoupled z_2, z_3 subsystem (11.10)–(11.11). Show that this subsystem has a critical problem of stability for its origin $(z_2, z_3) = (0, 0)$.

(c) Show that the origin $(z_2, z_3) = (0, 0)$ of the decoupled z_2, z_3 subsystem (11.10)–(11.11) is unstable. *Hint*: Try $V(z_2, z_3) = \frac{1}{2}(z_2^2 + z_3^2)$, and show that $\nabla V(z_2, z_3) \cdot f(z_2, z_3) > 0$ for some nonzero points (z_2, z_3) on the line $z_3 = -z_2$ arbitrarily near the origin, where $f(z_2, z_3)$ is the vector field in (11.10)–(11.11). Then consult the instability theorem of [15](Theorem 5.3, page 195).

Exercise 11.7 Show that the linear approximation (at the origin) of the normal form (11.12) coincides with the normal form of the linearization (at the origin) of the original system (11.1)–(11.2). *Hint*: Show that the relative degree of the original system and its linearization are the same. For this purpose, use expressions of the form

$$\begin{aligned} f(x) &= Ax + f_2(x), \quad Df_2(0) = 0, \\ g(x) &= b + g_1(x), \quad g_1(0) = 0, \\ h(x) &= cx + h_2(x), \quad Dh_2(0) = 0. \end{aligned}$$

Show that

$$\begin{aligned} cA^k b = L_g L_f^k h(0) &= 0 \quad \text{for all } k \le r - 2, \\ cA^{r-1} b = L_g L_f^{r-1} h(0) &\ne 0. \end{aligned}$$

Then complete the argument.

Exercise 11.8 Suppose that the linear approximation of a nonlinear system has uncontrollable modes associated only with eigenvalues on the imaginary axis, and that the zero dynamics has asympototically stable origin. Explain why the zero dynamics origin is asymptotically stable due to the nonlinear terms and not due to the linearization.

Exercise 11.9 Consider the system of Example 11.5, but now with $y = x_2$ as the output.

(a) Determine the normal form in a neighborhood of the origin.

(b) Discuss asymptotic stabilization of this system, based on a zero dynamics analysis. If possible, find an asymptotically stabilizing feedback.

Exercise 11.10 Determine, if possible, a stabilizing feedback control for the planar system

$$\begin{aligned} \dot{x} &= x - x^3 + \xi, \\ \dot{\xi} &= u, \end{aligned}$$

by using a zero dynamics analysis for the output $y = \xi - \alpha(x)$, where $\dot{x} = x - x^3 + \alpha(x)$ has an asymptotically stable equilibrium at $x = 0$. Write the normal form explicitly.

Exercise 11.11 Determine, if possible, a stabilizing feedback control for the planar system

$$\begin{aligned}\dot{x}_1 &= -x_1^2 x_2,\\ \dot{x}_2 &= u,\end{aligned}$$

by using a zero dynamics analysis for the output $y = x_2 - \alpha(x_1)$, where $\dot{x}_1 = -x_1^2\alpha(x_1)$ has an asymptotically stable equilibrium at $x_1 = 0$. Write the normal form explicitly.

Exercise 11.12 *Bounded Input-Bounded State*
In the normal form (11.12), suppose that, instead of the feedback (11.21), we apply the new feedback

$$u = \frac{1}{a(\xi,\eta)}(-b(\xi,\eta) - k_0 z_1 - k_1 z_2 - \cdots - k_{r-1} z_r + v),$$

where v is a new reference input. This gives the closed loop system

$$\begin{aligned}\dot{\xi} &= A\xi + bv,\\ \dot{\eta} &= q(\xi,\eta),\end{aligned}$$

where $b = [0 \ \cdots \ 0 \ 1]^T$. Observe that when $v = 0$, this closed loop reduces to (11.22)–(11.23). Suppose that the k_i are chosen such that the origin of (11.22)–(11.23) is asymptotically stable. Show that for each $\epsilon > 0$ there exist numbers $\delta(\epsilon) > 0$ and $M(\epsilon) > 0$ such that, if $\|x(0)\| < \delta$ and $|v(t)| < M$ for all $t \geq 0$, then $\|x(t)\| < \epsilon$ for all $t \geq 0$. *Hint*: Use the theorem on total stability.

Exercise 11.13 *On Example 11.11*
Suppose the altitude error dynamics (11.29) is replaced by the model (11.33). Write the new system equations as well as the closed loop equations that result from applying the same form of feedback control (11.30). Show, by an analysis similar to that of Example 11.11, that the same feedback control asymptotically stabilizes the origin of the closed loop.

Exercise 11.14 *Two-Dimensional SISO Controller Form*
The system

$$\begin{aligned}\dot{x}_1 &= x_2,\\ \dot{x}_2 &= u,\\ y &= c_1 x_1 + c_2 x_2\end{aligned}$$

has relative degree one if and only if $c_2 \neq 0$. Assuming $c_2 \neq 0$, analyze the stability properties of the zero dynamics.

Exercise 11.15 *3D Linear SISO Controller Forms*
Discuss the stability properties of the zero dynamics subsystem (11.34) in Example 11.13. Under what conditions on the c_i is the origin of the zero dynamics asymptotically stable? Under what conditions on the c_i is the origin of the zero dynamics merely Lyapunov stable? *Hint*: For a detailed answer, one might organize three cases: (i) all three c_i are nonzero, (ii) only c_1, c_3 are nonzero, (iii) only c_2, c_3 are nonzero.

11.7 NOTES AND REFERENCES

The zero dynamics concept for nonlinear systems was introduced in the paper by C. I. Byrnes and A. Isidori, A frequency domain philosophy for nonlinear systems, *Proceedings of the IEEE Conf. Dec. Contr.*, 23: 1569–1573, 1984. This chapter follows [47] (Sections 4.1, 4.3–4.4).

The assumption that $A(x_0)$ in (11.27) is nonsingular (the system has a vector relative degree) leads to a useful local normal form, as well as a concept of zero dynamics, for MIMO systems; for a complete general development, see [47] (pages 219–226).

This chapter emphasized the problem of determining initial conditions and inputs that allow exact tracking of *zero* output. This type of analysis can be extended to discuss the more general question of *exact output tracking* of a general nonzero output function [47]. *Asymptotic* output tracking, as opposed to *exact* tracking of an output trajectory, generally requires some stability analysis of time-varying systems. For more discussion of output tracking problems, see [47].

The airplane flight model in Example 11.10 is from [91] (page 90). Example 11.11 follows the motivating example in [24] (Example 1.1); this paper studies the stabilization of systems in the Byrnes-Isidori normal form (see [47]) under the assumption that the linearization has a center subspace corresponding to real zero eigenvalues only. Two earlier studies, [4] and [11], used center manifold theory to design stabilizing feedback.

Chapter Twelve

Feedback Linearization of Single-Input Nonlinear Systems

In this chapter we consider the question of equivalence, by means of local coordinate change and state feedback, between an n-dimensional single-input nonlinear system and an n-dimensional single-input controllable linear system. The solution of this equivalence problem is one of the fundamental results of geometric nonlinear control theory. There are connections here among topics in differential equations, linear algebra, analysis, and the basic differential-geometric concepts required for a special case of the Frobenius theorem.

12.1 INTRODUCTION

Let the matrix triple (c, A, b) denote a SISO linear system with state space $\mathbf{R}^n$; thus, A is $n \times n$, b is $n \times 1$, and c is $1 \times n$. In an earlier chapter we characterized the single-input single-output linear systems

$$\dot{x} = Ax + bu, \tag{12.1}$$

$$y = cx \tag{12.2}$$

that can be transformed by a nonsingular linear transformation, $z = Tx$, to a linear controllable system given by

$$\dot{z} = Pz + du$$
$$:= \begin{bmatrix} 0 & 1 & 0 & \cdot\;\cdot\;\cdot & 0 \\ 0 & 0 & 1 & \cdot\;\cdot\;\cdot & \cdot \\ \cdot & \cdot & \cdot & \cdot\;\cdot\;\cdot & \cdot \\ \cdot & \cdot & \cdot & \cdot\;\cdot\;\cdot & 1 \\ -k_n & -k_{n-1} & -k_{n-2} & \cdot\;\cdot\;\cdot & -k_1 \end{bmatrix} z + \begin{bmatrix} 0 \\ 0 \\ \cdots \\ 0 \\ 1 \end{bmatrix} u. \tag{12.3}$$

By an additional feedback transformation, $u = Kx + v$, where K is a $1 \times n$ matrix and v a new reference input, such systems are equivalent to the particularly simple form, $z_1^{(n)} = v$. For convenience we restate this characterization here as Theorem 12.1.

Theorem 12.1 *The single-input linear system* (12.1) *can be transformed by a nonsingular linear transformation, $z = Tx$, to the companion system* (12.3), *if and only if* rank $[b\ Ab\ \cdots A^{n-1}b] = n$, *that is,* (12.1) *is controllable; and the transformation* T *is uniquely determined by*

$$T = \begin{bmatrix} \tau \\ \tau A \\ \cdots \\ \tau A^{n-1} \end{bmatrix}, \tag{12.4}$$

where τ *is the unique* $1 \times n$ *solution of*

$$\tau\,[b\ Ab\ \cdots\ A^{n-1}b] = [0\ \cdots\ 0\ 1] =: d^T. \tag{12.5}$$

Given the SISO linear system (12.1)–(12.2), *there exists a nonsingular transformation* $z = Tx$ *that transforms the system to the companion system* (12.3) *with output* $y = z_1 = cx$, *if and only if*

$$\text{rank} \begin{bmatrix} c \\ cA \\ \cdots \\ cA^{n-1} \end{bmatrix} = n,$$

and

$$c\,[b\ Ab\ \cdots\ A^{n-1}b] = [0\ \cdots\ 0\ 1] = d^T.$$

The transformation matrix T *is unique and is given by* (12.4) *with* $\tau = c$.

The main goal of this chapter is an extension of Theorem 12.1 for the simplest form of single-input nonlinear systems, the control-affine systems given by

$$\dot{x} = f(x) + g(x)u.$$

The exposition uses ideas from the theory of ordinary differential equations, linear algebra, analysis, and the differential-geometric concepts of the Frobenius theorem.

We begin with a calculation that recapitulates the essential calculations that establish Theorem 12.1. Suppose that (c, A, b) satisfies $cA^k b = 0$ for $0 \le k \le n-2$, and $cA^{n-1}b \ne 0$; that is, the system has relative degree n. Consider the following product of $n \times n$ matrices:

$$\begin{bmatrix} c \\ cA \\ \vdots \\ cA^{n-1} \end{bmatrix} \begin{bmatrix} b & Ab & \cdots & A^{n-1}b \end{bmatrix} = \begin{bmatrix} 0 & \cdots & \cdots & 0 & cA^{n-1}b \\ 0 & & & * & * \\ \vdots & & & & \vdots \\ 0 & * & & & \vdots \\ cA^{n-1}b & * & \cdots & \cdots & * \end{bmatrix}. \tag{12.6}$$

Since $cA^{n-1}b \ne 0$, the matrix on the right-hand side is nonsingular, and thus both matrices on the left side must be nonsingular as well; nonsingularity on the right in (12.6) implies observability of (c, A) and controllability of (A, b). A geometric interpretation of the zeros on the right-hand side of (12.6) is that the nullspace of the linear functional $y = cx$ is the $(n-1)$ dimensional space given by span $\{b, Ab, \ldots, A^{n-2}b\}$. Similar geometric considerations in this chapter involve families of vector fields (distributions) and their annihilating family of covector fields, or differential 1-forms. A key result is Proposition 11.1 (we need the special case when the relative degree is $r = n$).

We are interested in determining exactly when the single-input nonlinear system $\dot{x} = f(x) + g(x)u$ is equivalent to a controllable linear system (A, b) by means of a local coordinate transformation and state feedback transformation. If such a transformation is possible, one can exploit the special control-theoretic properties possessed by linear controllable systems in order to analyze a nonlinear system locally, using linear control methods.

Example 12.1 Let $x \in \mathbf{R}^3$ and $u \in \mathbf{R}$. We would like to have computable conditions to determine if the system

$$\dot{x} = f(x) + g(x)u := \begin{bmatrix} \exp(-x_2) \\ x_1 \\ \frac{1}{2}x_1^2 \end{bmatrix} + \begin{bmatrix} 1 \\ 0 \\ 0 \end{bmatrix} u$$

is equivalent to a controllable linear system (A, b), and therefore equivalent to the third order scalar equation, $y^{(3)} = v$. $\triangle$

12.2 INPUT-STATE LINEARIZATION

Consider the single-input single-output system described by

$$\dot{x} = f(x) + g(x)u, \tag{12.7}$$

where f and g are smooth vector fields defined in an open set $D \subset \mathbf{R}^n$. As usual, the term *smooth* means that the components of f and g are continuously differentiable as often as required in our discussion. Equation (12.7) may be augmented with an output equation

$$y = h(x), \tag{12.8}$$

where h is a smooth real-valued function defined on D. We are interested in necessary and sufficient conditions under which system (12.7) is equivalent to a linear controllable system; such an equivalence, if possible, generally requires feedback transformations in addition to state coordinate change.

Recall that a *local coordinate change* about the point x_0 is defined by a mapping $z = T(x)$, where $DT(x_0)$ is nonsingular. A *regular feedback transformation* for a single-input system is defined by $u = \alpha(x) + \beta(x)v$, where α and β are smooth real-valued functions defined in a neighborhood of x_0, such that $\beta(x_0) \neq 0$. The variable v is called the new *reference input*. The general idea is that feedback by $u = \alpha(x)$ simplifies the system equations (for example, by cancelling nonlinearities when possible), and then subsequent controls v are available to control the system dynamics in some desirable way.

The *feedback linearization problem*, or *input-state linearization problem*, is the problem of finding a local coordinate change, $z = T(x)$, and a regular feedback transformation, $u = \alpha(x) + \beta(x)v$, with $\beta(x) \neq 0$ near x_0, such that the tranformed version of system (12.7), given by

$$\dot{z} = DT(T^{-1}(z))f(T^{-1}(z)) + DT(T^{-1}(z))g(T^{-1}(z))u =: Az + bu, \quad (12.9)$$

is a time invariant controllable linear system in z. The term input-state linearization serves to distinguish this problem from the problem of simply linearizing the input-output behavior, which is called *input-output linearization* and which can be achieved, for example, using any function with well-defined relative degree; recall Example 11.3.

12.2.1 Relative Degree n

Given system (12.7), output functions (12.8) having relative degree n at x_0 are especially useful. We have seen that linear output functions of relative degree n play a central role in the study of controllability and observability of linear systems; see Theorem 12.1. Suppose that an output $y = h(x)$ has relative degree n at x_0. Then y and the first $n-1$ derivatives of y along solutions of (12.7) yield the equations

$$\begin{bmatrix} y(t) \\ \dot{y}(t) \\ \vdots \\ y^{(n-1)}(t) \end{bmatrix} = \begin{bmatrix} h(x(t)) \\ L_f h(x(t)) \\ \vdots \\ L_f^{n-1} h(x(t)) \end{bmatrix} + \begin{bmatrix} 0 \\ \vdots \\ 0 \\ L_g L_f^{n-1} h(x(t)) \end{bmatrix} u. \quad (12.10)$$

Suppose that $u = 0$; then, since the differentials of the functions $L_f^k h(x)$, $0 \le k \le n-1$, are linearly independent at each point in a neighborhood of x_0, we can invoke the inverse function theorem to solve for $x(t)$ in terms of $y^{(k)}(t)$, $0 \le k \le n-1$. Thus, relative degree n implies a local observability property of the system. For our purposes, however, the most convenient result is the special case of Proposition 11.1 covering relative degree n, which we state next; note that we are simply setting the relative degree r equal to n in Proposition 11.1.

Proposition 12.1 *If system (12.7)–(12.8) has relative degree n in the open set U containing x_0, then, for all $x \in U$,*

(1) *the covectors $dh(x)$, $dL_f h(x)$, ..., $dL_f^{n-1} h(x)$ are linearly independent;*
(2) *the vectors $g(x)$, $ad_f g(x)$, ..., $ad_f^{n-1} g(x)$ are linearly independent.*

Proof. See the proof of Proposition 11.1 for relative degree $r=n$. For convenient reference in this chapter, we display the relevant matrix calculation

$$\begin{bmatrix} dh(x) \\ dL_f h(x) \\ \vdots \\ dL_f^{n-1}h(x) \end{bmatrix} \begin{bmatrix} g(x) & ad_f g(x) & \ldots & ad_f^{n-1}g(x) \end{bmatrix}$$

$$= \begin{bmatrix} 0 & 0 & \cdots & 0 & L_{ad_f^{n-1}g}h(x) \\ 0 & & & \star & \star \\ \cdots & & \cdots & & \cdots \\ 0 & \star & \cdots & \cdots & \star \\ L_g L_f^{n-1}h(x) & \star & \cdots & \star & \star \end{bmatrix}, \tag{12.11}$$

yielding a lower-right triangular $n \times n$ matrix, by the assumption of relative degree n. □

If the system (h, f, g) has relative degree n at x_0, then, by Proposition 12.1 (a), the vector function $[h(x)\ L_f h(x)\ \cdots\ L_f^{n-1}h(x)]^T$ has nonsingular Jacobian matrix at x_0; thus, the functions $L_f^k h(x)$, $0 \le k \le n-1$, qualify as the component functions in a nonlinear coordinate transformation defined locally around x_0. The presence of the nonzero function $L_g L_f^{n-1}h(x)$ as a coefficient of u prevents us from obtaining a linear, controllable companion form by coordinate transformation alone. However, a coordinate transformation together with regular state feedback does produce the simple form, $y^{(n)} = v$; we need only define

$$u = \frac{1}{L_g L_f^{n-1}h(x)}(-L_f^{n-1}h(x) + v), \tag{12.12}$$

where v is the new reference input. Therefore the existence of a function $h(x)$ such that the system (h, f, g) has relative degree n at x_0 is a sufficient condition for transforming (12.7) to a linear controllable system.

12.2.2 Feedback Linearization and Relative Degree n

Given a controllable linear system $\dot{x} = Ax + bu$, a coordinate change and feedback transformation can be used to produce the linear system

$$\dot{z} = Nz + dv := \begin{bmatrix} 0 & 1 & 0 & \cdot & \cdot & \cdot & 0 \\ 0 & 0 & 1 & \cdot & \cdot & \cdot & \cdot \\ \cdot & \cdot & \cdot & \cdot & \cdot & \cdot & \cdot \\ \cdot & \cdot & \cdot & \cdot & \cdot & \cdot & 1 \\ 0 & 0 & 0 & \cdot & \cdot & \cdot & 0 \end{bmatrix} z + \begin{bmatrix} 0 \\ 0 \\ \vdots \\ 0 \\ 1 \end{bmatrix} v, \tag{12.13}$$

where N is the nilpotent block with ones on the superdiagonal and zeros elsewhere, and $d = [0 \cdots 0\ 1]^T$. Thus we take the feedback linearization problem to mean the problem of achieving the form (12.13).

Our first goal is to show that the feedback linearization problem is solvable if and only if there exists a function h of relative degree n at x_0 for (12.7). We have already discussed the sufficiency of relative degree n for feedback linearization. We now show that feedback linearization is possible only if there exists a function of relative degree n at x_0, namely, the first coordinate function T_1 of the required coordinate transformation T. Our ultimate goal in the chapter is to obtain computable geometric conditions for the existence of a relative degree n function for the system under consideration. From the practical point of view, of course, even when a relative degree n function exists, there is a very real difference between having a known relative degree n function and having to construct such a function.

We now show that the existence of a relative degree n function is necessary for feedback linearization. Thus, suppose that, for $x \in U$, the transformation $z = T(x)$ combined with regular feedback $u = \alpha(x) + \beta(x)v$ produces the linear system (12.13). From the definitions of the variables, this transformation can occur if and only if

$$\frac{\partial T}{\partial x}(x)(f(x) + g(x)u) = NT(x) + dv, \tag{12.14}$$

where equality holds for all u and v related by $u = \alpha(x) + \beta(x)v$. By considering $u = 0$ and $u = 1$, we conclude that (12.14) implies the two partial differential equations,

$$\frac{\partial T}{\partial x}(x)f(x) = N\,T(x) - d\alpha(x)\beta^{-1}(x), \tag{12.15}$$

$$\frac{\partial T}{\partial x}(x)g(x) = d\beta^{-1}(x). \tag{12.16}$$

If $T(x) = [T_1(x)\ T_2(x)\ \cdots\ T_n(x)]^T$, then

$$N\,T(x) - d\alpha(x)\beta^{-1}(x) = [T_2(x) \quad T_3(x) \cdots T_n(x) \quad -\alpha(x)\beta^{-1}(x)]^T,$$

and

$$d\beta^{-1}(x) = [0 \cdots 0 \beta^{-1}(x)]^T.$$

This allows us to display (12.15) in component detail as follows:

$$dT_k(x) \cdot f(x) = L_f T_k(x) = T_{k+1}(x) \quad \text{for } 1 \le k \le n-1, \tag{12.17}$$

and

$$dT_n(x) \cdot f(x) = -\alpha(x)\beta^{-1}(x). \tag{12.18}$$

Therefore, by (12.17), all the component functions of T are determined by the first component function, T_1, and (12.18) specifies the ratio $-\alpha(x)\beta^{-1}(x)$ in terms of T_1. We can also display (12.16) using (12.17) and Lie derivative expressions as follows:

$$dT_{k+1}(x) \cdot g(x) = L_g L_f^k T_1(x) = 0, \quad \text{for } 0 \le k \le n-2, \tag{12.19}$$

while

$$dT_n(x) \cdot g(x) = L_g L_f^{n-1} T_1(x) = \beta^{-1}(x) \neq 0. \tag{12.20}$$

Thus, by (12.19) and (12.20), feedback linearization requires that the first component function $T_1(x)$ has relative degree n. As noted, the rest of the transformation T is then defined by (12.17). The required feedback $u = \alpha(x) + \beta(x)v$ is determined by (12.20) and (12.18), which yield

$$\beta(x) = \frac{1}{L_g L_f^{n-1} T_1(x)} \quad \text{and} \quad \alpha(x) = -\frac{L_f^{n-1} T_1(x)}{L_g L_f^{n-1} T_1(x)}. \tag{12.21}$$

Note that an equilibrium point of interest, say x_0 such that $f(x_0) = 0$, can always be transformed to $z_0 = T(x_0) = 0$; in fact, by (12.17), we can accomplish this by simply requiring that $T_1(x_0) = 0$. Therefore the following result has been proved completely.

Theorem 12.2 (Feedback Linearization and Relative Degree n) *System (12.7) can be transformed by coordinate transformation $z = T(x)$ and state feedback $u = \alpha(x) + \beta(x)v$ in a neighborhood of x_0 to the linear controllable system (12.13) if and only if there exists a function T_1 having relative degree n with respect to (12.7) at the point x_0; then, a transformation T is determined by (12.17) by letting T_1 be the first component function, and the required feedback is then determined by (12.21).*

The next example illustrates the ideas of Theorem 12.2 and also the fact that a coordinate transformation to a linear system is not unique in the nonlinear case.

Example 12.2 We return to the system in Example 11.1, repeated here for convenience,

$$\dot{x}_1 = \sin x_2,$$
$$\dot{x}_2 = -x_1^2 + u,$$

and we ask if it is feedback linearizable in a neighborhood of the equilibrium $x_0 = 0$ of the unforced system. A candidate for the function T_1 satisfies the equations

$$L_g T_1(x) = 0, \quad T_2 = dT_1(x) \cdot f(x), \quad \text{and} \quad L_g T_2(x) \neq 0.$$

Since $g = [0 \quad 1]^T$, T_1 is independent of x_2, and therefore

$$T_2(x_1, x_2) = \frac{\partial T_1}{\partial x_1}(x_1) \sin x_2.$$

The condition (12.20) is

$$L_g T_2(x_1, x_2) = \frac{\partial T_2}{\partial x_2}(x_1, x_2) = \frac{\partial T_1}{\partial x_1}(x_1) \cos x_2 \neq 0,$$

which holds as long as $\cos x_2 \neq 0$ and $\partial T_1/\partial x_1 \neq 0$. One solution is $T_1(x) = x_1$, which we noted as having relative degree 2 in Example 11.1. The solution for T_1 is not unique; for example, $T_1(x) = x_1 - x_1^2$ also works near $x_0 = 0$ in fulfilling the conditions of Theorem 12.2. △

We can also use Theorem 12.2 to show that the system of Example 12.1 is not feedback linearizable in any open set.

Example 12.3 The system of Example 12.1 is

$$\dot{x} = f(x) + g(x)u := \begin{bmatrix} \exp(-x_2) \\ x_1 \\ \frac{1}{2}x_1^2 \end{bmatrix} + \begin{bmatrix} 1 \\ 0 \\ 0 \end{bmatrix} u.$$

The input vector field is $g = [1 \quad 0 \quad 0]^T$. Let h be a smooth real-valued function, and suppose that $L_g h(x) = 0$ for (x_1, x_2, x_3) in an open set U. Since

$$L_g h(x) = \frac{\partial h}{\partial x}(x) g(x) = \frac{\partial h}{\partial x_1}(x_1, x_2, x_3),$$

it follows that h is independent of x_1 in U, so $h = h(x_2, x_3)$ in U. If we also impose the condition that $L_g L_f h(x) = 0$ in U, then we must have

$$\begin{aligned} L_g L_f h(x) &= \frac{\partial}{\partial x_1}\left[\frac{\partial h}{\partial x_2}(x_2, x_3)\, x_1 + \frac{\partial h}{\partial x_3}(x_2, x_3)\, \frac{1}{2}x_1^2\right] \\ &= \frac{\partial h}{\partial x_2}(x_2, x_3) + x_1 \frac{\partial h}{\partial x_3}(x_2, x_3) = 0 \quad \text{for } x \in U. \end{aligned}$$

Because of the x_1 factor, it is not possible for this last equation to hold throughout the open set U when $h = h(x_2, x_3)$. Thus, there is no function having relative degree three for this system in any open set in R^3. Therefore the system is not feedback linearizable in any open set. △

12.3 THE GEOMETRIC CRITERION

The Appendix D material on "Distributions and the Frobenius Theorem" can be read profitably in conjunction with this section.

As we have seen, a real-valued function $T_1 = h$, having the property that the nonlinear SISO system (h, f, g) has relative degree n at x_0, is completely determined by a nontrivial solution of the system of partial differential equations (12.19), where the nontriviality condition is (12.20). When is this system of first order partial differential equations solvable? We are asking for computable conditions for solvability directly in terms of the vector fields f and g.

The Lie bracket operation on vector fields was helpful in Proposition 11.1 in achieving the normal form for zero dynamics analysis; we refer in particular to the vector fields $ad_f^k g$ for $1 \leq k \leq r-1$ in Proposition 11.1 (b); see also equation (12.11) for the case $r = n$. It is helpful to consider a geometric interpretation of the matrix product in (12.11). The following remarks could be directed at the general case of relative degree r with $1 \le r \le n$, by reference to (11.4); however, $r = n$ is our immediate interest. Thus, we focus on (12.11), and observe that for h of relative degree n, the nullspace of the differential $dh(x)$ is the $(n-1)$-dimensional subspace of R^n given by

$$\mathcal{G}_{n-2}(x) = \text{span}\,\{g(x), ad_f g(x), \ldots, ad_f^{n-2} g(x)\} \quad \text{for } x \in U. \qquad (12.22)$$

For each $x \in U$, we have a subspace $\mathcal{G}_{n-2}(x)$, and the collection of these smoothly varying subspaces, for $x \in U$, is called a *distribution* on U. In the linear case, a constant distribution such as span $\{b, Ab, \ldots, A^{n-2}b\}$ automatically entails the existence of a smooth annihilator codistribution, namely, the constant codistribution defined by span $\{w\}$, where the row vector w is such that $wA^k b = 0$ for $0 \le k \le n-2$.

By reference to (12.11), and with Theorem 12.2 on relative degree n in mind, it is clear that the nonsingularity of both factors on the left-hand side in (12.11) is necessary and sufficient for the solution of the feedback linearization problem. And, although not mentioned explicitly in the calculations for Proposition 12.1, two other necessary conditions for feedback linearization are the involutivity and constant dimension (nonsingularity) of the distribution $\mathcal{G}_{n-2}(x)$ in (12.22) in a neighborhood of the point x_0 of interest. Involutivity of a distribution is a property which guarantees that the distribution is locally annihilated (pointwise) by the differentials of smooth independent functions. In fact, the Frobenius theorem establishes the equivalence of this complete integrability property and the involutivity property, for nonsingular (smooth, constant dimension) distributions.

It is not difficult to show that involutivity and nonsingularity are necessary in order that $\mathcal{G}_{n-2}(x)$ is the tangent space at x of the level set within U of a function h with relative degree n. The argument for nonsingularity is simple: If h has relative degree n at x_0, then, by Proposition 12.1, the matrix $[g(x)\ ad_f g(x)\ \ldots\ ad_f^{n-2} g(x)]$ has rank n for $x \in U$, so the distribution $\mathcal{G}_{n-2}$ is nonsingular with constant dimension $n-1$ on U; moreover, $dh(x)[g(x)\ ad_f g(x)\ \ldots\ ad_f^{n-2} g(x)] = 0$ for $x \in U$ and therefore the annihilator of $\mathcal{G}_{n-2}$ is

$$\mathcal{G}_{n-2}^{\perp}(x) = \text{span}\,\{dh(x)\}, \quad \text{for } x \in U.$$

(Hence, $\mathcal{G}_{n-2}$ is integrable on U.) We now show that $\mathcal{G}_{n-2}$ is involutive in U. Using the Jacobi identity, for $0 \le i, j \le n-2$ and $x \in U$, we have

$$\begin{aligned} dh(x)\,[ad_f^i g, ad_f^j g](x) &= L_{[ad_f^i g, ad_f^j g]} h(x) \\ &= L_{ad_f^i g} L_{ad_f^j g} h(x) - L_{ad_f^j g} L_{ad_f^i g} h(x) \\ &= 0, \end{aligned}$$

since h has relative degree n at x_0. Hence $[ad_f^i g, ad_f^j g](x) \in \mathcal{G}_{n-2}$ for all x in U. By definition, any two vector fields belonging to $\mathcal{G}_{n-2}$ are linear combinations, with smooth coefficients, of the ones we just dealt with, and one can show that, for smooth vector fields $\xi(x)$, $\eta(x)$ and smooth functions $a(x)$, $b(x)$ on U,

$$[a\xi, b\eta](x) = a(x)b(x)[\xi, \eta](x) + (L_\xi b(x))a(x)\eta(x) - (L_\eta a(x))b(x)\xi(x). \tag{12.23}$$

(See Exercise 12.1.) Hence, $\mathcal{G}_{n-2}$ is involutive in U.

We can now state and prove the main theorem of this chapter, which gives computable geometric conditions for feedback linearization.

Theorem 12.3 *Let f and g be smooth vector fields defined in some open set containing the point x_0. The single-input single-output nonlinear system*

$$\dot{x} = f(x) + g(x)u$$

is feedback linearizable in a neighborhood of x_0 if and only if the following two conditions hold:

(a) *the matrix $[g(x_0)\ ad_f g(x_0)\ \cdots\ ad_f^{n-1} g(x_0)]$ has rank n*

(b) *the distribution $\mathcal{G}_{n-2} = \text{span}\,\{g, ad_f g, \ldots, ad_f^{n-2} g\}$ is involutive and nonsingular with dimension $n-1$ in a neighborhood of x_0.*

Proof. Necessity of (i) and (ii). We must show that the conjunction of (i) and (ii) is implied by the existence of a function h such that (h, f, g) has relative degree n at x_0. Necessity of (i) is covered by Proposition 12.1, and necessity of (ii) was established in the paragraph preceding the theorem statement.

Sufficiency of (i) and (ii). By (i) and (ii), the distribution $\mathcal{G}_{n-2}$ is nonsingular with dimension $n-1$ and involutive in a neighborhood U of x_0; and, we may choose U such that

$$\text{rank}\,[g(x)\quad ad_f g(x)\quad \cdots\quad ad_f^{n-1} g(x)] = n \quad \text{for all } x \in U.$$

By the Frobenius theorem, $\mathcal{G}_{n-2}$ is completely integrable in U. Thus, in particular, given $x_0 \in U$, there exists a smooth function h defined in a neighborhood U_0 of x_0 such that $\text{span}\,\{dh(x)\} = \mathcal{G}_{n-2}^\perp(x)$ for $x \in U_0$ and such that $dh(x)\, ad_f^{n-1} g(x) \neq 0$ for $x \in U_0$. Thus, the system (h, f, g) has relative degree n in U_0. Hence, by Theorem 12.2, the single-input system $\dot{x} = f(x) + g(x)u$ is feedback linearizable in U_0. This completes the proof of the sufficiency of (i) and (ii) and completes the proof of the theorem. □

Theorem 12.3 involves a special codimension one case of the Frobenius theorem, where $\dim \mathcal{G}_{n-2}^\perp = 1$ and $\mathcal{G}_{n-2}$ is the distribution in (12.22). The major part of the proof of Theorem 12.3 lies in showing that involutivity is a sufficient integrability condition (solvability condition) for the system of partial differential equations (12.19).

Note that in (ii) of Theorem 12.3, the nonsingularity of $\mathcal{G}_{n-2}$ follows from the rank condition (i) in some open set U about x_0; so (ii) could be replaced by (ii)′ the distribution $\mathcal{G}_{n-2}$ is involutive in a neighborhood of x_0.

We can return to Example 12.1 to illustrate the fact that the lack of involutivity for $\mathcal{G}_{n-2}$ is tied to its lack of complete integrability.

Example 12.4 The distribution (12.22) for the three-dimensional system in Example 12.1 is

$$\mathcal{G}_{n-2}(x) = \mathcal{G}_1(x) = \text{span}\,\{g(x), ad_f g(x)\},$$

where

$$f(x) = \begin{bmatrix} \exp(-x_2) \\ x_1 \\ \frac{1}{2}x_1^2 \end{bmatrix} \quad \text{and} \quad g(x) = \begin{bmatrix} 1 \\ 0 \\ 0 \end{bmatrix}.$$

By direct calculation, we have

$$ad_f g(x) = -\begin{bmatrix} 0 & -e^{-x_2} & 0 \\ 1 & 0 & 0 \\ x_1 & 0 & 0 \end{bmatrix} \begin{bmatrix} 1 \\ 0 \\ 0 \end{bmatrix} = \begin{bmatrix} 0 \\ 1 \\ x_1 \end{bmatrix},$$

and then

$$[g, ad_f g](x) = \begin{bmatrix} 0 & 0 & 0 \\ 0 & 0 & 0 \\ 1 & 0 & 0 \end{bmatrix} \begin{bmatrix} 1 \\ 0 \\ 0 \end{bmatrix} = \begin{bmatrix} 0 \\ 0 \\ 1 \end{bmatrix}.$$

Form the matrix

$$\begin{bmatrix} g(x) & ad_f g(x) & [g, ad_f g](x) \end{bmatrix} = \begin{bmatrix} 1 & 0 & 0 \\ 0 & 1 & 0 \\ 0 & x_1 & 1 \end{bmatrix},$$

and observe that its rank is everywhere equal to 3. Hence, the vector $[g, ad_f g](x)$ is not an element of $\mathcal{G}_1(x)$ for any x. Therefore the distribution $\mathcal{G}_1$ is not involutive in any open set. It is interesting to compare this situation with Example 12.3, and observe that the calculations of that example also show that the distribution defined by span $\{g, ad_f g\}$ is not completely integrable in any open set in $\mathbf{R}^3$.

We now show that condition (i) does hold, thus showing that the rank condition (i) is not sufficient for feedback linearization. Calling on the previous calculations, we find

$$ad_f^2 g(x) = [f, ad_f g](x) = \begin{bmatrix} 0 & 0 & 0 \\ 0 & 0 & 0 \\ 1 & 0 & 0 \end{bmatrix} \begin{bmatrix} e^{-x_2} \\ x_1 \\ \frac{1}{2}x_1^2 \end{bmatrix} - \begin{bmatrix} 0 & -e^{-x_2} & 0 \\ 1 & 0 & 0 \\ x_1 & 0 & 0 \end{bmatrix} \begin{bmatrix} 0 \\ 1 \\ x_1 \end{bmatrix},$$

and therefore

$$\begin{bmatrix} g(x) & ad_f g(x) & ad_f^2 g(x) \end{bmatrix} = \begin{bmatrix} 1 & 0 & e^{-x_2} \\ 0 & 1 & 0 \\ 0 & x_1 & e^{-x_2} \end{bmatrix}.$$

Since $\det[g(x)\ ad_f g(x)\ ad_f^2 g(x)] = e^{-x_2} \neq 0$, the rank condition (i) of Theorem 12.3 is satisfied at every point. $\triangle$

It is possible to relate condition (i) of Theorem 12.3 to the controllability of the Jacobian linearization at x_0 of the nonlinear system. Suppose $f(x_0) = 0$, so that x_0 is an equilibrium for $\dot{x} = f(x)$. The rank condition (i) implies that the linearization at x_0 defined by

$$\dot{x} = Ax + bu := \frac{\partial f}{\partial x}(x_0)\, x + g(x_0)u \tag{12.24}$$

is controllable, that is, rank $[b \; Ab \; \cdots \; A^{n-1}b] = n$. (See Exercise 12.2.) Thus, the system of Example 12.1 has a controllable linearization, as shown in Example 12.4, but it is not feedback linearizable in a neighborhood of any point, because the involutivity condition (ii) of Theorem 12.3 does not hold at any point.

A return to Example 11.1 helps in understanding feedback linearization for planar systems.

Example 12.5 For convenience we repeat the equations of Example 11.1 here:

$$\begin{aligned} \dot{x}_1 &= \sin x_2, \\ \dot{x}_2 &= -x_1^2 + u. \end{aligned}$$

Let $x_0 = 0$. Conditions (i) and (ii) of Theorem 12.3 are easily checked. The rank condition (i) holds, since

$$\text{rank}\,[g \quad ad_f g](0) = \text{rank} \begin{bmatrix} 0 & 1 \\ 1 & 0 \end{bmatrix} = 2.$$

Condition (ii) holds, since span $\{g(x)\}$ = span $\{[0 \; 1]^T\}$ is a nonsingular one-dimensional distribution. △

Corollary 12.1 *A planar single-input system $\dot{x} = f(x) + g(x)u$, where f and g are C^1 and $f(0) = 0$, is locally feedback linearizable near $x = 0$ if and only if the linear approximation at the origin*

$$\dot{x} = Ax + bu := \frac{\partial f}{\partial x}(0)x + g(0)u$$

is a controllable linear system.

Proof. Suppose that the planar system is feedback linearizable near the origin. By (i) of Theorem 12.3, we have

$$\text{rank}\;[g(0)\;[f,g](0)] = 2. \tag{12.25}$$

Hence, $b = g(0) \neq 0$, and since $f(0) = 0$,

$$[f,g](0) = Dg(0)f(0) - Df(0)g(0) = -Df(0)g(0) = -Ab.$$

Thus, (i) implies that b and Ab are linearly independent, so the pair (A, b) is controllable.

Conversely, if the pair (A,b) is controllable, then (12.25) holds. Since $g(0) \neq 0$, $g(x) \neq 0$ for all x in a neighborhood of $x=0$. Thus, the distribution span $\{g\}$ is a nonsingular one-dimensional distribution in a neighborhood of the origin. Theorem 12.3 implies that the system is locally feedback linearizable near $x=0$. □

The next example concerns the system equations for a single-link robotic manipulator.

Example 12.6 The dynamical equations for a single-link, flexible-joint mechanism with negligible damping are

$$\begin{aligned} I\ddot{q}_1 + MGL\sin q_1 + k(q_1 - q_2) &= 0, \\ J\ddot{q}_2 - k(q_1 - q_2) &= u, \end{aligned}$$

where q_1 and q_2 are angular positions, I and J are moments of inertia, k is a spring constant, M is a mass, G is the gravitational constant, L is a distance, and u is a motor torque input. By writing $x = [q_1 \quad \dot{q}_1 \quad q_2 \quad \dot{q}_2]^T$, the four-dimensional state equations can be written as

$$\dot{x} = f(x) + g(x)u := \begin{bmatrix} x_2 \\ -a\sin x_1 - b(x_1 - x_3) \\ x_4 \\ c(x_1 - x_3) \end{bmatrix} + \begin{bmatrix} 0 \\ 0 \\ 0 \\ d \end{bmatrix} u, \tag{12.26}$$

using these positive constants: $a=(MGL)/I$, $b=k/I$, $c=k/J$, and $d=1/J$. The unforced system has an equilibrium at $x=0$. To determine whether this system is feedback linearizable near the origin, we check conditions (i), (ii) of Theorem 12.3. First, compute

$$\frac{\partial f}{\partial x}(x) = \begin{bmatrix} 0 & 1 & 0 & 0 \\ -a\cos x_1 - b & 0 & b & 0 \\ 0 & 0 & 0 & 1 \\ c & 0 & -c & 0 \end{bmatrix}.$$

Appropriate bracket calculations lead to the matrix needed in condition (i):

$$[g \quad ad_f g \quad ad_f^2 g \quad ad_f^3 g] = \begin{bmatrix} 0 & 0 & 0 & -bd \\ 0 & 0 & bd & 0 \\ 0 & -d & 0 & cd \\ d & 0 & -cd & 0 \end{bmatrix}.$$

Since this matrix has rank 4, condition (i) holds at every point. The involutivity condition (ii) holds for this four-dimensional system, since $\mathcal{G}_{n-2} = \mathcal{G}_2 = \text{span}\,\{g, ad_f g, ad_f^2 g\}$ is the span of constant vector fields. By Theorem 12.3, system (12.26) is feedback linearizable, and the construction of the function T_1 with relative degree four can be attempted. In fact, since we know that the nullspace of $dT_1(x)$ must be the linear subspace defined by span $\{g, ad_f g, ad_f^2 g\}$, we take T_1 to be a linear function of x_1

alone: $T_1(x) = x_1$. The complete coordinate transformation $T(x)$ is then obtained from (12.17), yielding

$$z = T(x) = \begin{bmatrix} T_1(x) \\ T_2(x) \\ T_3(x) \\ T_4(x) \end{bmatrix} = \begin{bmatrix} x_1 \\ x_2 \\ -a \sin x_1 - b(x_1 - x_3) \\ -a x_2 \cos x_1 - b(x_2 - x_4) \end{bmatrix}.$$

By defining the feedback transformation $u = \alpha(x) + \beta(x)v$ according to (12.21), the equations for z are given by (12.13). Equivalently, $z_1^{(4)} = v$. $\triangle$

There is another set of conditions on the spanning vector fields in (12.22) which are necessary and sufficient for feedback linearization; these conditions were not discussed in the argument for Proposition 12.1 (or Proposition 11.1). The conditions in question are the involutivity and nonsingularity of each of the distributions $\mathcal{G}_k$ defined for $0 \leq k \leq n-1$ as follows:

$$\begin{aligned} \mathcal{G}_0 &:= \text{span}\,\{g\}, \\ \mathcal{G}_1 &:= \text{span}\,\{g, ad_f g\}, \\ \cdots &\quad \cdots \\ \mathcal{G}_k &:= \text{span}\,\{g, ad_f g, \ldots, ad_f^k g\}, \\ \cdots &\quad \cdots \\ \mathcal{G}_{n-2} &:= \text{span}\,\{g, ad_f g, \ldots, ad_f^{n-2} g\}, \\ \mathcal{G}_{n-1} &:= \text{span}\,\{g, ad_f g, \ldots, ad_f^{n-1} g\}. \end{aligned} \tag{12.27}$$

In fact, we have the following result.

Theorem 12.4 *The single-input system $\dot{x} = f(x) + g(x)u$ is feedback linearizable in a neighborhood of x_0 if and only if each of the distributions $\mathcal{G}_k$, for $0 \leq k \leq n-1$, is involutive and nonsingular with $\dim \mathcal{G}_k = k+1$ in a neighborhood of x_0.*

Proof. Assume (i) and (ii) of Theorem 12.3 hold in the open set containing x_0. By (i), the distribution $\mathcal{G}_{n-1}$ is nonsingular in U, and it must be involutive since it has dimension n. Condition (i) also implies that $\mathcal{G}_k$ has constant dimension $k+1$ in U for $0 \leq k \leq n-1$. Since (12.11) holds, we conclude that for $x \in U$ and for each k with $0 \leq k \leq n-2$, the differentials $dh(x), \ldots, d(L_f^k h)$ are linearly independent, and satisfy

$$\begin{aligned} dh(x) &\in \mathcal{G}_{n-2}^{\perp}(x), \\ dh(x), d(L_f h)(x) &\in \mathcal{G}_{n-3}^{\perp}(x), \\ \cdots &\quad \cdots \\ dh(x), \ldots, d(L_f^k h)(x) &\in \mathcal{G}_{n-2-k}^{\perp}, \\ \cdots &\quad \cdots \\ dh(x), \ldots, d(L_f^{n-2} h)(x) &\in \mathcal{G}_0^{\perp}. \end{aligned}$$

Thus, each distribution $\mathcal{G}_k$ is completely integrable in U, and hence, by the Frobenius theorem, each $\mathcal{G}_k$ is involutive in U. This completes the proof of the only if part.

Conversely, suppose that each distribution $\mathcal{G}_k$ for $0 \leq k \leq n-1$ is involutive in U and satisfies $\dim \mathcal{G}_k(x) = k+1$ for $x \in U$. Then (i) certainly holds: $\dim \mathcal{G}_{n-1}(x_0) = n$, and so does (ii), since $\mathcal{G}_{n-2}$ is involutive in U. □

It is clear that Theorem 12.3 (and Theorem 12.4) highlight the very special nature of systems that are fully input-state linearizable. Nevertheless, certain system classes covered by the theorem can be very useful; see, for example, the triangular form system in Exercise 12.6, which can be viewed as a known perturbation of a controllable linear system.

On the other hand, as shown by Proposition 11.1 in the preceding chapter, even if a system is not fully input-state linearizable as in Theorem 12.3, but an output function with relative degree $r < n$ is known, then the system can still be transformed to a partially linear normal form which is quite useful in many control problems. Probably the most important point to make here is that the circle of ideas discussed in this chapter and the preceding one can be applied to a wide range of problems beyond those considered in this book.

12.4 LINEARIZING TRANSFORMATIONS

Suppose the system defined by the vector field pair (f, g) is locally feedback linearizable about x_0, and set $v_i(x) = ad_f^{i-1} g(x)$ for $1 \leq i \leq n-1$. Then $\text{span}\,\{v_1, \ldots, v_{n-1}\} = \text{span}\,\{g, \ldots, ad_f^{n-2} g\} = \mathcal{G}_{n-2}$ is involutive in a neighborhood of x_0. We outline here the argument that involutivity implies complete integrability, following the proof of the Frobenius theorem, as this argument explains how one can construct a locally linearizing coordinate change for a system that is locally feedback linearizable.

First, it is possible to augment the vector fields v_i, $1 \leq i \leq n-1$, by a smooth vector field v_n defined near x_0 such that

$$\text{span}\,\{v_1, \ldots, v_{n-1}, v_n\}(x) = \mathbf{R}^n$$

for all x in an open set about x_0. (If the hyperplanes given by $\mathcal{G}_{n-2}(x)$ vary smoothly with x, then so does a choice of unit normal vector, for example.)

For $1 \leq i \leq n$, let $\phi_t^{v_i}(x) = \phi^{v_i}(t, x)$ be the time t solution mapping for the vector field v_i. With the point x_0 fixed as above, the required coordinate change can be constructed from the composition of flow mappings given by

$$F(z) = F(z_1, \ldots, z_n) := \phi_{z_1}^{v_1} \circ \phi_{z_2}^{v_2} \circ \cdots \circ \phi_{z_{n-1}}^{v_{n-1}} \circ \phi_{z_n}^{v_n}(x_0). \tag{12.28}$$

By the local existence and uniqueness theorem for autonomous ordinary differential equations, the function F is well defined in a neighborhood of the origin $z = (z_1, \ldots, z_n) = (0, \ldots, 0) \in \mathbf{R}^n$, and F takes values in $\mathbf{R}^n$.

Application of the chain rule and evaluation at $z = (z_1, \ldots, z_n) = (0, \ldots, 0)$ give

$$\frac{\partial F}{\partial z_i}(0) = v_i(x_0) \quad \text{for } 1 \le i \le n. \tag{12.29}$$

By hypothesis, the vectors $v_i(x_0)$, $1 \le i \le n$, are linearly independent. By the inverse function theorem, there exists an $\epsilon > 0$ such that F restricted to $B_\epsilon(0) \subset \mathbf{R}^n$ is a local diffeomorphism onto its image $F(U)$; that is, $F : U \to \mathbf{R}^n$ is smooth, one-to-one, and has a smooth inverse mapping $F^{-1} : F(U) \to U$. The z_i coordinates of a point z near the origin $z = 0$ are exactly the travel times required along the solution flows of the vector fields v_i in order to reach the point z.

Write the local inverse of F in the form

$$F^{-1}(x) = \begin{bmatrix} \lambda_1(x) \\ \vdots \\ \lambda_n(x) \end{bmatrix}.$$

Now consider (12.29) together with the identity

$$\left(\frac{\partial F^{-1}}{\partial x}\right)_{z=F^{-1}(x)} \left(\frac{\partial F}{\partial z}\right)_{x=F(z)} = I_{n\times n}. \tag{12.30}$$

A key part of the proof of the Frobenius theorem shows that involutivity of $\mathcal{G}_{n-2}$ implies that $d\lambda_n(x)$ spans $\mathcal{G}_{n-2}^{\perp}(x)$, by showing that the first $n-1$ columns of $(\partial F/\partial z)_{x=F(z)}$ form a basis of $\mathcal{G}_{n-2}(x)$ at any $x \in U$; then (12.30) implies that span $\{d\lambda_n(x)\} = \mathcal{G}_{n-2}^{\perp}(x)$ for $x \in U$. In addition, λ_n satisfies the nontriviality condition required for relative degree n, because (12.30) and (12.29) imply

$$d\lambda_n(x_0)\frac{\partial F}{\partial z_n}(0) = d\lambda_n(x_0)v_n(x_0) = 1.$$

Therefore λ_n is a function of relative degree n in U.

The conclusion we want to emphasize here is that the linearizing transformation F is constructed from the solution flows as in (12.28). The next example illustrates the construction of a linearizing coordinate change and feedback transformation for a planar system.

Example 12.7 We illustrate the construction of a linearizing mapping $x = F(z)$ and its inverse $z = F^{-1}(x)$, as discussed above. The example also illustrates the local nature of a linearizing transformation and the local nature of a solution of the system of partial differential equations (12.19).

Consider the system

$$\begin{aligned} \dot{x}_1 &= x_2 + x_1^2 u, \\ \dot{x}_2 &= u. \end{aligned}$$

We have $f(x) = [x_2\ 0]^T$ and $g(x) = [x_1^2\ 1]^T$. The linearization at the origin is a controllable linear system, so this system is feedback linearizable in a neighborhood U of $x = 0$. We now identify such a neighborhood U. The distribution span $\{g(x)\}$ is involutive, and

$$\det[g(x)\ ad_f g(x)] = \det \begin{bmatrix} x_1^2 & 2x_1x_2 - 1 \\ 1 & 0 \end{bmatrix} = 1 - 2x_1x_2,$$

so that rank $[g(x)\ ad_f g(x)] = 2$ provided $1 - 2x_1x_2 \neq 0$. Thus the system is feedback linearizable in a neighborhood of any point such that $x_1x_2 \neq \frac{1}{2}$.

To construct the mapping F as in the proof of the Frobenius theorem (Theorem D.4), we can augment $g =: v_1$ with any vector field $w(x)$ such that span $\{v_1, w\}(x) = \mathbf{R}^2$. Set $w(x) = [1\ 0]^T$. Now consider the two systems

$$\dot{x} = v_1(x) = g(x) = \begin{bmatrix} x_1^2 \\ 1 \end{bmatrix}, \qquad \dot{x} = w(x) = \begin{bmatrix} 1 \\ 0 \end{bmatrix}.$$

The flows for these vector fields, which are used to construct the mapping F, are obtained by direct integration, using initial condition $x_0 = (x_{10}, x_{20})$. The flow mappings are as follows:

$$\phi_{z_1}^{v_1} = \begin{bmatrix} \dfrac{x_{10}}{1 - x_{10}z_1} \\ z_1 + x_{20} \end{bmatrix}, \qquad \phi_{z_2}^{w} = \begin{bmatrix} z_2 + x_{10} \\ x_{20} \end{bmatrix}.$$

Note that the flow $\Phi_{z_1}^{v_1}$ is not globally defined, since it is not defined for all times z_1. The mapping F is defined by

$$\begin{aligned} x = F(z) &= \phi_{z_1}^{v_1} \circ \phi_{z_2}^{w}(x_0) \\ &= \phi_{z_1}^{v_1}\left(\begin{bmatrix} z_2 + x_{10} \\ x_{20} \end{bmatrix}\right) \\ &= \begin{bmatrix} \dfrac{z_2 + x_{10}}{1 - (z_z + x_{10})z_1} \\ z_1 + x_{20} \end{bmatrix}. \end{aligned}$$

The inverse mapping is obtained, by straightforward algebraic manipulations, as

$$z = F^{-1}(x) = \begin{bmatrix} \lambda_1(x) \\ \lambda_2(x) \end{bmatrix} = \begin{bmatrix} \dfrac{x_2 - x_{20}}{\dfrac{x_1}{1 + x_1(x_2 - x_{20})} - x_{10}} \end{bmatrix}.$$

Then the relation

$$\left[\frac{\partial F^{-1}}{\partial x}\right]_{x=F(z)} \left[\frac{\partial F}{\partial z}\right]_z = I_{2\times 2}$$

implies that $\lambda_2(x)$ is a solution of the partial differential equation

$$d\lambda_2(x) \cdot v_1(x) = d\lambda_2(x) \cdot g(x) = 0.$$

Observe that λ_2, which we are assured has relative degree two, is given by

$$\lambda_2(x) = \frac{x_1}{1 + x_1(x_2 - x_{20})} - x_{10},$$

and it is not globally defined, as a consequence of the flow $\phi_{z_1}^{v_1}$ not being globally defined. (See Exercise 12.7.) △

12.5 EXERCISES

Exercise 12.1 *The Bracket Identity* (12.23)
Show that, for smooth vector fields $\xi(x)$, $\eta(x)$ defined for x in an open set U, and smooth functions $a(x)$, $b(x)$ on U,

$$[a\xi, b\eta](x) = a(x)b(x)[\xi, \eta](x) + (L_\xi b(x))a(x)\eta(x) - (L_\eta a(x))b(x)\xi(x).$$

Hint: Use the Jacobi identity, together with the following characterization of equality of two vector fields on U: Vector fields $v(x)$, $w(x)$ defined on U are equal if and only if for every smooth function λ on U, $L_v\lambda(x) = L_w\lambda(x)$ for all $x \in U$.

Exercise 12.2 *On Controllability of the Linearization*
Consider the system $\dot{x} = f(x) + g(x)u$ with $f(x_0) = 0$, and its linearization at x_0 given by

$$\dot{x} = Ax + bu := \frac{\partial f}{\partial x}(x_0)\, x + g(x_0)u.$$

Show that if $\text{rank}\,[g(x_0)\;\; ad_f g(x_0)\;\; \cdots\;\; ad_f^{n-1} g(x_0)] = n$, then the linearization is controllable. *Hint*: Write $f(x) = Ax + f_2(x)$, where $\frac{\partial f_2}{\partial x}(x_0) = 0$, and $g(x) = b + g_1(x)$, where $g_1(x_0) = 0$. Then compute $ad_f^k g(x_0)$ for $k = 0, \ldots, n-1$.

Exercise 12.3 Consider the system

$$\begin{aligned}
\dot{x}_1 &= x_2 + x_3^3,\\
\dot{x}_2 &= x_3,\\
\dot{x}_3 &= u.
\end{aligned}$$

(a) Show that the Jacobian linearization at the origin is a controllable linear system and the output $y = h(x) = x_1$ has relative degree three for that linearization.
(b) Show that the output function $y = h(x) = x_1$ does not have a well-defined relative degree at the origin for the nonlinear system.
(c) Show in two different ways that there is no smooth function $h(x)$ having relative degree three at the origin, hence this nonlinear system is not feedback linearizable in any neighborhood of the origin.

Exercise 12.4 Consider the robotic manipulator system in Example 12.6. Show how a feedback control can be designed that makes the link position $x_1 = z_1$ exactly track a prespecified trajectory $x_1 = z_1 = \sigma(t)$. *Hint:* The problem is equivalent to inverting the system, in the sense that the output $y(t) = x_1 = \sigma(t)$ is given, and it is required to specify the input control $u = \alpha(x) + \beta(x)v$ that achieves the given output.

Exercise 12.5 Show that the system

$$\begin{aligned}\dot{x}_1 &= x_1^3 + x_2^5 + 3x_3^3,\\ \dot{x}_2 &= x_3,\\ \dot{x}_3 &= u\end{aligned}$$

is not feedback linearizable in any neighborhood of the origin.

Exercise 12.6 *Systems in Triangular Form*
Show that a three dimensional system in the special triangular form,

$$\begin{aligned}\dot{x}_1 &= x_2 + \phi_1(x_1),\\ \dot{x}_2 &= x_3 + \phi_2(x_1, x_2),\\ \dot{x}_2 &= \phi_3(x_1, x_2, x_3) + u,\end{aligned}$$

in which ϕ_1, ϕ_2, ϕ_3 are smooth functions such that $\phi_i(0) = 0$, $1 \le i \le 3$, is locally feedback linearizable about the origin.

Exercise 12.7 *On Example 12.7*
Show directly that the function

$$\lambda_2(x_1, x_2) = \frac{x_1}{1 + x_1(x_2 - x_{20})} - x_{10},$$

constructed in Example 12.7, has relative degree two at the point (x_{10}, x_{20}) provided $2x_{10}x_{20} \neq 1$.

Exercise 12.8 Determine whether the system

$$\begin{aligned}\dot{x}_1 &= x_2 + \gamma(x_1, x_2)u,\\ \dot{x}_2 &= u\end{aligned}$$

is locally feedback linearizable about the origin, where $\gamma(x_1, x_2)$ is a smooth function with $\gamma(0, 0) = 0$. If so, determine a linearizing coordinate change and feedback transformation. What happens if $\gamma = \gamma(x_2)$?

Exercise 12.9 Consider the system

$$\begin{aligned}\dot{x}_1 &= x_2 + x_1 x_2,\\ \dot{x}_2 &= u.\end{aligned}$$

(a) Show that this system is locally feedback linearizable about the origin and determine a linearizing coordinate change and feedback transformation. Are your linearizing transformations globally defined? Is the system globally feedback linearizable?
(b) Show that there is no smooth feedback $u = \alpha(x_1, x_2)$ with $\alpha(0,0) = 0$ that makes the origin globally asymptotically stable.

Exercise 12.10 Consider the system

$$\begin{aligned}\dot{x}_1 &= -x_1^5 - x_2^3,\\ \dot{x}_2 &= x_2 + u.\end{aligned}$$

(a) Show that this system is not feedback linearizable about the origin.
(b) Show that there is a smooth feedback $u = \alpha(x_1, x_2)$ that makes the origin globally asymptotically stable.

Exercise 12.11 Show that the system

$$\begin{aligned}\dot{x}_1 &= x_2 + x_1 x_3,\\ \dot{x}_2 &= x_3.\\ \dot{x}_3 &= u.\end{aligned}$$

is feedback linearizable in a neighborhood of the origin. Find a linearizing coordinate transformation and feedback. *Hint*: Note that $y = h(x) = x_1$ does not have relative degree three at the origin. Also note that the system is not in the triangular form of Exercise 12.6.

12.6 NOTES AND REFERENCES

For more on feedback linearization, see [47]; for some engineering perspective, see [58] and [78]. Many additional examples of feedback linearization, as well as the problem of global feedback linearization, are discussed in R. Marino and P. Tomei, *Nonlinear Control Design: Geometric, Adaptive, and Robust*, Prentice-Hall Europe, London, 1995; several exercises for this chapter were taken from, or influenced by, their text; specifically, Exercises 12.5–12.6, 12.8–12.9. Reference [64] is an interesting survey of two decades of research on feedback linearization, its development from early ideas, and its influence on the development of nonlinear systems theory. Among many other topics, both [64] and the text by Marino and Tomei discuss the linearization of nonlinear systems by coordinate transformation alone.

The original formulation and solution of the feedback linearization problem appears in [17] for single-input systems with the restricted class of feedback transformations $u = \alpha(x) + \beta v$ having constant β. For early applications of Lie brackets in the study of reachability problems see [37] and the references therein. Some additional interesting historical remarks appear in [51] (pages 1–2); and [51] (Part I: Reachable Sets and Controllability) discusses controllability from a geometric point of view.

Information on controllability and observability for nonlinear systems appears in [47], [79], and [91]; see also [38].

Example 12.6 is from [58].

Chapter Thirteen

An Introduction to Damping Control

In this chapter we continue to study systems of the form

$$\dot{x} = f(x) + g(x)u, \tag{13.1}$$

and, for single-input systems, we explore some implications and limitations of the condition

$$\text{rank}\,[g(x_0)\ ad_f g(x_0)\ \cdots\ ad_f^{n-1} g(x_0)] = n, \tag{13.2}$$

which played an important role in the feedback linearization result in Theorem 12.3.

The first section of the chapter uses (13.2) to introduce the idea of *damping control*; the section features a result on global asymptotic stabilization by damping control. We note that in some references, damping control as discussed here is called *Jurdjevic-Quinn control*, or L_gV *control*, where $V(x)$ is a positive definite function. Damping control has applications beyond its connection with the particular global stabilization result given here.

The last section of the chapter is a convenient place to contrast some aspects of nonlinear systems with the familiar situation of linear systems. When $f(x_0) = 0$, (13.2) implies the controllability rank condition for the matrix pair $(A, b) := (Df(x_0), g(x_0))$ of the linearization at x_0. (See Exercise 12.2.) This section includes a definition of controllability for (13.1) by adapting the statement of Definition 4.3 to define a local controllability property for nonlinear systems. This section highlights two essential contrasting facts for the nonlinear case:

(a) The rank condition (13.2) does not imply local controllability of (13.1).
(b) Controllability of (13.1) does not imply smooth stabilizability.

While these are essentially negative results, they have stimulated much research on controllability issues for nonlinear systems.

13.1 STABILIZATION BY DAMPING CONTROL

In this section we show that, despite the general limitations outlined in items (a) and (b) of the introduction, the rank condition (13.2) can be used to establish a global stabilization result for a very special class of nonlinear systems useful in applications.

Consider system (13.1) where f is linear, $f(x) = Ax$. Further, suppose that all eigenvalues of A are pure imaginary, and suppose in addition that all the eigenvalues are simple, so that each eigenvalue has geometric multiplicity equal to its algebraic multiplicity. Under these conditions, the origin of $\dot{x} = f(x) = Ax$ is Lyapunov stable, but the question of asymptotic stabilization of the origin for (13.1) is a critical problem which cannot be resolved by the Jacobian linearization at the origin, given by $\dot{x} = Ax + g(0)u$. However, nonlinear feedback may be considered in order to asymptotically stabilize the equilibrium at the origin. To see how the stabilization might be done, in the next example we consider a linear harmonic oscillator with a scalar control.

Example 13.1 Consider the system

$$\begin{bmatrix} \dot{x}_1 \\ \dot{x}_2 \end{bmatrix} = \begin{bmatrix} 0 & 1 \\ -1 & 0 \end{bmatrix} x + \begin{bmatrix} g_1(x) \\ g_2(x) \end{bmatrix} u.$$

Take $V(x) = \frac{1}{2}x^T x$ as a candidate Lyapunov function. Along solutions $x(t)$ we have

$$\dot{V}(x) = \frac{1}{2}\dot{x}^T x + \frac{1}{2}x^T \dot{x} = x^T \dot{x} = x^T (Ax + g(x)u) = x^T g(x)u,$$

where we used the fact that A is skew-symmetric: $A^T = -A$. Note that this calculation may also be written as $\dot{V}(x) = L_g V(x)u$. Motivated by the desire to make $\dot{V}(x) < 0$ for $x \neq 0$, a natural choice for the feedback is to set

$$u = k(x) := -x^T g(x) = -L_g V(x).$$

Since $k(0) = 0$, the origin is still an equilibrium of the closed loop system, and we have

$$\dot{V}(x) = -(L_g V(x))^2 = -(x^T g(x))^2 \leq 0.$$

The set where $\dot{V}(x) = 0$ is the set where $x^T g(x) = 0$, that is, the set where the feedback $k(x) = 0$. If the only forward trajectory contained in this set is the origin itself, then the invariance theorem implies that $x = 0$ is globally asymptotically stable. Thus far, we have a general feedback form which makes $V(x) = \frac{1}{2}x^T x$ a true Lyapunov function. Without further information, however, this only allows us to say that the origin is stable after the feedback, and it was stable before the feedback. We now ask for conditions on the input vector field $g(x)$ so that the invariance theorem applies.

Let $x(t)$ be a nonzero forward solution of $\dot{x} = Ax + g(x)k(x)$, with $x(0) = x_0 \neq 0$, such that

$$\dot{V}(x(t)) = 0.$$

Then $k(x(t)) = 0$, and therefore $\dot{x}(t) = Ax(t)$, that is, $x(t)$ must be a solution of this linear system. Consequently, $x(t) = e^{tA}x_0$. The identity $x^T(t)g(x(t)) \equiv 0$ can be written as $x_0^T e^{-tA} g(e^{tA}x_0) \equiv 0$, since $A^T = -A$.

We have

$$\begin{aligned}\frac{d}{dt}(x_0^T e^{-tA} g(e^{tA}x_0)) &= x_0^T(-Ae^{-tA}g(e^{tA}x_0) + e^{-tA}Dg(e^{tA}x_0)Ae^{tA}x_0)\\ &= x_0^T(e^{-tA}Dg(e^{tA}x_0)Ae^{tA}x_0 - Ae^{-tA}g(e^{tA}x_0))\\ &= x_0^T e^{-tA}[Dg(e^{tA}x_0)Ae^{tA}x_0 - Ag(e^{tA}x_0)]\\ &= 0. \end{aligned} \tag{13.3}$$

Evaluation at $t = 0$ gives

$$0 = x_0^T \frac{d}{dt}\left[e^{-tA}g(e^{tA}x_0)\right]\Big|_{t=0} = x_0^T\,[f,g]\,(x_0), \quad \text{where } f(x) = Ax.$$

Thus, any nonzero forward solution $x(t)$ such that $\dot V(x(t)) = 0$ must satisfy the two orthogonality conditions at $x(0) = x_0$ given by

$$x_0^T\,[g(x_0) \quad [f,g](x_0)] = 0.$$

Now, if (13.2) holds at every nonzero x_0 for this planar system ($n = 2$), we have a contradiction of the assumption that $x_0 \neq 0$ for the solution $x(t)$; hence $x(t) \equiv 0$. Thus, under the assumption that (13.2) holds at every x_0, the invariance theorem applies, and we conclude that the origin is asymptotically stabilized by the feedback $u = -L_gV(x)$. △

The essential ideas of the analysis of Example 13.1 are as follows. Recalling that $V(x) = \frac{1}{2}x^Tx$ and $f(x) = Ax$, we had

$$\dot V(x(t)) = 0 \Longrightarrow (L_fV(x(t)) = 0 \text{ and } L_gV(x(t)) = 0)\,;$$

in addition,

$$(\,(13.2) \text{ and } \dot V(x(t)) = 0\,) \Longrightarrow x(t) \equiv 0.$$

Thus, (13.2) enabled the application of the invariance theorem when using the *damping control* defined by $u = -L_gV(x)$. In addition, for reference below in the proof of Theorem 13.1, it is helpful to note that the calculation in (13.3) actually yields

$$\frac{d}{dt}[k(x(t))] = \frac{d}{dt}(x^T(t)g(x(t))) = x^T(t)\,[f,g](x(t)) \tag{13.4}$$

for all t, where $k(x) = -x^Tg(x)$.

The proof of the next theorem generalizes the argument used in Example 13.1. It is important to realize that the calculations of the example were facilitated by the fact that A was skew-symmetric. However, note that if A is an $n \times n$ matrix with n simple, pure imaginary eigenvalues, then A is similar to a skew-symmetric matrix, namely, the real Jordan form of A.

Theorem 13.1 (V. Jurdjevic–J. P. Quinn)
Suppose $A \in \mathbf{R}^{n\times n}$ has n simple, pure imaginary eigenvalues, and let $f(x) = Ax$. If g is any vector field on $\mathbf{R}^n$ such that the pair (f,g) satisfies (13.2)

at all $x \neq 0$*, then the origin of the system* $\dot{x} = Ax + g(x)u$ *is globally asymptotically stabilized by the feedback*

$$u = -x^T (S^{-1})^T S^{-1} g(x),$$

where S *is such that* $J := S^{-1}AS$ *is the real Jordan form of* A*. In particular, if* A *is skew-symmetric, then we may take* $u = -x^T g(x)$*.*

Proof. After a linear change of coordinates, $x = Sz$, we may work with the real Jordan form J of A, which is skew-symmetric, $J = S^{-1}AS$. The transformed system is

$$\dot{z} = S^{-1}ASz + S^{-1}g(Sz)u =: \hat{f}(z) + \hat{g}(z).$$

The drift vector field in z is $\hat{f}(z) = S^{-1}ASz$ and the input vector field in z is $\hat{g}(z) = S^{-1}g(Sz)$. Since $x = Sz$, we have

$$\begin{aligned}
[\hat{f}, \hat{g}](z) &= D\hat{g}(z)\hat{f}(z) - D\hat{f}(z)\hat{g}(z) \\
&= (S^{-1}Dg(Sz)S)\, S^{-1}ASz - S^{-1}Ag(Sz) \\
&= S^{-1}[Dg(Sz)ASz - Ag(Sz)] \\
&= S^{-1}[f, g](Sz) \\
&= S^{-1}[f, g](x).
\end{aligned}$$

Thus, we may write

$$ad_{\hat{f}}\hat{g}(z) = S^{-1} ad_f g(x), \quad \text{where} \quad x = Sz.$$

Now suppose that

$$ad^k_{\hat{f}}\hat{g}(z) = S^{-1} ad^k_f g(x), \quad \text{where} \quad x = Sz.$$

We want to show that this identity holds with k replaced by $k+1$. Note that

$$\begin{aligned}
D(ad^k_{\hat{f}}\hat{g})(z) &= D(S^{-1} ad^k_f g \circ S)(z) \\
&= S^{-1} D(ad^k_f g \circ S)(z) \\
&= S^{-1} D(ad^k_f g)(Sz)S.
\end{aligned}$$

Thus, we have

$$\begin{aligned}
ad^{k+1}_{\hat{f}}\hat{g}(z) &= D(ad^k_{\hat{f}}\hat{g})(z)\hat{f}(z) - D\hat{f}(z) ad^k_{\hat{f}}\hat{g}(z) \\
&= S^{-1} ad^k_f g(Sz) S (S^{-1}ASz) - (S^{-1}AS) S^{-1} ad^k_f g(Sz) \\
&= S^{-1}\left[ad^k_f g(x) Ax - A ad^k_f g(x)\right] \\
&= S^{-1}[f, ad^k_f g](x), \quad \text{where} \quad x = Sz,
\end{aligned}$$

as we wished to show. Consequently, the pair $(\hat{f}, \hat{g})$ satisfies (13.2) at $z_0 = S^{-1}(x_0)$ for each $z_0 \neq 0$; in fact,

$$[\hat{g}(z)\ ad_{\hat{f}}\hat{g}(z)\ \cdots\ ad^{n-1}_{\hat{f}}\hat{g}(z)] = S^{-1}[g(Sz)\ ad_f g(Sz)\ \cdots\ ad^{n-1}_f g(Sz)].$$

Let $V = \frac{1}{2}z^T z$, and set

$$u = k(z) := -z^T \hat{g}(z) = -L_{\hat{g}} V(z).$$

Since $k(0) = 0$, the origin is an equilibrium of the closed loop z system, and we have

$$\dot{V}(z) = -(L_{\hat{g}} V(z))^2 = -(z^T \hat{g}(z))^2 \leq 0.$$

The set where $\dot{V}(z) = 0$ is the set where $k(z) = z^T \hat{g}(z) = 0$. We want to show that the only forward solution of the closed loop system contained in this set is the origin itself; since all forward solutions are bounded, the invariance theorem then implies that $z = 0$ is globally asymptotically stable, and the theorem follows.

Let $z(t)$, $z(0) = z_0 \neq 0$, be a forward solution of $\dot{z} = Jz + \hat{g}(z)k(z)$, such that

$$\dot{V}(z(t)) = 0.$$

Then $k(z(t)) = 0$, and therefore $z(t)$ is a solution of the linear system $\dot{z} = Jz$. Hence, $z(t) = e^{tJ} z_0$. The identity $k(z(t)) = z^T(t)\hat{g}(z(t)) \equiv 0$ can be written as $z_0^T e^{-tJ} \hat{g}(e^{tJ} z_0) \equiv 0$, since $J^T = -J$. Now, as we saw in (13.4), we can write

$$\begin{aligned} \frac{d}{dt}[k(z(t))] &= z_0^T \frac{d}{dt}[e^{-tJ} \hat{g}(e^{tJ} z_0)] \\ &= z_0^T [-Je^{-tJ} \hat{g}(e^{tJ} z_0) + e^{-tJ} D\hat{g}(e^{tJ} z_0) J e^{tJ} z_0] \\ &= z_0^T e^{-tJ} [-J\hat{g}(e^{tJ} z_0) + D\hat{g}(e^{tJ} z_0) J e^{tJ} z_0] \\ &= z_0^T e^{-tJ} [\hat{f}, \hat{g}](z(t)) \equiv 0. \end{aligned} \tag{13.5}$$

Evaluation at $t = 0$ gives

$$z_0^T [\hat{f}, \hat{g}](z_0) = z_0^T ad_{\hat{f}} \hat{g}(z_0) = 0$$

Now suppose that, for some k,

$$\frac{d^k}{dt^k}[k(z(t))] = z_0^T e^{-tJ} ad_{\hat{f}}^k \hat{g}(z(t)). \tag{13.6}$$

Differentiate (13.6) with respect to t to get

$$\begin{aligned} \frac{d^{k+1}}{dt^{k+1}}[k(z(t))] &= z_0^T \frac{d}{dt}[e^{-tJ} ad_{\hat{f}}^k \hat{g}(z(t))] \\ &= z_0^T [-Je^{-tJ} ad_{\hat{f}}^k \hat{g}(z(t)) + e^{-tJ} D(ad_{\hat{f}}^k \hat{g})(z(t)) J z(t)] \\ &= z_0^T e^{-tJ} [-J ad_{\hat{f}}^k \hat{g}(z(t)) + D(ad_{\hat{f}}^k \hat{g})(z(t)) J z(t)] \\ &= z_0^T e^{-tJ} [\hat{f}, ad_{\hat{f}}^k \hat{g}](z(t)) \\ &= z_0^T e^{-tJ} ad_{\hat{f}}^{k+1} \hat{g}(z(t)). \end{aligned}$$

Thus, (13.6) holds for $1 \leq k \leq n-1$, and evaluation at $t = 0$ for $1 \leq k \leq n-1$ yields

$$z_0^T [\hat{g} \quad ad_{\hat{f}} \hat{g} \quad \cdots \quad ad_{\hat{f}}^{n-1} \hat{g}](z_0) = 0.$$

Since the pair $(\hat{f}, \hat{g})$ satisfies (13.2) at $z_0 = S^{-1}(x_0)$ for each $z_0 \neq 0$, we must have $z_0 = 0$. This contradicts the assumption that $z(t)$ was a nonzero solution such that $\dot{V}(z(t)) \equiv 0$; therefore the only such solution is $z(t) \equiv 0$. By the invariance theorem, the origin $z = 0$ is globally asymptotically stable, and hence so is $x = 0$. Note that, in the original x coordinates, the stabilizing feedback is $u = -z^T \hat{g}(z) = -x^T (S^{-1})^T S^{-1} g(x)$, and a Lyapunov function is given by $V(x) = \frac{1}{2} x^T (S^{-1})^T S^{-1} x$. Finally, note that if A itself is skew-symmetric, then no coordinate transformation is required and the argument above shows that we may take $u = -x^T g(x)$. □

The next example illustrates the application of Theorem 13.1, and shows the construction of a globally stabilizing damping control as in the theorem.

Example 13.2 Consider the system

$$\begin{bmatrix} \dot{x}_1 \\ \dot{x}_2 \end{bmatrix} = \begin{bmatrix} 0 & 1 \\ -1 & 0 \end{bmatrix} x + \begin{bmatrix} 0 \\ x_1 \end{bmatrix} u.$$

The rank condition (13.2) holds at every point x_0 in the plane except at the origin, and Theorem 13.1 applies. As a stabilizing feedback, we may take $u = k(x) = -L_g V(x)$ where $V(x) = \frac{1}{2} x^T x$; thus, we take $u = -x_2 x_1$. This choice results in the closed loop system

$$\begin{aligned} \dot{x}_1 &= x_2, \\ \dot{x}_2 &= -x_1 - x_2 x_1^2. \end{aligned}$$

Along the solutions of this closed loop system, we have $\dot{V}(x) = -x_2^2 x_1^2 \leq 0$, and $\dot{V}(x) = 0$ on the union of the coordinate axes. However, as the proof of Theorem 13.1 shows, the invariance principle applies and ensures the asymptotic stability of the origin. △

The next example shows that the special structure for the A matrix is important in Theorem 13.1, by showing that the rank condition (13.2) by itself is not sufficient for smooth stabilization.

Example 13.3 Consider the system

$$\begin{bmatrix} \dot{x}_1 \\ \dot{x}_2 \end{bmatrix} = \begin{bmatrix} 0 & 1 \\ -1 & 1 \end{bmatrix} \begin{bmatrix} x_1 \\ x_2 \end{bmatrix} + \begin{bmatrix} 0 \\ x_1 \end{bmatrix} u.$$

It is straightforward to check that, with $n = 2$, (13.2) is satisfied at every nonzero x in the plane. However, the eigenvalue assumption of Theorem 13.1 does not hold here, since A has eigenvalues

$$\lambda = \frac{1}{2} \pm i \frac{\sqrt{3}}{2}.$$

We claim that the system is not stabilizable by any C^1 feedback $u = k(x_1, x_2)$ with $k(0,0) = 0$. To see this, note that for any choice of $k(x)$, the Jacobian matrix at $x = 0$ of the vector field

$$f(x) + g(x)k(x) = \begin{bmatrix} x_2 \\ -x_1 + x_2 + x_1 k(x_1, x_2) \end{bmatrix}$$

is given by

$$\begin{bmatrix} 0 & 1 \\ -1 & 1 \end{bmatrix}.$$

Hence, the eigenvalues are unchanged by the feedback, and the claim follows by Theorem 8.2 (b). △

A variation of Theorem 13.1 appears in Exercise 13.1, where the assumption that A has simple, pure imaginary eigenvalues is replaced by the assumption that A has n distinct real eigenvalues having nonpositive real part.

Theorem 13.1 (and the result of Exercise 13.1) provide useful results on certain special critical problems of smooth stabilization, where the Jacobian linearization at the origin is $\dot{x} = Ax + bu$, where $b = g(0)$, and the pair (A, b) is not stabilizable.

Damping Control with Nonlinear Drift Vector Fields: Examples

Consider a single-input single-output system of the form

$$\dot{x} = f(x) + g(x)u, \tag{13.7}$$

$$y = L_g V(x)\,. \tag{13.8}$$

The systems covered by Theorem 13.1 have $f(x) = Ax$ and the stabilizing feedback has the form $u = -y = -L_g V(x)$.

We consider some familiar examples, with nonlinear drift vector field f, using the damping control $u = -L_g V(x) = -y$. In the first example, Theorem 13.1 does not apply, but damping control is still effective.

Example 13.4 Consider the undamped nonlinear spring system with control,

$$\begin{aligned} \dot{x}_1 &= x_2, \\ \dot{x}_2 &= -x_1^3 + u, \\ y &= x_2\,. \end{aligned}$$

It is easy to verify that $V(x) = \frac{1}{4}x_1^4 + \frac{1}{2}x_2^2$ is a strict global Lyapunov function for the nonlinear unforced system. Moreover, the output is $y = h(x) = x_2 = L_g V(x)$, where $V(x)$ is exactly this Lyapunov function. Thus we have a system of the form (13.7)–(13.8). Note that the feedback $u = -y = -L_g V(x) = -x_2$ provides a friction term which speeds the asymptotic approach of any solution to the origin. △

A similar construction occurs in the nonlinear undamped pendulum system, which is critically stable.

Example 13.5 Consider the controlled undamped pendulum system

$$\begin{aligned}\dot{x}_1 &= x_2,\\ \dot{x}_2 &= -\sin x_1 + u,\\ y &= x_2.\end{aligned}$$

The total mechanical energy of the unforced system is

$$V(x_1, x_2) = \frac{1}{2}x_2^2 + (1 - \cos x_1).$$

The output is $y = x_2 = L_gV(x)$, and the feedback $u = -x_2 = -L_gV(x)$ asymptotically stabilizes the origin. We know this from previous experience, but note that the rate of change of energy along trajectories of the closed loop $\dot{x} = f(x) - h(x)g(x)$ is given by

$$L_fV(x) - [h(x)]^2 = 0 - [h(x)]^2 \leq 0.$$

We would like to apply the invariance theorem to deduce asymptotic stability of the origin. The set where $L_fV(x) - [h(x)]^2 = 0$ consists precisely of the points where $h(x) = 0$ (since $L_fV(x) = 0$) and this is the entire x_1-axis. In this system the only solution that remains on the x_1-axis for all time is the equilibrium $x = 0$. Thus, by the invariance theorem, the origin is asymptotically stabilized by $u = -L_gV(x)$. △

The last two examples suggest that damping control can be an effective form of feedback for some systems having a nonlinear drift vector field. Evidently, in a case where the inequality $L_fV(x) \leq 0$ signals only stability of $x = 0$ for the unforced system, we would like to determine, if possible, whether one of the following conditions holds:

(a) $h(x) = L_gV(x) \neq 0$ whenever $L_fV(x) = 0$;
(b) the only solution $x(t)$ that satisfies the equation $L_fV(x) - [h(x)]^2 = 0$ near the origin is $x = 0$.

In case (a), we would have $L_fV(x) - [h(x)]^2 < 0$ for $x \neq 0$, and asymptotic stability of $x = 0$ would follow immediately. Example 13.5 involved case (b). In the absence of detailed knowledge about the set where $L_fV(x) = 0$, Exercise 13.3 gives sufficient conditions for stabilization by the damping control $u = -L_gV(x)$, based on the ideas discussed so far in the text.

13.2 CONTRASTS WITH LINEAR SYSTEMS: BRACKETS, CONTROLLABILITY, STABILIZABILITY

Since the rank condition (13.2) implies controllability of the Jacobian linearization of (13.1) at x_0 (when $f(x_0) = 0$), it is natural to ask if (13.2) implies a local controllability property of the nonlinear system (13.1) on some connected neighborhood of x_0. We now show that this conjecture is false if we use the same controllability concept as for linear systems.

The Rank Condition (13.2) Does Not Imply Controllability

We need to give a definition of local controllability for nonlinear systems by adapting the statement of Definition 4.3. For simplicity, we consider only scalar control functions. Thus, let $\mathcal{U}$ be a set of admissible control functions, for example, the set of locally integrable functions $u : [0,\infty) \to \mathbf{R}$. Given a control function $u(\cdot)$, write $x(t, v, u(\cdot))$ for the unique solution of (13.1) with initial condition $x(0) = v$. The next definition is the same as that used for controllability of linear systems, but now with reference to the nonlinear system (13.1).

Definition 13.1 *Let U be an open connected subset of $\mathbf{R}^n$. The nonlinear system* (13.1) *is* controllable *on $U \subset \mathbf{R}^n$ if for any pair of points x^1, x^2 in U, there exist a finite time $T > 0$ and an admissible control function $u(\cdot) \in \mathcal{U}$ such that $x(T, x^1, u(\cdot)) = x^2$.*

We give two examples that indicate the difficulty in specifying Lie bracket conditions for controllability of nonlinear systems. These examples provide merely an initial appreciation for this difficulty in order to bring out the contrast with the situation for linear systems. The first example shows that controllability generally requires hypotheses stronger than (13.2).

Example 13.6 Consider the system

$$\begin{aligned} \dot{x}_1 &= (x_2 + 1)^2, \\ \dot{x}_2 &= u, \end{aligned}$$

where

$$f(x) = \begin{bmatrix} (x_2+1)^2 \\ 0 \end{bmatrix} \quad \text{and} \quad g(x) = \begin{bmatrix} 0 \\ 1 \end{bmatrix}.$$

Since $x_1(t)$ must be nondecreasing, it is clear that this system is not controllable on any connected open neighborhood of the origin. However, the rank condition (13.2) holds at every point x_0 in a neighborhood of the origin, since

$$[f, g](x) = \begin{bmatrix} 2(x_2+1) \\ 0 \end{bmatrix}$$

implies that

$$\operatorname{rank} [g(x_0) \quad [f, g](x_0)] = 2$$

for all $x_0 \neq -1$. △

Example 13.6 might prompt the following question. Suppose that some larger collection of iterated brackets involving the vector fields f and g spans the full tangent space (the whole plane $\mathbf{R}^2$) at x_0. Does it follow that the system (13.1) is controllable on some connected neighborhood U of x_0? The answer to this question is also negative, in general. In the following example, additional directions, generated by Lie brackets beyond those appearing in (13.2), are still not sufficient to guarantee controllability of the system on a neighborhood of x_0, even when these directions span the full tangent space at all points in a neighborhood of x_0.

Example 13.7 Consider the system

$$\dot{x}_1 = x_2^2,$$
$$\dot{x}_2 = u.$$

Note that controllability fails because, as in the previous example, we always have $\dot{x}_1 \geq 0$. For this system the drift vector field $f(x) = [x_2^2 \quad 0]^T$ satisfies $f(0) = 0$. We find that

$$[f, g](x) = -Df(x)g(x) = -\begin{bmatrix} 0 & 2x_2 \\ 0 & 0 \end{bmatrix} \begin{bmatrix} 0 \\ 1 \end{bmatrix} = \begin{bmatrix} -2x_2 \\ 0 \end{bmatrix}.$$

Since $[f, g](0) = 0$, (13.2) does not hold in any neighborhood of $x_0 = 0$. (The linearization at the origin is not controllable.) However, further brackets can be calculated, for example, $[\,[f, g], g\,]$. We find that

$$[\,[f, g], g\,](x) = -D[f, g](x)g(x) = -\begin{bmatrix} 0 & -2 \\ 0 & 0 \end{bmatrix} \begin{bmatrix} 0 \\ 1 \end{bmatrix} = \begin{bmatrix} 2 \\ 0 \end{bmatrix}.$$

Hence, it is true that

$$\operatorname{span}\{\, g,\ [f, g],\ [\,[f, g], g\,]\,\}(x) = \mathbf{R}^2 \quad \text{for all } x.$$

Thus, using iterated brackets beyond the collection in (13.2), it is possible to span all of $\mathbf{R}^2$ at each point in a neighborhood of the origin, yet the system fails to be controllable on any neighborhood of the origin. △

The preceding examples indicate that the determination of Lie bracket conditions for controllability is an interesting and difficult subject for nonlinear systems. To read more about the subject of nonlinear controllability and reachability properties, see the Notes and References for this chapter.

We now turn to another major contrast with linear systems.

Controllability Does Not Imply Smooth Stabilizability

We give a simple example which is cited often.

Example 13.8 *An Example of D. Aeyels*
Consider the system

$$\dot{x}_1 = x_1 + x_2^3,$$
$$\dot{x}_2 = u.$$

Note that the linearization at $x_0 = 0$ is not stabilizable, since the linearization pair (A, b) at the origin is

$$(A, b) := \left(\begin{bmatrix} 1 & 0 \\ 0 & 0 \end{bmatrix}, \begin{bmatrix} 0 \\ 1 \end{bmatrix} \right),$$

and the eigenvalue 1 cannot be changed by C^1 state feedback $u = \alpha(x_1, x_2)$ satisfying $\alpha(0,0) = 0$. By Corollary 8.1, the origin cannot be asymptotically stabilized by C^1 state feedback $u = \alpha(x_1, x_2)$ satisfying $\alpha(0,0) = 0$. However, it can be shown that this system is controllable on $\mathbf{R}^2$ (Exercise 13.4). For more information on this example, see the resources listed in the Notes and References for the chapter. △

The phenomenon in Example 13.8 might seem surprising because it says that the existence of open loop controls that can move a system from any point x^1 to any point x^2 in finite time does not imply the existence of a useful navigator (a smooth feedback controller $u = \alpha(x)$) that can accomplish even local stabilization of the origin by making instantaneous decisions based only on the current state.

13.3 EXERCISES

Exercise 13.1 Prove the following theorem:

THEOREM. Consider the system $\dot{x} = Ax + g(x)u$. Assume that the rank condition (13.2) holds for $x_0 \neq 0$, with $f(x) = Ax$, and assume that $A \in \mathbf{R}^{n \times n}$ has n distinct real eigenvalues having nonpositive real part. Then, with $V(x) = \frac{1}{2}x^T x$, the feedback $u = -L_g V(x)$ globally asymptotically stabilizes $x = 0$.

Hint: Under the eigenvalue assumption on A, any vector $x \in \mathbf{R}^n$ can be written as a linear combination $x = c_1 v_1 + \cdots + c_n v_n$ where the vectors v_i are eigenvectors of A.

Exercise 13.2 Consider the system

$$\dot{x} = Ax + g(x)u := \begin{bmatrix} 0 & 1 & 0 \\ -1 & 0 & 0 \\ 0 & 0 & -2 \end{bmatrix} x + \begin{bmatrix} x_2 \\ -x_1 \\ 1 \end{bmatrix} u.$$

(a) Show that the rank condition (13.2) does not hold for the pair $f(x) = Ax$ and $g(x)$.
(b) Show that the Jacobian linearization at the origin, $\dot{x} = Ax + bu$, with $b = g(0)$, is not stabilizable. *Hint*: Apply the PBH test.
(c) Show that Brockett's necessary condition for asymptotic stabilization holds. *Hint*: With $F(x, u) = Ax + g(x)u$, determine the rank of the 3×4 Jacobian matrix $DF(x, u)$ at $(x, u) = (0, 0)$.

Exercise 13.3 Consider the system $\dot{x} = f(x) + g(x)u$, $x \in \mathbf{R}^n$, where $f(0) = 0$ and $g(0) \neq 0$.

(a) Suppose there exists a positive definite function $V(x)$ on a neighborhood U of $x = 0$ such that the following two conditions hold:

(i) $L_f V(x) \leq 0$ for $x \in U$.

(ii) The function $h(x) := L_g V(x)$ has the property that

$$\text{rank} \begin{bmatrix} dh(0) \\ d(L_f h)(0) \\ \cdots \\ d(L_f^{n-1} h)(0) \end{bmatrix} = n.$$

Show that the feedback $u = -h(x) = -L_g V(x)$ makes the origin an asymptotically stable equilibrium. *Hint*: It may be helpful to write out the argument for the case of a linear system, where $f(x) = Ax$, $g(x) = b$, and $V(x) = x^T P x$ for a real symmetric positive definite P.

(b) Show that the result of part (a) applies to the system of Example 13.5 but not to the system of Example 13.4. (Use the V functions given in those examples.)

Exercise 13.4 Consider the system in Example 13.8.

(a) Show that the system is controllable on $\mathbf{R}^2$. *Hint*: Consider first a transfer-of-state from the origin to a given point (x_1, x_2).

(b) *A Project*. For a new approach, see the paper by Y. Sun, Necessary and sufficient condition for global controllability of planar affine nonlinear systems. *IEEE Trans. Aut. Contr.*, AC-52:1454-1460, 2007.

13.4 NOTES AND REFERENCES

Theorem 13.1 is from [52]. The text [91] (pages 240–241) has a result on damping control with multivariable input. Also, [91] (Section 8.5) relates damping controls to optimal control by showing that a globally stabilizing damping control is optimal with respect to some cost function. The connection between damping control, passivity, and optimal control is exploited for feedback design in [86], and the connection has roots in the earlier works [55] and [76].

The discussion of Lie brackets and controllability in this chapter was influenced, in part, by [79] (Section 3.1). The first published example of a nonlinear system that is controllable on $\mathbf{R}^n$ but not stabilizable by smooth feedback is a two-dimensional system in [52] (it is the planar system in Exercise 8.7 (b)). For more information on Example 13.8 and related material on controllability and stabilization for nonlinear systems, see the papers: D. Aeyels, Local and global controllability for nonlinear systems, *Syst. Contr.*

Lett. 5:19–26, 1984; and D. Aeyels, Stabilization of a class of nonlinear systems by a smooth feedback control, *Syst. Contr. Lett.* 5:289–294, 1985, as well as the recent paper listed in Exercise 13.4 (b).

Exercise 13.3 (a) was taken from R. Marino and P. Tomei, *Nonlinear Control Design: Geometric, Adaptive, and Robust*, Prentice-Hall Europe, London, 1995.

In this chapter, we considered only the strong form of controllability on a set $U \subseteq \mathbf{R}^n$ which requires control-directed steering between any pair of points in U in some finite time. The controllability question is a fundamental one, as it deals with an easily stated and intuitively appealing feature of a system. So it has been studied in search of insight for the case of nonlinear systems, and many controllability concepts have been studied. See the book by A. Bacciotti, *Local Stabilizability of Nonlinear Control Systems*, World Scientific, Singapore, 1992 (pages 34–39) for some discussion of controllability and stabilizability with many examples, including Example 13.8. General text references for controllability and observability of nonlinear systems include [47], [79], [91], and [105], and some fundamental papers are [38], [63], [88], [87], [98], [93], [94], and [96]. The extensive review article [95] offers insight on the application of differential geometric ideas in control theory.

Controllability notions find application especially in the area of geometric control of mechanical systems. See [18], [51], or [13]. See [51] (Part I: Reachable Sets and Controllability) for an introduction to controllability. These texts rely heavily on differential geometric concepts, primarily those of the Lie algebraic structure of families of vector fields. The strong modern connection between mechanics and differential geometry goes back at least to [1] (see the second edition [2]).

Chapter Fourteen

Passivity

This chapter contains a development of passive systems. Passive systems, and the idea of rendering a system passive by a feedback transformation, play a central role in the study of designs for feedback stabilization of nonlinear systems. We study systems with the same number of inputs and outputs,

$$\dot{x} = f(x) + g(x)u, \tag{14.1}$$

$$y = h(x), \tag{14.2}$$

where $x \in \mathbf{R}^n$, $y \in \mathbf{R}^m$, and $u \in \mathbf{R}^m$. We assume that the vector fields f and g, and the function h, are at least continuously differentiable.

14.1 INTRODUCTION TO PASSIVITY

We begin with a discussion of resistor and inductor circuit elements and a simple mass-spring system. From these examples, we draw out the essential feature of a passive system—its inability to increase its own energy—and we then define passive state space systems.

14.1.1 Motivation and Examples

The passivity concept has its roots in the analysis of passive circuit elements and circuit networks. Resistors and inductors are examples of passive elements. Recall that *power* is defined as the time rate of change of energy. For passive circuit elements, any gain in power must be due to the application of an external energy source such as a battery. Active elements—transistors, for example—can exhibit power gain because they can amplify an input signal to produce an output signal having greater power than the input.

PASSIVE CIRCUIT ELEMENTS. In order to fix some terminology and to motivate the definition of a passive system, we consider a resistor or an inductor circuit element with voltage v between two fixed points in the circuit and current i through the path between these points. Power is the time rate of change of energy in the element (the rate at which energy is either absorbed or spent), and we write

$$p(t) = \frac{dw(t)}{dt},$$

where p is power, w is energy, and t is time. Equivalently, we may write

$$w(t) = w(0) + \int_0^t p(s)\, ds.$$

The power consumed by a resistor or inductor is the work done per unit time, and it is given by

$$p(t) = v(t)\, i(t);$$

the units involved in this formula are

$$[p] = [v]\,[i] = \frac{[\text{work}]}{[\text{charge}]}\,\frac{[\text{charge}]}{[\text{time}]} = \frac{[\text{work}]}{[\text{time}]}.$$

Voltage is measured in volts, current is in amps, and power is in watts or, equivalently, joules per second [44]. The energy absorbed or stored by the element at time t is given by

$$w(t) = w(0) + \int_0^t v(s)i(s)\, ds. \tag{14.3}$$

A passive element is one for which $w(t) \geq 0$ for all t. If the element delivered more energy that it absorbed up to some instant t_0, then we would have $w(t_0) < 0$. As noted, a passive element is one that cannot deliver more energy than it has received.

In a resistor or an inductor, either the current i or the voltage drop v may be considered as the input, and the other variable is then the designated output. Consequently, with designated input u and output y, (14.3) implies that

$$w(t) = w(0) + \int_0^t u(s)y(s)\, ds. \tag{14.4}$$

We also have

$$\dot{w}(t) = u(t)y(t),$$

which is the differential form of (14.4). △

Simple mass-spring systems exhibit a similar differential relation involving energy, input and output.

A Mass-Spring System. The linear mass-spring equation is $m\ddot{x} + b\dot{x} + kx = u$, and the associated system, with velocity as output, is

$$\begin{aligned}
\dot{x}_1 &= x_2, \\
\dot{x}_2 &= -\frac{k}{m}x_1 - \frac{b}{m}x_2 + \frac{1}{m}u, \\
y &= x_2.
\end{aligned}$$

The total energy function is

$$V(x) = V(x_1, x_2) = \frac{1}{2}mx_2^2 + \frac{k}{2}x_1^2.$$

Let $x(t) = (x_1(t), x_2(t))$ be a solution. Differentiating $V(x(t))$ with respect to t, we find that

$$\frac{d}{dt}V(x_1(t), x_2(t)) = -bx_2^2(t) + x_2(t)u(t) \le u(t)y(t).$$

Integration of both sides of this inequality over the interval $0 \le s \le t$ gives

$$V(x_1(t), x_2(t)) - V(x_1(0), x_2(0)) \le \int_0^t u(s)y(s)\, ds$$

for all $t \ge 0$ for which the solution is defined.

Since V is positive definite, the inequality $\frac{d}{dt}V(x(t)) \le uy$ implies that, for the unforced system (with $u = 0$), the origin is Lyapunov stable.

Suppose now that the output is the displacement $y = x_1$. It can be shown that with this new output, the mass-spring system does not satisfy a dissipation inequality $\frac{d}{dt}S(x(t)) \le yu$ for any nonnegative function S. See Exercise 14.1 for a sample calculation. $\triangle$

Using the dynamical analogies between linear mass-spring systems and linear RLC circuit equations of the form $L\ddot{q} + R\dot{q} + \frac{1}{C}q = u(t)$, where q is the charge on the capacitor, a similar discussion could be given for an RLC circuit with the applied voltage u as the input and the loop current $i := \frac{dq}{dt}$ as the output.

In the next section we begin to develop the passivity concept, and its connections with stability, in state space terms.

14.1.2 Definition of Passivity

Given system (14.1)–(14.2), we define the instantaneous *power supply rate*, or *supply* $w(u, y)$, by the inner product

$$w(u, y) := y^T u = u^T y, \tag{14.5}$$

where $u, y \in \mathbf{R}^m$. Recall that the admissible control functions $u(\cdot)$ are in the space $\mathcal{U} := \{u : [0, \infty) \to \mathbf{R}^m : u(\cdot) \text{ is locally integrable}\}$.

Definition 14.1 (Passive System and Storage)
The system (14.1)–(14.2) is passive *if there exists a positive semidefinite function $S(x)$, called the* storage *function, such that $S(0) = 0$ and*

$$S(x(t)) - S(x_0) \le \int_0^t y^T(s)u(s)\, ds, \quad x(0) = x_0 \tag{14.6}$$

for all admissible inputs $u(\cdot)$ and all solutions $x(\cdot)$ for all t at which they are defined.

Inequality (14.6) is called the *dissipation inequality*. Rewrite the dissipation inequality in the form

$$\frac{S(x(t+\delta)) - S(x(t))}{\delta} \le \frac{1}{\delta} \int_t^{t+\delta} y^T(s) u(s)\, ds \,,$$

where $\delta > 0$ and the solution $x(s)$ is well defined on $t \le s \le t+\delta$. Now take the limit of both sides of this inequality as $\delta \to 0$, to obtain

$$\frac{d}{dt} S(x(t)) \le y^T(t) u(t) = u^T(t) y(t). \tag{14.7}$$

This inequality holds for all admissible inputs $u(\cdot)$ and all solutions $x(\cdot)$ for all t at which they are defined. Inequality (14.7) is called the *differential form of the dissipation inequality.*

We have seen above that the mass-spring system, with input u and velocity output $y = x_2$, is passive with the total energy as a storage function. However, with position output $y = x_1$, the mass-spring system is not passive.

Does a storage function S of a passive system always qualify as a Lyapunov function for the unforced system? The answer is no, because Definition 14.1 does not require that the storage be positive definite. It is possible to have $S(x) = 0$ for some $x \ne 0$, and then the storage function is not a Lyapunov function. The next example has a simple illustration.

Example 14.1 The two-dimensional double integrator system

$$\begin{aligned} \dot{x}_1 &= x_1, \\ \dot{x}_2 &= u, \\ y &= x_2 \end{aligned}$$

is passive with storage $S(x) = \frac{1}{2} x_2^2$, since $\frac{d}{dt} S(x(t)) = x_2(t) u(t) = y(t) u(t)$. But S is not positive definite, and when $u = 0$ the origin is unstable. Thus, passivity may still allow an unstable equilibrium at $x = 0$ for the unforced system. $\triangle$

Despite examples like Example 14.1, there are indeed strong links between passivity and stability. One of our observations about the mass-spring system can be stated more generally.

Lemma 14.1 (Positive Definite Storage and Lyapunov Stability)
If (14.1)–(14.2) is passive with a C^1 positive definite storage function, then the origin is Lyapunov stable for the unforced system and the storage function is a Lyapunov function.

Proof. The statement follows from (14.7); when $u = 0$, (14.7) implies that, along solutions $x(t)$ of $\dot{x} = f(x)$,

$$\frac{d}{dt} V(x(t)) = L_f V(x(t)) \le 0.$$

If V is positive definite then V qualifies as a Lyapunov function and the equilibrium $x = 0$ is Lyapunov stable for the unforced system. □

Suppose that the linear system $\dot{x} = Ax + Bu$, $y = Cx$, is passive, with a quadratic storage function $V(x) = \frac{1}{2}x^T Px$, with $P \geq 0$. Then the differential form of the dissipation inequality is

$$\frac{d}{dt}V(x(t)) = x^T(t)(A^T P + PA)x(t) \leq 0 \quad \text{for all } x.$$

The linear mass-spring system had quadratic storage. The existence of quadratic storage for passive linear systems is proved below in Theorem 14.2.

In order to study more links between passivity and stability, we need a characterization of passive systems in state-space terms, and that is the subject of the next section.

14.2 THE KYP CHARACTERIZATION OF PASSIVITY

The definition of passivity relates the change in storage along solutions to the total supply, which is given by the integral of the supply rate, and the supply rate $y^T u$ involves only the input and output. However, passivity can be characterized directly and completely in terms of the storage and the triple (h, f, g) of equations (14.1)–(14.2). The main goal of this section is the KYP characterization of passivity in state-space terms.

The KYP Property and Nonlinear Passive Systems

The terminology in the next definition is convenient in describing the state-space characterization of passivity. The KYP abbreviation used here credits the work of R. E. Kalman, V. A. Yakubovich, and V. M. Popov, who developed passivity conditions for linear systems in terms of a transfer function description. The KYP conditions were extended in [39] and thus they are sometimes called the Hill-Moylan conditions. The work of D. J. Hill and P. J. Moylan built upon the foundations of dissipative systems developed by J. C. Willems in [101], [102].

Definition 14.2 *The system* (14.1)–(14.2) *has the* KYP property *if there exists a* C^1 *nonnegative function* $V : \mathbf{R}^n \mapsto \mathbf{R}$, *with* $V(0) = 0$, *such that, for all* $x \in \mathbf{R}^n$,

$$L_f V(x) \leq 0 \tag{14.8}$$

and

$$L_g V(x) = h^T(x). \tag{14.9}$$

Note that (14.9) is equivalent to $(L_g V)^T(x) = h(x)$, and in some calculations the condition may be written this way.

We now show that a system is passive with storage V if and only if it has the KYP property with the same function V.

Theorem 14.1 (P. J. Moylan—Characterization of Passivity)
If the system (14.1)–(14.2) has the KYP property with a function V, then it is passive with storage function V. Conversely, if the system is passive with a C^1 storage function V, then it has the KYP property with the same V in (14.8)–(14.9).

Proof. If (14.1)–(14.2) has the KYP property with function V, then, along any solution $x(t)$, (14.8) and (14.9) imply that

$$\begin{aligned}\frac{d}{dt}V(x(t)) &= L_fV(x(t)) + L_gV(x(t))u(t)\\ &\leq y^T(t)u(t).\end{aligned}$$

Writing $x(0) = x_0$, and integrating over the interval $0 \leq s \leq t$, we have

$$V(x(t)) - V(x_0) = \int_0^t \frac{d}{ds}V(x(s))\,ds \leq \int_0^t y^T(s)u(s)\,ds,$$

which is the dissipation inequality for storage V and supply rate y^Tu. Therefore the system is passive with storage V.

Conversely, suppose (14.1)–(14.2) is passive with a C^1 storage function V. The differential form of the dissipation inequality says that

$$\frac{d}{dt}V(x(t)) = L_fV(x(t)) + L_gV(x(t))u(t) \leq y^T(t)u(t),$$

for all admissible inputs $u(\cdot)$ and for all solutions $x(\cdot)$ for all t at which the solution is defined. Set $u = 0$ to see that (14.8) holds for all x. By rearranging terms, we have

$$[L_gV(x(t)) - h^T(x(t))]u(t) \leq -L_fV(x(t)) \tag{14.10}$$

for any $u(t)$ and $x(t)$. Suppose (14.9) does not hold for some $x_0 = x(0)$; then, we are free to choose $u_0 = u(0)$ such that the left-hand side of (14.10) is greater than $-L_fV(x(0))$ at $t = 0$, giving a contradiction. Therefore (14.9) holds for all x. Hence, the system has the KYP property with V being the storage. □

There is an additional observation from the dissipation inequality, this time involving the zero dynamics subsystem of a passive system.

Lemma 14.2 (Positive Definite Storage and Zero Dynamics)
If (14.1)–(14.2) is passive with a C^1 positive definite storage function, then the zero dynamics subsystem has a Lyapunov stable equilibrium at its origin.

Proof. Let $V(x)$ be the assumed positive definite storage for (14.1)–(14.2). We must consider solutions of the closed loop system generated by the zero dynamics feedback $u_{zd} = \alpha(x)$. Any solution $x(t)$ of the zero dynamics lies within the manifold defined locally near $x = 0$ by $h(x) = 0$; along such a solution, the second of the KYP conditions must hold, that is, $(L_gV)^T(x(t)) = h(x(t))$, and therefore $L_gV(x(t)) = 0$. Using the KYP conditions and the fact that $y = h(x(t)) = 0$ for this solution of $\dot{x} = f(x) + g(x)\alpha(x)$, we have

$$\frac{d}{dt}V(x(t)) = L_fV(x(t)) + L_gV(x(t))\alpha(x) \leq 0.$$

Hence, the restriction of V to the manifold $\{x : h(x) = 0\}$ in a neighborhood of $x = 0$ is a Lyapunov function for the origin of the zero dynamics subsystem, which is therefore stable by Theorem 8.1 (a). □

The KYP Property and Linear Passive Systems

It is convenient to state a complete result for linear systems based on Theorem 14.1, because it is possible to show that the storage function of a passive linear system must be quadratic. Let us first restate the KYP property for the case of a quadratic storage function, $V(x) = \frac{1}{2}x^TPx$, where P is symmetric positive semi-definite and the passive linear system is

$$\dot{x} = Ax + Bu, \tag{14.11}$$

$$y = Cx. \tag{14.12}$$

It is straightforward to show that the KYP conditions (14.8)–(14.9), with $V(x) = \frac{1}{2}x^TPx$, $f(x) = Ax$ and $g(x) = B$, are equivalent, respectively, to

$$A^TP + PA \leq 0 \tag{14.13}$$

and

$$PB = C^T. \tag{14.14}$$

The next result specializes Theorem 14.1 to the case of linear time invariant systems. The one essential difference in the statement, as compared with Theorem 14.1, is that a passive *linear* system has a *quadratic* storage function.

Theorem 14.2 (KYP and Quadratic Storage)
If the linear system (14.11)–(14.12) *has the KYP property with quadratic function* $V(x) = \frac{1}{2}x^TPx$, *then* $P \geq 0$, P *satisfies* (14.13)–(14.14), *and the system is passive with quadratic storage* $V(x) = \frac{1}{2}x^TPx$.

Conversely, if system (14.11)–(14.12) *is passive with a* C^1 *storage function* V, *then it is possible to write* $V(x) = \frac{1}{2}x^TPx$ *where* P *is symmetric,* $P \geq 0$, *and* P *satisfies* (14.13)–(14.14).

Proof. The proof is recommended as Exercise 14.3. □

Theorem 14.2 helps to complete the discussion of the double integrator system.

Example 14.2 Consider again the double integrator system

$$\dot{x}_1 = x_2,$$
$$\dot{x}_2 = u,$$
$$y = x_2.$$

We have seen already that this system is passive with storage $V(x_1, x_2) = \frac{1}{2}x_2^2$. In order to illustrate Theorem 14.2, we ask if the KYP conditions (14.13)–(14.14) are satisfied for a quadratic storage function

$$V(x) = \frac{1}{2}(\alpha x_1^2 + 2\beta x_1 x_2 + \gamma x_2^2) = \frac{1}{2}x^T P x,$$

where

$$P = \begin{bmatrix} \alpha & \beta \\ \beta & \gamma \end{bmatrix}.$$

It is easy to check that $PB = C^T$ implies that $\beta = 0$ and $\gamma = 1$. In addition, the requirements that $P \geq 0$ and $A^T P + PA \leq 0$ imply that $\alpha = 0$. Therefore

$$V(x) = \frac{1}{2}x^T \begin{bmatrix} 0 & 0 \\ 0 & 1 \end{bmatrix} x = \frac{1}{2}x_2^2$$

is the only possible quadratic storage function for this passive system, and by Theorem 14.2, it is the only possible C^1 storage function. △

14.3 POSITIVE DEFINITE STORAGE

Consider the system

$$\dot{x} = f(x) + g(x)u, \tag{14.15}$$
$$y = h(x) \tag{14.16}$$

and the associated unforced system, $\dot{x} = f(x)$. In this section we show that an observability condition on the unforced system, namely *zero-state observability* (Definition 14.4 below), implies that the storage of a passive system (14.15)–(14.16) is positive definite.

The concepts described in the next two definitions are familiar for linear time invariant systems.

Definition 14.3 (Zero-State Detectability)
System (14.15)–(14.16) *is* zero-state detectable (ZSD) *if, for any solution $x(t)$ of the unforced system that is defined for all $t \geq 0$, we have*

$$y(t) = h(x(t)) = 0 \text{ for all } t \geq 0 \quad \Longrightarrow \quad \lim_{t\to\infty} x(t) = 0.$$

The linear system $\dot{x} = Ax + Bu$, $y = Cx$ is zero-state detectable if and only if the pair (C, A) is detectable.

Definition 14.4 (Zero-State Observability)
System (14.15)–(14.16) *is* zero-state observable (ZSO) *if, for any solution* $x(t)$ *of the unforced system that is defined for all* $t \geq 0$, *we have*

$$y(t) = h(x(t)) = 0 \ for \ all \ t \geq 0 \quad \Longrightarrow \quad x(t) \equiv 0 \quad for \ t \geq 0.$$

The linear system $\dot{x} = Ax + Bu$, $y = Cx$ is zero-state observable if and only if the pair (C, A) is observable. (See Exercise 14.4.) Clearly, for any system of the form (14.15)–(14.16), zero-state observability implies zero-state detectability.

Whether system (14.15)–(14.16) is passive or not, there is a concept of *available storage* with supply rate $y^T u$.

Definition 14.5 (Available Storage with Supply $y^T u$)
The available storage at x *of system* (14.15)–(14.16), *with supply function* $y^T u$, *is*

$$V_a(x) = \sup_{x_0 = x, u(\cdot) \in \mathcal{U}, t \geq 0} \left\{ -\int_0^t y^T(s) u(s)\, ds \right\}, \tag{14.17}$$

where $y(\cdot)$ *depends on* $u(\cdot)$, *the initial condition* $x(0) = x_0 = x$, *and the resulting solution* $x(\cdot)$ *of* (14.15).

Passive systems can be characterized in terms of the available storage.

Theorem 14.3 (Available Storage and Passivity)
The available storage $V_a(x)$ *of system* (14.15)–(14.16) *is finite for all* x *if and only if the system is passive. Moreover,* $0 \leq V_a(x) \leq V(x)$ *for passive systems with storage* V *and* V_a *is itself a possible storage function.*

Proof. This is covered by a more general result in [101] (page 328). However, for the *if* part, that passivity implies finite available storage at each x, see Exercise 14.5. □

We now show that for passive systems, zero-state observability implies positive definite storage.

Proposition 14.1 *If* (14.15)–(14.16) *is passive with storage* V *and zero-state observable, then the storage* V *is positive definite.*

Proof. By Theorem 14.3, the available storage $V_a(x)$ is finite at each x, and $V(x) \geq V_a(x) \geq 0$ for all x. By definition,

$$V_a(x) = \sup_{x_0 = x, u(\cdot) \in \mathcal{U}, t \geq 0} \left\{ -\int_0^t y^T(s) u(s)\, ds \right\}.$$

Given $x_0 \neq 0$, if we show that there is an input-output pair $u(s)$, $y(s)$ such that $-y^T(s)u(s) > 0$ on some interval $[0, \delta)$, then we will have $V_a(x_0) > 0$, and therefore $V(x_0) > 0$.

Suppose $h(x_0) \neq 0$. Let $y(t) = h(\phi_t^f(x_0))$, where $\phi_t^f(x_0)$ solves $\dot{x} = f(x)$ with $\phi_0^f(x_0) = x_0$. Now let $u = \bar{u}(t) = -y(t)$ in the system $\dot{x} = f(x)+g(x)u$; using initial condition $x(0) = x_0$, denote the solution by $\bar{x}(t)$, and the resulting output by $\bar{y}(t) := h(\bar{x}(t))$. Note that $\bar{y}(0) = h(\bar{x}(0)) = h(x_0) = y(0)$. Thus, for the input-output pair $\bar{u}(s) = -y(s)$, $\bar{y}(s)$, we have

$$-\bar{y}^T(0)\bar{u}(0) = y^T(0)y(0) = h^T(x_0)h(x_0) > 0.$$

By continuity of h and the solutions, there is a $\delta > 0$ such that $-\bar{y}^T(s)\bar{u}(s) = h(\bar{x}(s))h(\phi_s^f(x_0)) > 0$ on the interval $0 \leq s < \delta$. Thus, $V_a(x_0) > 0$, provided $h(x_0) \neq 0$.

Now suppose $x_0 \neq 0$ and $h(x_0) = 0$. By zero-state observability, we cannot have $y(t) = h(\phi_t^f(x_0)) = 0$ for all $t \geq 0$. Thus, there exists $\tau > 0$ such that $y(\tau) = h(\phi_\tau^f(x_0)) \neq 0$. We can now deal with the point $\phi_\tau^f(x_0)$ much as we did with the point x_0 when we had $h(x_0) \neq 0$. We just need to define $\bar{u}(s) = 0$ for $0 \leq s < \tau$, so that the output of $\dot{x} = f(x) + g(x)\bar{u}(t)$ with initial condition x_0 at time 0 is $\bar{y}(s) = h(\phi_s^f(x_0))$ for $0 \leq s \leq \tau$; and define $\bar{u}(\tau) = -h(\phi_\tau^f(x_0))$. Hence, we have $\bar{y}^T(s)\bar{u}(s) = 0$ for $0 \leq s < \tau$, and

$$-\bar{y}^T(\tau)\bar{u}(\tau) = [h(\phi_\tau^f(x_0))]^2 > 0.$$

The input $\bar{u}(t)$ can be chosen continuous for $t \geq \tau$, and thus the output $\bar{y}(t)$ will be continuous for all $t \geq 0$. Hence, there exists a $\delta > 0$ such that $-\bar{y}^T(s)\bar{u}(s) > 0$ for $0 \leq s < \delta$, and we again conclude that $V_a(x_0) > 0$, hence $V(x_0) > 0$. □

The ZSO condition in Proposition 14.1 cannot be weakened to ZSD and still guarantee positive definite storage. To see this, consider the system

$$\begin{aligned}\dot{x}_1 &= -x_1,\\ \dot{x}_2 &= u,\\ y &= x_2,\end{aligned}$$

which is zero-state detectable and passive with storage $V(x) = \frac{1}{2}x_2^2$, but the storage is not positive definite. The zero-state detectability condition can help to enable applications of the invariance theorem. For example, in situations where we want to use the storage as a valid Lyapunov function for the unforced system $\dot{x} = f(x)$, and for asymptotic stability arguments as well, it may be necessary to assume both ZSD and positive definite storage.

Proposition 14.2 *Suppose that (C, A, B) defines a passive linear system with positive definite quadratic storage $V = \frac{1}{2}x^TPx$, and assume that B has full rank equal to m. Then the system has uniform relative degree one and the origin of the zero dynamics subsystem is Lyapunov stable.*

Proof. The statement about the origin of the zero dynamics subsystem being Lyapunov stable follows from Lemma 14.2. In order to show that the system has uniform relative degree one, start with the second of the KYP conditions, $B^T P = C$, and multiply it on the right by B, to get

$$CB = B^T PB.$$

Clearly, the $m \times m$ matrix CB is symmetric. Since $P > 0$ and rank $B = m$, CB is positive definite. Differentiating the output once gives $\dot{y} = CAx + CBu$, with CB nonsingular; hence, the system has uniform relative degree one. □

A comment on the proof of Proposition 14.2 is in order. In the proof, the assumptions that $P > 0$ and $B^T P = C$ imply that C has full rank. On the other hand, under the assumption that C has full rank and $PB = C^T$, it is straightforward to see that B has full rank. Note that uniform relative degree one is equivalent to CB nonsingular. Thus, we have the following statement: If a linear system (C, A, B) satisfies the KYP conditions (14.13)–(14.14) with $P > 0$, and at least one of the matrices B, C is known to have full rank, then the other has full rank as well, and the system has uniform relative degree one and Lyapunov stable zero dynamics.

The next example shows that relative degree one and Lyapunov stable zero dynamics cannot guarantee passivity.

Example 14.3 Consider the system

$$\begin{aligned} \dot{x}_1 &= 0, \\ \dot{x}_2 &= x_1 + u, \\ y &= x_2, \end{aligned}$$

which has uniform relative degree one. The zero dynamics feedback is $u_{zd} = -x_1$, and the zero dynamics system is $\dot{x}_1 = 0$, which has Lyapunov stable origin. We now show that the KYP conditions cannot hold for any symmetric P. For symmetric P, the condition $PB = C^T$ implies that

$$P = \begin{bmatrix} p_1 & 0 \\ 0 & 1 \end{bmatrix}, \quad \text{where } p_1 \text{ is arbitrary.}$$

Note that $P \geq 0$ if and only if $p_1 \geq 0$. But for any p_1, we have

$$A^T P + PA = \begin{bmatrix} 0 & 1 \\ 1 & 0 \end{bmatrix},$$

which is indefinite rather than negative semidefinite, so the KYP conditions cannot hold for any symmetric P. Therefore the system is not passive. △

We now give a nonlinear version of Proposition 14.2; the assumptions are analogous to those of Proposition 14.2, except for a mild additional smoothness assumption on the storage function. Recall that the system has uniform relative degree one at the origin if $L_g h(0)$ is nonsingular.

Proposition 14.3 *If the system (14.15)–(14.16) is passive with a C^2 positive definite storage function such that $\frac{\partial^2 V}{\partial x^2}(0)$ is nonsingular, and $g(0)$ has full rank equal to m, then the system has uniform relative degree one at the origin and the zero dynamics subsystem has Lyapunov stable origin.*

Proof. The zero dynamics has Lyapunov stable origin by Lemma 14.2. We must show that the $m \times m$ matrix $L_g h(0)$ is nonsingular. Let $V(x)$ be a C^2 positive definite storage function for the system. The KYP conditions are that

$$L_f V(x) \le 0,$$
$$g^T(x) \left[\frac{\partial V}{\partial x}(x) \right]^T = h(x)$$

for all x. Differentiate the second of these with respect to x, to get

$$g^T(x) \left[\frac{\partial^2 V}{\partial x^2}(x) \right]^T + \left[\frac{\partial g}{\partial x}(x) \right]^T \left[\frac{\partial V}{\partial x}(x) \right]^T = \frac{\partial h}{\partial x}(x). \tag{14.18}$$

Then multiply on the right by $g(x)$, to get

$$g^T(x) \left[\frac{\partial^2 V}{\partial x^2}(x) \right]^T g(x) + \left[\frac{\partial g}{\partial x}(x) \right]^T \left[\frac{\partial V}{\partial x}(x) \right]^T g(x) = \frac{\partial h}{\partial x}(x) g(x). \tag{14.19}$$

Since V is a positive definite storage function, it has a local minimum at $x = 0$, and since V is differentiable, $\frac{\partial V}{\partial x}(0) = 0$. Evaluate (14.19) at $x = 0$ to get

$$g^T(0) \frac{\partial^2 V}{\partial x^2}(0) g(0) = L_g h(0).$$

By hypothesis, the symmetric matrix $\frac{\partial^2 V}{\partial x^2}(0)$ is nonsingular (in fact, positive definite); hence, we have

$$\frac{\partial^2 V}{\partial x^2}(0) = R^T R$$

for a positive definite matrix R (Lemma 2.5), and therefore

$$L_g h(0) = g^T(0) R^T R g(0)$$

is positive definite. Since g is smooth and V is C^2, $L_g h(x)$ is nonsingular in a neighborhood of $x = 0$. Therefore the system has uniform relative degree one at the origin. □

In the proof, note that evaluation of (14.18) at $x = 0$ gives the rank m matrix

$$g^T(0) R^T R = \frac{\partial h}{\partial x}(0).$$

14.4 PASSIVITY AND FEEDBACK STABILIZATION

This section includes two feedback stabilization results, one for linear passive systems, and an analogous result for nonlinear passive systems. The simplicity of the stabilizing feedback in these results is one of the reasons why passivity is an appealing property. The section ends with a theorem on the equivalence of detectability and stabilizability for passive linear systems with positive definite storage.

Consider the system

$$\dot{x} = Ax + Bu, \tag{14.20}$$
$$y = Cx. \tag{14.21}$$

There is an especially simple form of stabilizing feedback for passive systems that are zero-state detectable.

Proposition 14.4 *Suppose the linear system* (14.20)–(14.21) *is passive with positive definite quadratic storage* $V(x) = \frac{1}{2}x^T Px$. *If the system is also zero-state detectable, then setting* $u = -y = -Cx$ *renders the origin of the closed loop system* $\dot{x} = (A - BC)x$ *globally asymptotically stable.*

Proof. Zero-state detectability for the linear system means that the pair (C, A) is detectable. Set $u = -y = -Cx$, and consider the closed loop system $\dot{x} = (A - BC)x$. We compute that

$$\begin{aligned}\frac{\partial V}{\partial x}(x)((A - BC)x) &= x^T((A - BC)^T P + P(A - BC))x \\ &= x^T(A^T P + PA)x - 2x^T PBCx \\ &= x^T(A^T P + PA)x - 2x^T C^T Cx.\end{aligned}$$

Since each of the quadratic forms on the right is negative semidefinite, we conclude that $V = \frac{1}{2}x^T Px$ is a Lyapunov function and $x = 0$ is stable for the closed loop. All trajectories of the closed loop approach the largest forward invariant subset M contained in the set where $x^T(A^T P + PA)x = 0$ and $Cx = 0$. A solution $x(t)$ of $\dot{x} = (A - BC)x$ which lies in M must satisfy $Cx(t) = 0$ for all $t \geq 0$. Hence, $x(t)$ is a solution of $\dot{x}(t) = Ax(t)$, and detectability implies that $x(t) \to 0$ as $t \to \infty$. Thus $V(x(t)) \to V(0) = 0$. But $V(x(t))$ must be constant for a solution $x(t)$ in M, so $V(x(t)) \equiv 0$. Since V is positive definite, we have $x(t) = 0$ for solutions in M. Thus, $M = \{0\}$. By the invariance theorem, the origin is an asymptotically stable equilibrium of the linear system $\dot{x} = (A - BC)x$. □

What can happen if the zero-state detectability assumption is dropped?

Example 14.4 Consider the linear system

$$\begin{aligned}\dot{x}_1 &= x_2, \\ \dot{x}_2 &= -x_1 + u, \\ \dot{x}_3 &= 0, \\ y &= x_2.\end{aligned}$$

This system is not detectable, since we can have constant $x_3 = c \neq 0$ when $y \equiv 0$. It is passive with positive definite storage given by $V(x_1, x_2, x_3) = \frac{1}{2}(x_1^2 + x_2^2 + x_3^2)$. Note that if $u = -y = -x_2$, then the closed loop coefficient matrix is

$$\begin{bmatrix} 0 & 1 & 0 \\ -1 & -1 & 0 \\ 0 & 0 & 0 \end{bmatrix},$$

which is not Hurwitz. △

There is a direct generalization of Proposition 14.4 to nonlinear passive systems. We consider a nonlinear system having the form

$$\dot{x} = f(x) + g(x)u, \tag{14.22}$$

$$y = h(x). \tag{14.23}$$

Proposition 14.5 *Suppose system (14.22)–(14.23) has $f(0) = 0$ and is passive with a positive definite, radially unbounded C^1 storage function V. If the system is also zero-state detectable, then setting $u = -y = -h(x)$ renders the origin of the closed loop system $\dot{x} = f(x) - g(x)h(x)$ globally asymptotically stable.*

Proof. The proof will follow closely the pattern of argument for Proposition 14.4. By hypothesis, the system is passive with a positive definite storage function $V(x)$. Hence, for all x,

$$\frac{\partial V}{\partial x}(x)f(x) \leq 0 \quad \text{and} \quad \frac{\partial V}{\partial x}(x)g(x) = h^T(x).$$

With $u = -y = -h(x)$, and $x = x(t)$ any solution of the closed loop system, we have

$$\frac{d}{dt}V(x) = \frac{\partial V}{\partial x}(x)\left[f(x) - g(x)h(x)\right] \leq -h^T(x)h(x) \leq 0.$$

Therefore V is a Lyapunov function for the closed loop system, and the origin is stable. Since V is radially unbounded, all forward trajectories are bounded. By the invariance theorem, all solutions $x(t)$, $t \geq 0$, approach the largest invariant set M contained in the set where $\frac{\partial V}{\partial x}(x)f(x) = 0$ and $h^T(x)h(x) = 0$. Suppose $x(t)$, $t \geq 0$, is a solution in M. Then $h(x(t)) = 0$ for all $t \geq 0$, and therefore $u = -y = -h(x(t)) = 0$ along $x(t)$. So $x(t)$ is a solution of the unforced system $\dot{x} = f(x)$ with $h(x(t)) = 0$ for all $t \geq 0$. Zero-state detectability implies that $x(t) \to 0$ as $t \to \infty$. Therefore

$V(x(t)) \to V(0) = 0$. But $V(x(t))$ must be constant for a solution $x(t)$ in M. Thus, $V(x(t)) = 0$ for $t \geq 0$. Since V is positive definite, $x(t) = 0$ for $t \geq 0$. Hence, $M = \{0\}$, and the invariance theorem implies that the origin of the closed loop system is globally asymptotically stable. $\square$

Again, without zero-state detectability in Proposition 14.5, stabilization by negative output feedback may not be possible. Before considering an example to illustrate this point, note that the ZSD property for a nonlinear system is independent of whether the linearization at the origin is ZSD (see Exercise 14.6). Consequently, we cannot use the previous Example 14.4 as our illustration. Consider the nonlinear system in the next example.

Example 14.5 Consider the system

$$\begin{aligned}\dot{x}_1 &= x_2^2,\\ \dot{x}_2 &= -x_1 x_2 + u,\\ y &= x_2.\end{aligned}$$

This system is passive with positive definite storage $V(x_1, x_2) = \frac{1}{2}(x_1^2 + x_2^2)$. It is not zero-state detectable, since the nonzero constant solutions $(x_1, 0) = (c, 0) \neq (0, 0)$ of the unforced system give identically zero output. Negative output feedback cannot asymptotically stabilize the origin. In fact, no state feedback can stabilize the origin, because in the first equation, $\dot{x}_1 = x_2^2 \geq 0$ for all x_2. Observe also that Brockett's necessary condition for asymptotic stabilization fails, due to the first equation. $\triangle$

There is a direct connection here with $L_g V$ control. Suppose that $f(x)$ is a smooth vector field on $\mathbf{R}^n$, $f(0) = 0$, and there exists a smooth, real-valued positive definite function $V(x)$ such that $L_f V(x) \leq 0$ for all x. If the system

$$\begin{aligned}\dot{x} &= f(x) + g(x)u, \qquad u \in \mathbf{R},\\ y &= L_g V(x)\end{aligned}$$

is zero-state detectable, then the feedback $u = -y = -L_g V(x)$ asymptotically stabilizes the equilibrium at $x = 0$; note that the system is passive with storage function $V(x)$, since

$$\frac{\partial V}{\partial x}(x)(f(x) + g(x)u) = L_f V(x) + L_g V(x)u \leq yu.$$

Thus, Proposition 14.5 applies.

The final result of this section shows that stabilizability and detectability are equivalent for passive linear systems with positive definite storage.

Theorem 14.4 *A passive linear system with a positive definite storage function is stabilizable if and only if it is detectable. Moreover, if the system is stabilizable, then the origin is globally asymptotically stabilized by negative output feedback,* $u = -y = -Cx$.

Proof. Denote the passive linear system by (C, A, B). By hypothesis, there is a positive definite P such that $A^T P + PA \leq 0$ and $C = B^T P$, where $V(x) = \frac{1}{2}x^T Px$ is the storage function. We must show the equivalence of detectability of (C, A) and stabilizability of (A, B).

If (C, A) is detectable, then the system is globally asymptotically stabilized by negative output feedback $u = -y = -Cx$, by Proposition 14.4. Thus the pair (A, B) is stabilizable.

Conversely, suppose that (A, B) is stabilizable. Let $u = -Cx = -B^T Px$. We show first that this feedback asymptotically stabilizes the origin. Along solutions of the closed loop system $\dot{x} = (A - BC)x$, we have

$$\begin{aligned} \frac{d}{dt}V(x) &= \frac{\partial V}{\partial x}(x)[(A - BC)x] \\ &= \frac{1}{2}x^T(A^T P + PA)x - x^T PBB^T Px \\ &\leq -x^T PBB^T Px \\ &= -||Cx||_2^2. \end{aligned} \tag{14.24}$$

Thus, $x = 0$ is Lyapunov stable for the closed loop. By the stability of the origin and the linearity of the closed loop, every solution $x(t)$ of $\dot{x} = (A - BC)x$ is bounded for $t \geq 0$. By the invariance theorem, the ω-limit set of a solution $x(t)$ is contained in the set

$$E := \{x : B^T Px = 0 \text{ and } x^T(A^T P + PA)x = 0\}.$$

Now suppose $x(t)$ is a solution of $\dot{x} = (A - BC)x$ in E. Since

$$B^T Px(t) \equiv 0, \tag{14.25}$$

$x(t)$ must be a solution of $\dot{x} = Ax$. Since $x^T(t)(A^T P + PA)x(t) \equiv 0$, the symmetry and negative semidefiniteness of $A^T P + PA$ implies that $(A^T P + PA)x(t) \equiv 0$, hence

$$-A^T Px(t) \equiv PAx(t). \tag{14.26}$$

Differentiation of (14.25) with respect to t gives $B^T PAx(t) \equiv 0$, and, inductively, $B^T PA^k x(t) \equiv 0$ for all $k \geq 0$. We may also apply (14.26) inductively to get

$$B^T(A^T)^k Px(t) \equiv 0 \quad \text{for } k \geq 0.$$

Taking transposes of the last expression for each k, we arrive at

$$x^T(t)\, P\, [B \quad AB \quad \cdots \quad A^k B] \equiv 0 \quad \text{for } k \geq 0. \tag{14.27}$$

Since (A, B) is stabilizable, there is a linear coordinate change $x = Qz$, where $z = [z_1^T \; z_2^T]^T$, such that

$$\begin{bmatrix} \dot{z}_1 \\ \dot{z}_2 \end{bmatrix} = Q^{-1}AQz + Q^{-1}Bu := \begin{bmatrix} A_{11} & A_{12} \\ 0 & A_{22} \end{bmatrix} \begin{bmatrix} z_1 \\ z_2 \end{bmatrix} + \begin{bmatrix} B_1 \\ 0 \end{bmatrix} u,$$

where the pair (A_{11}, B_1) is controllable and all eigenvalues of A_{22} have negative real part. There is a compatible partition of the matrix

$$Q^T PQ = \begin{bmatrix} P_{11} & P_{12} \\ P_{12}^T & P_{22} \end{bmatrix}.$$

It is straightforward to check that for each $k \geq 0$,

$$Q^{-1}A^k B = \begin{bmatrix} A_{11}^k B_1 \\ 0 \end{bmatrix}.$$

Therefore the constraint (14.27) implies that

$$\left[z_1^T(t) \; z_2^T(t)\right] \begin{bmatrix} P_{11} & P_{12} \\ P_{12}^T & P_{22} \end{bmatrix} \begin{bmatrix} B_1 & A_{11}B_1 & \cdots & A_{11}^{r-1}B_1 \\ 0 & 0 & \cdots & 0 \end{bmatrix} \equiv 0,$$

since z_1 is in $\mathbf{R}^r$. Equivalently, we have

$$\left(z_1^T(t)P_{11} + z_2^T(t)P_{12}^T\right) \left[\; B_1 \quad A_{11}B_1 \quad \cdots \quad A_{11}^{r-1}B_1 \;\right] \equiv 0.$$

Since the pair (A_{11}, B_1) is controllable, we must have

$$z_1^T(t)P_{11} + z_2^T(t)P_{12}^T \equiv 0,$$

and, since P is positive definite, we have

$$z_1^T(t) \equiv -z_2^T(t)P_{12}^T P_{11}^{-1}.$$

Since $z_2(t) \to 0$ as $t \to \infty$, we also have $z_1(t) \to 0$ as $t \to \infty$, and therefore $x(t) = Qz(t) \to 0$ as $t \to \infty$. But $V(x(t))$ must be constant for solutions in E, hence $V(x(t)) = 0$ for all t, and therefore $x(t) \equiv 0$. By the invariance theorem, every solution of $\dot{x} = (A - BC)x$ converges to the origin as $t \to \infty$, so the system is stabilized by $u = -Cx$. It remains to show that (C, A) is detectable. Suppose that $x(t)$ is a solution of $\dot{x} = Ax$ for which $y = Cx(t) = 0$ for $t \geq 0$. Then $x(t)$ is also a solution of $\dot{x} = (A - BC)x$, hence, $x(t) \to 0$ as $t \to \infty$ by the argument above. Therefore the pair (C, A) is detectable. □

14.5 FEEDBACK PASSIVITY

We have seen that passive systems with positive definite storage have some stability and stabilization properties that are conceptually quite simple. Thus it should be useful to be able to recognize when a system can be made passive with positive definite storage by a feedback transformation.

Definition 14.6 (Feedback Passivity)
The system (14.1)–(14.2) *is* feedback passive with positive definite storage *if there is a regular feedback transformation* $u = \alpha(x)+\beta(x)v$*, with nonsingular* $m \times m$ $\beta(x)$*, such that the transformed system given by*

$$\dot{x} = f(x) + g(x)\alpha(x) + g(x)\beta(x)v, \tag{14.28}$$

$$y = h(x) \tag{14.29}$$

satisfies the KYP conditions for a positive definite $V(x)$*; that is, for every* x*,*

$$L_{f+g\alpha}V(x) \leq 0$$

and

$$L_{g\beta}V(x) = h^T(x).$$

The system (14.1)–(14.2) *is* locally feedback passive with positive definite storage *if there is a regular feedback transformation defined near the origin such that the feedback-transformed system satisfies the KYP conditions in an open set about the origin.*

For all smooth $n \times m$ matrix functions $g(x)$, $m \times m$ matrix functions $\gamma(x)$, and real-valued functions $V(x)$, we have

$$L_{g\gamma}V(x) = \frac{\partial V}{\partial x}(x)g(x)\gamma(x) = [L_gV(x)]\gamma(x).$$

Thus the KYP conditions on the feedback transformed system are equivalent, respectively, to

$$L_fV(x) + L_gV(x)\alpha(x) \leq 0 \quad \text{for all } x,$$

and

$$L_gV(x)\beta(x) = h^T(x) \quad \text{for all } x.$$

For reference later in this section, it is convenient to state Definition 14.6 explicitly for the linear system case, as follows.

Definition 14.7 (Feedback Passivity of a Linear System)
The linear system defined by the matrix triple (C, A, B) *is* feedback passive with positive definite storage *if there is a regular feedback transformation* $u = Kx+\beta v$*,* β *nonsingular, such that the feedback transformed triple* $(C, A+BK, B\beta)$ *satisfies the KYP conditions for some* $P > 0$*, that is,*

$$(A + BK)^T P + P(A + BK) \leq 0$$

and

$$PB\beta = C^T.$$

Invariance of Relative Degree One and Zero Dynamics

Characterizations of feedback passivity depend on the invariance under feedback transformations of the properties of uniform relative degree one and stable zero dynamics. We proceed now to establish the invariance of these two properties.

COMPUTATION OF THE ZERO DYNAMICS. Lemma 14.2 gave a coordinate-independent proof that the zero dynamics of a passive system has asymptotically stable origin under the assumption that the storage is a C^1 positive definite function. We now compute the zero dynamics system in more detail. Recall that the relative degree one condition at $x = 0$, which is guaranteed by Proposition 14.3 for a passive system with C^2 positive definite storage such that $\frac{\partial^2 V}{\partial x^2}(0)$ is nonsingular, implies that a local change of coordinates of the form $(z, \xi) = (T(x), h(x)) =: S(x)$ gives the normal form

$$\begin{aligned} \dot{z} &= q(z,\xi) + \gamma(z,\xi)u, \\ \dot{\xi} &= a(z,\xi) + b(z,\xi)u, \\ y &= \xi\,, \end{aligned} \tag{14.30}$$

in which the coefficient $b(z,\xi) = L_g h(x) = L_g h(S^{-1}(z,\xi))$ is invertible near the origin. The constraint of zero output, $y = \xi = 0$, is maintained by the zero dynamics feedback

$$u_{zd} = -b^{-1}(z,0)a(z,0),$$

which is defined in a neighborhood of $z = 0$. The resulting zero dynamics subsystem is defined in a neighborhood of $z = 0$ and is described by the differential equation

$$\dot{z} = f_{zd}(z) := q(z,0) - \gamma(z,0)\, b^{-1}(z,0)a(z,0). \tag{14.31}$$

Under the assumption of Proposition 14.3, that the system has a C^2 positive definite storage function, the zero dynamics subsystem has Lyapunov stable origin with a C^2 Lyapunov function. That is, the equilibrium $z = 0$ for (14.31) is Lyapunov stable and the storage function is a C^2 Lyapunov function. (In [19] and [86], a system with $L_g h(0)$ nonsingular is called *weakly minimum phase* if the origin of the zero dynamics subsystem is stable with a C^2 Lyapunov function.)

We now show that uniform relative degree one and stable zero dynamics are properties invariant under regular feedback transformations $u = \alpha(x) + \beta(x)v$, where $\beta(x)$ is nonsingular for x near zero. It will then follow from Lemma 14.2 and Proposition 14.3 that these two conditions are necessary for feedback passivity of smooth nonlinear systems if we require C^2 storage for the feedback-transformed system.

INVARIANCE OF UNIFORM RELATIVE DEGREE ONE. To see the invariance of uniform relative degree one at the origin, consider the normal form (14.30) after application of the feedback transformation

$$u = \alpha(x) + \beta(x)v = \alpha \circ S^{-1}(z,\xi) + \beta \circ S^{-1}(z,\xi)v.$$

The output is $y = \xi$, therefore one differentiation of the output gives

$$\dot{\xi} = a(z,\xi) + b(z,\xi)\alpha \circ S^{-1}(z,\xi) + b(z,\xi)\beta \circ S^{-1}(z,\xi)v.$$

Since $b(z,\xi) = L_g h(S^{-1}(z,\xi))$ and $\beta(S^{-1}(z,\xi))$ are both nonsingular at $(z,\xi) = (0,0)$, the transformed system also has uniform relative degree one at the origin.

Invariance of the Zero Dynamics. To see the invariance of the zero dynamics subsystem, note the changes in the terms in system (14.30) due to the feedback. After application of the feedback transformation, system (14.30) becomes (with arguments suppressed)

$$\begin{aligned} \dot{z} &= q + \gamma\alpha + \gamma\beta v, \\ \dot{\xi} &= a + b\alpha + b\beta v, \\ y &= \xi\,. \end{aligned} \tag{14.32}$$

The zero dynamics feedback is

$$v = -(b\beta)^{-1}(a + b\alpha).$$

The zero dynamics subsystem is therefore

$$\dot{z} = q + \gamma\alpha - \gamma\beta(b\beta)^{-1}(a + b\alpha)\,,$$

where all terms on the right side are evaluated at $(z, 0)$. Simplifying the right-hand side gives the zero dynamics of the transformed system as

$$\dot{z} = q(z,0) - \gamma(z,0)b^{-1}(z,0)a(z,0)\,,$$

which is exactly the same as (14.31).

In the remainder of this section, we characterize feedback passivity, first for linear systems, and then for nonlinear systems.

14.5.1 Linear Systems

If a linear system is passive with positive definite storage, then the storage is quadratic and, hence, C^2; moreover, by Proposition 14.2, the system has relative degree one and the origin of the zero dynamics is Lyapunov stable. The key idea of feedback passivity is that the latter two properties are invariant under regular feedback transformations. The next proposition characterizes feedback passive linear systems.

Proposition 14.6 (Feedback Passivity of Linear Systems)
Suppose B has full rank equal to m. The linear system defined by the matrix triple (C, A, B) is feedback passive with positive definite storage, $V(x) = \frac{1}{2}x^T P x$, if and only if it has relative degree one and the origin of the zero dynamics subsystem is Lyapunov stable.

Proof. If the system is feedback passive with positive definite storage, then the transformed system has relative degree one and stable zero dynamics, by Proposition 14.2. Since these two properties are invariant under regular feedback transformations, the original system has them as well.

Suppose that the system has relative degree one and the origin of the zero dynamics is Lyapunov stable. Since the output is $y = Cx$, we can write the system in the normal form

$$\begin{aligned} \dot{\xi} &= Q_{11}\xi + Q_{12}y, \\ \dot{y} &= Q_{21}\xi + Q_{22}y + CBu. \end{aligned}$$

Since CB is invertible, we can define the regular feedback transformation $u = -(CB)^{-1}(Q_{21}\xi + Q_{22}y) + (CB)^{-1}v$, which puts the system into the form

$$\dot{\xi} = Q_{11}\xi + Q_{12}y, \tag{14.33}$$
$$\dot{y} = v, \tag{14.34}$$

which has relative degree one with the same output y. Since the zero dynamics has a stable origin, there is a symmetric positive definite matrix P_{11} such that

$$Q_{11}^T P_{11} + P_{11}Q_{11} \leq 0. \tag{14.35}$$

(See Exercise 3.12.) In (14.34) we define the feedback transformation

$$v = -Q_{12}^T P_{11}\xi + \bar{v},$$

which transforms (14.33)–(14.34) into

$$\dot{\xi} = Q_{11}\xi + Q_{12}y, \tag{14.36}$$
$$\dot{y} = -Q_{12}^T P_{11}\xi + \bar{v}. \tag{14.37}$$

The function

$$S(\xi, y) := \frac{1}{2}\xi^T P_{11}\xi + \frac{1}{2}y^T y$$

is positive definite. Using (14.35), differentiation of $S(\xi, y)$ along solutions of (14.36)–(14.37) gives

$$\begin{aligned} \frac{d}{dt}S(\xi, y) &= \xi^T P_{11}(Q_{11}\xi + Q_{12}y) + y^T(-Q_{12}^T P_{11}\xi + \bar{v}) \\ &= \frac{1}{2}\xi^T(Q_{11}^T P_{11} + P_{11}Q_{11})\xi + \xi^T P_{11}Q_{12}y - y^T Q_{12}^T P_{11}\xi + y^T\bar{v} \\ &\leq y^T\bar{v}. \end{aligned}$$

This is the required dissipation inequality for (14.36)–(14.37) with positive definite storage given by S. Thus, the original system is feedback passive. □

We consider two examples to illustrate Proposition 14.6.

Example 14.6 The system

$$\begin{aligned}\dot{x}_1 &= x_2,\\ \dot{x}_2 &= u,\\ y &= x_2,\end{aligned}$$

denoted (c, A, b), has relative degree one and Lyapunov stable zero dynamics. From Example 14.2, we know that the only storage function for this passive system is $\frac{1}{2}x_2^2$, which is not positive definite. By Proposition 14.6, this system is feedback passive with a positive definite storage function. In fact, it is straightforward to check that the feedback transformation $u = Kx + \beta v = [k_1 \quad k_2]x + \beta v$ produces a system that satisfies $Pb = c^T$ with positive definite P if and only if P is diagonal, $P = \text{diag}\,[p_{11}, p_{22}]$, and $\beta = 1/p_{22}$. We have

$$(A + bK)^T P + P(A + bK) = \begin{bmatrix} 0 & p_{11} + k_1 p_{22} \\ p_{11} + k_1 p_{22} & 2k_2 p_{22} \end{bmatrix},$$

which is negative semidefinite if and only if $k_1 = -\, p_{11}/p_{22}$. After the application of the feedback transformation, we have the system

$$\begin{aligned}\dot{x}_1 &= x_2,\\ \dot{x}_2 &= -\frac{p_{11}}{p_{22}}x_1 + k_2 x_2 + \frac{1}{p_{22}}v,\\ y &= x_2.\end{aligned}$$

With $V(x) = \frac{1}{2}x^T P x = \frac{1}{2}(p_{11}x_1^2 + p_{22}x_2^2)$, we get

$$\begin{aligned}\frac{d}{dt}V(x_1, x_2) &= p_{11}x_1\dot{x}_1 + p_{22}x_2\dot{x}_2\\ &= p_{11}x_1x_2 + p_{22}x_2\left(-\frac{p_{11}}{p_{22}}x_1 + k_2x_2 + \frac{1}{p_{22}}v\right)\\ &= p_{22}k_2x_2^2 + x_2 v.\end{aligned}$$

Since we must have $p_{22} > 0$, the dissipation inequality for V is satisfied if we choose $k_2 \le 0$, which we are free to do. Thus, the feedback-transformed system is passive with positive definite storage. In particular, we may take $P = I$; then we have $k_1 = -1$ and $\beta = 1$, with a free choice of $k_2 \le 0$. $\triangle$

For the next illustration we return to the system of Example 14.3.

Example 14.7 The system (c, A, b) is

$$\begin{aligned}\dot{x}_1 &= 0,\\ \dot{x}_2 &= x_1 + u,\\ y &= x_2,\end{aligned}$$

which is not passive, but it does have relative degree one and Lyapunov stable zero dynamics. By Proposition 14.6, the system is feedback passive. We now look for a feedback transformation $u = Kx + \beta v = [k_1 \quad k_2]x + \beta v$

that produces a passive system with positive definite storage. The feedback-transformed system is

$$\begin{aligned}\dot{x}_1 &= 0,\\ \dot{x}_2 &= (1+k_1)x_1 + k_2x_2 + \beta v,\\ y &= x_2.\end{aligned}$$

If P is symmetric, then $Pb = c^T$ implies that $P = \text{diag}\,[p_{11}, p_{22}]$, with p_{11} arbitrary and $p_{22} = 1/\beta$. In addition, $P > 0$ if and only if $p_{11} > 0$ and $p_{22} > 0$ ($\beta > 0$). We have

$$A^TP + PA = \begin{bmatrix} 0 & (1+k_1)p_{22} \\ (1+k_1)p_{22} & 2k_2p_{22} \end{bmatrix},$$

and for this matrix to be negative semidefinite, it is necessary and sufficient that $k_1 = -1$ and $k_2 \le 0$. With these choices, the feedback-transformed system is

$$\begin{aligned}\dot{x}_1 &= 0,\\ \dot{x}_2 &= k_2x_2 + \frac{1}{p_{22}}v,\\ y &= x_2.\end{aligned}$$

The storage must be $V(x) = \frac{1}{2}x^TPx$, with $p_{11} > 0$ and $p_{22} > 0$, but we verify the dissipation inequality by computing that, indeed,

$$\frac{d}{dt}V(x_1, x_2) = p_{11}x_1\dot{x}_1 + p_{22}x_2\dot{x}_2 = p_{22}k_2x_2^2 + x_2v \le yu$$

for any choice of $k_2 \le 0$ and $p_{22} > 0$. Therefore the feedback-transformed system is passive with positive definite storage function given by

$$V(x_1, x_2) = \frac{1}{2}p_{11}x_1^2 + \frac{1}{2}\frac{1}{\beta}x_2^2,$$

for any choice of $p_{11} > 0$ and $\beta > 0$. △

Two-dimensional SISO feedback passive linear systems are characterized in Exercise 14.10. We now consider three-dimensional SISO controller forms.

Example 14.8 (Feedback Passivity of 3D SISO Controller Forms)
Consider the three-dimensional single-input single-output controller form

$$\begin{aligned}\dot{x}_1 &= x_2,\\ \dot{x}_2 &= x_3,\\ \dot{x}_3 &= a_{31}x_1 + a_{32}x_2 + a_{33}x_3 + u,\\ y &= c_1x_1 + c_2x_2 + c_3x_3.\end{aligned}$$

The normal form and zero dynamics equations for these systems are developed in Example 11.13; however, in this example we only need the characteristic polynomial for the zero dynamics, and this polynomial is repeated below.

The relative degree one condition is equivalent to $c_3 \neq 0$. Example 11.13 showed that the case where $c_1 = c_2 = 0$ has unstable zero dynamics, so we continue on with $c_1^2 + c_2^2 \neq 0$. The two-dimensional zero dynamics system has characteristic equation (11.35), which we repeat here as

$$\lambda^2 + \frac{c_2}{c_3}\lambda + \frac{c_1}{c_3} = 0. \tag{14.38}$$

Based on (14.38), the question of whether the system is feedback passive can be answered for all cases of the output coefficients c_1, c_2, c_3 with $w = c_1^2 + c_2^2 \neq 0$ and $c_3 \neq 0$. See Exercise 14.11. △

14.5.2 Nonlinear Systems

In this subsection, we show that there are geometric constraints on feedback passive nonlinear systems of the form

$$\dot{x} = f(x) + g(x)u, \tag{14.39}$$

$$y = h(x), \tag{14.40}$$

namely, that (14.39)–(14.40) has uniform relative degree one at the origin and Lyapunov stable zero dynamics with a C^2 Lyapunov function.

Proposition 14.6 stated a criterion for feedback passivity of linear systems. The next theorem does the same for feedback passivity of nonlinear systems; however, we should note that Theorem 14.5 gives necessary and sufficient conditions for *local* feedback passivity of a nonlinear system, because in general we are only guaranteed a transformation to a normal form in some neighborhood of the origin.

Theorem 14.5 (Feedback Passivity of Nonlinear Systems)
Suppose that rank $g(0) = m$, *that is,* $g(0)$ *has full column rank. Then system (14.39)–(14.40) is feedback passive with a* C^2 *positive definite storage function if and only if it has uniform relative degree one at* $x = 0$ *and the origin of the zero dynamics is stable with a* C^2 *Lyapunov function that depends only on the state* z *of the zero dynamics system.*

Proof. The *only if* part follows from Lemma 14.2 and Proposition 14.3 applied to the feedback-transformed system and the invariance under regular feedback of uniform relative degree one and stable zero dynamics.

Suppose (14.39)–(14.40) has uniform relative degree one at $x = 0$ and the origin of the zero dynamics is stable with a C^2 Lyapunov function. The normal form coordinates may be chosen in such a way that the coefficient $\gamma(z, \xi)$ in system (14.30) is equal to zero (Proposition 11.2). Assuming that such a choice of coordinates has been made, we consider the normal form given by

$$\begin{aligned} \dot{z} &= q(z,\xi), \\ \dot{\xi} &= a(z,\xi) + b(z,\xi)u, \\ y &= \xi\,, \end{aligned} \tag{14.41}$$

and then the zero dynamics subsystem is $\dot{z} = q(z, 0)$. By the chain rule and the fundamental theorem of calculus, we may write

$$q(z,\xi) - q(z,0) = \int_0^1 \frac{\partial q}{\partial \xi}(z, s\xi)\, \xi \, ds.$$

Letting

$$p(z,\xi) := \int_0^1 \frac{\partial q}{\partial \xi}(z, s\xi)\, ds,$$

we may write the first equation of system (14.41) as

$$\dot{z} = q(z,0) + p(z,\xi)\xi. \tag{14.42}$$

By the hypothesis on the zero dynamics subsystem, there exists a C^2 Lyapunov function $W(z)$ such that $L_{q(z,0)}W(z) \le 0$, and therefore

$$\frac{d}{dt}W(z) = L_{q(z,0)}W(z) + (L_{p(z,\xi)}W(z))\xi \le (L_{p(z,\xi)}W(z))\xi\,. \tag{14.43}$$

Define the feedback transformation

$$u(z,\xi) = b^{-1}(z,\xi)(-a(z,\xi) - (L_{p(z,\xi)}W(z))^T + v). \tag{14.44}$$

Substitution of (14.44) into (14.41) yields the system

$$\begin{aligned} \dot{z} &= q(z,\xi), \\ \dot{\xi} &= -(L_{p(z,\xi)}W(z))^T + v, \\ y &= \xi\,. \end{aligned}$$

Let

$$S(z,\xi) := W(z) + \frac{1}{2}\xi^T\xi\,;$$

then S is positive definite since $W(z)$ is. Differentiate, using (14.43), to get

$$\frac{d}{dt}S(z,\xi) = \frac{d}{dt}W(z) + \xi^T\dot{\xi} \le L_{p(z,\xi)}W(z)\xi - \xi^T(L_{p(z,\xi)}W(z))^T + \xi^T v = \xi^T v.$$

Thus, the feedback (14.44) transforms (14.41) into a passive system with the C^2 positive definite storage function S. □

14.6 EXERCISES

Exercise 14.1 Consider the mass-spring system with $k = m = 1$, $b = 0$ and with position as output,

$$\begin{aligned}\dot{x}_1 &= x_2,\\ \dot{x}_2 &= -x_1 + u,\\ y &= x_1.\end{aligned}$$

Suppose the input is $u(t) = \sin \omega t$.

(a) Show that if $(x_1(t), x_2(t))$ is the solution with initial conditions $x_1(0) = 0$, $x_2(0) = 0$, then

$$x_1(t) = -\frac{\omega}{1-\omega^2}\sin t + \frac{1}{1-\omega^2}\sin \omega t.$$

(b) Show that the integral form of the dissipation inequality (14.6) cannot hold for any nonnegative function $S(x_1, x_2)$ with $S(0,0) = 0$. *Hint:* Choose a frequency ω and the upper limit of the integral so that (14.6) is violated.

Exercise 14.2 Consider the system

$$\begin{aligned}\dot{x}_1 &= x_2,\\ \dot{x}_2 &= u,\\ y &= x_1.\end{aligned}$$

Show in two ways that this system is not passive with C^1 storage, by considering (1) the KYP property, and (2) the relative degree.

Then do the same for the linear mass-spring system with position output.

Exercise 14.3 Write out a complete proof of Theorem 14.2, using the proof of Theorem 14.1 as a guide.

Exercise 14.4 Show that a linear system defined by the matrix triple (C, A, B) is zero-state detectable if and only if (C, A) is detectable, and it is zero-state observable if and only if (C, A) is observable.

Exercise 14.5 *Passivity Implies Finite Available Storage*
Show that if (14.15)–(14.16) is passive with storage V, then the available storage $V_a(x)$ is finite for each x, and $0 \le V_a(x) \le V(x)$.

Exercise 14.6 Consider a nonlinear system $\dot{x} = f(x) + g(x)u$, $y = h(x)$ with $f(0) = 0$ and $h(0) = 0$, and its linearization at the origin, given by

$(C, A, B) := (\frac{\partial h}{\partial x}(0), \frac{\partial f}{\partial x}(0), g(0))$. Show by examples that the property of zero-state detectability for the nonlinear system does not imply, and is not implied by, detectability of the pair (C, A).

Exercise 14.7 Consider the control-affine system $\dot{x} = f(x) + g(x)u$ with output $y = h(x)$. Determine whether these implications are true or false. Explain.

(a) If the system is passive and zero-state observable (ZSO), then it is stabilizable by output feedback.
(b) If the system is linear, passive with positive definite storage, and detectable, then it is stabilizable by output feedback.

Exercise 14.8 Consider the linear system of oscillators,

$$\dot{x} = \begin{bmatrix} 0 & 1 & 0 & 0 \\ -1 & 0 & 0 & 0 \\ 0 & 0 & 0 & 1 \\ 0 & 0 & -1 & 0 \end{bmatrix} x + \begin{bmatrix} 0 \\ 1 \\ 0 \\ 1 \end{bmatrix} u$$

with output $y = x_2 + x_4$.

(a) Show that the system is passive with positive definite storage.
(b) Is the system zero-state detectable?
(c) Is the system asymptotically stabilized by $u = -y = -x_2 - x_4$?

Exercise 14.9 *Stabilization with Output Feedback*
Under the hypotheses of Proposition 14.5, show that the origin can be globally asymptotically stabilized using output feedback of the form $u = -\phi(y)$, where $\phi(y)$ is any C^1 function such that $\phi(0) = 0$ and $y^T \phi(y) > 0$ for all nonzero y.

Exercise 14.10 *Feedback Passive Linear Systems: 2D SISO Case*
Consider the two-dimensional SISO linear system in controller form,

$$\begin{aligned} \dot{x}_1 &= x_2, \\ \dot{x}_2 &= a_{21}x_1 + a_{22}x_2 + u, \\ y &= c_1 x_1 + c_2 x_2. \end{aligned}$$

Use Proposition 14.6 to show that the system is feedback passive with positive definite storage if and only if $c_2 \neq 0$ and $c_1 c_2 \geq 0$. *Hint*: For simplicity, and without loss of generality, assume that the coefficients $a_{2j} = 0$ for $j = 1, 2$. Verify that a transformation to normal form is obtained by setting

$$\begin{bmatrix} \xi \\ \eta \end{bmatrix} = \begin{bmatrix} c_1 & c_2 \\ 1 & 0 \end{bmatrix} \begin{bmatrix} x_1 \\ x_2 \end{bmatrix}.$$

Exercise 14.11 *Feedback Passive Linear Systems: 3D Controller SISO Case*
For the three-dimensional controller form SISO linear system,

$$\begin{aligned}\dot{x}_1 &= x_2,\\ \dot{x}_2 &= x_3,\\ \dot{x}_3 &= a_{31}x_1 + a_{32}x_2 + a_{33}x_3 + u,\\ y &= c_1x_1 + c_2x_2 + c_3x_3,\end{aligned}$$

assume that $c_3 \neq 0$ (the relative degree one condition) and $w := c_1^2 + c_2^2 \neq 0$. Show that the system is feedback passive with positive definite storage if and only if all the nonzero output coefficients have the same sign. *Hint*: For simplicity, and without loss of generality, assume that the coefficients $a_{3j} = 0$ for $j = 1, 2, 3$.

Exercise 14.12 Assume that the single-input system $\dot{x} = f(x) + g(x)u$ is globally defined on $\mathbf{R}^n$, $f(0) = 0$, and $x = 0$ is globally asymptotically stabilizable by smooth state feedback. Show that there are a feedback transformation $u = \alpha(x) + v$ and a smooth function $h(x)$ such that the closed loop system

$$\begin{aligned}\dot{x} &= f(x) + g(x)\alpha(x) + g(x)v,\\ y &= h(x)\end{aligned}$$

is passive with a positive definite and radially unbounded storage function.

Exercise 14.13 Show that passivity of the linearization of a system does not imply passivity of the full nonlinear system. *Hint*: For a two-dimensional example, use the double integrator system as the linearization at the origin.

Exercise 14.14 *Passivity of the Pendulum-on-a-Cart*
Consider the inverted pendulum-on-a-cart system in Exercise 7.7. This exercise shows that this nonlinear system is passive when the output is taken to be the velocity of the cart.

(a) Show that the pair of second order equations of motion can be written in the form

$$M(q)\ddot{q} + C(q, \dot{q})\dot{q} + G(q) = F,$$

where

$$q := \begin{bmatrix}\xi\\ \theta\end{bmatrix}, \quad M(q) := \begin{bmatrix}M + m & ml\cos\theta\\ ml\cos\theta & ml^2\end{bmatrix},$$

$$C(q, \dot{q}) := \begin{bmatrix}0 & -ml\dot{\theta}\sin\theta\\ 0 & 0\end{bmatrix}, \quad G(q) := \begin{bmatrix}0\\ -mgl\sin\theta\end{bmatrix}, \quad F := \begin{bmatrix}u\\ 0\end{bmatrix}.$$

(b) Show that $M(q)$ is symmetric positive definite, and that

$$\frac{d}{dt}[M(q)] - 2C(q,\dot{q}) = \begin{bmatrix} 0 & ml\dot{\theta}\sin\theta \\ -ml\dot{\theta}\sin\theta & 0 \end{bmatrix}.$$

(c) Define the energy function

$$E(q,\dot{q}) := \frac{1}{2}\dot{q}^T M(q)\dot{q} - mgl(1-\cos\theta),$$

and show that the system is passive with input u, output $y = \dot{\xi}$, and storage E.

(d) Show that the full nonlinear system, with $y = \dot{\xi}$ as output, is not zero-state detectable near the origin. *Hint*: The angle θ is restricted by the mechanical set-up to the interval $-\pi < \theta < \pi$; however, consider solutions $(q(t), \dot{q}(t))$ with initial condition θ_0 near zero that produce identically zero output on some nontrivial time interval.

14.7 NOTES AND REFERENCES

This chapter is guided by [19] and [86]. The terminology, statement, and proof of Theorem 14.1 are from [19]; this result appeared earlier in [76] (Theorem 1), in a form that includes direct feedthrough models, where the output equation takes the form $y = h(x) + j(x)u$. Passive systems form a subclass of the important larger class of *dissipative systems*; the foundations of dissipative systems appear in the papers [101], [102].

Definition 14.5 of available storage is from [101]. Theorem 14.3 is a special case of [101] (Theorem 1, page 328). Proposition 14.1 is from [39] (Lemma 1); in this reference, the zero-state observability property is called "zero-state detectability."

Theorem 14.4 and its proof follow [48] (pages 72–74). The result in Proposition 14.6 is included in [83] (Proposition 2). Theorem 14.5 and the proof given here are from [86].

See [44] for information on circuit elements.

Exercise 14.12 is from [19] (Proposition 4.14). The converse of the statement in Exercise 14.12 holds with some additional hypotheses; see [19].

Exercise 14.14 was extracted from the analysis in the paper by R. Lozano and I. Fantoni, Passivity based control of the inverted pendulum, in *Perspectives in Control: Theory and Applications* edited by D. Normand-Cyrot, Springer, London, 1998.

Passivity and feedback passivation play a central role in the recursive feedback designs for stabilization in [86].

Chapter Fifteen

Partially Linear Cascade Systems

In this chapter we consider cascade systems in which a controlled subsystem is linear. These systems take the form

$$\dot{x} = \hat{F}(x, \xi), \tag{15.1}$$

$$\dot{\xi} = A\xi + Bu, \tag{15.2}$$

where $x \in \mathbf{R}^n$ and $\xi \in \mathbf{R}^\nu$. The main distinguishing feature of such a composite system is that the control u enters only the ξ-subsystem, and thus (15.2) is called the *driving system*, while (15.1) is the *driven system* of the cascade. In (15.2), the driving linear subsystem can be more general than the chain of integrators subsystem that appeared in the normal form in our study of zero dynamics.

The standing assumptions for the chapter are as follows. The mapping $\hat{F}$ satisfies $\hat{F}(0,0) = 0$, which guarantees that the unforced system (with $u = 0$) has an equilibrium at $(x, \xi) = (0, 0)$. The origin $x = 0$ of the undriven x-subsystem given by $\dot{x} = \hat{F}(x, 0)$ is assumed to be globally asymptotically stable. The ξ-subsystem is assumed to be stabilizable, that is, the pair (A, B) is stabilizable.

The main goal of the chapter is to develop sufficient conditions for global stabilizability of the cascade system. First we consider local stabilization by linear *partial-state feedback*, that is, linear feedback that uses the state of the driving subsystem alone, and then we develop an important result on global stabilization by *full-state feedback*, $u = u(x, \xi)$. Occasionally, we use convenient abbreviations: LAS (for "local asymptotic stability," "local asymptotic stabilization," or "locally asymptotically stable"), and, similarly, GAS for the global property.

15.1 LAS FROM PARTIAL-STATE FEEDBACK

We have seen by example that even when the decoupled subsystems of a cascade are globally asymptotically stable, the cascade itself need not be GAS. This situation has immediate consequences for stabilization of cascades when using *partial-state feedback*. Consider the next example.

Example 15.1 Consider the system

$$\dot{x} = -x + x^2\xi,$$

$$\dot{\xi} = u.$$

With feedback $u = -\xi$, the closed loop cascade has asymptotically stable origin by Corollary 9.2. But it is straightforward to check that the hyperbola $x\xi = 2$ is invariant for the closed loop system, by the following calculation:

$$\frac{d}{dt}(x\xi) = \dot{x}\xi + x\dot{\xi} = (-x + x^2\xi)\xi + x(-\xi) = -2x\xi + x^2\xi^2 = -x\xi(2 - x\xi).$$

Clearly, the invariance of the curve $x\xi = 2$ precludes global asymptotic stability for $(x, \xi) = (0, 0)$ after this feedback. (The basin of attraction for the origin of the closed loop is the region $x\xi < 2$; see Example 9.3.) △

It may seem plausible that a faster rate of convergence of $\xi(t)$ to the origin might overcome the growth represented by the interconnection term $x^2\xi$ in Example 15.1.

Example 15.2 Consider again the system

$$\begin{aligned} \dot{x} &= -x + x^2\xi, \\ \dot{\xi} &= u. \end{aligned}$$

Set $u = -k\xi$ and work with the closed loop system. Much as in Example 15.1, one can check that the hyperbola $x\xi = k + 1$ is invariant for the closed loop system, since we have

$$\frac{d}{dt}(x\xi) = \dot{x}\xi + x\dot{\xi} = (-x + x^2\xi)\xi + x(-k\xi) = -x\xi(k + 1 - x\xi).$$

In fact, it can be shown that the basin of attraction of the origin is the region $x\xi < k+1$. By taking larger and larger values of $k > 0$, it is possible to include any given initial condition (x_0, ξ_0) in the basin of attraction of the origin of the closed loop defined by $u = -k\xi$. However, there is no linear partial-state feedback that makes the origin GAS. (See also Exercise 9.6.) △

There is a standard way of describing the result of Example 15.2. Let Z be a compact subset of the plane. Since Z is closed and bounded, it is contained in some closed ball centered at the origin; hence, there is a value of $k > 0$ such that Z is contained in the region $x\xi < k + 1$, and thus Z is contained in the basin of attraction for the closed loop system defined by the feedback $u = -k\xi$. It is worthwhile to identify this property of *semiglobal stabilizability* in the next definition.

Definition 15.1 *A system $\dot{x} = f(x, u)$, where $x \in \mathbf{R}^n$ and $f(0, 0) = 0$, is* semiglobally stabilizable *(to the origin) if, given any compact subset Z of $\mathbf{R}^n$, there exists a feedback control $u = k(x)$ such that the origin is an asymptotically stable equilibrium of $\dot{x} = f(x, k(x))$ and the set Z is contained in its basin of attraction.*

The next example shows that linear partial-state feedback can fail to achieve semiglobal stabilization.

Example 15.3 Consider the system

$$\begin{aligned}\dot{x} &= -x + x^2\xi_1,\\ \dot{\xi}_1 &= \xi_1 + \gamma\xi_2 + u,\\ \dot{\xi}_2 &= -\xi_2,\end{aligned}$$

where γ is an arbitrary constant. The driving linear subsystem is certainly stabilizable. However, for any choice of linear partial-state feedback $u = -k_1\xi_1 - k_2\xi_2$ which asymptotically stabilizes $(\xi_1, \xi_2) = (0, 0)$, there are initial conditions that yield unbounded solutions for (x, ξ). In fact, reference to Example 15.2 shows that we only have to choose initial conditions of the form $(x_0, \xi_{10}, 0)$ with the product $x_0\xi_{10}$ sufficiently large, specifically, $x_0\xi_{10} \geq k_1$. $\triangle$

The presence of the interconnection term $x^2\xi_1$ is the obstacle to global asymptotic stabilization by linear partial-state feedback in the previous example. However, a general interconnection term presents no obstacle to *local* asymptotic stabilization of the cascade, based on the local asymptotic stability result for cascades given in Corollary 9.2.

In the next section we set up some notation for the interaction of the driving variable ξ with the driven variable x in (15.1). We also consider a little further the obstruction to global stabilization presented by the interconnection term.

15.2 THE INTERCONNECTION TERM

The examples so far have shown that the GAS assumption for $x = 0$ in $\hat{F}(x, 0)$ and the stabilizability assumption on the pair (A, B) cannot guarantee global asymptotic stabilization of (15.1)–(15.2) using partial-state feedback from ξ alone. Global stabilization of the cascade depends, in general, on the way in which the driving state ξ enters (15.1). In order to describe the interconnection of the subsystems, write

$$f(x) := \hat{F}(x, 0)$$

and

$$F(x, \xi) := \hat{F}(x, \xi) - \hat{F}(x, 0) = \hat{F}(x, \xi) - f(x).$$

Then the cascade (15.1)–(15.2) has the form

$$\dot{x} = f(x) + F(x, \xi), \tag{15.3}$$

$$\dot{\xi} = A\xi + Bu, \tag{15.4}$$

where $F(x, 0) = 0$. The next examples will show some of the possible dynamic effects of the interconnection term $F(x, \xi)$.

Example 15.4 Consider the system

$$\begin{aligned}\dot{x} &= -x + x\xi,\\ \dot{\xi} &= u,\end{aligned}$$

in which the interconnection term $F(x,\xi) := x\xi$ is linear in the driven variable x. If $u = -k\xi$ with $k > 0$, then direct integration of $\dot{x} = (-1+e^{-kt}\xi_0)x$ shows that the origin is a globally asymptotically stable equilibrium for the closed loop system. △

In addition to the standing assumptions of this chapter, global asymptotic stabilization of the cascade (15.1)–(15.2) by partial-state feedback generally requires a severe restriction on the growth of the interconnection term in the driven variable x as well as growth restrictions on a radially unbounded Lyapunov function for $\dot{x} = f(x)$; see [86] (pages 128–132).

If $\|F(x,\xi)\|$ grows faster than linearly with $\|x\|$, then the dependence of the interconnection term on ξ can be crucial, as we saw in Example 15.3 where semiglobal stabilizability was ruled out.

The next example provides some insight into the interaction of the driving variable ξ with nonlinear growth in x in the interconnection term $F(x,\xi)$.

Example 15.5 Consider the system

$$\begin{aligned}\dot{x} &= -x^3 - x^3\xi_2,\\ \dot{\xi}_1 &= \xi_2,\\ \dot{\xi}_2 &= u.\end{aligned}$$

Suppose we use linear feedback to stabilize the origin of the (ξ_1,ξ_2) subsystem. For large $k > 0$, we can achieve subsystem eigenvalues equal to $-k, -k$ using the feedback $u = -k^2\xi_1 - 2k\xi_2$, since the characteristic polynomial for the closed loop linear driving system is then $(\lambda+k)^2 = \lambda^2 + 2k\lambda + k^2$. If we choose initial conditions $\xi_1(0) = 1$, $\xi_2(0) = 0$, then the matrix exponential solution (or a direct verification) shows that

$$\xi_1(t) = (1+kt)e^{-kt} \quad \text{and} \quad \xi_2(t) = -k^2te^{-kt}.$$

For the present discussion, the important feature of $\xi_2(t)$ is that it has a peak absolute value of k/e, which occurs at time $t_p = 1/k$. The larger the value of k, the higher the peak absolute value, and the shorter the time interval required for the peak to occur. The driven equation for x is given by

$$\dot{x} = -x^3 - x^3\xi_2(t) = (-1 + k^2te^{-kt})x^3. \tag{15.5}$$

For large $k > 0$ the coefficient of x^3 on the right-hand side of (15.5) becomes large positive during the peaking time interval $0 \le t \le 1/k$. Eventually, for even larger positive times t, this coefficient will become negative, so it appears that $\dot{x}$ can decrease eventually. However, the peaking behavior of ξ_2 can cause a finite escape time in the solution for $x(t)$ before the stabilizing influence of the approach of $\xi_2(t) \to 0$ ever occurs. To see this, use separation of variables to write

$$\int_0^t x^{-3}\,dx = \int_0^t (-1 - \xi_2(s))\,ds$$
$$= \int_0^t (-1 + k^2 s e^{-ks})\,ds.$$

A direct integration followed by algebraic rearrangement yields

$$\frac{1}{[x(t)]^2} = \frac{1}{[x(0)]^2} + 2(t - 1 + (1 + kt)e^{-kt}),$$

and therefore

$$[x(t)]^2 = \frac{1}{\dfrac{1}{[x(0)]^2} + 2(t - 1 + (1 + kt)e^{-kt})}.$$

If the integration is valid on the time interval $0 \leq s \leq t_p = 1/k$, then we have

$$[x(t_p)]^2 = [x(1/k)]^2 = \frac{1}{\dfrac{1}{[x(0)]^2} + 2(\dfrac{1}{k} - 1 + 2e^{-1})}.$$

Multiply the top and bottom of the right-hand side by $[x(0)]^2 ke$ to get

$$[x(t_p)]^2 = \frac{[x(0)]^2 ke}{ke + 2e[x(0)]^2 - 2ke[x(0)]^2 + 4k[x(0)]^2}$$
$$= \frac{[x(0)]^2 ke}{k(e + 2[x(0)]^2(2 - e)) + 2e[x(0)]^2}.$$

From the last expression we have that $[x(0)]^2 > \frac{e}{2(e-2)}$ and sufficiently large $k > 0$ imply $[x(t_p)]^2 < 0$, a contradiction. Thus the solution $x(t)$ cannot exist up to time $t_p = 1/k$ for these initial conditions and k values. The conclusion is that the peaking behavior of $\xi_2(t)$ prevents semiglobal stabilization of the cascade using partial-state feedback, which drives ξ to zero more rapidly. $\triangle$

In essence, the *peaking phenomenon* of Example 15.5 involves an early transient peak in the driving solution, which, combined with nonlinear growth in the interconnection term, can be destabilizing for some initial conditions and can even cause finite escape times. As we have seen in these examples, the attempt to use feedback to increase the rate of convergence of the driving variable may or may not expand the closed loop basin of attraction in all directions — and such a strategy may even shrink the basin of attraction in certain directions as the driving rate of convergence increases. Consider the next example.

Example 15.6 In the cascade

$$\begin{aligned}\dot{x} &= -x + x^2\xi_2,\\ \dot{\xi}_1 &= \xi_1^3 + u,\\ \dot{\xi}_2 &= -\xi_1\end{aligned}$$

we may set $u = -k\xi_1 + k^2\xi_2$ to produce eigenvalues $\lambda = -\frac{k}{2} \pm i\frac{\sqrt{3}k}{2}$ for the linearization at the origin of the closed loop driving system given by

$$\begin{aligned}\dot{\xi}_1 &= -k\xi_1 + k^2\xi_2 + \xi_1^3,\\ \dot{\xi}_2 &= -\xi_1.\end{aligned}$$

We now relate this closed loop to the reversed-time Van der Pol system of Example 8.9, and examine how the basin changes as the feedback parameter $k > 0$ increases. Define the scaling transformation of time and space variables

$$\tau = kt, \quad \eta_1 = k^{-\frac{1}{2}}\xi_1, \quad \eta_2 = k^{\frac{1}{2}}\xi_2.$$

The system for η_1, η_2 is easily verified to be

$$\begin{aligned}\frac{d\eta_1}{d\tau} &= -\eta_1 + \eta_2 + \frac{1}{3}\eta_1^3,\\ \frac{d\eta_2}{d\tau} &= -\eta_1.\end{aligned}$$

This is the system of Example 8.9. There is a number $\gamma > 3$ such that the region $\eta_1^2 + \eta_2^2 > \gamma$ lies outside the basin of attraction of the equilibrium $(\eta_1, \eta_2) = (0, 0)$. Thus, in the original ξ_1, ξ_2 coordinates, the region

$$\frac{1}{k}\xi_1^2 + k\xi_2^2 > \gamma$$

lies outside the basin of attraction of the equilibrium $(\xi_1, \xi_2) = (0, 0)$. In particular, if $|\xi_2(0)| > \sqrt{\gamma/k}$ and $\xi_1(0) = 0$, then the solution $(\xi_1(t), \xi_2(t))$ is unbounded in forward time. Consequently, as $k \to \infty$ and the eigenvalues $\lambda = -\frac{k}{2} \pm i\frac{\sqrt{3}k}{2}$ are shifted further to the left in the complex plane, the basin for $(\xi_1, \xi_2) = (0, 0)$ shrinks to zero along the ξ_1-axis. $\triangle$

For more on the peaking phenomenon, see the Notes and References for this chapter.

In the next section we consider conditions on the interconnection term that allow for global asymptotic stabilization using full-state feedback, rather than partial-state feedback from the driving variable alone.

15.3 STABILIZATION BY FEEDBACK PASSIVATION

Using the system representation in (15.3)–(15.4), the standing assumptions on the cascade system are expressed as follows:

(LS) The pair (A, B) is stabilizable

(NLS) The origin $x = 0$ is a globally asymptotically stable equilibrium for $\dot{x} = f(x)$.

We want to find conditions on the interconnection term $F(x, \xi)$ in (15.3) that are sufficient for global asymptotic stabilizability of the cascade (15.3)–(15.4). In particular, when $\|F(x, \xi)\|$ grows faster than linearly in $\|x\|$, we attempt to factor out of F a linear function $y = C\xi$, and thus write

$$F(x, \xi) = \psi(x, \xi)C\xi = \psi(x, \xi)y,$$

where the factor $y = C\xi$ is viewed as the output of the linear ξ-subsystem. This allows us to write (15.3)–(15.4) as

$$\dot{x} = f(x) + \psi(x, \xi)y, \tag{15.6}$$

$$\dot{\xi} = A\xi + Bu. \tag{15.7}$$

Such a factorization of F might be done in more than one way, as examples will show below.

Obtaining a Direct Sum Lyapunov Function. Recall that the system has m inputs and m outputs. By assumption (LS), there exists linear feedback $u = K\xi$ such that $A + BK$ is Hurwitz. If we apply full-state feedback of the form $u = K\xi + v(x, \xi)$, then the closed loop for (15.6)-(15.7) is

$$\dot{x} = f(x) + \psi(x, \xi)C\xi, \tag{15.8}$$

$$\dot{\xi} = (A + BK)\xi + Bv(x, \xi). \tag{15.9}$$

If the feedback $v(x, \xi)$ satisfies the condition $v(0, \xi) = 0$, then, under the standing assumptions, each of the decoupled subsystems $\dot{x} = f(x)$ and $\dot{\xi} = (A + BK)\xi$ has a globally asymptotically stable origin. This fact alone does not guarantee the GAS property for the origin of the composite closed loop system, because (15.8)–(15.9) is a fully interconnected system; however, the observation provides reasonable motivation to search for a direct sum Lyapunov function. We are seeking conditions on the factorization $\psi(x, \xi)C\xi$ that allow for a choice of stabilizing feedback $v(x, \xi)$ for the composite system.

Let $V(x)$ be a globally defined and radially unbounded Lyapunov function for the equilibrium $x = 0$ of $\dot{x} = f(x)$, which exists by the converse Lyapunov theorem (Theorem 8.7 (b)). In addition, let $\frac{1}{2}\xi^T P\xi$ be a Lyapunov function for $\xi = 0$ in $\dot{\xi} = (A + BK)\xi$. As a Lyapunov function candidate, we consider a *direct sum* composite function

$$W(x, \xi) = V(x) + \frac{1}{2}\xi^T P\xi. \tag{15.10}$$

Along solutions of (15.8)–(15.9), we have

$$\dot{W}(x,\xi) = \nabla V(x)\cdot f(x) \tag{15.11}$$
$$+\nabla V(x)\cdot \psi(x,\xi)y$$
$$+\frac{1}{2}\xi^T((A+BK)^TP + P(A+BK))\xi \tag{15.12}$$
$$+\xi^T PBv(x,\xi).$$

It is possibly a good exercise at this point to look at the second and fourth terms, and ask if a choice for the feedback $v(x,\xi)$ presents itself, maybe after some further assumption. Recall that $y = C\xi$. In order to make $\dot{W}$ negative semidefinite, it is sufficient to make

$$\nabla V(x)\cdot \psi(x,\xi)C\xi + \xi^T PBv(x,\xi) = 0.$$

By making the additional assumption that $PB = C^T$, a choice for $v(x,\xi)$ presents itself, namely

$$v(x,\xi) = -\nabla V(x)\cdot \psi(x,\xi).$$

(Note that we do have $v(0,\xi) = 0$ then, since V is smooth and has a minimum at $x = 0$.) With this choice of $v(x,\xi)$, we have

$$\dot{W}(x,\xi) = \nabla V(x)\cdot f(x) + \xi^T((A+BK)^TP + P(A+BK))\xi \le 0$$

for all (x,ξ).

By our choices above, on the right-hand side of the expression for $\dot{W}(x,\xi)$, the term (15.11) is negative definite, and the term (15.12) is negative semidefinite. (The matrix $(A+BK)^TP + P(A+BK)$ need not be negative definite, due to the additional condition on P that $PB = C^T$; for an example, see Exercise 15.1.) Under these conditions $\dot{W}(x,\xi)$ is negative semidefinite. Since W is radially unbounded, we would like to apply Theorem 8.5 to conclude that $(x,\xi) = (0,0)$ is globally asymptotically stable, but we must verify that the largest invariant set M contained in the set where $\dot{W}(x,\xi) = 0$ is $M = \{(0,0)\}$.

Before we state a formal result, let us first summarize the assumptions we have made. Under the assumptions (LS), (NLS), and given the factorization in (15.8), the additional conditions that yield the composite Lyapunov function (15.10) are that the linear system defined by the triple (C,A,B) is feedback passive using the feedback transformation $u = K\xi + v$, with positive definite storage given by $\frac{1}{2}\xi^T P\xi$, while $A+BK$ is Hurwitz. Note that (15.12) is negative semidefinite (the first of the KYP conditions) and also that $PB = C^T$ (the other KYP condition). It remains to show that, under these conditions, the origin of system (15.6)–(15.7) is globally asymptotically stabilized by full-state feedback given by

$$u = K\xi + v(x,\xi) = K\xi - \nabla V(x)\cdot \psi(x,\xi),$$

where $\psi(x,\xi)$ is from the interconnection term in (15.8).

We now state a result for a single-input single-output system, and then we give some examples to illustrate it.

Proposition 15.1 *Suppose* (15.3)–(15.4) *is single-input single-output, that is,* $m=1$. *Suppose, in addition, that conditions* (LS) *and* (NLS) *hold, and that* $F(x,\xi)$ *in* (15.3) *can be factored as* $F(x,\xi) = \psi(x,\xi)c\xi$, *where the linear system defined by the triple* (c, A, b) *and having scalar output* $y = c\xi$ *is feedback equivalent to a passive linear system under the feedback* $u = K\xi + v$, *and satisfies the KYP conditions*

$$(A + bK)^T P + P(A + bK) = -Q$$

and

$$Pb = c^T,$$

with K *such that* $A+bK$ *is Hurwitz,* $P > 0$, *and* $Q \geq 0$. *If* $V(x)$ *is a globally defined and radially unbounded Lyapunov function for the origin of* $\dot{x} = f(x)$, *then the origin of system* (15.3)–(15.4) *is made globally asymptotically stable by full-state feedback given by*

$$u = K\xi - \nabla V(x) \cdot \psi(x,\xi).$$

Proof. As indicated prior to the proposition statement, it only remains to show that the largest invariant set M contained in the set where $\dot{W}(x,\xi) = 0$ is $M = \{(0,0)\}$. When this is established, we can invoke Theorem 8.5 to conclude that $(x,\xi) = (0,0)$ is globally asymptotically stable. We leave this argument for the single-input single-output case as an exercise; the main ideas are covered in the proof of the more general result in Theorem 15.1 below. □

We now apply the full-state feedback of Proposition 15.1 to Example 15.1.

Example 15.7 The system is

$$\begin{aligned} \dot{x} &= -x + x^2\xi, \\ \dot{\xi} &= u, \end{aligned}$$

where we take $y = \xi$, $f(x) = -x$, and $\psi(x,\xi) = x^2$. We identify $A = 0$, $b = 1$, and $c = 1$ for the scalar linear subsystem. Thus, the conditions for the (scalar) Q and P in this case are

$$KP + PK = -Q \quad \text{and} \quad P = c^T = 1.$$

We may choose $Q = 1$, and then $K = -\frac{1}{2}$. We use the natural choice of Lyapunov function $V(x) = \frac{1}{2}x^2$ for the undriven nominal system $\dot{x} = -x$. Our composite Lyapunov function is

$$W(x,\xi) := V(x) + \frac{1}{2}\xi^T P\xi = \frac{1}{2}x^2 + \frac{1}{2}\xi^2.$$

In accordance with Proposition 15.1, we let

$$v(x,\xi) = -\nabla V(x) \cdot \psi(x,\xi) = -(x)x^2 = -x^3.$$

With this choice, the feedback $u = K\xi + v(x,\xi) = -\frac{1}{2}\xi - x^3$ gives the closed loop system

$$\begin{aligned}\dot{x} &= -x + \xi x^2,\\ \dot{\xi} &= -\frac{1}{2}\xi - x^3.\end{aligned}$$

Along solutions of this closed loop, we have

$$\dot{W}(x,\xi) = -x^2 - \frac{1}{2}\xi^2 \le 0,$$

and $\dot{W}(x,\xi) < 0$ for $(x,\xi) \neq (0,0)$. Moreover, W is radially unbounded: $W(x,\xi) \to \infty$ as $\|(x,\xi)\| \to \infty$. Thus, the origin $(0,0)$ is globally asymptotically stable. $\triangle$

For another illustration of Proposition 15.1, we return to Example 15.3.

Example 15.8 We repeat the system equations here:

$$\begin{aligned}\dot{x} &= -x + x^2\xi_1,\\ \dot{\xi}_1 &= \xi_1 + \gamma\xi_2 + u,\\ \dot{\xi}_2 &= -\xi_2,\end{aligned}$$

where γ is an arbitrary constant. Choose $y = \xi_1$ as the output. Note that the linear subsystem, denoted (c, A, b), is not controllable, but it is stabilizable. The linear driving system with this output has relative degree one. Moreover, the zero dynamics are given by

$$\dot{\xi}_2 = -\xi_2,$$

with Lyapunov stable, even asymptotically stable, origin $\xi_2 = 0$. Therefore Proposition 15.1 applies (by Proposition 14.6 on feedback passivity). It can be verified that, with $K = [-2 \quad 2]$ the matrix $A + bK$ is Hurwitz with eigenvalues $-1, -1$. We use the natural choice of Lyapunov function $V(x) = \frac{1}{2}x^2$ for the undriven nominal system $\dot{x} = -x$. To proceed with the feedback construction in Proposition 15.1, the condition $Pb = c^T$ requires that the symmetric matrix P have the form

$$P = \begin{bmatrix} 1 & 0 \\ 0 & p_{22} \end{bmatrix},$$

and P is positive definite provided $p_{22} > 0$. The other KYP condition is that

$$(A + bK)^T P + P(A + bK) = \begin{bmatrix} -2 & 2+\gamma \\ 2+\gamma & -2p_{22} \end{bmatrix} \le 0,$$

and this will hold if we take $p_{22} \geq \frac{(2+\gamma)^2}{4}$. With this choice, the composite Lyapunov function is

$$W(x, \xi_1, \xi_2) = V(x) + \frac{1}{2}\xi^T P\xi = \frac{1}{2}x^2 + \frac{1}{2}\xi_1^2 + \frac{1}{2}p_{22}\xi_2^2.$$

The stabilizing feedback is $u = -2\xi_1 + 2\xi_2 + v(x, \xi_1, \xi_2)$, where

$$v(x, \xi_1, \xi_2) = -\nabla V(x) \cdot \psi(x, \xi) = -x(x^2) = -x^3.$$

Choose p_{22} such that $4p_{22} > (\gamma + 2)^2$. Since W is radially unbounded, the feedback $u = -2\xi_1 + 2\xi_2 - x^3$ makes $(x, \xi_1, \xi_2) = (0, 0, 0)$ globally asymptotically stable. △

With the statement of Proposition 15.1 in mind, it is recommended that Exercise 15.1 and Exercise 15.2 be worked out. Exercise 15.1 in particular shows clearly how the condition $PB = C^T$ can impose negative *semidefiniteness* on the term (15.12) (with the same P) rather than negative definiteness, with $A + BK$ Hurwitz.

In order to extend Proposition 15.1 to the general case of m inputs and m outputs, we assume that $F(x, \xi)$ in (15.3) can be factored as $\psi(x, \xi)C\xi$ with an appropriate vector output $y = C\xi$. Note that, if we write $y = C\xi = [y_1 \ \cdots \ y_m]^T$ and

$$\psi(x, \xi)_{n\times m} = [f_1(x, \xi) \quad \cdots \quad f_m(x, \xi)],$$

where each $f_i : \mathbf{R}^n \times \mathbf{R}^\nu \to \mathbf{R}^n$, then we have

$$\psi(x, \xi)C\xi = \sum_{i=1}^{m} y_i f_i(x, \xi). \tag{15.13}$$

Theorem 15.1 *Suppose that (LS) and (NLS) hold, and that the interconnection term $\psi(x, \xi)y = \psi(x, \xi)C\xi$ in (15.6) is written as in (15.13). Suppose further that the linear system defined by (C, A, B) is feedback passive using the feedback $u = K\xi + v$ and that*

$$(A + BK)^T P + P(A + BK) = -Q$$

and

$$PB = C^T,$$

where $A + BK$ is Hurwitz, $P > 0$, and $Q \geq 0$. Then the origin of (15.6)–(15.7) is globally asymptotically stabilizable by full-state feedback. Moreover, if $V(x)$ is a globally defined and radially unbounded Lyapunov function for the origin of $\dot{x} = f(x)$, then a globally stabilizing feedback is given by $u = K\xi + v(x, \xi)$, where

$$v(x, \xi) = \begin{bmatrix} v_1(x, \xi) \\ \cdots \\ v_m(x, \xi) \end{bmatrix} = \begin{bmatrix} -\nabla V(x) \cdot f_1(x, \xi) \\ \cdots \\ -\nabla V(x) \cdot f_m(x, \xi) \end{bmatrix}. \tag{15.14}$$

Proof. Let K be a feedback matrix and let $P > 0$ and $Q \geq 0$ be symmetric matrices such that the KYP conditions hold; that is,

$$(A + BK)^T P + P(A + BK) = -Q$$

and

$$PB = C^T,$$

where $A+BK$ is Hurwitz. We consider feedback of the form $u = K\xi + v(x, \xi)$, where $v(x, \xi)$ is to be defined. By hypothesis, we may use (15.13) to write

$$\dot{x} = f(x) + \sum_{i=1}^{m} y_i f_i(x, \xi)$$

and $\dot{x} = f(x)$ has a globally asymptotically stable equilibrium at $x = 0$. By a converse Lyapunov theorem (Theorem 8.7 (b)), there is a Lyapunov function $V(x)$ for $\dot{x} = f(x)$ with the properties that $V(x) > 0$ for $x \neq 0$, $V(0) = 0$, $\nabla V(x) \cdot f(x) < 0$ for all $x \neq 0$, and $V(x)$ is radially unbounded. We consider the candidate Lyapunov function

$$W(x, \xi) = V(x) + \frac{1}{2}\xi^T P \xi,$$

and full-state feedback of the form $u = K\xi + v(x, \xi)$ yielding the closed loop system

$$\dot{x} = f(x) + \sum_{i=1}^{m} y_i f_i(x, \xi), \tag{15.15}$$

$$\dot{\xi} = (A + BK)\xi + Bv(x, \xi). \tag{15.16}$$

Write $\hat{A} = A + BK$. Along solutions of (15.15)-(15.16) we have

$$\begin{aligned} \dot{W}(x, \xi) = {} & \nabla V(x) \cdot f(x) + \sum_{i=1}^{m} y_i \nabla V(x) \cdot f_i(x, \xi) \\ & + \frac{1}{2}\xi^T (\hat{A}^T P + P\hat{A})\xi + \xi^T PBv(x, \xi). \end{aligned}$$

Now use the KYP conditions to write $\hat{A}^T P + P\hat{A} = -Q$ and $\xi^T PB = \xi^T C^T = y^T$. Thus,

$$\begin{aligned} \dot{W}(x, \xi) = {} & \nabla V(x) \cdot f(x) + \sum_{i=1}^{m} y_i \nabla V(x) \cdot f_i(x, \xi) \\ & - \frac{1}{2}\xi^T Q \xi + y^T v(x, \xi). \end{aligned}$$

We can guarantee that $\dot{W}(x, \xi)$ is negative semidefinite by choosing

$$v_i(x, \xi) = -\nabla V(x) \cdot f_i(x, \xi),$$

because then we have

$$\dot{W}(x,\xi) = \nabla V(x) \cdot f(x) - \frac{1}{2}\xi^T Q \xi \leq 0 \quad \text{for all } (x,\xi).$$

The function $W(x,\xi)$ is smooth, positive definite, and radially unbounded.

In order to show global asymptotic stability of $(x,\xi) = (0,0)$, we will show that if $(x(t),\xi(t))$ is a solution of (15.15)–(15.16) such that

$$\dot{W}(x(t),\xi(t)) = 0, \tag{15.17}$$

then $(x(t),\xi(t)) = (0,0)$ for all t. First, if (15.17) holds, then $x(t) = 0$, since $\nabla V(x) \cdot f(x)$ is negative definite and $-Q$ is negative semidefinite. With $x(t) = 0$, the feedback $v(x,\xi)$ vanishes along the solution $(x(t),\xi(t))$, because

$$v(0,\xi(t)) = -\nabla V(0) \cdot f_1(0,\xi(t))$$

and $\nabla V(0) = 0$ since V has a minimum at $x = 0$. Therefore, by (15.16), $\xi(t)$ is a solution of

$$\dot{\xi} = (A + BK)\xi.$$

We also have

$$W(x(t),\xi(t)) = \frac{1}{2}\xi(t)^T P \xi(t) \quad \text{for all } t.$$

Since $A + BK$ is Hurwitz, $\xi(t) \to 0$ as $t \to \infty$. Since W is constant along the solution $(x(t),\xi(t)) = (0,\xi(t))$ and $W(0,\xi(t)) \to 0$ as $t \to \infty$, we have $\xi(t)^T P \xi(t) = 0$ for all t. But P is positive definite, so $\xi(t) = 0$ for all t. By the invariance theorem, the origin of (15.15)–(15.16) is globally asymptotically stable. □

It is interesting to note that the term $f(x)$ in (15.6) may be replaced by a term that is dependent on ξ, that is, a term of the form $f_0(x,\xi)$ with $f_0(0,\xi) = 0$ for all ξ, assuming that $f_0(x,\xi)$ is naturally stabilizing. Here is an appropriate definition to address these situations.

Definition 15.2 *Suppose $f_0(x,\xi)$ is C^1 on $\mathbf{R}^n \times \mathbf{R}^\nu$ and $f_0(0,\xi) = 0$ for all ξ in $\mathbf{R}^\nu$. We say that the equilibrium $x = 0$ is* globally asymptotically stable for $\dot{x} = f_0(x,\xi)$, uniformly in ξ, *if there exists a function $V : \mathbf{R}^n \to \mathbf{R}$ which is a strict Lyapunov function for the equilibrium $x = 0$ of $\dot{x} = f_0(x,\xi)$, where ξ is considered as a parameter; that is, V is C^1, positive definite, and radially unbounded, and*

$$(L_{f_0}V)(x,\xi) := \nabla V(x) \cdot f_0(x,\xi) < 0 \quad \textit{for all } x \neq 0 \textit{ and all } \xi.$$

The next theorem is the main result of the paper [61].

Theorem 15.2 *Consider the cascade*

$$\dot{x} = f_0(x,\xi) + \psi(x,\xi)C\xi, \tag{15.18}$$

$$\dot{\xi} = A\xi + Bu, \tag{15.19}$$

where $f_0(0,\xi)=0$ *for all* ξ, *and* $x=0$ *is globally asymptotically stable for* $\dot{x}=f_0(x,\xi)$, *uniformly in* ξ. *If the linear system* (C,A,B) *is feedback passive using the feedback transformation* $u=K\xi+v$, *with*

$$(A+BK)^T P + P(A+BK) = -Q$$

and

$$PB = C^T,$$

where $A+BK$ *is Hurwitz,* $P>0$, *and* $Q\geq 0$, *then the origin of* (15.18)–(15.19) *is globally asymptotically stabilizable by full-state feedback. In particular, if* $\psi(x,\xi)C\xi$ *is written in the form* (15.13), *then the feedback construction for* $u=K\xi+v(x,\xi)$ *of Theorem 15.1 is globally stabilizing, where* $v(x,\xi)$ *is given by* (15.14).

Proof. The proof follows directly the pattern of argument for Theorem 15.1; the details are left as Exercise 15.3. □

Theorem 15.2 allows us to handle examples like the following one, which requires an appropriate choice of the term $f_0(x,\xi)$.

Example 15.9 Consider the system

$$\begin{aligned}\dot{x} &= -x - x^3\xi_2^2 + x^2\xi_1,\\ \dot{\xi}_1 &= \xi_1 + \gamma\xi_2 + u,\\ \dot{\xi}_2 &= -\xi_2.\end{aligned}$$

If we choose $f_0(x)=-x$, then $F(x,\xi):=-x^3\xi_2^2+x^2\xi_1$ cannot be written in the factored form $\psi(x,\xi)C\xi$ using a linear output of the driving system. Thus, set $f_0(x,\xi):=-x-x^3\xi_2^2$, and note that $f_0(0,\xi)=0$ for all ξ. The function $V(x)=\frac{1}{2}x^2$ has rate of change along the solutions of $\dot{x}=f_0(x,\xi)$ given by

$$x\dot{x} = -x^2 - x^4\xi_2^2 < 0, \quad \text{for all } x\neq 0 \text{ and all } \xi.$$

Therefore $x=0$ is asymptotically stable for $\dot{x}=f_0(x,\xi)=-x-x^3\xi_2^2$, uniformly in ξ, and $V(x)$ is a radially unbounded Lyapunov function. Thus, we can take $x^2\xi_1$ as the interconnection term, with $y=\xi_1$ as the output of the driving system. Now proceed as before, and, in accordance with the proof of Theorem 15.1, construct the feedback

$$u = -2\xi_1 + 2\xi_2 - x^3$$

as a globally asymptotically stabilizing control for the origin of this cascade. Note that the term $-x^3\xi_2^2$ is already stabilizing due to the uniformity condition. Thus the same choice of $v(x,\xi)=-x^3$ as in Example 15.8 still works. △

There are several interesting consequences of the main results. To develop them, we need the next lemma.

Lemma 15.1 *Suppose $f : \mathbf{R}^n \times \mathbf{R}^\nu \to \mathbf{R}^n$ is a smooth function and $f(0,0) = 0$. Then there exist smooth functions $f_1(x,\xi), \ldots, f_\nu(x,\xi)$ such that*

$$f(x,\xi) = f(x,0) + \sum_{i=1}^{\nu} \xi_i f_i(x,\xi)$$

where $f_i(x,0) = \frac{\partial f}{\partial \xi_i}(x,0)$ for each i. In fact, we may take $f_i(x,\xi) := \int_0^1 \frac{\partial f}{\partial \xi_i}(x,t\xi)\,dt$.

Proof. Fix x and ξ, and consider $h(t) = f(x,t\xi)$ for $0 \le t \le 1$. We have

$$h(1) - h(0) = \int_0^1 \frac{dh}{dt}(t)\,dt = \int_0^1 \frac{\partial f}{\partial \xi}(x,t\xi)\xi\,dt.$$

Expressing the integrand as a sum over components, and using $h(1) = f(x,\xi)$, $h(0) = f(x,0)$, yields

$$\begin{aligned} f(x,\xi) &= f(x,0) + \int_0^1 \sum_{i=1}^{\nu} \frac{\partial f}{\partial \xi_i}(x,t\xi)\,\xi_i\,dt \\ &= f(x,0) + \sum_{i=1}^{\nu} \xi_i \int_0^1 \frac{\partial f}{\partial \xi_i}(x,t\xi)\,dt. \end{aligned}$$

Thus, we may define $f_i(x,\xi) := \int_0^1 \frac{\partial f}{\partial \xi_i}(x,t\xi)\,dt$. The functions f_i so defined are smooth, in fact, differentiable of order (at least) one less than f. □

The next result appears to involve a condition on the driving system alone. However, using Lemma 15.1, it is seen to be covered by Theorem 15.1.

Corollary 15.1 *Suppose that $f : \mathbf{R}^n \times \mathbf{R}^\nu \to \mathbf{R}^n$ is a smooth function and $x = 0$ is a globally asymptotically stable equilibrium for $\dot{x} = f(x,0)$. If $B_{\nu\times m}$ has rank ν, then the system*

$$\begin{aligned} \dot{x} &= f(x,\xi), \\ \dot{\xi} &= A\xi + Bu \end{aligned}$$

is globally asymptotically stabilizable by state feedback $u(x,\xi)$.

Proof. By Lemma 15.1, there are functions $f_1(x,\xi), \ldots, f_\nu(x,\xi)$ such that

$$f(x,\xi) = f(x,0) + \sum_{i=1}^{\nu} \xi_i f_i(x,\xi).$$

Since $f(x,0)$ is independent of ξ, the uniformity in ξ for the Lyapunov function for the equilibrium $x=0$ of $\dot{x} = f(x,0)$ is not an issue. Consider now the driving system. By hypothesis, $\nu \le m$. If $\nu < m$, then we can choose linearly independent columns of B that form a $\nu \times \nu$ invertible input matrix, and we need only work with the corresponding components of the input u;

thus, without loss of generality, we simply assume that $m = \nu$ and B is invertible. Let $\alpha < 0$, and apply the feedback transformation

$$u = Kx + B^{-1}v; \quad K := -B^{-1}A + B^{-1}\alpha I.$$

The resulting equation for ξ is

$$\dot{\xi} = (A + BK)\xi + BB^{-1}v = (\alpha I)\xi + v.$$

Define the output $y := \xi = [\xi_1 \quad \cdots \quad \xi_\nu]^T$. The closed loop has coefficient matrices given by

$$\hat{A} = \alpha I \ (\alpha < 0), \qquad \hat{B} = I, \quad \hat{C} = I.$$

The KYP property is easy to verify: $P\hat{B} = \hat{C}^T$ says that $PI = I$, so we can take $P = I$, and then $\hat{A}P + P\hat{A} = 2\alpha I \leq 0$, as required. By Theorem 15.1, there exists a smooth, globally asymptotically stabilizing full-state feedback $v(x, \xi)$. The globally stabilizing feedback for the original cascade is then $u(x, \xi) = [-B^{-1}A + B^{-1}\alpha I]x + B^{-1}v(x, \xi)$. □

The next corollary assumes not that the origin of $\dot{x} = f(x, 0)$ is asymptotically stable, but rather that it is stabilizable by smooth feedback. This corollary provides another proof that augmentation of a stabilizable system by an integrator results in a stabilizable system.

Corollary 15.2 *Suppose $f : \mathbf{R}^n \times \mathbf{R} \to \mathbf{R}^n$ is C^1 and suppose that $\dot{x} = f(x, u)$ is globally asymptotically stabilizable by a C^1 feedback $u = k(x)$. Then the cascade*

$$\begin{aligned} \dot{x} &= f(x, \xi), \\ \dot{\xi} &= v \end{aligned}$$

is globally asymptotically stabilizable by a C^1 feedback $v(x, \xi)$.

Proof. Let $\eta := \xi - k(x)$ and write $g(x, \eta) := f(x, k(x) + \eta)$. We want to express the system in terms of the variables (x, η). Differentiation of η with respect to t gives

$$\dot{\eta} = v - Dk(x) \cdot \dot{x} = v - Dk(x) \cdot g(x, \eta).$$

This suggests a feedback transformation to define the new input variable $w := -Dk(x) \cdot g(x, \eta) + v$, giving the system

$$\begin{aligned} \dot{x} &= g(x, \eta), \\ \dot{\eta} &= w. \end{aligned}$$

By the definiton of η and the GAS property of the origin of $\dot{x} = f(x, k(x))$, we have that $x = 0$ is a GAS equilibrium for $\dot{x} = g(x, 0) = f(x, k(x))$. Therefore the transformed system satisfies the hypotheses of Corollary 15.1, and there exists a globally asymptotically stabilizing feedback $w(x, \eta)$. The resulting stabilizing feedback for the original system is

$$v(x,\xi) := Dk(x) \cdot g(x,\eta) + w(x,\eta); \quad \eta = \xi - k(x). \tag{15.20}$$

This completes the proof. □

The proof of Corollary 15.2 actually works for any dimension vector ξ, not just real ξ; that is, we could have $f : \mathbf{R}^n \times \mathbf{R}^\nu \to \mathbf{R}^n$. The argument for Corollary 15.2 also gives a different proof of the earlier result on the preservation of stabilization when adding a single scalar integrator (Proposition 10.1, which was proved by a center manifold argument). The feedback constructions in these two results are identical. See also Example 10.6.

An induction argument using Corollary 15.2 establishes the next result involving a chain of integrators as the driving system.

Corollary 15.3 *Let $f : \mathbf{R}^n \times \mathbf{R} \to \mathbf{R}$ be smooth and suppose that the origin $x = 0$ of $\dot{x} = f(x,u)$ is globally asymptotically stabilizable by a C^1 feedback $u = k(x)$. Then, for any positive integer ν, the origin of the cascade*

$$\begin{aligned} \dot{x} &= f(x,\xi_1), \\ \dot{\xi}_1 &= \xi_2, \\ \dot{\xi}_2 &= \xi_3, \\ &\cdots \\ \dot{\xi}_{\nu-1} &= \xi_\nu, \\ \dot{\xi}_\nu &= v \end{aligned}$$

is globally asymptotically stabilizable by smooth feedback $v(x,\xi)$, where $\xi = [\xi_1 \quad \cdots \quad \xi_\nu]^T$.

Nonlinear Driving Systems

The main idea in the previous results is that a driving linear subsystem is made passive by linear feedback. Theorem 15.2 is an example of stabilization by *feedback passivation*. Additional results are available using this approach in cascades with a nonlinear, feedback passive driving system. These results depend on conditions describing when a system can be made passive by a feedback transformation. For linear systems, these feedback passivation conditions are given in Proposition 14.6; for nonlinear systems, Theorem 14.5 states the feedback passivity criterion.

At this point, it is possible to formulate a result on the stabilization of a cascade with nonlinear driving system, using feedback passivation ideas along the lines of Proposition 15.1. Instead of stating a formal result, we present one simple example here.

Example 15.10 Consider the system

$$\begin{aligned}\dot{x} &= -x + x^2 y,\\ \dot{\xi}_1 &= \xi_2,\\ \dot{\xi}_2 &= -\sin\xi_1 + u,\\ y &= \xi_2.\end{aligned}$$

The interconnection term is $\psi(x,\xi) := x^2$. The nonlinear driving system, with state (ξ_1,ξ_2) and output $y = \xi_2$, is already passive with storage given by the energy $E(\xi_1,\xi_2) = \frac{1}{2}\xi_2^2 + (1-\cos\xi_1)$. The driving system is feedback passive since it has relative degree one and Lyapunov stable zero dynamics. Let $u = -\xi_2 + v$ to provide damping. Note that $E(\xi_1,\xi_2)$ is the total mechanical energy for the closed loop ξ system given by

$$\begin{aligned}\dot{\xi}_1 &= \xi_2,\\ \dot{\xi}_2 &= -\sin\xi_1 - \xi_2,\end{aligned}$$

and that E is a positive definite storage function for this system. Let $W(x,\xi) = \frac{1}{2}x^2 + E(\xi_1,\xi_2)$, where $V(x) = \frac{1}{2}x^2$ is a global Lyapunov function for $\dot{x} = -x$. We want to choose v such that $\dot{W}(x,\xi) \le 0$. We have

$$\begin{aligned}\dot{W}(x,\xi_1,\xi_2) &= x\dot{x} + (\sin\xi_1)\dot{\xi}_1 + \xi_2\dot{\xi}_2,\\ &= -x^2 + x^3\xi_2 + \xi_2\sin\xi_1 - \xi_2\sin\xi_1 - \xi_2^2 + \xi_2 v,\\ &= -x^2 + x^3\xi_2 - \xi_2^2 + \xi_2 v.\end{aligned}$$

By choosing $v = -x^3$ (which has the form $v = -\nabla V(x)\cdot\psi(x,\xi)$), we have

$$\dot{W}(x,\xi_1,\xi_2) = -x^2 - \xi_2^2 \le 0.$$

Moreover, $\dot{W}(x,\xi_1,\xi_2) = 0$ if and only if $x = 0$ and $\xi_2 = 0$. The differential equations for the closed loop imply that a forward solution $(x(t),\xi_1(t),\xi_2(t))$ near the origin can remain in the set where $\dot{W}(x,\xi_1,\xi_2) = 0$ only if

$$(x(t),\xi_1(t),\xi_2(t)) \equiv (0,0,0).$$

By the invariance theorem, the feedback

$$u = -\xi_2 - x^3$$

locally asymptotically stabilizes the origin. Note that there are multiple equilibria for the closed loop under $u = -\xi_2 + v = -\xi_2 - x^3$, so the origin cannot be globally asymptotically stable. △

In Example 15.10, the system $\dot{x} = f(x) = -x$ had globally asymptotically stable origin $x = 0$, and, after the preliminary feedback $u = -\xi_2$, the (ξ_1,ξ_2) system (from v to y) was passive with storage $E(\xi_1,\xi_2)$. It was also zero-state observable in a local sense for solutions (x,ξ_1,ξ_2) near the origin, but not zero-state detectable in a global sense because there are nonzero solutions of the form $(x,\xi_1,\xi_2) = (0,c,0)$, c constant, that produce identically zero output (due essentially to the multiple equilibria, in this case).

We consider one result for a cascade system of the form

$$\dot{x} = f(x) + \psi(x,\xi)y, \tag{15.21}$$
$$\dot{\xi} = a(\xi) + b(\xi)u, \tag{15.22}$$
$$y = h(\xi). \tag{15.23}$$

The next result, taken from [86], relaxes the GAS assumption on $\dot{x} = f(x)$ and replaces it with a weaker stability-plus-global-boundedness assumption, plus a zero-state detectability assumption on the system after a feedback transformation.

Theorem 15.3 *Suppose that the system* (15.21)–(15.22) *with output given by* (15.23) *is globally defined for* $(x,\xi) \in \mathbf{R}^n \times \mathbf{R}^\nu$. *Assume there exists a smooth, positive definite, and radially unbounded function* $V(x)$ *such that* $L_f V(x) \leq 0$ *for all* x, *and the subsystem defined by* $(h(\xi), a(\xi), b(\xi))$ *is passive with a smooth, positive definite, and radially unbounded storage function* $S(\xi)$. *Then* (15.21)–(15.22) *with output* (15.23) *is made passive by the feedback transformation*

$$u = -(L_\psi V)^T(x,\xi) + v, \tag{15.24}$$

and

$$W(x,\xi) = V(x) + S(\xi)$$

is a storage function. Moreover, if the closed loop system with input v, *output* y, *and state* (x,ξ) *is zero-state detectable, then the additional feedback* $v := -ky = -kh(\xi)$, $k > 0$, *makes* $(x,\xi) = (0,0)$ *globally asymptotically stable.*

Proof. The proof is left to Exercise 15.8. □

15.4 INTEGRATOR BACKSTEPPING

This section provides an introduction to an important approach to feedback design called *integrator backstepping*, or simply *backstepping*. In addition to proving the basic lemma that describes this approach, we indicate briefly the importance of backstepping for recursive feedback designs.

Consider the cascade system

$$\dot{x} = f(x) + g(x)\xi, \tag{15.25}$$
$$\dot{\xi} = u, \tag{15.26}$$

where $x \in \mathbf{R}^n$, $\xi \in \mathbf{R}$, $f(0) = 0$, and there exists a smooth feedback $\alpha(x)$ such that $\dot{x} = f(x) + g(x)\alpha(x)$ has $x = 0$ as a globally asymptotically stable equilibrium. We wish to find a feedback control $u = u(x,\xi)$ that stabilizes the origin $(x,\xi) = (0,0)$.

In some cases, the problem might be approached with a zero dynamics analysis, using the output $y = \xi - \alpha(x)$ and investigating whether the zero dynamics subsystem is asymptotically stable; such an approach would involve differentiating the control $\alpha(x)$ in order to maintain the relation $y = \xi - \alpha(x) = 0$. This zero dynamics approach will work in some cases; for examples, see Exercises 11.10–11.11.

Here we consider a different approach. By a change of variable, the cascade (15.25)–(15.26) is reformulated to allow a known stabilizing control for (15.25), say $\xi = \alpha(x)$ (where ξ is considered the control variable for equation (15.25)), to be "backstepped" through an integrator. The resulting feedback u is a function of both x and ξ. A simple example illustrates the procedure.

Example 15.11 Consider the cascade

$$\begin{aligned} \dot{x} &= x + x^3 + \xi, \\ \dot{\xi} &= u, \end{aligned}$$

with scalar variables x, ξ, and $f(x) = x + x^3$. The origin $x = 0$ of $\dot{x} = x + x^3 + v$ is stabilized by the feedback $v = \alpha(x) = -2x$, which gives

$$\dot{x} = -x + x^3.$$

A Lyapunov function is $V(x) = \frac{1}{2}x^2$ with

$$x\dot{x} = -x^2 + x^4,$$

which shows local asymptotic stability of $x = 0$ after the feedback. To proceed with the change of variable, set

$$z := \xi - \alpha(x) = \xi + 2x.$$

Substitution into the original system gives the equivalent system in the variables x, z:

$$\begin{aligned} \dot{x} &= -x + x^3 + z, \\ \dot{z} &= u - \frac{\partial \alpha}{\partial x}(x)\dot{x} = u + 2(-x + x^3 + z). \end{aligned}$$

For a composite Lyapunov function, let $V_c(x, z) = V(x) + \frac{1}{2}z^2 = \frac{1}{2}x^2 + \frac{1}{2}z^2$, and compute

$$\begin{aligned} \dot{V}_c(x, z) = x\dot{x} + z\dot{z} &= -x^2 + x^4 + xz + zu + 2z(-x + x^3 + z) \\ &= -x^2 + x^4 + z[x + u + 2(-x + x^3 + z)]. \end{aligned}$$

Thus, if we take

$$u = -x - 2(-x + x^3 + z) - kz, \quad k > 0,$$

then

$$\dot{V}_c(x, z) = -x^2 + x^4 - kz^2 < 0 \quad \text{for } 0 < |x| < 1, \quad z \neq 0,$$

which shows that $(x, z) = (0, 0)$ is locally asymptotically stable. Since $z = \xi - \alpha(x)$ and $\alpha(0) = 0$, the origin $(x, \xi) = (0, 0)$ of the original cascade is locally asymptotically stable.

At the first step above, we might have chosen instead the stabilizing feedback $v = \alpha_0(x) = -2x - x^3$ which makes the origin $x = 0$ for the first equation globally asymptotically stable. Then, with $z = \xi - \alpha_0(x) = \xi + 2x + x^3$, we have the equivalent system

$$\begin{aligned} \dot{x} &= -x + z, \\ \dot{z} &= u - \frac{\partial \alpha_0}{\partial x}(x)\dot{x} = u + (2 + 3x^2)(-x + z). \end{aligned}$$

Letting $V_c(x, z) = V(x) + \frac{1}{2}z^2 = \frac{1}{2}x^2 + \frac{1}{2}z^2$ for this new system, we have

$$\dot{V}_c(x, z) = -x^2 + z[x + u + (2 + 3x^2)(-x + z)].$$

Thus, by choosing

$$u = -x - (2 + 3x^2)(-x + z) - kz, \quad k > 0,$$

we have

$$\dot{V}_c(x, z) = -x^2 - kz^2 < 0, \quad \text{for } (x, z) \neq (0, 0).$$

Thus, $(x, \xi) = (0, 0)$ is GAS. $\triangle$

In the backstepping approach, the difference $\xi - \alpha(x)$ is not required to be maintained at zero value; instead, we think of this difference as an error variable to be driven to zero by an appropriate choice of feedback. The fundamental procedure is known as *integrator backstepping.* The next lemma formulates the global version of the procedure for a cascade as in (15.25)–(15.26).

Lemma 15.2 (Integrator Backstepping)
Suppose that f and g are globally defined, smooth vector fields on $\mathbf{R}^n$. Suppose there exist a C^1 feedback control $u = \alpha(x)$, with $\alpha(0) = 0$, and a C^1 positive definite and radially unbounded function $V(x)$ such that

$$\frac{\partial V}{\partial x}(x)[f(x) + g(x)\alpha(x)] < 0 \quad \text{for } x \neq 0. \tag{15.27}$$

Then the origin $(x, \xi) = (0, 0)$ of the augmented system

$$\dot{x} = f(x) + g(x)\xi, \tag{15.28}$$

$$\dot{\xi} = u, \tag{15.29}$$

where $u \in \mathbf{R}$, is globally asymptotically stabilized by the feedback

$$u = -k(\xi - \alpha(x)) + \frac{\partial \alpha}{\partial x}(x)[f(x) + g(x)\xi] - \frac{\partial V}{\partial x}(x)g(x), \tag{15.30}$$

and the function

$$V(x) + \frac{1}{2}[\xi - \alpha(x)]^2 \tag{15.31}$$

is a Lyapunov function.

Proof. Let $z := \xi - \alpha(x)$. Rewrite (15.28)–(15.29) as

$$\dot{x} = f(x) + g(x)(\alpha(x) + z) \tag{15.32}$$

$$\dot{z} = u - \frac{\partial \alpha}{\partial x}(x)[f(x) + g(x)(\alpha(x) + z)]. \tag{15.33}$$

Using (15.27), the derivative of (15.31) along solutions of (15.32)–(15.33) satisfies the following inequality (suppressing the argument x throughout):

$$\begin{aligned}
&\frac{\partial V}{\partial x}(f + g(\alpha + z)) + z\left[u - \frac{\partial \alpha}{\partial x}(f + g(\alpha + z))\right] \\
&= \frac{\partial V}{\partial x}(f + g\alpha) + z\left[u - \frac{\partial \alpha}{\partial x}(f + g(\alpha + z)) + \frac{\partial V}{\partial x}g\right] \\
&< z\left[u - \frac{\partial \alpha}{\partial x}(f + g(\alpha + z)) + \frac{\partial V}{\partial x}g\right].
\end{aligned} \tag{15.34}$$

The right-hand side in (15.34) consists only of the z-dependent terms. The easiest choice that makes the right-hand side in (15.34) purely z-dependent and negative definite in z is given by the formula (15.30), which makes the quantity in brackets equal to $-kz$. The right side of (15.34) is then $-kz^2$. Thus the derivative of (15.31) along solutions is strictly negative for $(x, z) \neq (0, 0)$. Hence, $(x, z) = (0, 0)$ is asymptotically stable for (15.32)–(15.33), and since $\xi = \alpha(x) + z$, we conclude that $(x, \xi) = (0, 0)$ is asymptotically stable for (15.28)–(15.29) with u given by (15.30). □

Possibly the most important observation to make about Lemma 15.2 is that formula (15.30) is the easiest choice to make and there are other choices that might be better in specific problems. In some problems it may be possible to relax the assumptions of positive definiteness of $V(x)$ or negative definiteness in (15.27), provided one can choose a control $\alpha(x)$ and a feedback according to (15.30) (or some modification of it) such that asymptotic stability of $(x, \xi) = (0, 0)$ for the closed loop may be concluded, perhaps by the invariance theorem. For additional examples of backstepping see the Exercises for this chapter.

Several extensions of the basic idea of Lemma 15.2 are possible. First, consider a system having the form

$$\dot{x} = f(x) + g(x)\xi_1, \tag{15.35}$$

$$\dot{\xi}_1 = f_1(x, \xi_1) + g_1(x, \xi_1)u, \tag{15.36}$$

where $\xi_1 \in \mathbf{R}$, $g_1(0, 0) \neq 0$ and $f_1(0, 0) = 0$. Using the preliminary feedback

$$u = -\frac{f_1(x, \xi_1)}{g_1(x, \xi_1)} + \frac{1}{g_1(x, \xi_1)}v,$$

which is defined in a neighborhood of $(x, \xi_1) = (0, 0)$, we arrive at a system of the form (15.25)–(15.26) with input v instead of u. Backstepping can then be used to define v to stabilize the origin of the cascade. After a stabilizing control for system (15.35)–(15.36) has been determined this way, one can consider an augmented system (we now write ξ_2 in place of u in (15.36))

$$\dot{x} = f(x) + g(x)\xi_1, \tag{15.37}$$
$$\dot{\xi}_1 = f_1(x, \xi_1) + g_1(x, \xi_1)\xi_2, \tag{15.38}$$
$$\dot{\xi}_2 = f_2(x, \xi_1, \xi_2) + g_2(x, \xi_1, \xi_2)u, \tag{15.39}$$

where $\xi_2 \in \mathbf{R}$, $g_2(0,0,0) \neq 0$, $f_2(0,0,0) = 0$, and there is a known stabilizing control $\xi_2 = \alpha_1(x, \xi_1)$ for the first two equations with ξ_2 considered as the control. This reasoning can be extended inductively to produce a recursive feedback design for a system having the form

$$\begin{aligned}
\dot{x} &= f(x) + g(x)\xi_1, \\
\dot{\xi}_1 &= f_1(x, \xi_1) + g_1(x, \xi_1)\xi_2, \\
\dot{\xi}_2 &= f_2(x, \xi_1, \xi_2) + g_2(x, \xi_1, \xi_2)\xi_3, \\
\cdots & \quad \cdots \\
\dot{\xi}_{k-1} &= f_{k-1}(x, \xi_1, \xi_2, \ldots, \xi_{k-1}) + g_{k-1}(x, \xi_1, \xi_2, \ldots, \xi_{k-1})\xi_k, \\
\dot{\xi}_k &= f_k(x, \xi_1, \xi_2, \ldots, \xi_k) + g_k(x, \xi_1, \xi_2, \ldots, \xi_k)u,
\end{aligned} \tag{15.40}$$

where $\xi_i \in \mathbf{R}$, $f_i(x, \xi_1, \ldots, \xi_i)$ satisfies $f_i(0,0,\ldots,0) = 0$, and $g_i(x, \xi_1, \xi_2, \ldots, \xi_i)$ satisfies $g_i(0,0,0,\ldots,0) \neq 0$, for each $1 \leq i \leq k$. Systems of the form (15.40) are called *strict-feedback systems* in [65].

Suppose the feedback $v = \alpha(x)$ for the seed system $\dot{x} = f(x) + g(x)v$ is globally asymptotically stabilizing with a known global Lyapunov function, and suppose further that all the functions f_i and g_i are globally defined and, for each $1 \leq i \leq k$, $g_i(x, \xi_1, \xi_2, \ldots, \xi_i) \neq 0$ for all $(x, \xi_1, \xi_2, \ldots, \xi_i)$. Then the recursive backstepping procedure of [65] produces a globally asymptotically stabilizing feedback $u = u(x, \xi_1, \ldots, \xi_k)$ for system (15.40). Moreover, the intermediate feedbacks, such as $\alpha(x)$, $\alpha_1(x, \xi_1)$, provide the tools to construct, in a step-by-step manner, a Lyapunov function for the closed loop system.

An augmentation of the system in Example 15.11 illustrates the recursive design.

Example 15.12 Consider the system

$$\dot{x} = x + x^3 + \xi_1, \tag{15.41}$$
$$\dot{\xi}_1 = \xi_2, \tag{15.42}$$
$$\dot{\xi}_2 = u. \tag{15.43}$$

We extend the globally stablizing construction from Example 15.11. We had $\alpha_0(x) = -2x - x^3$, and, with $z_1 := \xi_1 - \alpha_0(x)$, we concluded Example 15.11 with the feedback given by

$$-x - (2 + 3x^2)(-x + z_1) - kz_1, \quad k > 0.$$

We take $k = 1$ and write this feedback as

$$\alpha_1(x, z_1) = -x - (2 + 3x^2)(-x + z_1) - z_1. \tag{15.44}$$

This is the feedback that, when substituted for ξ_2 considered as a control for (15.41)–(15.42), starts the next step of backstepping. The associated Lyapunov function is given by $V(x, z_1) = \frac{1}{2}x^2 + \frac{1}{2}z_1^2$. In order to see the structure involved, we can view the system for (x, z_1, ξ_2):

$$\begin{aligned} \dot{x} &= -x + z_1, \\ \dot{z}_1 &= \xi_2 + (2 + 3x^2)(-x + z_1), \\ \dot{\xi}_2 &= u. \end{aligned}$$

Now, write $z_2 := \xi_2 - \alpha_1(x, z_1)$, and substitute the known $\alpha_1(x, z_1)$ to get the control system in (x, z_1, z_2):

$$\dot{x} = -x + z_1, \tag{15.45}$$

$$\dot{z}_1 = -x - z_1 + z_2, \tag{15.46}$$

$$\dot{z}_2 = u - \frac{\partial \alpha_1}{\partial(x, z_1)}(x, z_1)[f(x, z_1) + g(x, z_1)(\alpha_1(x, z_1) + z_2)], \tag{15.47}$$

where the functions $f(x, z_1)$ and $g(x, z_1)$ are identified from (15.45)–(15.46) as the vector fields

$$f(x, z_1) = \begin{bmatrix} -x + z_1 \\ -x - z_1 \end{bmatrix}, \quad g(x, z_1) = \begin{bmatrix} 0 \\ 1 \end{bmatrix}.$$

We now define u in (15.47), following the formula (15.30) with $k = 1$, and we find that (suppressing some arguments (x, z_1))

$$u = -z_2 + \frac{\partial \alpha_1}{\partial(x, z_1)}[f + g(\alpha_1 + z_2)] - z_1. \tag{15.48}$$

The choice (15.48) yields the following closed loop system in (x, z_1, z_2):

$$\begin{aligned} \dot{x} &= -x + z_1, \\ \dot{z}_1 &= -x - z_1 + z_2, \\ \dot{z}_2 &= -z_1 - z_2. \end{aligned}$$

It is not difficult to see that the function

$$\frac{1}{2}x^2 + \frac{1}{2}z_1^2 + \frac{1}{2}z_2^2$$

is a globally defined Lyapunov function. △

Recursive backstepping design is not limited to the strict-feedback systems (15.40). The many interesting and important variations described in [65] allow larger classes of systems to be addressed. The backstepping approach allows flexibility in feedback design, depending on problem specifics. For an interesting design example which illustrates the basic tools and shows the ingenuity and flexibility required in specific problems, see the discussion of jet engine stall and surge in [65] (pages 66–72).

15.5 EXERCISES

Exercise 15.1 Show that the system

$$\begin{aligned} \dot{x} &= -x^3 + x^3\xi_2, \\ \dot{\xi}_1 &= \xi_2, \\ \dot{\xi}_2 &= u \end{aligned}$$

is globally asymptotically stabilizable by full-state feedback.

Exercise 15.2 Consider the system

$$\begin{aligned} \dot{x} &= -x^3 + x^3\xi_2, \\ \dot{\xi}_1 &= 0, \\ \dot{\xi}_2 &= \xi_1 + u. \end{aligned}$$

(a) Verify that the origin of the unforced system is unstable.
(b) Show that the origin can be made Lyapunov stable by smooth feedback, but there is no state feedback control that makes it asymptotically stable.
(c) Note that the driving system with output $y = \xi_2$ is feedback passive (Example 14.7). Why does Proposition 15.1 not apply in this case?

Exercise 15.3 Verify that the proof of Theorem 15.1 adapts to provide a proof of the more general Theorem 15.2.

Exercise 15.4 Consider the system

$$\begin{aligned} \dot{x} &= -x^3 + (\xi_1 + \xi_2)x^3 - \xi_1^2 x^3, \\ \dot{\xi}_1 &= \xi_2, \\ \dot{\xi}_2 &= u. \end{aligned}$$

Show that $(x, \xi_1, \xi_2) = (0, 0, 0)$ is globally asymptotically stabilizable by smooth state feedback, and find a stabilizing feedback.

Exercise 15.5 Consider the system

$$\begin{aligned} \dot{x} &= -x^3 + (\xi_2 + \xi_3)x^3 - \xi_2^2 x^3, \\ \dot{\xi}_1 &= \xi_2, \\ \dot{\xi}_2 &= \xi_3, \\ \dot{\xi}_3 &= u. \end{aligned}$$

Determine whether the origin $(x, \xi_1, \xi_2, \xi_3) = (0, 0, 0, 0)$ is locally or globally asymptotically stabilizable by smooth state feedback, and find a stabilizing feedback.

Exercise 15.6 Apply the feedback construction of Corollary 15.1 to the system

$$\begin{aligned} \dot{x} &= -x + x^2(\xi_1 + \xi_2), \\ \dot{\xi}_1 &= \xi_2 + u_2, \\ \dot{\xi}_2 &= \xi_1 + u_1 + u_2, \end{aligned}$$

and thus find a feedback $u(x, \xi_1, \xi_2) = (u_1(x, \xi_1, \xi_2),\ u_2(x, \xi_1, \xi_2))$ that globally asymptotically stabilizes the origin $(x, \xi_1, \xi_2) = (0, 0, 0)$.

Exercise 15.7 Consider the system

$$\begin{aligned} \dot{x} &= -x + x^2\xi_2, \\ \dot{\xi}_1 &= \xi_2, \\ \dot{\xi}_2 &= \xi_1 + u. \end{aligned}$$

Find a feedback $u(x, \xi_1, \xi_2)$ that globally asymptotically stabilizes the origin $(x, \xi_1, \xi_2) = (0, 0, 0)$.

Exercise 15.8 Prove Theorem 15.3.

Exercise 15.9 Consider the system

$$\begin{aligned} \dot{x} &= -x^3 + \xi_1\xi_3 x^3, \\ \dot{\xi}_1 &= \xi_2, \\ \dot{\xi}_2 &= \xi_3, \\ \dot{\xi}_3 &= u. \end{aligned}$$

(a) Show that neither of the output choices $y = \xi_1$, $y = \xi_3$ yields a feedback passive driving system.

(b) Rewrite the first equation as

$$\dot{x} = -x^3 - \xi_1^2 x^3 + \xi_1(\xi_1 + \xi_3)x^3$$

and let $y = \xi_1 + \xi_3$. Show that the linear driving system with output y is feedback passive.

(c) Show that the system is globally asymptotically stabilizable by smooth state feedback.

Exercise 15.10 *Using Backstepping and Zero Dynamics Analysis*
Consider the system

$$\begin{aligned}\dot{x}_1 &= x_1^2 x_2,\\ \dot{x}_2 &= u.\end{aligned}$$

Examine both backstepping and a zero dynamics analysis to determine an asymptotically stabilizing feedback control.

Exercise 15.11 *Backstepping Again*
Use backstepping to design an asymptotically stabilizing feedback for the system

$$\begin{aligned}\dot{x} &= x^3 + x\xi,\\ \dot{\xi} &= u.\end{aligned}$$

Exercise 15.12 *Backstepping Using Existing Stabilizing Terms*
Use the backstepping approach to asymptotically stabilize the origin of the nonlinear system

$$\begin{aligned}\dot{x}_1 &= x_2 - x_1^5\\ \dot{x}_2 &= u\end{aligned}$$

without cancellation of the stabilizing term $-x_1^5$.

Exercise 15.13 *More Backstepping*
The planar system

$$\begin{aligned}\dot{x} &= x^2\xi,\\ \dot{\xi} &= u\end{aligned}$$

has uncontrollable linearization at the origin, so it is not feedback linearizable. Find a globally stabilizing feedback control, using both feedback passivation and backstepping. In particular, show that (15.30) can be used with $\alpha(x) = 0$, even though this choice does not asymptotically stabilize $x = 0$ when substituted for ξ in the first equation.

15.6 NOTES AND REFERENCES

The primary references for this chapter are the paper [61] and the monograph [86] (Chapter 4). The main theorem of [61] (Theorem 2.1) appears here as Theorem 15.2. The results of Corollaries 15.1, 15.2, 15.3 also appear in [61], but they were obtained earlier; see the references in [61].

The discussion of the interconnection term follows [86] (Chapter 4) and [48] (Chapter 10). The full analysis of the peaking phenomenon (illustrated in Example 15.5) is complex; detailed analyses of the peaking phenomenon in nonlinear systems appear in the monograph [86] (Section 4.5) and in the earlier paper [97]. The discussion in Example 15.6, showing that a basin can shrink in certain directions under high gain feedback, is adapted from [86] (Example 6.4, page 244).

The motivation for the passivity condition on the linear subsystem triple (C, A, B) in Proposition 15.1 and Theorem 15.1 was the desire to get a direct sum Lyapunov function, $W(x, \xi) = V(x) + \xi^T P \xi$; this approach appears in [83]. It should be noted that for more general cascade systems, a composite Lyapunov function may require a cross term which depends on both the driving states and the driven states; see [86] (Chapter 5).

Theorem 15.3 is taken from [86] (page 139). For more on stabilization by feedback passivation for cascades with nonlinear driving system, see [19] and [86] (Chapter 4).

The integrator backstepping Lemma 15.2 follows [65], where many additional examples will be found. The Bode lecture [60] has an interesting overview of backstepping. There is also a backstepping lemma for a seed system having the form $\dot{x} = f(x, \xi)$, which is nonaffine in ξ; see [91] (page 242).

Backstepping is one of the foundations of the recursive feedback designs for stabilization described in [86]; this reference includes discussion of the connections between feedback passivation and backstepping.

Chapter Sixteen

Input-to-State Stability

This chapter is an introduction to the concept of *input-to-state stability* (ISS) and its importance for problems of stability and stabilization. The ISS concept was introduced in [89], and it has proven to be a very effective tool in the study of stabilization problems, as shown for example in [65] and [48]. The ISS concept also provides an extension to control systems of some classical Lyapunov theorems for uncontrolled systems, namely, the direct and converse Lyapunov theorems on asymptotic stability.

The presentation in this chapter concentrates on the definition of input-to-state stability, the use of ISS Lyapunov functions to determine input-to-state stability, and applications to cascade systems. The intention is to provide some motivation for further study of this important concept.

This chapter uses the language of comparison functions from Appendix E on **Comparison Functions and a Comparison Lemma**.

16.1 PRELIMINARIES AND PERSPECTIVE

We begin by recalling the issue of boundedness for the driven trajectories in a cascade system (Theorem 9.2). The input-to-state stability concept provides a condition for boundedness of the driven trajectories that does not depend on the linearization of the driven system. However, we can provide some motivation for the ISS definition by first considering a few facts about Hurwitz linear systems.

Estimates for a Hurwitz Linear System

Consider the linear system

$$\dot{x} = Ax + Bu,$$

where the admissible inputs $u(t)$ are piecewise continuous bounded functions for $t \geq 0$. The variation of parameters formula gives the general solution

$$x(t) = e^{tA}x(0) + \int_0^t e^{(t-s)A}Bu(s)\,ds. \tag{16.1}$$

Now suppose that A is Hurwitz. Then there exist real constants $\alpha > 0$ and $M > 0$ such that

$$\|e^{(t-s)A}\| \leq Me^{-\alpha(t-s)} \quad \text{for } t \geq s.$$

Using this norm estimate of $e^{(t-s)A}$ together with straightforward norm estimates applied to the variation of parameters formula (16.1), we obtain (Exercise 16.1)

$$\|x(t)\| \leq Me^{-\alpha t}\|x(0)\| + \frac{M\|B\|}{\alpha} \sup_{0 \leq s \leq t} \|u(s)\| \quad \text{for } t \geq 0. \tag{16.2}$$

Most discussions of input-to-state stability use the language of comparison functions. In terms of comparison functions, the estimate (16.2) can be expressed as

$$\|x(t)\| \leq \beta(\|x(0)\|, t) + \gamma\left(\sup_{0 \leq s \leq t} \|u(s)\|\right) \tag{16.3}$$

where β is of class $\mathcal{KL}$ and γ is of class $\mathcal{K}$. (See Definition E.2 and Definition E.1, respectively, for this terminology.) In this instance, from (16.2) we have

$$\beta(r, s) = Mre^{-\alpha s}$$

and

$$\gamma(b) = \frac{M\|B\|}{\alpha} b.$$

Sometimes we want to consider a different initial time t_0. If $u(t)$ is a real- or vector-valued function bounded on $[t_0, \infty)$, with the initial time t_0 given, then it is straightforward to show that we may write (using the same functions β and γ)

$$\|x(t)\| \leq \beta(\|x(t_0)\|, t - t_0) + \gamma\left(\sup_{t_0 \leq t} \|u(t)\|\right) \quad \text{for } t \geq t_0, \tag{16.4}$$

where $x(t)$ is the unique solution of $\dot{x} = Ax + Bu(t)$ with initial condition $x(t_0) = x_0$.

There are several important consequences of the estimate (16.4); for later reference, we list them here:

(a) When $u = 0$, $\|x(t)\| \to 0$ as $t \to \infty$. This is immediate from the first bounding term on the right in (16.4), which is monotone decreasing to zero for any given $x(t_0)$.

(b) For any initial condition $x(t_0)$ and any input $u(t)$ bounded on $[t_0, \infty)$, the norm $\|x(t)\|$ is bounded on $[t_0, \infty)$ as well. This property is called *bounded-input bounded-state (BIBS) stability*. Moreover, for a given bounded input $u(t)$, there is an ultimate bound (as $t \to \infty$) on $\|x(t)\|$ which is no greater than a class $\mathcal{K}$ function of

$$\sup_{t \geq t_0} \|u(t)\|.$$

This estimate for an ultimate bound for a solution $x(t)$ is more revealing about asymptotic behavior than $\sup_{t\geq 0}\|x(t)\|$, which takes into account early transient peaks in the solution; such transient peaks are embedded in the constant M in (16.2).

(c) If $u(t)\to 0$ as $t\to\infty$, then we call $u(t)$ a *converging input*. If $u(t)$ is a converging input, then for any $t_0 > 0$ and any initial condition $x(t_0)$, the solution $x(t)$ is also converging; that is, $x(t)\to 0$ as $t\to\infty$. This property is called the *converging-input converging-state* (CICS) property implied by (16.4).

We would like a condition on a nonlinear system that guarantees properties (a)–(c), but with no assumptions on the linearization. The key is the type of estimate indicated in (16.4). First, we show that properties (a)–(c) are not generally to be expected for nonlinear control systems, and then we show how an estimate such as (16.4) can be obtained for a simple nonlinear system for which the linearization at $x = 0$ is not Hurwitz.

Review of a Nonlinear Example

We consider nonlinear systems of the form

$$\dot{x} = f(x) + g(x)u. \tag{16.5}$$

However, to begin, it will be helpful to recall previous experience with a planar cascade system.

Example 16.1 Recall the planar system

$$\begin{aligned}\dot{x} &= -x + x^2 z,\\ \dot{z} &= -z\end{aligned}$$

from Example 9.3. We think of z as the input to the x equation which has the form (16.5). We first show that properties (b), (c) do not hold in general for the x equation with input z, even if (a) holds. Note that $x = 0$ is a globally asymptotically stable equilibrium of $\dot{x} = -x$.

In the equation $\dot{x} = -x + x^2 u$, the origin is GAS when $u = 0$. Let $u \equiv 1$, and take any $x(0) = x_0$ such that $x_0 > 2$. The corresponding solution $x(t)$ has a finite escape time. For instance, if $x_0 = 3$, the solution can be verified to be

$$x(t) = \frac{-3}{2e^t - 3},$$

and $|x(t)| \to \infty$ as $t \to \ln\frac{3}{2}$. Thus, a bounded input u can yield an unbounded solution $x(t)$, including solutions with finite escape time. Thus, the BIBS stability and boundedness properties of item (b) above do not hold in general for nonlinear systems.

The same equation also serves to show that converging inputs do not necessarily imply converging solution trajectories. Consider the effect of converging inputs u in the equation $\dot{x} = -x + x^2u$. Take any initial values with $x(0)u(0) = x_0u_0 > 2$, and let $u(t) = u_0e^{-t}$. Then $x(t)$ becomes unbounded in finite time; in fact, $|x(t)| \to \infty$ as $t \to \frac{1}{2}\ln\frac{x_0u_0}{x_0u_0-2}$. (Compare with Exercise 9.5.) △

In this discussion, we have used previous knowledge of a particular cascade only to make the counterexample to the global properties (b), (c) more easily transparent. The BIBS and CICS properties do not hold in a general global sense for (16.5), for *all* bounded inputs and *all* converging inputs, independently of initial conditions for x.

If the linearization of (16.5) is not Hurwitz, then the variation of parameters formula generally will not yield useful bounds that would establish long-term bounds for $\|x(t)\|$ or show the converging behavior of $\|x(t)\|$, if present. To see this, let us write

$$f(x) = \frac{\partial f}{\partial x}(0)x + f_2(x) =: Ax + f_2(x), \text{ where } \frac{\partial f_2}{\partial x}(0) = 0.$$

If $x(t)$ is a solution of (16.5) for a given input $u(\cdot)$, then we can use variation of parameters to write the equivalent integral equation

$$x(t) = e^{tA}x(0) + \int_0^t e^{(t-s)A}[f_2(x(s)) + g(x(s))u(s)]\,ds.$$

However, if A is not Hurwitz, then generally we cannot get useful bounds from this, even if we knew bounds on $\|f_2(x)\|$ and $\|g(x)\|$. The following scalar example may illustrate the situation more concretely.

Example 16.2 Consider $\dot{x} = -x^3 + e^{-t}$. The linearization of the unforced equation at $x = 0$ is $\dot{x} = 0$, and it provides no help in determining boundedness properties of solutions. The equivalent integral equation for the forced equation is

$$x(t) = x(t_0) + \int_{t_0}^t [-x(s)^3 + e^{-s}]\,ds.$$

It is possible to show that solutions $x(t)$ remain bounded for $t \geq t_0$. (We will do this later on below.) However, the integral formula itself cannot provide an estimate for a bound on $|x(t)|$ as $t \to \infty$, even if the fact of boundedness were known. Also, the integral formula cannot help to decide whether every solution of $\dot{x} = -x^3 + e^{-t}$ satisfies $x(t) \to 0$ as $t \to \infty$. (This property does indeed hold.) △

Obtaining Bounds for a Nonlinear System

We now show how Lyapunov analysis can be used to determine specific bounds for solutions of

$$\dot{x} = f(t, x), \tag{16.6}$$

where $f : [0, \infty) \times D \to \mathbf{R}^n$, $D \subset \mathbf{R}^n$ is an open set, and f is piecewise continuous in t and locally Lipschitz in x.

Example 16.3 Consider the scalar system $\dot{x} = -x^3 + b$, where b is a real constant. Because of the cubic term, we expect that solutions will have a decrease in $|x|$ for large x. Let t_0 be a fixed real number; we will consider the rate of change of $V(x) = \frac{1}{2}x^2$ along solutions $x(t)$ for $t \ge t_0$. We have

$$\dot{V}(x) = \frac{d}{dt}V(x) = x\dot{x} = -x^4 + xb.$$

If $|x| > |b|^{\frac{1}{3}}$, then $|x|^3 > |b|$, hence $x^4 = |x|^4 > |x||b| > xb$, and we have $\dot{V}(x) < 0$. Therefore all solutions remain bounded for $t \ge t_0$, and thus all solutions are defined on $[t_0, \infty)$. It is useful to have a negative definite upper bound for $\dot{V}(x)$ for sufficiently large $|x|$. That is, we want to have $\dot{V}(x) \le -W(x)$ for sufficiently large $|x|$, where $W(x)$ is a class $\mathcal{K}$ function. For this purpose, we can use part of the negative definite term $-x^4$ to identify such an upper bound. Letting $0 < \theta < 1$, we may write

$$\begin{aligned} \frac{d}{dt}V(x) &= -(1-\theta)x^4 - \theta x^4 + xb, \\ &\le -(1-\theta)x^4, \end{aligned} \tag{16.7}$$

provided we have $-\theta x^4 + xb \le 0$, and the latter condition is guaranteed if

$$|x| > \left(\frac{|b|}{\theta}\right)^{\frac{1}{3}}.$$

For x satisfying the latter inequality, we have the desired bound on $\dot{V}(x)$ with $W(x) = (1-\theta)x^4$. Now take $x(t_0)$ with $V(x(t_0)) = c$ and $|x(t_0)| = \sqrt{2c} > (\frac{|b|}{\theta})^{\frac{1}{3}}$. Choose $0 < \epsilon < c$ such that $\sqrt{2\epsilon} > (\frac{|b|}{\theta})^{\frac{1}{3}}$. Then the set $\Omega_{[\epsilon,c]} = \{x : \epsilon \le V(x) \le c\}$ is contained in the set where (16.7) holds. We now show that the solution with initial condition $x(t_0)$ reaches the set $\Omega_\epsilon = \{x : V(x) \le \epsilon\}$ in finite time independent of t_0. To see this, let

$$M = \min_{x \in \Omega_{[\epsilon,c]}} W(x) > 0;$$

the minimum exists and is positive since $W(x)$ is continuous and positive on the compact set $\Omega_{[\epsilon,c]}$. Thus, for as long as $x(t)$ is in $\Omega_{[\epsilon,c]}$, we have

$$\frac{d}{dt}V(x(t)) \le -M.$$

Integration from t_0 to t gives

$$V(x(t)) \le V(x(t_0)) - M(t - t_0) \le c - M(t - t_0) \quad \text{for } x(t) \in \Omega_{[\epsilon,c]}.$$

Consequently, $V(x(t))$ must decrease to a value of ϵ within the time interval $[t_0, t_0 + \frac{c-\epsilon}{M}]$, and the solution $x(t)$ will remain within the set Ω_ϵ for all $t \ge t_0 + \frac{c-\epsilon}{M}$. In fact, the argument just given shows that for any initial condition with $|x(t_0)| \le \sqrt{2c}$, the solution $x(t)$ will satisfy $|x(t)| \le \sqrt{2\epsilon}$ for all $t \ge t_0 + \frac{c-\epsilon}{M}$, and the finite time to reach this bound depends only on c and ϵ, and not on t_0. Moreover, since the system is defined for all real x, the argument above applies to arbitrarily large initial conditions which belong to the set $\Omega_c = \{x : V(x) \le c\}$ for some number $c > 0$. △

In the next example we allow for time dependence in the differential equation and study the effect of a converging input.

Example 16.4 Consider the equation $\dot{x} = -x^3 + e^{-t}$. Note that the input term e^{-t} has upper bound $b = e^{-t_0}$ for $t \ge t_0$, and by choosing $t_0 > 0$ sufficiently large we can have b as small positive as we wish. As in Example 16.3, it is not difficult to show that all solutions exist on the interval $[0, \infty)$. It is a recommended exercise to repeat the analysis of Example 16.3 in order to show the following: If $|x(t_0)| = \sqrt{2c}$ and $0 < \epsilon < c$ with $\sqrt{2\epsilon} > (\frac{|b|}{\theta})^{\frac{1}{3}}$, then

$$\frac{d}{dt}V(x(t)) \le -M := -\min_{x \in \Omega_{[\epsilon,c]}} W(x),$$

where $W(x) = (1-\theta)x^4$. Moreover, the solution $x(t)$ reaches the set Ω_ϵ within the finite time $T = \frac{c-\epsilon}{M}$. Noting that $\epsilon > 0$ is arbitrary, subject only to $c > \epsilon > \frac{1}{2}(\frac{|b|}{\theta})^{\frac{2}{3}}$, we can also state the following: Given any $\delta > 0$, there exists a finite time $T = T(c, \delta)$ such that any solution $x(t)$ with $|x(t_0)| \le \sqrt{2c}$ satisfies

$$|x(t)| \le \left(\frac{|b|}{\theta}\right)^{\frac{1}{3}} + \delta \quad \text{for all } t \ge t_0 + T(c, \delta).$$

Of course, we are thinking of δ subject to the restriction that $(\frac{|b|}{\theta})^{\frac{1}{3}} + \delta < \sqrt{2c}$; otherwise, we may take $T(c, \delta) = 0$.

Observe now that the number $b = \sup_{t \ge t_0} e^{-t}$ may be taken arbitrarily small by appropriate choice of t_0. Moreover, the system is defined for all real x, so the argument above can be applied to arbitrarily large initial conditions $|x(t_0)|$ by taking c sufficiently large. As an exercise one can now argue that every solution of $\dot{x} = -x^3 + e^{-t}$ must satisfy $x(t) \to 0$ as $t \to \infty$. △

16.2 STABILITY THEOREMS VIA COMPARISON FUNCTIONS

In the analysis of Examples 16.3–16.4, we translated between values of V and values of $\|x\|$ as needed. In general, these translations are accomplished systematically by means of comparison functions. Suppose $V(x)$ is continuous and positive definite in an open set D containing the origin, and $B_d(0) = \{x : \|x\| < d\} \subset D$. By Lemma E.2, there exist class $\mathcal{K}$ functions α_1 and α_2 such that

$$\alpha_1(\|x\|) \le V(x) \le \alpha_2(\|x\|) \quad \text{for } x \in D. \tag{16.8}$$

When $V(x) = \frac{1}{2}\|x\|^2$, we may take $\alpha_i(r) = \frac{1}{2}r^2$ for $i = 1, 2$. We now illustrate the role of comparison functions in proving the basic Lyapunov theorems on stability and asymptotic stability for autonomous systems

$$\dot{x} = f(x), \tag{16.9}$$

where x is in $\mathbf{R}^n$, $f(0) = 0$, and f is Lipschitz in x on an open set containing the origin.

The next theorem is a restatement of Theorem 8.1 in the language of comparison functions, including a global result stated here in part (c).

Theorem 16.1 (Stability Theorems via Comparison Functions)
Let $V : B_d(0) \to \mathbf{R}$ be a C^1 positive definite function, and let $\alpha_1(\cdot)$, $\alpha_2(\cdot)$ be class $\mathcal{K}$ functions defined on $[0, d)$ such that

$$\alpha_1(\|x\|) \le V(x) \le \alpha_2(\|x\|) \tag{16.10}$$

for $\|x\| < d$. Then the following results hold.

(a) *If*
$$\frac{\partial V}{\partial x}(x) f(x) \le 0 \quad \text{for all } \|x\| < d, \tag{16.11}$$
then the equilibrium $x = 0$ of (16.9) is stable.

(b) *If there is a class $\mathcal{K}$ function $\alpha_3(\cdot)$ defined on $[0, d)$ such that*
$$\frac{\partial V(x)}{\partial x} f(x) \le -\alpha_3(\|x\|) \quad \text{for all } \|x\| < d, \tag{16.12}$$
then the equilibrium $x = 0$ of (16.9) is asymptotically stable.

(c) *If (16.12) holds with $d = \infty$ and the lower bound $\alpha_1(\cdot)$ in (16.10) is a class $\mathcal{K}_\infty$ function, then the equilibrium $x = 0$ of (16.9) is globally asymptotically stable.*

Proof. (a) If (16.11) holds, and $\|x(0)\| < d$, then as long as the solution $x(t)$ is defined, we have

$$V(x(t)) \le V(x(0)).$$

Let $\epsilon > 0$, and choose $\delta = \alpha_2^{-1}(\alpha_1(\epsilon))$. Observe that, by (16.10), if $\|x(0)\| < \delta$, then for as long as the solution is defined, we have

$$\alpha_1(\|x(t)\|) \le V(x(t)) \le V(x(0)) \le \alpha_2(\|x(0)\|) \le \alpha_2(\delta) = \alpha_1(\epsilon).$$

Since α_1 is an increasing function, $\|x(t)\| \le \epsilon$. Since the solution remains bounded, it exists for all $t \ge 0$. Thus, Lyapunov stability is proved.

(b) Let $\|x_0\| < d$ and let $x(t)$ be the solution with $x(0) = x_0$. Write $V(t) := V(x(t))$. Using (16.10) and (16.12), we have

$$\frac{d}{dt}V(t) \le -\alpha_3\big(\alpha_2^{-1}(V(t))\big) =: -\alpha(V(t)),$$

where we have set $\alpha(\cdot) = \alpha_3 \circ \alpha_2^{-1}(\cdot)$. If we knew that the class $\mathcal{K}$ function $\alpha(\cdot)$ was locally Lipschitz, then we would be in a position to invoke

Lemma E.5. However, we may always replace $\alpha(\cdot)$, if needed, by a locally Lipschitz class $\mathcal{K}$ function $\bar{\alpha}(\cdot)$ that satisfies $\alpha(r) \geq \bar{\alpha}(r)$. So without loss of generality, we simply assume that $\alpha(\cdot)$ is locally Lipschitz. Lemma E.5 then implies that the solution of

$$\dot{y} = -\alpha(y)$$

with initial condition $y(0) = V(0) = V(x(0))$ satisfies

$$y(t) = \beta(V(0), t)$$

for some class $\mathcal{KL}$ function $\beta(\cdot,\cdot)$. Then, by the comparison lemma (Lemma E.4),

$$V(x(t)) = V(t) \leq \beta(V(0), t),$$

and by (16.10) we conclude that

$$\|x(t)\| \leq \alpha_1^{-1}\big(\beta(V(x(0)), t)\big) \leq \alpha_1^{-1}\big(\beta(\alpha_2(\|x_0\|), t)\big).$$

By Lemma E.1, the right-hand side of the last inequality is a class $\mathcal{KL}$ function of $(\|x_0\|, t)$; hence, the stability and attractivity of the equilibrium $x = 0$ follow immediately.

(c) Under the assumptions of part (c), the system is globally defined and the function $V(x)$ is radially unbounded. Thus the arguments of part (b) apply to arbitrary initial conditions $x(0) = x_0$ in $\mathbf{R}^n$.

This completes the proof of Theorem 16.1. □

Converse Lyapunov results, for example Theorem 8.7, may also be formulated in the language of comparison functions. See Exercise 16.11.

16.3 INPUT-TO-STATE STABILITY

We consider control systems of the form

$$\dot{x} = f(x, u), \tag{16.13}$$

where $f : \mathbf{R}^n \times \mathbf{R}^m \to \mathbf{R}^n$ is locally Lipschitz in x on $\mathbf{R}^n \times \mathbf{R}^m$ and $f(0,0) = 0$. The admissible input functions $u(\cdot)$ are elements of the set L_∞^m defined by

$$L_\infty^m := \big\{u : [0,\infty) \to \mathbf{R}^m \; : \; u(\cdot) \text{ is piecewise continuous and bounded}\big\}.$$

The set L_∞^m is a normed vector space with norm given by

$$\|u(\cdot)\|_\infty := \sup_{t \geq 0} \|u(t)\|.$$

Definition 16.1 (Input-to-State Stability)
The system (16.13) *is* input-to-state stable (ISS) *if there exist a class $\mathcal{KL}$ function $\beta(\cdot,\cdot)$ and a class $\mathcal{K}$ function $\gamma(\cdot)$ such that, for any input $u(\cdot) \in L_\infty^m$ and any x_0 in $\mathbf{R}^n$, the solution $x(t)$ with initial condition $x_0 = x(0)$ exists*

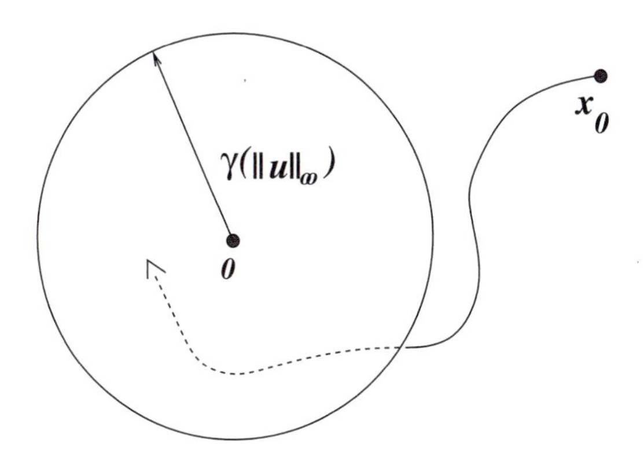

Figure 16.1 Input-to-state stability—Definition 16.1.

for all $t \geq 0$ *and satisfies*

$$\|x(t)\| \leq \beta(\|x_0\|, t) + \gamma(\|u(\cdot)\|_\infty) \quad \textit{for all } t \geq 0. \tag{16.14}$$

The class $\mathcal{K}$ function γ in (16.14) is called an *ISS gain function.* (See Figure 16.1.) Note that, for any $t \geq 0$, the solution value $x(t)$ depends on the input values $u(s)$ only for $0 \leq s \leq t$; thus, the second term on the right side of (16.14) can be replaced by

$$\gamma\left(\sup_{0 \leq s \leq t} \|u(s)\|\right).$$

Also, an equivalent concept results if the estimate (16.14) is replaced by

$$\|x(t)\| \leq \max\left\{\beta(\|x_0\|, t), \gamma(\|u(\cdot)\|_\infty)\right\} \quad \text{for all } t \geq 0, \tag{16.15}$$

for some class $\mathcal{KL}$ function $\beta(\cdot, \cdot)$ and class $\mathcal{K}$ function $\gamma(\cdot)$.

It is immediate from the estimate (16.14) of Definition 16.1 that an ISS system has the bounded-input bounded-state (BIBS) property. In addition, by setting $u = 0$ in (16.14), we have

$$\|x(t)\| \leq \beta(\|x_0\|, t) \quad \text{for all } t \geq 0$$

for any solution $x(t)$ of the unforced equation $\dot{x} = f(x, 0)$. Since the class $\mathcal{KL}$ function $\beta(\cdot, \cdot)$ is strictly decreasing in the second variable when the first variable is fixed, the origin of the unforced system is globally asymptotically stable. It can also be shown that an ISS system has the converging-input converging-state (CICS) property; see Exercise 16.2.

The next example shows that global asymptotic stability of the origin of $\dot{x} = f(x)$ does not imply input-to-state stability.

Example 16.5 The scalar system $\dot{x} = -x - x^3 u$, defined for $x \in \mathbf{R}$, is not input-to-state stable. For example, consider the bounded input $u = -1$ and initial conditions $|x_0| > 1$; the corresponding solutions satisfy the equation $\dot{x} = -x + x^3 = -x(1 - x^2)$ and therefore $|x(t)|$ grows without bound as t

increases. Similar comments apply to the scalar system $\dot{x} = -x + x^3 v$ with input v. △

The next definition highlights the features that were present in the function $V(x)$ used in Examples 16.3–16.4.

Definition 16.2 (ISS Lyapunov Function)
A C^1 function $V : \mathbf{R}^n \to \mathbf{R}$ is called an ISS Lyapunov function *for system* (16.13) *if there exist class $\mathcal{K}_\infty$ functions α_1, α_2, α_3, and a class $\mathcal{K}$ function χ such that*

$$\alpha_1(\|x\|) \le V(x) \le \alpha_2(\|x\|) \quad \textit{for all } x \in \mathbf{R}^n \tag{16.16}$$

and

$$\|x\| \ge \chi(\|u\|) \Longrightarrow \frac{\partial V}{\partial x}(x) f(x,u) \le -\alpha_3(\|x\|). \tag{16.17}$$

It should be emphasized that Definition 16.1 specifies a global property and Definition 16.2 specifies a globally defined ISS Lyapunov function.

The next result says that the existence of an ISS Lyapunov function for a system of the form (16.13) implies the ISS property for that system.

Theorem 16.2 *If $V : \mathbf{R}^n \to \mathbf{R}$ is an ISS Lyapunov function for system* (16.13)*, and* (16.16)–(16.17) *hold for appropriate bounding functions with notation as in Definition 16.2, then* (16.13) *is input-to-state stable with ISS gain function given by $\gamma = \alpha_1^{-1} \circ \alpha_2 \circ \chi$.*

Proof. Suppose $V(x)$ is an ISS Lyapunov function for (16.13), and that (16.16)–(16.17) hold for appropriate bounding functions with notation as in Definition 16.2. Setting $u = 0$ in (16.17) yields the inequality

$$\frac{\partial V}{\partial x}(x) f(x,0) \le -\alpha_3(\|x\|) \quad \text{for all } x \in \mathbf{R}^n.$$

Therefore $V(x)$ is a globally defined Lyapunov function in the usual sense for the system $\dot{x} = f(x,0)$. Then, as in the proof of Theorem 16.1, there exists a class $\mathcal{KL}$ function $\beta(\cdot,\cdot)$, which depends only on the composition $\alpha := \alpha_3 \circ \alpha_2^{-1}$, such that, for any initial state $x(0) = x_0$, the solution of $\dot{x} = f(x,0)$ satisfies

$$\|x(t)\| \le \beta(\|x_0\|, t) \quad \text{for all } t \ge 0. \tag{16.18}$$

Now let $u(\cdot) \in L_\infty^m$ be a nonzero input function. Since $u(\cdot)$ is bounded, we may write

$$M := \|u(\cdot)\|_\infty < \infty.$$

Let

$$c := \alpha_2(\chi(M)),$$

and observe that, by the inequality on the right-hand side of (16.16), the set

$$\Omega_c := \{x : V(x) \le c\}$$

contains the ball $B_{\chi(M)}$; that is,

$$B_{\chi(M)} \subseteq \Omega_c.$$

Thus, we have $\|x\| \geq \chi(M)$ for each point x on the boundary of Ω_c. Therefore, at any $t \geq 0$ for which the point $x(t)$ of a solution of $\dot{x} = f(x, u(t))$ is on the boundary of Ω_c, we have $\|x(t)\| \geq \chi(M) \geq \chi(\|u(t)\|)$. Thus, by (16.17), we have

$$\frac{\partial V}{\partial x}(x(t))f(x(t), u(t)) \leq -\alpha_3(\|x(t)\|) < 0$$

at any $t \geq 0$ for which the point $x(t)$ is on the boundary of Ω_c. Furthermore, for the same input $u(\cdot)$, and any initial condition $\bar{x}(0)$ in Ω_c, the solution $\bar{x}(t)$ of (16.13) is defined for all $t \geq 0$ and remains in Ω_c for all $t \geq 0$. We may estimate such a solution $\bar{x}(\cdot)$ in forward time as follows. By the left-hand inequality of (16.16), we have

$$\|\bar{x}(t)\| \leq \alpha_1^{-1}(V(\bar{x}(t)) \leq \alpha_1^{-1}(c) = \alpha_1^{-1}(\alpha_2(\chi(M))) \quad \text{for all } t \geq 0.$$

Now write

$$\gamma(r) := \alpha_1^{-1} \circ \alpha_2 \circ \chi(r),$$

and observe that for all $t \geq 0$, $\bar{x}(t)$ satisfies

$$\|\bar{x}(t)\| \leq \gamma(\|u(\cdot)\|_\infty).$$

Consequently, for any forward solution $\bar{x}(t)$ of $\dot{x} = f(x, u(t))$ with initial condition $\bar{x}(0)$ in Ω_c, we have $\|\bar{x}(\cdot)\|_\infty \leq \gamma(\|u(\cdot)\|_\infty)$. Thus, for these solutions, the estimate (16.14) of the ISS definition certainly holds, with the class $\mathcal{KL}$ function $\beta(\cdot, \cdot)$ being the one in (16.18), which exists by virtue of the GAS property for the unforced system $\dot{x} = f(x, 0)$.

Now suppose that the initial condition $x(0) = x_0$ satisfies $V(x_0) > c$, and let $x(t)$ be the corresponding solution of $\dot{x} = f(x, u(t))$. As long as $V(x(t)) > c$, we have $\|x(t)\| \geq \chi(\|u(t)\|)$, and therefore by (16.17) we have

$$\frac{d}{dt}V(x(t)) = \frac{\partial V}{\partial x}(x(t))\, f(x(t), u(t)) \leq -\alpha_3(\|x(t)\|) < 0.$$

Thus, as long as $V(x(t)) > c$, the function $V(x(t))$ is decreasing, which implies that $\|x(t)\|$ remains bounded in forward time, in fact, $\|x(t)\| \leq \alpha_1^{-1}(V(x_0))$, and therefore $x(t)$ is defined for all $t \geq 0$. Moreover, the solution $x(t)$ must reach the set where $V(x) = c$ in finite time. To see this, let

$$\mathcal{A} := \{x \,:\, c \leq V(x) \leq V(x_0)\},$$

and note that $\mathcal{A}$ is a compact set. ($\mathcal{A}$ is closed since $V(x)$ is continuous, and $\mathcal{A}$ is bounded because of the estimate (16.16), which uses class $\mathcal{K}_\infty$ bounding functions; that is, $V(x)$ is radially unbounded.) The function $V(x(t))$ will decrease while the solution remains in $\mathcal{A}$. The function $\alpha_3(\|x\|)$ is positive definite and continuous, so it has a minimum value in the compact set $\mathcal{A}$, and this minimum value must be positive. Thus we may write

$$0 < k := \min_{x \in \mathcal{A}} \alpha_3(\|x\|).$$

It follows that $-\alpha_3(\|x\|) \le -k$ for all $x \in \mathcal{A}$, and hence that

$$\frac{\partial V}{\partial x}(x(t))\, f(x(t), u(t)) \le -\alpha_3(\|x\|) \le -k$$

as long as the solution $x(t)$ remains in $\mathcal{A}$. Integration over the time interval $0 \le s \le t$ gives

$$V(x(t)) \le V(x_0) - kt.$$

By this last inequality, the solution must reach the boundary of $\mathcal{A}$ where $V(x) = c$ within the time interval $[0, (V(x_0) - c)/k]$. Let $T > 0$ be the first time at which we have $V(x(T)) = c$; of course, $T \le (V(x_0) - c)/k$. Then straightforward comparison estimates show that, for all $t \in [0, T)$, we have

$$\frac{d}{dt}V(x(t)) \le -\alpha_3(\alpha_2^{-1}(V(x(t)))) = -\alpha_3 \circ \alpha_2^{-1}(V(x(t))).$$

Consequently, for all $t \in [0, T)$ we have

$$\|x(t)\| \le \beta(\|x_0\|, t),$$

where $\beta(\cdot, \cdot)$ is the same class $\mathcal{KL}$ function as in (16.18). Since we have

$$\|x(t)\| \le \gamma(\|u(\cdot)\|_\infty) \quad \text{for all } t \ge T,$$

we may conclude that

$$\|x(t)\| \le \beta(\|x_0\|, t) + \gamma(\|u(\cdot)\|_\infty) \quad \text{for all } t \ge 0,$$

and this is (16.14) in the definition of input-to-state stability. □

There is a converse of Theorem 16.2 for systems of the form (16.13). That is, input-to-state stability of (16.13) implies the existence of an ISS Lyapunov function; see the Notes and References for this chapter. Thus, for (16.13), the ISS property is equivalent to the existence of an ISS Lyapunov function.

Theorem 16.2 (and its converse) extend the classical direct and converse Lyapunov theorems on asymptotic stability of uncontrolled systems (classical ODE systems), to the class of controlled systems of the form (16.13). More precisely, the ISS Lyapunov function guaranteed by Theorem 16.2 (or its converse) serves as a classical Lyapunov function for the globally asymptotically stable equilibrium $x = 0$ of the unforced system, $\dot{x} = f(x, 0)$.

We now apply Theorem 16.2 in some examples.

Example 16.6 Consider again the scalar system $\dot{x} = -x^3 + u$. We use Theorem 16.2 to show that the system is input-to-state stable, and to identify the ISS gain function γ. The origin is globally asymptotically stable for the unforced equation, $\dot{x} = -x^3$, with the globally defined and radially unbounded Lyapunov function $V(x) = \frac{1}{2}x^2$. We compute the rate of change of $V(x)$ along solutions of the controlled system and obtain

$$\frac{d}{dt}V(x) = x\dot{x} = -x^4 + xu.$$

The second term is indefinite, but according to (16.17) and Theorem 16.2, we only need a negative definite bound for sufficiently large $|x|$. We can use part of the term $-x^4$ to dominate for large $|x|$. Letting θ be constant with $0 < \theta < 1$, we may write

$$\frac{d}{dt}V(x) = x\dot{x} = -(1-\theta)x^4 - \theta x^4 + xu \leq -(1-\theta)x^4,$$

provided we have

$$-\theta x^4 + xu \leq 0.$$

It is straightforward to check that the latter inequality is satisfied if

$$|x| \geq \left(\frac{|u|}{\theta}\right)^{\frac{1}{3}}.$$

By Theorem 16.2, the system is input-to-state stable with $\chi(r) = (\frac{r}{\theta})^{\frac{1}{3}}$ and with gain function $\gamma(r) = (\frac{r}{\theta})^{\frac{1}{3}}$, because for this $V(x)$ we may take $\alpha_1(r) = \frac{1}{2}r^2 = \alpha_2(r)$. △

Example 16.7 Consider the scalar system

$$\dot{x} = -x^5 + x^3 u.$$

The unforced system has $x = 0$ as a globally asymptotically stable equilibrium, and $V(x) = \frac{1}{2}x^2$ is a globally defined and radially unbounded Lyapunov function. The term $-x^5$ is dominant for sufficiently large $|x|$. Taking $0 < \theta < 1$, we have

$$\begin{aligned}\frac{d}{dt}V(x) &= -x^6 + x^4 u \\ &= -(1-\theta)x^6 - \theta x^6 + x^4 u \\ &\leq -(1-\theta)x^6,\end{aligned}$$

provided we have $-\theta x^6 + x^4 u \leq 0$, and the latter inequality will be satisfied if

$$|x| \geq \left(\frac{|u|}{\theta}\right)^{\frac{1}{2}}.$$

In summary, we have

$$\frac{d}{dt}V(x) \leq -(1-\theta)x^6 \quad \text{for all } |x| \geq \left(\frac{|u|}{\theta}\right)^{\frac{1}{2}} =: \chi(|u|).$$

By Theorem 16.2, this system is input-to-state stable with gain function $\gamma(r) = (\frac{r}{\theta})^{\frac{1}{2}}$.

By way of contrast, consider the scalar equation

$$\dot{x} = -x^3 + x^5 u,$$

which has unbounded solutions for some bounded inputs u. For example, consider the constant input $u = 1$. There are unbounded solutions of $\dot{x} = -x^3 + x^5$ for initial conditions $x(0) > 1$; hence, this scalar system is not input-to-state stable. △

In the next example, the ISS property is helpful even though there are restrictions on the state variable.

Example 16.8 Let $a > 0$. The scalar equation

$$\dot{R} = -aR^2 - aR(2\phi + \phi^2) \quad \text{for } R \geq 0$$

is input-to-state stable with input ϕ and state R. The nonnegative axis $R \geq 0$ is an invariant set for the differential equation. We have

$$\begin{aligned} \dot{R} &\leq -aR^2 + 2aR|\phi| - a\phi^2 R \\ &\leq -\frac{a}{2}R^2 - \frac{a}{2}R^2 + 2aR|\phi| \\ &\leq -\frac{a}{2}R^2 - \frac{a}{2}R(R - 4|\phi|). \end{aligned}$$

When $R > 4|\phi|$, a solution $R(t)$ decreases faster than the solution of $\dot{r} = -\frac{a}{2}r^2$ having the same initial condition. Thus, we may write

$$R(t) \leq \frac{R(0)}{1 + R(0)\frac{a}{2}t} + 4 \sup_{0 \leq s \leq t} |\phi(s)| \quad \text{for all } t \geq 0.$$

Hence, the scalar system is ISS with input ϕ and state R. △

16.4 ISS IN CASCADE SYSTEMS

The next theorem addresses the question of global asymptotic stability in a cascade connection of time invariant nonlinear systems. Input-to-state stability of the driven system of a cascade makes it relatively easy to address the boundedness issue highlighted by Theorem 9.2.

Theorem 16.3 *Let $f : \mathbf{R}^n \times \mathbf{R}^m \to \mathbf{R}^n$ and $g : \mathbf{R}^m \to \mathbf{R}^m$ be C^1 functions, with $f(0,0) = 0$ and $g(0) = 0$. Suppose that*

$$\dot{x} = f(x, z)$$

is input-to-state stable with input z, and $z = 0$ is a globally asymptotically stable equilibrium of

$$\dot{z} = g(z).$$

Then $(x, z) = (0, 0)$ *is a globally asymptotically stable equilibrium of the cascade system*

$$\dot{x} = f(x, z), \tag{16.19}$$
$$\dot{z} = g(z). \tag{16.20}$$

Proof. By hypothesis, every solution of (16.20) converges to $z = 0$ as $t \to \infty$. In addition, $x = 0$ is a globally asymptotically stable equilibrium of $\dot{x} = f(x, 0)$, by input-to-state stability of (16.19). The origin $(x, z) = (0, 0)$ is a stable equilibrium of the cascade, by Corollary 9.1. Thus, by Theorem 9.2, we need only show that, for each solution $z(t)$, every solution $x(t)$ of $\dot{x} = f(x, z(t))$ is defined and bounded for $t \geq 0$. But this is immediate from the ISS definition and the estimate (16.14) applied to (16.19) with input z. □

Alternatively, instead of invoking Theorem 9.2, one can prove Theorem 16.3 based on the CICS property of (16.19), which is implied by ISS. Since $z(t)$ is a converging input to (16.19), we have $x(t) \to 0$ as $t \to \infty$ for every solution $x(t)$ of $\dot{x} = f(x, z(t))$. (See Exercise 16.2.)

The ISS condition on (16.19) is a strong condition that guarantees global asymptotic stability of the origin of the cascade (16.19)–(16.20) when (16.20) is GAS. An easy example which shows that ISS of the driven system is not necessary for the GAS property in a cascade is the system given by $\dot{x} = -x + xz$, $\dot{z} = -z$.

Input-to-state stability is also helpful in more general composite systems.

Theorem 16.4 *Let* $f : \mathbf{R}^n \times \mathbf{R}^m \to \mathbf{R}^n$ *and* $g : \mathbf{R}^m \times \mathbf{R}^\mu \to \mathbf{R}^m$ *be* C^1 *functions. Suppose that*

$$\dot{x} = f(x, z)$$

is input-to-state stable with input z*, and*

$$\dot{z} = g(z, u)$$

is input-to-state stable with input u*. Then the following statements are true for the composite system*

$$\dot{x} = f(x, z), \tag{16.21}$$
$$\dot{z} = g(z, u). \tag{16.22}$$

(a) *The composite system* (16.21)–(16.22) *with input* u *and state* (x, z) *has the BIBS property and the CICS property.*

(b) *The composite system* (16.21)–(16.22) *is input-to-state stable with input* u *and state* (x, z)*.*

Proof. Of course (a) follows from (b); however, the argument for (b) requires development of some additional ideas, and we cite [48] (page 34) for its proof. Instead, we prove (a) directly from the hypotheses on the individual component systems.

If $u(t)$ is bounded for $t \geq 0$, then, by input-to-state stability of (16.22), for any initial condition $z(0) = z_0$ the solution $z(t)$ of $\dot{z} = g(z, u(t))$ exists and is bounded for all $t \geq 0$. Since (16.21) is ISS, the same is true of every solution $x(t)$ of $\dot{x} = f(x, z(t))$. This proves the BIBS property for the composite system with input u and state (x, z).

If $u(t) \to 0$ as $t \to \infty$, then, by the converging-input converging-state property of the ISS system (16.22), for any initial condition $z(0) = z_0$ the solution $z(t)$ of $\dot{z} = g(z, u(t))$ satisfies $z(t) \to 0$ as $t \to \infty$. Since (16.21) is ISS, every solution $x(t)$ of $\dot{x} = f(x, z(t))$ satisfies $x(t) \to 0$ as $t \to \infty$. This proves the CICS property for the composite system with input u and state (x, z). □

We end the section with some simple examples to illiustrate Theorem 16.4.

Example 16.9 The composite system

$$\begin{aligned} \dot{x} &= -x^3 + z, \\ \dot{z} &= -z^3 + u \end{aligned}$$

is input-to-state stable with scalar input u, because the scalar system $\dot{\xi} = -\xi^3 + v$ is ISS with input v. △

Example 16.10 With the help of Example 16.7, an application of Theorem 16.4 shows that the composite system

$$\begin{aligned} \dot{x} &= -x^5 + x^3 z, \\ \dot{z} &= -z^3 + u \end{aligned}$$

is input-to-state stable with scalar input u, and hence has the BIBS and CICS properties. △

It is beyond the scope of this book to explore these ideas further here, but interested readers can do so through the references.

16.5 EXERCISES

Exercise 16.1 Show that (16.2) follows from (16.1).

Exercise 16.2 *ISS Implies the CICS Property*
Suppose that $\dot{x} = f(x, u)$ is input-to-state stable, where $f : \mathbf{R}^n \times \mathbf{R}^m \to \mathbf{R}^n$. Use the estimate (16.14) to deduce that, if $u(t) \to 0$ as $t \to \infty$, then for any initial condition $x_0 = x(0)$, the solution $x(t) \to 0$ as $t \to \infty$.

Exercise 16.3 For each of these scalar systems, determine whether the input-to-state stability property holds:

(a) $\dot{x} = -(1 + u)x^5$;
(b) $\dot{x} = -x + x^2 u$;

(c) $\dot{x} = -x^3 - x^5 + u$;
(d) $\dot{x} = x^3 - x^5 + u$;
(e) $\dot{x} = -x^3 + x^5 + u$;
(f) $\dot{x} = -x^3 + x^2 u$.

Exercise 16.4 For each of these planar systems, determine whether the input-to-state stability property holds:
(a) $\dot{x}_1 = (-x_1 + 9x_2)(x_1^2 - 1), \qquad \dot{x}_2 = (-x_2 + u)(x_1^2 - 1)$;
(b) $\dot{x}_1 = -x_1 - x_2, \qquad \dot{x}_2 = x_1 + x_2^3 + u$;
(c) $\dot{x}_1 = -x_1 - x_1^3 x_2, \qquad \dot{x}_2 = -3x_1 + 2x_2 + u$;
(d) $\dot{x}_1 = -x_1^3 - x_1^2 x_2, \qquad \dot{x}_2 = -2x_2 + u$.

Exercise 16.5 Consider the scalar system

$$\dot{x} = -x - \frac{x^3}{1 + |u|}.$$

(a) Show that the unforced system has a GAS origin and $V(x) = \frac{1}{2}x^2$ is a Lyapunov function.
(b) Show that the system is ISS with gain function $\gamma \equiv 0$.

Exercise 16.6 Investigate input-to-state stability for these planar systems:

(a) $\dot{x}_1 = -x_1 - \dfrac{x_1^3}{1 + |x_2|}, \qquad \dot{x}_2 = -x_2^3 - x_2 u$;

(b) $\dot{x}_1 = -x_1 - \dfrac{x_1^3}{1 + |x_2|}, \qquad \dot{x}_2 = -x_2^3 + x_2^3 u$;

(c) $\dot{x}_1 = -x_1 + \dfrac{x_1^3}{1 + |x_2|}, \qquad \dot{x}_2 = -x_2 + x_2^3 + u$.

Hint: Note Exercise 16.5.

Exercise 16.7 Consider the scalar system

$$\dot{x} = ax^p + bx^q u,$$

where a, b are real constants and p and q are positive integers. Find conditions on a, b, p, q which are necessary and sufficient for the system to be input-to-state stable with input u. *Hint*: First, determine necessary conditions for ISS. Then show that those conditions are also sufficient for ISS.

Exercise 16.8 *Does Converging $z(t)$ Imply Bounded $x(t)$?*
Revisit the systems in Exercise 9.9 in the light of the ISS concept. In particular, consider the planar system

$$\begin{aligned} \dot{x} &= -x^3 + x^2 z, \\ \dot{z} &= -z^3 \end{aligned}$$

and show that $(x, z) = (0, 0)$ is globally asymptotically stable.

Exercise 16.9 *ISS can Help without Feedback Passivity*
Give an example of a globally asymptotically stabilizable three-dimensional cascade system of the form (15.6)–(15.7) with the following features:

(i) The linear driving system is two-dimensional and the triple (C, A, B) is not feedback passive.
(ii) The one-dimensional driven system is ISS with $y = C\xi$ as the input.

Is your example globally asymptotically stabilizable by partial-state feedback? full-state feedback?

Exercise 16.10 *Feedback Passivity can Help without ISS*
Give an example of a globally asymptotically stabilizable three-dimensional cascade system of the form (15.6)–(15.7) with the following features:

(i) The linear driving system is two-dimensional and the triple (C, A, B) is feedback passive.
(ii) The one-dimensional driven system with input $y = C\xi$ is not ISS.

Is your example globally asymptotically stabilizable by partial-state feedback? full-state feedback?

Exercise 16.11 *Converse Theorems via Comparison Functions*
Reformulate the results of the converse Lyapunov theorem (Theorem 8.7) in the language of comparison functions.

16.6 NOTES AND REFERENCES

In this chapter we considered the ISS property for time invariant systems, $\dot{x} = f(x, u)$. This is the case considered in the original paper [89] where the ISS concept was introduced. The presentation in this chapter owes a debt to [48]. The converse of Theorem 16.2, that input-to-state stability of $\dot{x} = f(x, u)$ implies the existence of an ISS Lyapunov function, is proved in E. D. Sontag and Y. Wang, On characterizations of the input-to-state stability property, *Syst. Contr. Lett.*, 24:351–359, 1995.

For a complete development of the ISS concept in the context of time varying systems, see the progression of ideas in [58] (Sections 4.5, 4.8–4.9).

The equation in Example 16.8 is taken from the jet engine surge and stall example in [65].

It is worth noting briefly that a system $\dot{x} = f(x, u)$ is ISS, with ISS Lyapunov function $V(x)$, if and only if [48] (page 21) there exist class $\mathcal{K}_\infty$ functions $\alpha_1(\cdot)$, $\alpha_2(\cdot)$, $\alpha(\cdot)$, and a class $\mathcal{K}$ function $\sigma(\cdot)$ such that

$$\alpha_1(\|x\|) \le V(x) \le \alpha_2(\|x\|) \quad \text{for all } x \in \mathbf{R}^n$$

and

$$\frac{\partial V}{\partial x}(x) f(x, u) \leq -\alpha(\|x\|) + \sigma(\|u\|) \quad \text{for all } x \in \mathbf{R}^n \quad \text{and all } u \in \mathbf{R}^m.$$

This last inequality is similar to the dissipation inequality for a passive system, except that here, instead of the function $w(u, y) = y^T u$ as supply rate, we have

$$w(u, y) = -\alpha(\|y\|) + \sigma(\|u\|), \quad \text{with } y = x,$$

and $V(x)$ has the role of storage function. In fact, ISS systems can be viewed as dissipative systems in the sense of [101], [102]. For more on the connection between ISS systems and general dissipative systems, see [48] (pages 42–43).

Chapter Seventeen

Some Further Reading

Mathematical control theory is conceptually rich and undergoing healthy development. Readers looking for a mathematical subject area with a thriving interaction between theory and applications are encouraged to explore further. In addition to the end-of-chapter Notes and References on specific topics, some readers may appreciate a brief look at further reading in mathematical control theory or in the area of stability and stabilization. The items listed here are research surveys, standard advanced texts and monographs, and even a few foundational papers. There is no claim of completeness; these resources are listed because they have been helpful to the author in gaining perspective on a vast subject area.

Some Research Surveys

There are several surveys on control theory that provide perspective on research trends since around 1980.

- First, [59], from 1985, describes four major areas of research emphasis during the 1980s.
- Second, [90], from 1990, focuses on nonlinear stabilization.
- The authors of [27], from 1995, state that their survey can be seen as a prolongation of the overview on stabilization in [90].
- Finally, [62], from 2001, describes a transition from earlier, more descriptive concepts of control to "the activation of these concepts into design tools and constructive procedures." The Selected Applications section has four representative examples that exhibit "the mutually enriching theory-applications transitions that have been common in recent developments of nonlinear control."

Texts, Monographs, and a Few Papers

STABILITY AND STABILIZATION. There are several excellent comprehensive texts on the analysis and feedback control of nonlinear systems. These texts all have much material and many references on stability and stabilization; see, in particular, [47], [58], [85], [91], [105], and the text by R. Marino and P. Tomei, *Nonlinear Control Design: Geometric, Adaptive, and Robust*, Prentice-Hall Europe, London, 1995.

The book by A. Bacciotti, *Local Stabilizability of Nonlinear Control Systems*, World Scientific Publishing, 1992, provides valuable insight and perspective. Some attention is given there to the question of *continuous* stabilizability as contrasted with *smooth* (C^1) stabilizability. (For example, a continuous (but not C^1) feedback stabilizer for Example 13.8 in the present text is given there; see page 37 and Example 6.8 on page 28 in A. Bacciotti's book.) In its general approach, Bacciotti's book emphasizes definitions, methods, and examples, rather than complete proofs of all results. It includes many examples of low-dimensional systems. See also the recent monograph by A. Bacciotti and L. Rosier, *Liapunov Functions and Stability in Control Theory*, Second Edition, Springer, New York, 2005, which discusses converse Lyapunov theorems, time varying as well as time invariant systems, and systems with discontinuous right-hand sides. For more on issues of discontinuities and disturbances, see E. D. Sontag, Stability and stabilization: Discontinuities and the effect of disturbances, in *Nonlinear Analysis, Differential Equations, and Control*, F. H. Clarke, R. J. Stern, and G. Sabidussi, editors, Kluwer, Dordrecht, 1999, pages 551–598.

The monograph [65] is a comprehensive presentation of backstepping and its applications to nonlinear and adaptive control designs.

The monograph [86] is aimed at constructive procedures for smooth stabilization of equilibria. It identifies a class of systems with special structure (the *interlaced systems*) for which smooth stabilization is possible with no growth restrictions on the nonlinear terms of the system. Passivity is the main tool employed, and important connections between passivity, backstepping, and optimal control play a central role.

The monograph [48] presents a self-contained and coordinated development of several design methods for achieving stabilization of nonlinear control systems either globally or on arbitrarily large domains, and in the presence of model uncertainties.

Numerical Methods in Linear Control. As mentioned in the introductory chapter, for numerical methods in linear control see the text by B. N. Datta, *Numerical Methods for Linear Control Systems*, Elsevier Academic Press, London, 2004.

Optimal Control. An early and fundamental paper is R. E. Kalman, Contribution to the theory of optimal control, *Bol. Soc. Matem. Mex.*, 5:102–119, 1960.

There are many excellent introductory texts devoted completely to optimal control, including the following: M. Athans and P. L. Falb, *Optimal Control: An Introduction to the Theory and its Applications*, McGraw-Hill, New York, 1966 (reprinted by Dover, New York, 2007); A. Locatelli, *Optimal Control: An Introduction*, Birkhäuser, Boston 2001; R. F. Stengel, *Stochastic Optimal Control*, John Wiley and Sons, 1986 (reprinted as *Optimal Control and Estimation*, Dover, New York, 1994); F. L. Lewis and V. L. Syrmos, *Optimal Control*, John Wiley, New York, second edition, 1995. And, as mentioned in the introductory chapter, there are three chapters on optimal control in [91]. See also the chapters on linear optimal control in [104].

ROBUST AND OPTIMAL CONTROL. The text [28] is a graduate text that uses convex optimization techniques, specifically linear matrix inequalities (LMIs), in an exposition of the major advances in the 1990s on robust control. See also the comprehensive text by Zhou, K. with J. C. Doyle and K. Glover, *Robust and Optimal Control*, Prentice-Hall, Englewood Cliffs, NJ 1996. See as well the graduate text on linear time invariant control systems by H. L. Trentelman, A. A. Stoorvogel, and M. Hautus, *Control Theory for Linear Systems*, Springer, London, 2001; this text has an emphasis on the geometric approach, as in [104].

PASSIVE SYSTEMS. We have noted earlier the main resources for the state-space results presented in this book. They are [19], [39], [76], [86]; in addition, see the foundational papers [101], [102] on the class of *dissipative systems* which includes passive systems. The text [58] includes a study of passivity using both state-space methods and transfer function analysis. For a look at linear passive systems that is close to its electrical engineering origins, see B. D. O. Anderson and S. Vongpanitlerd, *Network Analysis and Synthesis*, Prentice-Hall, 1973 (reprinted by Dover, New York, 2006).

DIFFERENTIAL GEOMETRIC CONTROL. Local and global decompositions of nonlinear systems that generalize the controllability and observability normal forms appear in [47] (Chapters 1–2). The global results in [47] (Chapter 2), in particular, depend on results from the differential geometric theory of distributions and their integral manifolds [36], [92]. See also [51].

Reference [47] presents important control methods that go well beyond the basic ideas discussed in Chapters 11–13 of the present text, and it includes a development of the required differential geometric concepts. See also reference [79], as well as the text by R. Marino and P. Tomei mentioned above, for the geometric approach to nonlinear systems. Much of the differential geometric approach to nonlinear systems generalizes the geometric approach to linear control theory based on invariant subspaces in [104]. Reference [79] includes material on Hamiltonian control systems.

Two other helpful introductions to differentiable manifolds can be mentioned. A very accessible introduction to manifolds and smooth mappings from one manifold to another is [31] (Chapter 1). The text [14] is a comprehensive graduate introduction to differentiable manifolds. Both [31] and [14] require only a solid background in undergraduate analysis and linear algebra.

Appendix A

Notation: A Brief Key

This listing should prove convenient for some frequently used notation. For certain constructions involving derivatives there are alternative standard notations which are explained here. In addition, index entries are available for many of the terms that appear in this listing.

$\mathbf{R}$	the real number field
$\mathbf{R}^n$	real n-dimensional space
$\mathbf{C}$	the complex number field
$\mathbf{C}^n$	complex n-dimensional space
$a := b$	a is defined by b
$a =: b$	b is defined by a
$\equiv$	identically equal to, as in $\cos^2 t + \sin^2 t \equiv 1$
$S_1 \implies S_2$	Statement S_1 implies statement S_2; equivalently, S_1 is sufficient for S_2 and S_2 is necessary for S_1.
$\|x\|_2$	$:= (x^T x)^{\frac{1}{2}}$, the Euclidean norm of the real vector x
$\|x\|$	usually this refers to $\sum_{j=1}^n \lvert x_j \rvert$, the absolute sum norm of a vector x
$\|A\|$	usually this refers to $\sum_{i=1}^m \sum_{j=1}^n \lvert A_{ij} \rvert$, the absolute sum norm of matrix A
$x(t) \to L$	denotes $\lim_{t \to c} x(t) = L$, where the limit point c is understood from context or mentioned explicitly, as in $x(t) \to L$ as $t \to \infty$.
sup	supremum (least upper bound)
inf	infimum (greatest lower bound)
$B_r(x_0)$	the set $\{x : \|x - x_0\| < r\}$, where the norm is understood
$\text{dist}\,(p, S)$	the distance from the point p to the set S; it is norm-dependent
$\dot{x}$	the first derivative of x with respect to t, $\frac{dx}{dt}$
$\ddot{x}$	the second derivative of x with respect to t, $\frac{d^2x}{dt^2}$
$x^{(j)}$	the j-th derivative of x with respect to t, $\frac{d^j x}{dt^j}$
x^T, A^T	the transpose of vector x, the transpose of matrix A
$P > 0$	matrix P is symmetric positive definite
$P \geq 0$	matrix P is symmetric positive semidefinite
$\text{Re}\, z$	is the real part of the complex number z
$\text{Im}\, z$	is the imaginary part of the complex number z
$\bar{z}$	is the conjugate of the complex number $z = a + ib$; thus, $\bar{z} = a - ib$

$O(\cdot)$	is an order of magnitude notation
$N(A)$	$:= \{x : Ax = 0\}$ is the null space (or kernel) of matrix A
$R(A)$	$:= \{y : y = Ax \text{ for some } x\}$ is the range space (column space) of matrix A
span S	the set of all finite linear combinations of vectors from S; thus, if $S = \{v_1, \ldots, v_k\}$, then span S is the set of all linear combinations of the vectors $v_1, \ldots, v_k$ with coefficients in the scalar field
$S^\perp$	is the orthogonal complement of a subspace (or a set) S, (where the ambient space is understood).
$\mathcal{D}^\perp(x)$	is the covector annihilator of the subspace $\mathcal{D}(x)$; $\mathcal{D}^\perp(x)$ is spanned by row vectors that form a basis of the orthogonal complement of $\mathcal{D}(x)$.
$\omega(\phi)$	is the ω-limit set of a function $\phi : [0, \infty) \to \mathbf{R}$.
$\frac{\partial f}{\partial x}(x)$	is the $m \times n$ Jacobian matrix of first order partial derivatives of a function $f : \mathbf{R}^n \to \mathbf{R}^m$, evaluated at x; the dimensions are understood from the context.
$Df(x)$	$:= \frac{\partial f}{\partial x}(x)$
$\frac{\partial V}{\partial x}(x)$	is the row gradient of a real-valued function $V(x)$; it is also the Jacobian matrix $DV(x) = (\nabla V(x))^T$ of V at x.
$L_f V(x)$	$:= \nabla V(x) \cdot f(x)$, is the Lie derivative of the function V along the vector field f. It denotes the quantity $\frac{d}{dt}V(\phi(t))\|_{t=0}$ when $\frac{d}{dt}\phi(t) = f(\phi(t))$ and $\phi(0) = x$.
$\dot{V}(x)$	$:= L_f V(x)$, when the autonomous vector field $f(x)$ is understood. Observe that $\dot{V}(x)$ is a state-dependent function, rather than an explicitly time-dependent function. The Lie derivative notation $L_f h(x)$ often has advantages when higher order derivatives of h are considered.
$dh(x)$	the differential of a real-valued function h at x. For computations, it can be identified with the row gradient $\nabla^T h(x) = (\nabla h(x))^T$; thus, note that $L_f h(x) = dh(x)\, f(x)$ for a vector field f.
SISO	single-input single-output
MIMO	multi-input multi-output
LTI	linear time invariant
$\triangle$	denotes the end of an example
$\square$	denotes the end of a proof
$[xx]$	citation to item number xx in the bibliography

Appendix B

Analysis in R and $\mathbf{R}^n$

This appendix is not a substitute for a course in advanced calculus. It is recommended that the reader have some exposure to undergraduate analysis at the rigorous "Advanced Calculus" level. Our main reference is the text [7]. See also [68]. Many questions that arise in the analysis of differential equations can be understood and resolved with an understanding of analysis in $\mathbf{R}^n$. In general, the appendix assumes an understanding of the integral of vector-valued functions, some experience with field axioms and order axioms for the real numbers, and the basic metric topological concepts required for a precise discussion of continuity and differentiability.

B.1 COMPLETENESS AND COMPACTNESS

We begin with the topological notions of open sets, accumulation points, closed sets, and bounded sets.

Let $\|\cdot\|$ be a norm on $\mathbf{R}^n$. The ball of radius $r > 0$ centered at the point a in $\mathbf{R}^n$ is the set $B_r(a) := \{x : \|x - a\| < r\}$. Recall that a subset A of $\mathbf{R}^n$ is *open* if every point in A is the center point of some ball which is contained entirely in A; that is, every point of A is an interior point of A. A point x in $\mathbf{R}^n$ is an *accumulation point* of a set $S \subset \mathbf{R}^n$ if every ball $B_r(x)$ centered at x contains some point of S distinct from x. A set S is closed if S contains all its accumulation points, and the closure of S, written $\mathrm{cl}(S)$, is the union of S and the set of its accumulation points. The closure $\mathrm{cl}(S)$ is the smallest closed set which contains S. S is closed if and only if $S = \mathrm{cl}(S)$. Also, a set S is closed if and only if its complement $\mathbf{R}^n - S = \{x \in \mathbf{R}^n : x \notin S\}$ is open. A subset B of $\mathbf{R}^n$ is *bounded* if $B \subset B_r(a)$ for some point a and some $r > 0$. So B is bounded if and only if it is contained in some ball.

The field axioms and order axioms for the real numbers can be found in [7] (pages 1–3). Some experience with these properties is assumed. A very important property of the real number field is the least upper bound property. Recall that a number m is an upper bound for a set S if $s \le m$ for every s in S. The concept of *least upper bound* is defined next.

Definition B.1 *[7]* (page 9) *Let S be a set of real numbers bounded above. A real number b is a* least upper bound *for S if it has these two properties:*

(a) *b is an upper bound for S.*
(b) *If $m < b$, then m is not an upper bound for S.*

It is easy to show that if a least upper bound for S exists, then it must be unique. As examples, the interval $S = (0, \pi]$ has the number π as least upper bound, and this is also the maximum of S. On the other hand, the interval $(0, \pi)$ has least upper bound π even though the interval has no maximum number. Standard terminology for the least upper bound of S is *supremum* (of S), written $\sup S$.

In addition to the field axioms and the usual order axioms, we have the following axiom known as the Completeness Axiom, or the Least Upper Bound Property of the real numbers.

THE COMPLETENESS AXIOM (LEAST UPPER BOUND PROPERTY). [7] (page 9) *Every nonempty subset S of the real numbers which is bounded above has a least upper bound in $\mathbf{R}$; that is, there is a real number b such that*

$$b = \sup S.$$

The set of real numbers may also be *constructed* from the set of rational numbers, so that the field and order properties of [7] (pages 1–3) hold; one can then deduce the theorem that $\mathbf{R}$ has the least upper bound property.

As a consequence of the least upper bound property, every nonempty subset S of the real numbers which is bounded *below* has a *greatest lower bound* in $\mathbf{R}$, with the obvious definition of greatest lower bound. In fact, it is not difficult to show that, if S is bounded below, then the set

$$-S := \{-s \, : \, s \in S\}$$

is bounded above, and we have

$$\inf S = -\sup(-S). \tag{B.1}$$

As an example, the interval $S = (\pi, 4]$ has the number π as greatest lower bound, even though S has no minimum number. Note also that $\inf(\pi, 4] = -\sup[-4, -\pi)$, in accordance with (B.1). Standard terminology for the greatest lower bound of S is *infimum* (of S), written $\inf S$.

The least upper bound property implies the Bolzano-Weierstrass theorem.

Theorem B.1 (Bolzano-Weierstrass)
If a bounded subset S of $\mathbf{R}^n$ contains infinitely many points, then there is at least one point in $\mathbf{R}^n$ which is an accumulation point of S.

Proof. See [7] (pages 54–56). □

The Bolzano-Weierstrass theorem may be applied to show that every Cauchy sequence in $\mathbf{R}^n$ must converge to a point of $\mathbf{R}^n$. We state and prove the result for the real number field only.

Theorem B.2 (Completeness of the Real Number Field)
Every Cauchy sequence of real numbers converges to a real number.

Proof. Let $\{x_n\}$ be a Cauchy sequence of real numbers. If there are only finitely many values of x_n, then the sequence is constant past some index value, so the sequence converges. Suppose now that the range of $n \mapsto x_n$ is infinite. Since the sequence is Cauchy, it is bounded: There is an N_1 such that if $n, m \geq N_1$, then $|x_n - x_m| < 1$. Let $M = \max_{1 \leq j \leq N}\{|x_j|\}$. Then for all k, we have

$$|x_k| \leq \max\{M, M+1\}.$$

By the Bolzano-Weierstrass theorem, the sequence $\{x_n\}$ has an accumulation point p. We want to show that $\lim_{n\to\infty} x_n = p$. Given $\epsilon > 0$, there is an N such that

$$n, m \geq N \Longrightarrow |x_n - x_m| < \frac{\epsilon}{2}.$$

Since p is an accumulation point of the sequence, there is an m such that

$$|x_m - p| < \frac{\epsilon}{2}.$$

Thus, if $n \geq N$, we have

$$|x_n - p| \leq |x_n - x_m| + |x_m - p| \leq \frac{\epsilon}{2} + \frac{\epsilon}{2} = \epsilon.$$

Since $\epsilon > 0$ was arbitrary, we have $\lim_{n\to\infty} x_n = p$. □

The completeness of $\mathbf{R}^n$ and $\mathbf{C}^n$, as well as the matrix spaces $\mathbf{R}^{n\times n}$ and $\mathbf{C}^{n\times n}$, is a consequence of the completeness of $\mathbf{R}$.

The concept of compactness can be defined in different ways. We will follow [7]. A collection of open sets whose union contains a set S is called an *open covering* (or an *open cover*) of S. Given an open cover of S, a *finite subcover* (of S) is a finite subcollection of the given cover that is also an open cover of S.

Definition B.2 (Compact Set)
A subset S of $\mathbf{R}^n$ is compact *provided that, whenever S is contained in a union of open subsets of $\mathbf{R}^n$, there is some finite subcollection of these open sets whose union also contains S.*

Thus, a set S is compact if and only if every open cover of S has a finite subcover.

Theorem B.3 (Compactness Characterization)
Let S be a subset of $\mathbf{R}^n$. Then the following statements are equivalent.

(a) *S is compact.*
(b) *S is closed and bounded.*
(c) *Every infinite subset of S has an accumulation point in S.*

Proof. See [7] (pages 59–60). □

Theorem B.4 *If $f : K \to \mathbf{R}$ is continuous on the compact set $K \subset \mathbf{R}^n$, then f attains its absolute maximum and absolute minimum values on K.*

Proof. See [7] (pages 82–83). □

B.2 DIFFERENTIABILITY AND LIPSCHITZ CONTINUITY

The basic concepts of limits of sequences and limits of functions, and the continuity of functions in a metric space such as $\mathbf{R}^n$, are assumed. For a development of these ideas for sequences in a metric space, and functions mapping one metric space to another metric space, see [7] (pages 70–79). In this book, we are interested in the metric spaces $\mathbf{R}^n$, where for a given n the metric distance $d(x, y)$ between two points $x = (x_1, \ldots, x_n)$ and $y = (y_1, \ldots, y_n)$ is defined by a *norm*, for example the Euclidean norm

$$\|x - y\|_2 = \sqrt{(x_1 - y_1)^2 + \cdots + (x_n - y_n)^2},$$

or some other convenient norm, such as

$$\|x - y\| = |x_1 - y_1| + \cdots + |x_n - y_n|.$$

A formal definition of norms is included in Chapter 2 (Mathematical Background).

We say that a function $f : \mathbf{R} \to \mathbf{R}$ is *locally integrable* if $\int_a^b |f(x)|\, dx$ exists for all real numbers a, b.

A function $f : \mathbf{R} \to \mathbf{R}$ is *piecewise continuous* if it is continuous except possibly at a set of points E such that any finite interval contains at most finitely many points of E, and the left- and right-hand limits of f exist at each point of E. If f is piecewise continuous, then the set E is at most countably infinite.

Let $f : U \to \mathbf{R}^m$, where U is an open set in $\mathbf{R}^n$. We say that f is *differentiable* at a point $x \in U$ if and only if there exists a linear mapping $T(x) : h \mapsto T(x)h$ of tangent vectors h based at x,

$$T(x)(\alpha_1 h_1 + \alpha_2 h_2) = \alpha_1 T(x)h_1 + \alpha_2 T(x)h_2 \quad \text{for all } h_1,\ h_2 \in \mathbf{R}^n,$$

satisfying the tangent estimate

$$\|f(x + h) - f(x) - T(x)h\| = o(\|h\|) \quad \text{as } \|h\| \to 0.$$

Recall that the term $o(\|h\|)$ means that $o(\|h\|)/\|h\| \to 0$ as $\|h\| \to 0$. The linearity property and tangent estimate uniquely determine the linear mapping $T(x)$. If f is differentiable at x, then all first order partial derivatives of f exist at x, and in fact, the matrix representation of the linear mapping $T(x)$ with respect to the standard bases in $\mathbf{R}^n$ and $\mathbf{R}^m$ is the Jacobian matrix of f at x whose entries are the first order partial derivatives of the components of f evaluated at x. To see this, let $A = [a_{ij}]$ be the matrix of

$T(x)$ in the standard bases. It must be shown that

$$a_{ij} = \frac{\partial f_i}{\partial x_j}(x) = \lim_{h \to 0} \frac{f_i(x + he_j) - f_i(x)}{h},$$

where e_j is the j-th standard basis vector in $\mathbf{R}^n$. Note that

$$\begin{aligned} 0 \leq & \left| \frac{f_i(x + he_j) - f_i(x)}{h} - a_{ij} \right| \\ \leq & \left\| \frac{f(x + he_j) - f(x)}{h} - Ae_j \right\| \\ = & \frac{\|f(x + he_j) - f(x) - A(he_j)\|}{\|he_j\|}. \end{aligned}$$

It follows that

$$\lim_{h \to 0} \left| \frac{f_i(x + he_j) - f_i(x)}{h} - a_{ij} \right| = 0,$$

so the partial derivative $\frac{\partial f_i}{\partial x_j}(x)$ exists and equals a_{ij}.

Exercise. Show that differentiability of f at x implies that f is continuous at x, that is, $\lim_{y \to x} f(y) = f(x)$.

Exercise. [7] (page 115) The mere existence of partial derivatives of a function at a point does not imply continuity of the function at that point. Use the example

$$f(x_1, x_2) = \begin{cases} x_1 + x_2 & \text{if } x_1 = 0 \text{ or } x_2 = 0; \\ 1 & \text{otherwise} \end{cases}$$

to illustrate this. Show that

$$\frac{\partial f}{\partial x_1}(0,0) = 1 = \frac{\partial f}{\partial x_2}(0,0),$$

but that $\lim_{(x_1,x_2) \to (0,0)} f(x_1, x_2) \neq f(0,0) = 0$.

We usually denote the Jacobian matrix at x by $Df(x)$:

$$Df(x) = \begin{bmatrix} \frac{\partial f_1}{\partial x_1}(x) & \frac{\partial f_1}{\partial x_2}(x) & \cdots & \frac{\partial f_1}{\partial x_n}(x) \\ \frac{\partial f_2}{\partial x_1}(x) & \frac{\partial f_2}{\partial x_2}(x) & \cdots & \frac{\partial f_2}{\partial x_n}(x) \\ \vdots & \vdots & & \vdots \\ \frac{\partial f_m}{\partial x_1}(x) & \frac{\partial f_m}{\partial x_2}(x) & \cdots & \frac{\partial f_m}{\partial x_n}(x) \end{bmatrix}.$$

The Jacobian matrix $Df(x)$ may also be viewed as

$$Df(x) = \begin{bmatrix} (\nabla f_1(x))^T \\ (\nabla f_2(x))^T \\ \vdots \\ (\nabla f_m(x))^T \end{bmatrix}.$$

We may also write $df(x)$ (lower case d) for the Jacobian matrix of a scalar-valued function f at x

$$df(x) := Df(x),$$

and call $df(x)$ the *differential* of f at x. We can view $df(x)$ as a linear mapping (or *linear functional*) that determines directional derivatives of f at x, in the direction of the vector h, according to the chain rule formula

$$\frac{d}{dt} f(x + th)|_{t=0} = df(x)h = \nabla f(x) \cdot h. \tag{B.2}$$

For $f : \mathbf{R}^n \to \mathbf{R}^m$, the chain rule formula reads

$$\frac{d}{dt} f(x + th)|_{t=0} = Df(x)h.$$

The idea of directional derivative of a function is useful in a different construction as well. Instead of a constant vector h, suppose $g(x)$ is a smooth vector field on U, and f is a real valued function on U. Then we define the *Lie derivative of f with respect to g*, by

$$L_g f(x) = df(x) \cdot g(x) = \nabla f(x) \cdot g(x), \quad x \in U.$$

Thus, at each x, the Lie derivative of f with respect to g gives the directional derivative of f at x in the direction of the vector $g(x)$.

For mappings $f : U \subseteq \mathbf{R}^n \to \mathbf{R}^m$, we have the following terminology. We will say that the mapping f is of class C^1 (or simply C^1), or *continuously differentiable* at x, if all first order partial derivatives of f are continuous at x. We say that f is C^1 on an open set $U \subseteq \mathbf{R}^n$ if all first order partial derivatives of f are continuous at each x in U. It can be shown that, if f is C^1 on an open neighborhood U about the point a, then f is differentiable at a in the sense that a well-defined tangent approximation mapping $T(a)$ exists. Similarly, we say that f is C^2 on U if all second order partial derivatives of f are continuous at each x in U. It is worthwhile to point out that the continuity of second order partials implies that all mixed second order partial derivatives may be computed in any order. Thus,

$$\frac{\partial^2 f}{\partial x_i \partial x_j}(x) = \frac{\partial^2 f}{\partial x_j \partial x_i}(x)$$

when the indicated partial derivatives are continuous in a neighborhood of the point x. This follows from the continuity of the partial derivatives and the usual differential mean value theorem for real valued differentiable functions [68] (Theorem 1.1, pages 372–373). It follows that if f is a C^2

real-valued function on a domain U, then the *Hessian matrix* of second order partial derivatives of f must be symmetric:

$$H(x) = \left[\frac{\partial^2 f}{\partial x_j \partial x_i}(x)\right] \Longrightarrow H^T(x) = H(x), \quad x \in U.$$

For some discussions in the text, a function is called *smooth* if it is sufficiently differentiable that all derivatives required for the current discussion exist and are continuous on the relevant domain. This terminology avoids the necessity of having to count derivatives and specify the exact order of differentiability every time.

The main goal for the remainder of this section is to show that continuous differentiability of $f(t, x)$ with respect to x implies that f satisfies a local Lipschitz condition in x. We begin by defining the local Lipschitz condition. (This is a repeat of Definition 2.14.)

Definition B.3 *Let D be an open set in $\mathbf{R}^{n+1}$. A function $f : D \mapsto \mathbf{R}^n$, denoted $f(t, x)$ with $t \in \mathbf{R}$ and $x \in \mathbf{R}^n$, is locally Lipschitz in x on D if for any point $(t_0, x_0) \in D$, there are an open ball $B_r(t_0, x_0)$ about (t_0, x_0) and a number L such that*

$$\|f(t, x_1) - f(t, x_2)\| \leq L\|x_1 - x_2\|$$

for all (t, x_1), (t, x_2) in $B_r(t_0, x_0) \cap D$.

A local Lipchitz condition is a strong type of continuity. Note that Definition B.3 applies if f does not depend explicitly on t, that is, $f : U \subseteq \mathbf{R}^n \to \mathbf{R}^m$.

The next result is an integral form of mean value theorem which follows from the one-variable Fundamental Theorem of Calculus and the simple chain rule formula (B.2).

Theorem B.5 (Mean Value Theorem)
Let U be an open set in $\mathbf{R}^n$ and let $x \in U$. Suppose $f : U \to \mathbf{R}^m$ is C^1, and let $h \in \mathbf{R}^n$. Suppose the line segment consisting of the points $x + th$ with $0 \leq t \leq 1$ is contained in U. Then

$$f(x + h) - f(x) = \int_0^1 Df(x + th)h\, dt = \int_0^1 Df(x + th)\, dt \cdot h.$$

Proof. Consider the real-valued function of t defined by $g(t) = f(x+th)$. By the chain rule, $g'(t) = Df(x + th)h$ for all t. By the Fundamental Theorem of Calculus, we have

$$g(1) - g(0) = \int_0^1 g'(t)\, dt.$$

The result follows, since $g(1) = f(x + h)$, $g(0) = f(x)$, and h can be taken outside the integral. □

Exercise. Show the component detail in the proof of Theorem B.5.

We will now show how a local Lipschitz condition for $f : U \subseteq \mathbf{R}^n \to \mathbf{R}^m$ follows from the fact that f is continuously differentiable on U. Let $x \in U$ and take any closed ball B about x which is contained within U. Each of the first order partial derivatives of any component of f is a continuous real-valued function on the compact set B, and therefore attains an absolute maximum and absolute minimum on B. In particular there exist numbers $m_{ij} > 0$ such that

$$\left|\frac{\partial f_i}{\partial x_j}(x)\right| \leq m_{ij} \quad \text{for all } x \in B.$$

Using the sum of absolute entries norm for $\|Df(x)\|$, we then have

$$\|Df(x)\| \leq \sum_{i=1}^{m}\sum_{j=1}^{n} m_{ij} =: L \quad \text{for all } x \in B.$$

By Theorem B.5, if x and y are any two points in B, then, writing $h = y - x$, we have

$$\|f(y) - f(x)\| = \left\| \int_0^1 Df(x+th)\cdot h\, dt \right\| \leq \int_0^1 \|Df(x+th)\cdot h\|\, dt.$$

We now use the fact that the sum of absolute entries norm for a matrix is compatible with the sum norm for vectors, to write

$$\|Df(x+th)\cdot h\| \leq \|Df(x+th)\|\ \|h\|,$$

hence

$$\|f(y) - f(x)\| \leq L\|y - x\|$$

for all x and y in B. Note that the Lipschitz constant L depends on the neighborhood B.

Similarly, if $f(t,x)$ is continuously differentiable in x on an open set $D \subseteq \mathbf{R}^{n+1}$, then f is locally Lipschitz in x on D.

In general, continuous differentiability of a vector field f implies a local Lipschitz condition in a neighborhood of each point of the domain, but not necessarily a global Lipschitz constant and global Lipschitz condition over the entire domain.

Example B.1 The Van der Pol system

$$\begin{aligned}\dot{x}_1 &= x_2,\\ \dot{x}_2 &= -x_1 + \epsilon(1 - x_1^2)x_2\end{aligned}$$

is defined on the whole plane. The Jacobian matrix of the vector field is

$$\begin{bmatrix} 0 & 1 \\ -1 - 2\epsilon x_1 x_2 & \epsilon(1 - x_1^2) \end{bmatrix}.$$

Clearly, the entries in the second row are not bounded in absolute value over the whole plane, so the matrix norm as defined above cannot be bounded by any finite L over the whole plane. Is it possible that some other choice of the matrix norm or the vector norm would work to produce a *global* Lipschitz constant? That is not possible, by the equivalence of norms in finite-dimensional vector spaces. △

The vector field in the next example is globally Lipschitz.

Example B.2 The equations of motion for a pendulum with friction are

$$\dot{x}_1 = x_2,$$
$$\dot{x}_2 = -\sin x_1 - x_2.$$

Although there is redundancy in the use of an entire axis for the angular variable x_1, we consider this system to be defined on the whole plane. The Jacobian matrix of the vector field at x is

$$\frac{\partial f}{\partial x}(x) = \begin{bmatrix} 0 & 1 \\ -\cos x_1 & -1 \end{bmatrix}.$$

Using the sum of absolute entries matrix norm, we have a global bound for the norm of the Jacobian, given by

$$\left\| \frac{\partial f}{\partial x}(x) \right\| \leq 3 \quad \text{for all } x.$$ △

Our reference for Taylor's theorem is the text [7].

Theorem B.6 (Taylor's Theorem: Single Variable Real-Valued Function) *Suppose $f : (a,b) \to \mathbf{R}$ has an n-th derivative $f^{(n)}(x)$ at each $x \in (a,b)$ and $f^{(n-1)}(x)$ is continuous on the closed interval $[a,b]$. Let $c \in [a,b]$. Then for every $x \in [a,b]$, $x \neq c$, there exists a point ξ strictly between x and c, such that*

$$f(x) = f(c) + \sum_{k=1}^{n-1} \frac{1}{k!} f^{(k)}(c)(x-c)^k + \frac{1}{n!} f^{(n)}(\xi)(x-c)^n. \tag{B.3}$$

Proof. See [7] (page 113). □

The last term in (B.3) is called the *remainder term*. In particular, if $f^{(n)}(x)$ is bounded on (a,b), say $|f^{(n)}(x)| \leq M$, then (B.3) implies that the remainder term is bounded by $\frac{M}{n!}|b-a|^n$ for all $x \in (a,b)$.

Let $f : U \to \mathbf{R}$ be defined on the open set $U \subset \mathbf{R}^n$ and suppose that all second order partial derivatives of f exist at $x \in U$. If $v = (v_1, \ldots, v_n)$ is any point in $\mathbf{R}^n$, then we write

$$f^{(2)}(x; v) := \sum_{i=1}^{n} \sum_{j=1}^{n} \frac{\partial^2 f}{\partial v_i \partial v_j}(x) v_j v_i.$$

If all third order partial derivatives of f exist at x, then we write

$$f^{(3)}(x;v) := \sum_{i=1}^{n}\sum_{j=1}^{n}\sum_{k=1}^{n} \frac{\partial^3 f}{\partial v_i \partial v_j \partial v_k}(x) v_k v_j v_i.$$

There is a similar definition for $f^{(m)}(x;v)$ involving the sum over all order m partial derivatives of f at x times the corresponding m-fold products of components of v.

Theorem B.7 (Taylor's Theorem: Vector Variable Real-Valued Function) *Suppose $f : U \to \mathbf{R}$ and all partial derivatives of f of order m exist at each point of the open set $U \subset \mathbf{R}^n$. If a and b are two points of U such that $a + t(b-a) \in U$ for $0 \le t \le 1$, then there is a point z of the form $z = a + \theta(b-a)$, where $0 < \theta < 1$, such that*

$$f(b) - f(a) = \sum_{k=1}^{m-1} \frac{1}{k!} f^{(k)}(a; b-a) + \frac{1}{m!} f^{(m)}(z; b-a).$$

Proof. Apply Theorem B.6 to the function $g(t) := f(a + t(b-a))$, which is defined on the interval $(-\delta, 1+\delta)$ for some $\delta > 0$ since U is an open set. Note that $g(1) - g(0) = f(b) - f(a)$. Apply the chain rule repeatedly to find that $g^{(m)}(t) = f^{(m)}(a + t(b-a); b-a)$. Since

$$g(1) - g(0) = \sum_{k=1}^{m-1} \frac{1}{k!} g^{(k)}(0) + \frac{1}{m!} g^{(m)}(\theta), \quad \text{where } 0 < \theta < 1,$$

the theorem follows. □

Appendix C

Ordinary Differential Equations

Chapter 2 (Mathematical Background) includes a statement of the basic existence and uniqueness theorem (Theorem 2.3) for solutions of a nonautonomous system

$$\dot{x} = f(t, x), \tag{C.1}$$

where $x \in \mathbf{R}^n$ and $f : \mathcal{I} \times D \to \mathbf{R}^n$ is a locally Lipschitz mapping of an interval times an open set $\mathcal{I} \times D \subseteq \mathbf{R} \times \mathbf{R}^n$. This appendix includes a proof of Theorem 2.3 based on the contraction mapping theorem in Banach spaces, which is also proved here. In addition, the extension (continuation) of solutions is considered in Theorem C.3 for autonomous systems, and the continuous dependence of solutions on initial conditions, and on right hand sides, is considered in Theorem C.4 and Theorem C.5, respectively.

Other texts where material on these issues may be found include [25], [33], [40], [41], [57], [81], and [91].

C.1 EXISTENCE AND UNIQUENESS OF SOLUTIONS

The purpose of this section is to prove the existence and uniqueness theorem (stated as Theorem 2.3 in the text). The theorem will be proved as an application of the contraction mapping theorem in Banach spaces.

Theorem C.1 (Contraction Mapping Theorem)
Let X be a complete normed vector space (a Banach space), and let Y be a nonempty closed subset of X. If $T : Y \to Y$ is a mapping and there exists a number $0 < k < 1$ such that

$$\|T(x) - T(y)\| \le k\|x - y\| \quad \text{for all } x, y \in Y,$$

then there is a unique point z in Y such that $T(z) = z$.

Proof. Let x_0 be any given point in Y, fixed for the argument to follow. Define

$$x_{n+1} = T(x_n) \quad \text{for } n = 0, 1, 2, \ldots.$$

Then, using the definition of the sequence x_n, we have

$$\|x_{n+1} - x_n\| = \|T(x_n) - T(x_{n-1})\| \le k\|x_n - x_{n-1}\| \le \cdots \le k^n\|x_1 - x_0\|.$$

Thus,

$$\begin{aligned}\|x_{n+m} - x_n\| &\leq \|x_{n+m} - x_{n+m-1}\| + \cdots + \|x_{n+1} - x_n\| \\ &\leq (k^{n+m-1} + \cdots + k^n)\|x_1 - x_0\| \\ &\leq k^n(1 + k + \cdots + k^{m-1})\|x_1 - x_0\| \\ &\leq k^n\Big(\frac{1-k^m}{1-k}\Big)\|x_1 - x_0\| \\ &\leq \Big(\frac{k^n}{1-k}\Big)\|x_1 - x_0\|.\end{aligned}$$

Since $0 < k < 1$, the points x_n form a Cauchy sequence in $Y \subseteq X$. Since X is complete, there is a point z in X such that $\|x_n - z\| \to 0$ as $n \to \infty$. In fact, since Y is closed, the limit point z must lie in Y. We want to show that $T(z) = z$. We have

$$\begin{aligned}\|z - T(z)\| &\leq \|z - x_n\| + \|x_n - T(z)\| \\ &\leq \|z - x_n\| + \|T(x_{n-1}) - T(z)\| \\ &\leq \|z - x_n\| + k\|x_{n-1} - z\|.\end{aligned}$$

Letting $n \to \infty$ on the right side gives $z = T(z)$. Now suppose that we also have $T(y) = y$ for some point y in Y. Then

$$\|y - z\| = \|T(y) - T(z)\| \leq k\|y - z\|.$$

Since $0 < k < 1$, we must have $y = z$. □

The result of the contraction mapping theorem is often formulated more generally in complete metric spaces; see, for example, A. Friedman, *Foundations of Modern Analysis*, Holt, Rinehart and Winston, New York, 1970. The proof is essentially the same. We merely note here that the closed subset Y of the theorem is a complete metric space (even if it is not a subspace, and hence not a Banach space in its own right) with the metric distance between two points x, y given by the norm of their difference.

In order to apply the contraction mapping theorem to establish existence and uniqueness of solutions of $\dot{x} = f(x)$, we need the result in the following exercise.

Exercise. Let $C_n[a, b]$ be the set of functions continuous on the interval $[a, b]$ and taking values in $\mathbf{R}^n$. Then $C_n[a, b]$ is a normed vector space over the real field, with the sup norm defined by

$$\|\phi\| := \sup_{a \leq t \leq b} \|\phi(t)\|,$$

where the norm on the right is any given, fixed norm on $\mathbf{R}^n$, for example, the absolute sum norm. In fact, the supremum may be replaced by the maximum, since the elements of $C_n[a, b]$ are continuous functions on $[a, b]$. Moreover, $C_n[a, b]$ is a Banach space, that is, every Cauchy sequence in

$C_n[a, b]$ converges in the sup norm to an element of $C_n[a, b]$.

If $\phi : [a, b] \to \mathbf{R}^n$ is continuous, then for any t_0 and t in $[a, b]$, we have

$$\left\| \int_{t_0}^{t} \phi(s)\, ds \right\| \leq \int_{t_0}^{t} \|\phi(s)\|\, ds,$$

where the norm on each side is the absolute sum norm.

We now apply Theorem C.1 to prove existence and uniqueness of solutions of initial value problems for locally Lipschitz systems $\dot{x} = f(x)$.

Theorem C.2 (Existence and Uniqueness)
Let D be an open set in $\mathbf{R}^{n+1}$. If $f : D \mapsto \mathbf{R}^n$ is locally Lipschitz on D, then, given any $x_0 \in D$ and any $t_0 \in \mathbf{R}$, there exists a $\delta > 0$ such that the initial value problem

$$\dot{x} = f(t, x), \quad x(t_0) = x_0 \tag{C.2}$$

has a unique solution $x(t, t_0, x_0)$ defined on the interval $[t_0 - \delta, t_0 + \delta]$; that is, if $z(t) := x(t, t_0, x_0)$, then

$$\frac{d}{dt} z(t) = f(t, z(t)) \quad \text{for } t \in [t_0 - \delta, t_0 + \delta]$$

and $z(t_0) = x(t_0, t_0, x_0) = x_0$.

Proof. Note first that a function $x(t)$ is a solution of the initial value problem (C.2) on the interval $[t_0 - \delta, t_0 + \delta]$ if and only if

$$x(t) = x_0 + \int_{t_0}^{t} f(s, x(s))\, ds \quad \text{for all} \quad t \in [t_0 - \delta, t_0 + \delta]. \tag{C.3}$$

Let $N_1 \subset D$ be an open set about (t_0, x_0) on which $\|f(t, x)\| \leq M$ for all $(t, x) \in N_1$, and such that

$$\|f(t, x_1) - f(t, x_2)\| \leq L\|x_1 - x_2\|$$

for any (t, x_1), (t, x_2) in N_1. Choose $\delta > 0$ such that the rectangular set

$$R = \{(t, x) : t_0 - \delta \leq t \leq t_0 + \delta, \ \|y - y_0\| \leq \delta M\}$$

is contained in N_1 and such that $\delta L < 1$. The space $C_n[t_0 - \delta, t_0 + \delta]$ is a Banach space with the sup norm. Consider the subset Y of $C_n[t_0 - \delta, t_0 + \delta]$ consisting of those functions $\phi(t)$ for which $\phi(t_0) = x_0$ and

$$\|\phi(t) - x_0\| \leq \delta M \quad \text{for all } t \in [t_0 - \delta, t_0 + \delta].$$

Then Y is a closed subset of $C_n[t_0 - \delta, t_0 + \delta]$ (Exercise). Define $T : Y \to C_n[t_0 - \delta, t_0 + \delta]$ by

$$(T\phi)(t) = x_0 + \int_{t_0}^{t} f(s, \phi(s))\, ds \quad \text{for } t \in [t_0 - \delta, t_0 + \delta].$$

Then $(T\phi)(t)$ is a continuous function on $[t_0 - \delta, t_0 + \delta]$. A fixed point of the mapping T must be a solution of the initial value problem (C.2), and, conversely, a solution of (C.2) is a solution of the integral equation (C.3) and therefore, by the choice of δ, this solution lies in the set Y, hence it is a fixed point of the mapping T. Thus, we want to show that $T : Y \to Y$. We have $(T\phi)(t_0) = x_0$, and

$$\begin{aligned}\|(T\phi)(t) - x_0\| &= \left\| \int_{t_0}^{t} f(s, \phi(s))\, ds \right\| \\ &\leq \sup_{t_0-\delta\leq t\leq t_0+\delta} \int_{t_0}^{t} \|f(s, \phi(s))\|\, ds \leq \delta M.\end{aligned}$$

Therefore $T\phi \in Y$. Moreover, T is a contraction mapping on Y, since, for each t, we have

$$\begin{aligned}\|(T\phi_1)(t) - (T\phi_2)(t)\| &= \left\| \int_{t_0}^{t} f(s, \phi_1(s))\, ds - \int_{t_0}^{t} f(s, \phi_2(s))\, ds \right\| \\ &\leq L\delta \max_{t_0-\delta\leq t\leq t_0+\delta} \|\phi_1(s) - \phi_2(s)\| \\ &= L\delta \|\phi_1 - \phi_2\|.\end{aligned}$$

Hence, using the sup norm,

$$\|T\phi_1 - T\phi_2\| \leq L\delta \|\phi_1 - \phi_2\|$$

and $L\delta < 1$. Thus, by Theorem C.1, T has a unique fixed point in Y which is the desired unique solution of (C.2) on $[t_0 - \delta, t_0 + \delta]$. □

C.2 EXTENSION OF SOLUTIONS

We now address the continuation (or extension) of solutions to a maximal interval of definition, and we include a result on the behavior of solutions at the boundaries of the maximal interval.

We assume that the vector field f in (C.1) is C^1, or at least locally Lipschitz in x. Each solution of (C.1) has a maximal open interval of existence J. For a given initial condition x_0, we may extend the solution satisfying $x(t_0) = x_0$, using the basic existence and uniqueness theorem in a step by step manner, until the solution is defined on an interval $(t_-(x_0), t_+(x_0))$ but on no larger open interval containing t_0. The endpoints of this maximal interval will depend on x_0, in general. For simplicity in notation, given an initial value x_0, we write

$$(t_-, t_+) := (t_-(x_0), t_+(x_0)).$$

This maximal interval of definition is an open interval, since solutions always exist on some interval starting from an initial point in the domain $\mathcal{I} \times D$ of the vector field. Thus, if $(\hat{t}, \hat{x})$ is any point in the domain $\mathcal{I} \times D$ where a solution $x(t)$ is defined, this solution can be continued in some interval about $\hat{t}$. The continuation of the solution may proceed as long as the curve $(t, x(t))$ remains within the domain $\mathcal{I} \times D$. Specific examples of maximal intervals of definition for initial value problems appear in the text.

Let f be locally Lipschitz on the open set $D \subseteq \mathbf{R}^n$ and let $\phi(t, x_0)$ be the solution of the autonomous system $\dot{x} = f(x)$ with $\phi(0, x_0) = x_0 \in D$. We also write $\phi_t(x_0) = \phi(t, x_0)$. Given x_0, let (t_-, t_+) be the maximal interval of definition of the solution $\phi_t(x_0)$. The set

$$\{\phi_t(x_0) \, : \, t_- < t < t_+\}$$

is called the *orbit through* x_0; and the set

$$\{\phi_t(x_0) \, : \, 0 \le t < t_+\}$$

is the *forward orbit through* x_0.

Lemma C.1 *Let $x_0 \in D$. For all t, s for which $t + s \in (t_-, t_+)$, we have*

$$\phi_{t+s}(x_0) = \phi_t \circ \phi_s(x_0) = \phi_t(\phi_s(x_0)), \tag{C.4}$$

or, equivalently, $\phi(t + s, x_0) = \phi(t, \phi(s, x_0))$.

Proof. Let s be fixed but arbitrary such that $\phi_s(x_0)$ is defined. Let $x(t) = \phi_{t+s}(x_0)$ and $y(t) = \phi_t(\phi_s(x_0))$. By definition, $x(t)$ and $y(t)$ are local solutions of $\dot{x} = f(x)$ defined for t near zero. The solution $x(t)$ has initial condition $x(0) = \phi_s(x_0)$, as does $y(t)$. By the local uniqueness theorem, $x(t) = y(t)$ for some interval of t values near zero. This argument applies for any choice of the point $\phi_s(x_0)$ on the orbit through x_0. This completes the proof. □

The collection of all time t mappings given by $\phi_t(x) = \phi(t, x)$, $x \in D$, is called the *flow* of the differential equation $\dot{x} = f(x)$, or the flow of the vector field f. For example, the flow of the differential equation $\dot{x} = x^2$ is given by

$$\phi_t(x) = \phi(t, x) = \frac{x}{1 - xt}.$$

The flow is defined for $t \in (-\infty, 1/x)$ if $x > 0$, for $t \in (1/x, \infty)$ if $x < 0$, and for all real t if $x = 0$. The property in Lemma C.1 is not difficult to verify directly for this flow; indeed, we have

$$\phi_t(\phi_s(x_0)) = \frac{\phi_s(x_0)}{1 - \phi_s(x_0)t} = \frac{x_0}{1 - x_0(t + s)} = \phi_{t+s}(x_0).$$

On the other hand, the equation $\dot{x} = tx^2$ has solution given by

$$\phi(t, t_0, x_0) = \frac{2x_0}{2 - x_0(t^2 - t_0^2)}, \quad \text{where} \quad \phi(t_0, t_0, x_0) = x_0.$$

Taking $t_0 = 0$, one can see that

$$\phi(t + s, 0, x_0) \neq \phi(t, s, \phi(s, 0, x_0)).$$

We now discuss the behavior of a solution $x(t)$ as time t approaches either endpoint of its maximal interval (t_-, t_+). We consider this question first for autonomous systems and the initial value problem

$$\dot{x} = f(x), \quad x(0) = x_0, \tag{C.5}$$

where $f : D \to \mathbf{R}^n$ and D is an open subset of $\mathbf{R}^n$. The domain of (C.5) is $\mathbf{R} \times D \subseteq \mathbf{R} \times \mathbf{R}^n$ since there is no explicit dependence of the right side on t. With no loss in generality we take the initial time to be $t_0 = 0$ in (C.5).

The next theorem says that if a solution $\phi_t(x_0)$ is not defined for all time, then it must leave all compact subsets of the domain D as $t \to t_-$ or as $t \to t_+$. References for the next theorem include [40] (page 171) and [81] (page 144).

Theorem C.3 (Behavior of Solutions at Time Boundaries)
Let D be open in $\mathbf{R}^n$ and $f : D \to \mathbf{R}^n$ a C^1 function.

(a) *Given x_0 in D, let (a, b) be the maximal open interval of definition of the solution $\phi_t(x_0)$. Let K be an arbitrary compact subset of D. If $b < \infty$, then there is a time t_K with $0 \le t_K < b$ such that $\phi_{t_K}(x_0) \notin K$. Similarly, if $a > -\infty$, then there is a time t_K with $a < t_K \le 0$ such that $\phi_{t_K}(x_0) \notin K$.*

(b) *If the vector field f is globally defined, that is $f : \mathbf{R}^n \to \mathbf{R}^n$, and $\|f(x)\|$ is bounded, then all solutions exist for all time t.*

Proof. (a) Let K be a compact subset of D. Since f is C^1, there are constants $M > 0$ and $L > 0$ such that $\|f(x)\| \le M$ and $\|f(x) - f(z)\| \le L\|x - z\|$ for $x, z \in K$. For as long as the solution $\phi_t(x_0)$ remains in K, we have

$$\|\phi_t(x_0) - \phi_s(x_0)\| \le M|t - s|.$$

Suppose that $\phi_t(x_0) \in K$ for $0 \le t < b$. Then the limit

$$\lim_{t \to b} \phi_t(x_0) =: \phi_b(x_0)$$

exists as a finite limit. By the existence and uniqueness theorem, there are a $\delta > 0$ and a solution defined on the interval $(b - \delta, b + \delta)$ which agrees with $\phi_b(x_0)$ at $t = b$. But this says that (a, b) is not the maximal open interval of definition of $\phi_t(x_0)$, a contradiction. Therefore the solution $\phi_t(x_0)$ must leave the set K before time $t = b$. The argument for the behavior as $t \to a$ is similar.

(b) If $\|f(x)\| \le M$ for all x, then we have $\|\phi_t(x_0) - x_0\| \le M|t|$ for all t for which the solution is defined. Hence, the solution must remain within the ball $B_R(x_0)$ for $|t| \le R/M$. This is true for any $R > 0$. By part (a), the solution $\phi_t(x_0)$ must have maximal open interval of definition $(-\infty, \infty)$. □

In particular, if there is a compact set K which is invariant for $\dot{x} = f(x)$, then every solution starting in K exists for all $t \geq 0$, and thus the system is forward complete on K.

A similar result holds for nonautonomous systems. Let D be an open set in $\mathbf{R}^{n+1}$ and $f : D \to \mathbf{R}^n$ a continuous function. If (a, b) is the maximal open interval of existence of a solution $x(t)$ of $\dot{x} = f(t, x)$, then $(t, x(t))$ approaches the boundary of D as $t \to a$ and as $t \to b$ [33] (page 17), that is, $(t, x(t))$ must leave every compact subset of D.

It may be useful conceptually to realize that every autonomous system $\dot{x} = f(x)$ can be time-reparameterized to be a complete system. Arc length s can be used as a new independent variable, using the definition

$$\frac{ds}{dt} = \sqrt{1 + \|f(x)\|^2}.$$

The new differential equation is

$$\frac{dx}{ds} = \frac{dx}{dt}\frac{dt}{ds} = \frac{1}{\sqrt{1 + \|f(x)\|^2}} f(x).$$

The right-hand side is now bounded and all solutions exist on the interval $-\infty < s < \infty$. We may think of the solution curves in the phase portrait for $\frac{dx}{dt} = f(x)$ being traversed at a different speed given by $\frac{dx}{ds}$, for $-\infty < s < \infty$.

C.3 CONTINUOUS DEPENDENCE

The main reference for this section is [40]. We begin with Gronwall's inequality, which is the key to showing that solutions of ordinary differential equations depend continuously on initial conditions and on the vector field.

Lemma C.2 (Gronwall's Inequality)
If $u(t)$ and $v(t)$ are nonnegative continuous functions on $[a, \beta)$ that satisfy

$$u(t) \leq M + \int_a^t v(s)u(s)\, ds \quad \text{for } t \geq a,$$

where $M \geq 0$ is constant, then

$$u(t) \leq M e^{\int_a^t v(s)\, ds} \quad \text{for } t \geq a.$$

Proof. Let $U(t) := M + \int_a^t v(s)u(s)\, ds$ for $a \leq t < \beta$. By assumption, $u(t) \leq U(t)$ for $t \geq a$. By the fundamental theorem of calculus,

$$\dot{U}(t) = v(t)u(t) \leq v(t)U(t),$$

where we have used the fact that $v(t) \geq 0$. Multiply both sides by $e^{-\int_a^t v(s)\, ds}$, and use the product rule, to obtain

$$\frac{d}{dt}\left[U(t)e^{-\int_a^t v(s)\, ds}\right] \leq 0.$$

Now integrate both sides from a to t to get

$$U(t)e^{-\int_a^t v(s)\,ds} - U(a) \leq 0.$$

Since $U(a) = M$ and $u(t) \leq U(t)$, we have

$$u(t) \leq U(t) \leq Me^{\int_a^t v(s)\,ds} \quad \text{for } t \geq a.$$

This completes the proof of Gronwall's inequality. □

In Gronwall's inequality we may have β finite or infinite; note that all statements in the proof hold for $a \leq t < \beta$.

Gronwall's inequality leads to the next result on continuous dependence of solutions on initial conditions.

Theorem C.4 (Continuous Dependence on Initial Conditions)
Suppose that in a neighborhood B of (t_0, x_0), $f(t,x)$ satisfies an estimate

$$\|f(t,x) - f(t,y)\| \leq K\|x-y\|.$$

Suppose $x^1(t)$ and $x^2(t)$ are solutions of $\dot{x} = f(t,x)$ with initial conditions $(t_0, x^1(t_0))$, $(t_0, x^2(t_0))$ in B. Then, for all $t \geq t_0$ for which these solutions exist and

$$\|f(t,x^1(t)) - f(t,x^2(t))\| \leq K\|x^1(t) - x^2(t)\|, \tag{C.6}$$

we have

$$\|x^1(t) - x^2(t)\| \leq \|x^1(t_0) - x^2(t_0)\|\, e^{K(t-t_0)}.$$

Proof. Write $a = x^1(t_0)$ and $b = x^2(t_0)$. For the solutions in question we have

$$x^1(t) = a + \int_{t_0}^t f(s, x^1(s))\,ds$$

and

$$x^2(t) = b + \int_{t_0}^t f(s, x^2(s))\,ds.$$

A straightforward norm estimate gives

$$\|x^1(t) - x^2(t)\| \leq \|a-b\| + \int_{t_0}^t \|f(s,x^1(s)) - f(s,x^2(s))\|\,ds.$$

As long as the Lipschitz estimate (C.6) holds, we have

$$\|x^1(t) - x^2(t)\| \leq \|a-b\| + \int_{t_0}^t K\|x^1(s) - x^2(s)\|\,ds.$$

Apply Gronwall's inequality with $u(t) := \|x^1(t) - x^2(t)\|$, $v(t) := K$ and $M := \|a-b\|$ to conclude that

$$\|x^1(t) - x^2(t)\| \leq \|x^1(t_0) - x^2(t_0)\|\, e^{K(t-t_0)},$$

for as long as the solutions exist and the estimate (C.6) holds. □

The planar system

$$\dot{x}_1 = -x_1 + x_1^2 x_2,$$
$$\dot{x}_2 = -x_2$$

provides a good example to help in understanding continuous dependence on initial conditions. In particular, consider the forward time behavior of solutions with initial condition in a small ball centered on any point on the hyperbola $x_1 x_2 = 2$. This hyperbola forms the boundary of the basin of attraction of the equilibrium $(x_1, x_2) = (0, 0)$; see Example 9.3 (where $x_1 = x$ and $x_2 = z$).

The next result is a statement about the continuous dependence of solutions on the right-hand side of the differential equation.

Theorem C.5 (Continuous Dependence on Right-Hand Sides)
Suppose $W \subset \mathbf{R} \times \mathbf{R}^n$ is an open set and $f : W \to \mathbf{R}^n$ and $g : W \to \mathbf{R}^n$ are Lipschitz in x on W. Suppose that for all $(t, x) \in W$,

$$\|f(t, x) - g(t, x)\| < \epsilon.$$

Let K be a Lipschitz constant in x for f on W; that is,

$$\|f(t, x) - f(t, y)\| \leq K\|x - y\| \quad \text{for all } (t, x) \in W.$$

If $x(t)$ and $y(t)$ are solutions of $\dot{x} = f(t, x)$, $\dot{y} = g(t, y)$, respectively, on an interval J, and $x(t_0) = y(t_0)$ for some $t_0 \in J$, then

$$\|x(t) - y(t)\| \leq \frac{\epsilon}{K}(e^{K|t-t_0|} - 1) \quad \text{for all } t \in J.$$

Proof. For $t \in J$ we have, using $x(t_0) = y(t_0)$,

$$x(t) - y(t) = \int_{t_0}^{t} (\dot{x}(s) - \dot{y}(s))\, ds = \int_{t_0}^{t} (f(s, x(s)) - g(s, y(s)))\, ds.$$

Hence,

$$\begin{aligned}
\|x(t) - y(t)\| &\leq \int_{t_0}^{t} \|f(s, x(s)) - g(s, y(s))\|\, ds \\
&\leq \int_{t_0}^{t} \|f(s, x(s)) - f(s, y(s))\|\, ds + \int_{t_0}^{t} \|f(s, y(s)) - g(s, y(s))\|\, ds \\
&\leq K \int_{t_0}^{t} \left(\|x(s) - y(s)\| + \frac{\epsilon}{K} \right) ds.
\end{aligned}$$

Letting $u(t) = \|x(t) - y(t)\|$, we have

$$u(t) + \frac{\epsilon}{K} \leq \frac{\epsilon}{K} + \int_{t_0}^{t} K\Big(u(s) + \frac{\epsilon}{K}\Big)\, ds.$$

With $v(s) := K$, Gronwall's inequality gives

$$u(t) + \frac{\epsilon}{K} \leq \frac{\epsilon}{K} e^{K(t-t_0)} \quad \text{for all } t \in J,$$

from which the result follows. □

It is possible to prove a result that combines continuous dependence on initial conditions and on right-hand sides in a single estimate; see [91] (page 486).

Appendix D

Manifolds and the Preimage Theorem; Distributions and the Frobenius Theorem

Introduction

This appendix is for readers who wish to know something more about manifolds in $\mathbf{R}^n$, including a definition, as well as standard ways of recognizing or producing manifolds. The definitions, terminology, and discussions of this appendix provide general background and a convenient reference for Chapters 10–12 of the text. The appendix includes a statement, proof, and discussion of the Frobenius theorem (Theorem D.4). This theorem is invoked in Proposition 11.2 as a key step in developing a local normal form for a nonlinear single-input single-output system having a well-defined relative degree at a point of interest. In Chapter 12 on feedback linearization, Frobenius theorem is invoked in the proof of the feedback linearization result in Theorem 12.3.

D.1 MANIFOLDS AND THE PREIMAGE THEOREM

Smooth Mappings on Subsets of $\mathbf{R}^n$

This presentation of manifolds and smooth mappings follows [31], and readers are encouraged to see that text for additional information as well as interesting graphic illustrations of manifolds.

Systems of differential equations in this text are defined on open subsets of Euclidean space $\mathbf{R}^n$. (For linear systems, $\dot{x} = Ax$, we sometimes consider complex solutions even when A is real, so we also consider subspaces of $\mathbf{C}^n$ for those systems.) Here, the main interest is in nonlinear systems, so we concentrate on the appropriate subsets of $\mathbf{R}^n$ only. Sometimes we want to focus attention on a proper subset of the initial domain, because the solution behavior is of particular interest on that subset. In order to use the techniques of differential calculus, these subsets need to be, in a precise sense, locally equivalent to open sets in a Euclidean space. We first consider, in the next paragraph, the proper equivalence relation on appropriate sets. The allowable subsets will certainly include linear subspaces of $\mathbf{R}^n$ and open sets in $\mathbf{R}^n$, but there are many more manifolds in $\mathbf{R}^n$ of lower dimension, that is, submanifolds of $\mathbf{R}^n$.

Open Domains. Suppose $F : U \to V$ is a mapping of open sets $U \subset \mathbf{R}^n$, $V \subset \mathbf{R}^m$. If all the partial derivatives of F of all orders exist (that is, F is *smooth*, or of differentiability class C^∞), and F is one-to-one and onto V, and in addition, all the partial derivatives of $F^{-1} : V \to U$ exist (so F^{-1} is also smooth, or C^∞), we say that F is a diffeomorphism and that U and V are diffeomorphic. Diffeomorphism thus provides an equivalence relation on open subsets of Euclidean spaces. (An equivalence relation on open subsets involving F and F^{-1} of class C^r could also be defined.)

If F is a smooth invertible mapping, the smoothness of the inverse F^{-1} is needed to ensure that solutions x of the equation $F(x) = y$ depend in a smooth manner on y. The fact that F is smooth and invertible does not imply that F^{-1} is smooth, as the next example shows.

Example D.1 Consider the mapping $F : (-\epsilon, \epsilon) \to \mathbf{R}$ given by $y = F(x) = x^3$. F is one-to-one and onto the image interval $(-\epsilon^3, \epsilon^3)$, but $F^{-1}(y) = \sqrt[3]{y}$ is not smooth, not even continuously differentiable, since F^{-1} is not differentiable at $y = 0$. △

More General Domains. Sometimes we need to consider mappings with domains which are not open in $\mathbf{R}^n$. As examples, a proper linear subspace of $\mathbf{R}^n$ is not open in $\mathbf{R}^n$, and a closed disk of unit radius in the x, y plane is not an open subset of $\mathbf{R}^3$. When the domain X of a mapping $F : X \to \mathbf{R}^m$ is not an open set, there may be boundary points of X at which some directional derivatives of F cannot be defined. We will say that a mapping $F : X \to \mathbf{R}^m$ of an arbitrary subset of $\mathbf{R}^n$ is smooth (C^∞) if, for each point $x \in X$ there is an open set $U \subset \mathbf{R}^n$ containing x and a smooth mapping $\hat{F} : U \to \mathbf{R}^m$ that extends F, that is, $\hat{F}(x) = F(x)$ for $x \in U \cap X$. We say that a smooth (C^∞) mapping $F : X \to Y$ of subsets of (possibly different) Euclidean spaces is a *diffeomorphism* if it is one-to-one and onto Y and the inverse mapping $F^{-1} : Y \to X$ is also smooth (C^∞).

Manifolds of Dimension k. It is convenient to have the notion of a k-dimensional manifold in Euclidean space $\mathbf{R}^n$. If S is a subset of $\mathbf{R}^n$, then S is a *k-dimensional manifold* if it is locally diffeomorphic to $\mathbf{R}^k$, that is, every point $x \in S$ has a neighborhood U contained in S (U is *relatively open* in S, meaning $U = \tilde{U} \cap S$ for some open set $\tilde{U}$ in $\mathbf{R}^n$) such that U is diffeomorphic to an open set V in $\mathbf{R}^k$.

If S is a k-dimensional manifold, then we write $\dim S = k$. A diffeomorphism $\phi : V \to U$ is a *parametrization* of the neighborhood $U \subset S$, and the inverse $\phi^{-1} : U \to V$ is a *coordinate system* on U. If we write the mapping ϕ^{-1} using coordinate functions as $\phi^{-1} = (x_1, x_2, \dots, x_k)$, the k smooth functions $x_1, \dots, x_k$ are called local coordinates on U. This allows us to write a typical point of S in local coordinates as the point $(x_1, \dots, x_k)$. An example will be given below.

It follows from the definition just given that the subset $S = \mathbf{R}^n$ is a manifold of dimension n. So we also refer to a k-dimensional manifold in $\mathbf{R}^n$ as a *submanifold* of $\mathbf{R}^n$ (for $k \le n$). We define the *codimension* of a k-dimensional manifold S in $\mathbf{R}^n$ to be $\operatorname{codim} S = n - k = \dim \mathbf{R}^n - \dim S$. Of course, the codimension depends on the ambient space under consideration.

In order to discuss examples of manifolds, it is helpful to use the inverse function theorem and the implicit function theorem.

The Inverse and Implicit Function Theorems

If U and V are subsets of Euclidean spaces, diffeomorphisms $f : U \to V$ are usually required to be continuously differentiable of class C^r for some finite r, that is, the diffeomorphisms have continuous partial derivatives of all orders less than or equal to r. The reason for this is that the manifolds in this text appear as the zero set of a set of known functions, say $\{f_1, \ldots, f_\nu\}$, that is, the solution set of the system of equations

$$\begin{aligned} f_1(x) &= 0, \\ f_2(x) &= 0, \\ &\cdots \\ f_\nu(x) &= 0, \end{aligned}$$

where x is in $\mathbf{R}^n$. We need only the fact that the resulting solution set can be locally parametrized by a mapping which is C^r, with a C^r inverse, for some r. This results in manifolds in Euclidean space that are at least as smooth as class C^r for some r appropriate to the problem under consideration. In practice, one determines that the functions involved are *sufficiently differentiable* so that the desired statements are valid. The main tools needed for such determinations are the inverse function theorem and the implicit function theorem. We recall here the statements of these important theorems.

Theorem D.1 (Inverse Function Theorem)
Let X be an open subset of $\mathbf{R}^n$ and $F : X \to \mathbf{R}^n$ a C^r mapping. Let x_0 be a point in X. If the Jacobian matrix $DF(x_0)$ is nonsingular, then there exists an open set U about x_0, $U \subset X$, such that $V := F(U)$ is open in $\mathbf{R}^n$ and the restriction of F to U is a C^r diffeomorphism onto V.

In the inverse function theorem we may have $r = \infty$, in which case the restriction of F to U is a C^∞ diffeomorphism onto V.

Theorem D.2 (Implicit Function Theorem)
Let $X \subset \mathbf{R}^k$ and $Y \subset \mathbf{R}^n$ be open sets. Let $F : X \times Y \to \mathbf{R}^n$ be a C^r mapping, write F in component form as $F = (f_1, \ldots, f_n)$, and write the points of $X \times Y$ as $(x_1, \ldots, x_k, y_1, \ldots, y_n)$. Let

$$F(x_0, y_0) = 0$$

for some point $(x_0, y_0) \in X \times Y$ *and suppose that the square Jacobian matrix*

$$\frac{\partial F}{\partial y}(x_0, y_0) = \begin{bmatrix} \frac{\partial f_1}{\partial y_1}(x_0, y_0) & \cdots & \frac{\partial f_1}{\partial y_n}(x_0, y_0) \\ \cdots & \cdots & \cdots \\ \frac{\partial f_n}{\partial y_1}(x_0, y_0) & \cdots & \frac{\partial f_n}{\partial y_n}(x_0, y_0) \end{bmatrix}$$

is nonsingular. Then there exist open neighborhoods $U_1 \subset X$ *of* x_0 *and* $U_2 \subset Y$ *of* y_0 *and a unique* C^r *mapping* $G : U_1 \to U_2$ *such that*

$$F(x, G(x)) = 0$$

for all $x \in U_1$.

In the implicit function theorem we may have $r = \infty$, in which case the unique mapping G is a C^∞ diffeomorphism.

Example D.2 *A Circle in the Plane*
The unit circle in the real plane is the solution set of the equation $x_1^2 + x_2^2 - 1 = 0$. We will use the implicit function theorem to show, by reference to the definition, that the circle is a one-dimensional manifold. Let $F(x_1, x_2) = x_1^2 + x_2^2 - 1$, so that the circle is the solution set of the equation $F(x_1, x_2) = 0$. We have

$$\frac{\partial F}{\partial x_2}(x_1, x_2) = 2x_2 \neq 0$$

at every point of the solution set other than $(\pm 1, 0)$. Clearly, in a neighborhood of any point except these two singular points, the solution set may be written as the *graph* of a unique mapping $x_1 \mapsto G(x_1)$. To be precise, in a neighborhood of (x_{10}, x_{20}) for $x_{20} > 0$, we have $G^+(x_1) = \sqrt{1 - x_2^2}$; and, in a neighborhood of (x_{10}, x_{20}) for $x_{20} < 0$, we have $G^-(x_1) = -\sqrt{1 - x_2^2}$. In either case, the projection mapping $(x_1, x_2) \mapsto x_1$ provides a local coordinate system on a portion of the circle, while the mapping $x_1 \mapsto (x_1, G^{\pm}(x_1))$ provides a parametrization of that portion.

Finally, in a neighborhood of the points $(\pm 1, 0)$, we have

$$\frac{\partial F}{\partial x_1}(\pm 1, 0) = \pm 2 \neq 0.$$

Thus, within the solution set, we can solve for x_1 in terms of x_2. The projection mapping $(x_1, x_2) \mapsto x_2$ provides a coordinate system near each point $(\pm 1, 0)$, while the functions $H^{\pm}(x_2) := \pm\sqrt{1 - x_2^2}$ can be used to define local parametrizations by means of the mappings $x_2 \mapsto (H^{\pm}(x_2), x_2) = (\pm\sqrt{1 - x_2^2}, x_2)$. $\triangle$

Example D.3 *A Cubic Curve*
In $\mathbf{R}^3$, consider the set S defined as the solution set of the two equations

$$x_1 = 0,$$
$$x_3 - x_2^3 = 0.$$

Geometrically, this set can be viewed as the *curve* $\mathcal{C}$ defined by $x_3 - x_2^3 = 0$ which sits within the *plane* defined by $x_1 = 0$. This curve is indeed a one-dimensional manifold in $\mathbf{R}^3$ of differentiability class C^1. (But this fact does not follow from any false parametrization of the curve using the formula $x_3 = x_2^3$ and its inverse. Why?) We will use the inverse function theorem to understand why $\mathcal{C}$ is a one-dimensional manifold in $\mathbf{R}^3$. We consider first a portion of the curve near the origin. We can map a neighborhood U of the point $(0, 0, 0)$ in $\mathcal{C}$ to $\mathbf{R}$ by the mapping $(x_1, x_2, x_3) \mapsto x_2$ for (x_1, x_2, x_3) in $\mathcal{C}$. To see that this defines a scalar coordinate system for the curve $\mathcal{C}$, we now apply the inverse function theorem to see that it gives the required C^1 diffeomorphism of a neighborhood $U \subset \mathcal{C}$. To do so, consider the mapping $z = F(x)$ defined by

$$z_1 = x_1,$$
$$z_2 = x_3 - x_2^3,$$
$$z_3 = x_2.$$

The Jacobian matrix of this mapping at a point $x = (x_1, x_2, x_3)$ is

$$DF(x) = \begin{bmatrix} 1 & 0 & 0 \\ 0 & -3x_2^2 & 1 \\ 0 & 1 & 0 \end{bmatrix},$$

and it is straightforward to check that $\det DF(x) \neq 0$ for all x, so that $DF(x)$ is nonsingular for all x. In particular it is nonsingular at $x = 0$. Thus, by the inverse function theorem, F is at least a C^1 diffeomorphism in a neighborhood of $x = 0$. Observe that it is possible to explicitly write the C^1 inverse of F as the mapping $(x_1, x_2, x_3) = F^{-1}(z) = (z_1, z_3, z_2 + z_3^3)$, that is,

$$x_1 = z_1,$$
$$x_2 = z_3,$$
$$x_3 = z_2 + z_3^3.$$

Note that the curve $\mathcal{C}$ is described in the z coordinate system by $\mathcal{C} = \{z : z_1 = 0, z_2 = 0\}$. Thus, $\mathcal{C}$ is parametrized by the mapping

$$\phi : z_3 \mapsto F^{-1}(0, 0, z_3) = (0, z_3, z_3^3) = (0, x_2, x_2^3),$$

defined on an interval about the origin of the real line, and $\mathcal{C}$ is provided a coordinate system by the mapping

$$\phi^{-1} : (x_1, x_2, x_3) \mapsto x_2 = z_3,$$

defined on a portion of the curve $\mathcal{C}$, a portion which contains the origin in $\mathbf{R}^3$, and is relatively open in $\mathcal{C}$. This justifies the statement that x_2 serves as a scalar coordinate for the curve $\mathcal{C}$ in a neighborhood of the origin.

Next, observe that similar constructions provide the required parametrizations and coordinate systems in a neighborhood of any other point of $\mathcal{C}$.

Finally, note that $\mathcal{C}$ has codimension 2 in $\mathbf{R}^3$: $\operatorname{codim} \mathcal{C} = 2$. △

The Preimage Theorem: Constructing and Recognizing Manifolds

The implicit function theorem helped us to see why the circle in Example D.2 is a manifold, by reference to the definition of manifold. And we used the inverse function theorem to show that the cubic curve $\mathcal{C}$ in Example D.3 was a one-dimensional manifold in $\mathbf{R}^3$. Both theorems can be useful. The implicit function theorem may also be used to define manifolds in $\mathbf{R}^n$, by means of the preimage theorem (Theorem D.3 below). The zero dynamics manifolds in Chapter 11 appear as the zero set of a finite set of known functions, say $\{f_1, \ldots, f_\nu\}$. The preimage theorem guarantees that the zero sets of interest there are submanifolds of $\mathbf{R}^n$.

Let $F : X \subset \mathbf{R}^{k+n} \to \mathbf{R}^n$. A value y in the range of F is a *regular value* of F if $DF(x)$ maps onto $\mathbf{R}^n$ for every point x such that $F(x) = y$. So y is a regular value if and only if $\operatorname{rank} DF(x) = n$ for every x such that $F(x) = y$. When rank $DF(x) = n$, there are n linearly independent columns (and n linearly independent rows) of $DF(x)$. By a relabeling of the components of x, these linearly independent columns may be permuted to be the last n columns. Then the implicit function theorem applies at each point x in the preimage set $F^{-1}(y)$.

In [31], the next theorem is called the preimage theorem.

Theorem D.3 (Preimage Theorem)
Let $F : X \subset \mathbf{R}^{k+n} \to \mathbf{R}^n$ be a smooth function on an open set X in $\mathbf{R}^{k+n}$. If y is a regular value of F, then the set

$$F^{-1}(y) = \{x \, : \, F(x) = y\}$$

is a k-dimensional manifold in $\mathbf{R}^{k+n}$.

Proof. Let y be a regular value, and let $x_0 \in F^{-1}(y)$. As noted above, we can relabel the components of x, if necessary, in order to write $x = (p, q)$ with $p \in \mathbf{R}^k$ and $q \in \mathbf{R}^n$, such that Theorem D.2 applies at $x_0 = (p_0, q_0)$. Thus, there exist open sets $U_1 \subset \mathbf{R}^k$ containing p_0 and $U_2 \subset \mathbf{R}^n$ containing q_0 and a unique smooth mapping $G : U_1 \to U_2$, such that $G(p_0) = q_0$ and $F(p, q) = 0$ for $(p, q) \in U_1 \times U_2$ if and only if $q = G(p)$. Thus we have

$$F(p, G(p)) = 0$$

for all $x=(p,q) \in U_1 \times U_2 \cap F^{-1}(y)$. Observe that the mapping $\phi : p \mapsto (p, G(p))$ is the required parametrization of the relatively open set $U_1 \times U_2 \cap F^{-1}(y)$, and $\phi^{-1} : (p,q) \mapsto p$ is a coordinate system on $U_1 \times U_2 \cap F^{-1}(y)$. □

According to the preimage theorem, if y is a regular value of $F : X \subset \mathbf{R}^{k+n} \to \mathbf{R}^n$, then the codimension of $F^{-1}(y)$ in the domain space $\mathbf{R}^{k+n}$ is n.

Note that the constructions in the proof of Theorem D.3 are exactly the constructions used in Example D.2 involving the circle in the plane. It is also instructive to view the cubic curve $\mathcal{C}$ in Example D.3 in the light of Theorem D.3.

Example D.4 The equations for the cubic curve $\mathcal{C}$ are

$$x_1 = 0,$$
$$x_3 - x_2^3 = 0,$$

which we write as $F(x) = 0$, $F : \mathbf{R}^3 \to \mathbf{R}^2$ being defined by the left side of this system of equations. Let us consider the solution set near the origin $x = (0,0,0)$. We compute that

$$DF(x) = \begin{bmatrix} 1 & 0 & 0 \\ 0 & -3x_2^2 & 1 \end{bmatrix}.$$

The first and last columns of $DF(x)$ are linearly independent, so we can take $q := (x_1, x_3)$, $p := x_2$, and, according to Theorem D.2 (as well as the proof of Theorem D.3), within some open set of the form $U_1 \times U_2 \subset \mathbf{R} \times \mathbf{R}^2 \equiv \mathbf{R}^3$, the solution set can be expressed as the *graph* of a mapping $G : U_1 \to U_2$. So, within $U_1 \times U_2$, the solutions of the system of equations must take the form $(p, G(p))$. In this case, we have $G(p) = G(x_2) = (0, x_2^3) = q$. A similar analysis applies in a neighborhood of any other point x of the solution set, since the first and last columns of $DF(x)$ are always linearly independent. Compare the construction here with the construction of the mappings ϕ and ϕ^{-1} of Example D.3. △

The preimage theorem provides a way to construct, or to recognize, manifolds. It is usually much easier to invoke Theorem D.3 than to construct local parametrizations, in order to simply recognize a manifold when it is presented to us. Here are two more examples of manifolds defined by means of Theorem D.3.

Example D.5 *Spheres in* $\mathbf{R}^n$
Let $(a_1, \dots, a_n)$ be a point of $\mathbf{R}^n$ and d a nonzero real number. Consider the mapping $f : \mathbf{R}^n \to \mathbf{R}$ given by

$$f(x) = (x_1 - a_1)^2 + \cdots + (x_n - a_n)^2 - d^2.$$

The Jacobian matrix of f at x is $Df(x) = df(x) = [2(x_1 - a_1) \; \cdots \; 2(x_n - a_n)]$. It follows that $y = 0$ is a regular value of f. By Theorem D.3, every sphere in $\mathbf{R}^n$ of nonzero radius is an $(n-1)$-dimensional manifold in $\mathbf{R}^n$. △

Example D.6 *A Zero Dynamics Manifold*
Consider a nonlinear system on $\mathbf{R}^n$ given by $\dot{x} = f(x) + g(x)u$ with output $y = h(x)$, and $f(0) = 0$. Assume that h has relative degree $r < n$ at $x_0 = 0$. The set

$$Z = \{x : h(x) = L_f h(x) = \cdots = L_f^{r-1} h(x) = 0\},$$

intersected wih a suitable neighborhood U of $x_0 = 0$, can be rendered invariant by a uniquely defined feedback, the zero dynamics feedback. The set $Z \cap U$ is a manifold in $\mathbf{R}^n$ of dimension $n - r$. It is the intersection with U of the preimage of the origin under the mapping

$$x \mapsto (h(x), L_f h(x), \ldots, L_f^{r-1} h(x))$$

from $\mathbf{R}^n$ to $\mathbf{R}^r$. This mapping has full rank r at $x_0 = 0$ by Proposition 11.1 (a). △

Many other examples of manifolds can be given by the application of the Preimage Theorem. See [31] for more information.

D.2 DISTRIBUTIONS AND THE FROBENIUS THEOREM

This section is guided by [47], [85], and [91], and readers should also consult these references for more on distributions and Frobenius' theorem; in particular, [91] (pages 164–176) contains a detailed development of distributions with exercises.

The space $\mathbf{R}^n$ is the n-dimensional space of real column n-vectors. Recall that the dual space $(\mathbf{R}^n)^*$ of $\mathbf{R}^n$ is the vector space of all linear functionals on $\mathbf{R}^n$; we may identify the dual space of $\mathbf{R}^n$ with the n-dimensional space of row vectors; a row vector w is identified with the linear functional $v \mapsto wv$, $v \in \mathbf{R}^n$. A *vector field* on $\mathbf{R}^n$ is an assignment of a vector at each point $x \in \mathbf{R}^n$, and a *covector field* is an assignment of a covector (linear functional) $\omega(x)$ at each point $x \in \mathbf{R}^n$. If these assignments are accomplished by smooth functions defined on an open set, then we refer to a *smooth vector field* and *smooth covector field*.

We now give the definitions of *distribution* (in the differential geometric sense), *codistribution*, and the *annihilator (codistribution)* of a given distribution. We also discuss distributions defined as the span of a set of smooth vector fields and codistributions defined as the span of a set of smooth covector fields.

Definition D.1 (Distributions, Codistributions, Annihilators, Spans)
Let U be an open set in $\mathbf{R}^n$.

- *A* smooth distribution *on U is a smooth mapping $\mathcal{D} : x \mapsto \mathcal{D}(x)$ which assigns, for each $x \in U$, a subspace $\mathcal{D}(x)$ of $\mathbf{R}^n$.*
- *A* smooth codistribution *on U is a smooth mapping $\Omega : x \mapsto \Omega(x)$ which assigns, for each $x \in U$, a subspace $\Omega(x)$ of the dual space $(\mathbf{R}^n)^*$ of linear functionals on $\mathbf{R}^n$.*
- *If $\mathcal{D}(x)$ is a smooth distribution on U, then its* annihilator *is a codistribution on U that assigns, for each $x \in U$, a subspace $\Omega(x)$ of the dual space $(\mathbf{R}^n)^*$ of linear functionals on $\mathbf{R}^n$, written $\Omega = \mathcal{D}^\perp$, such that $\omega v = 0$ for every ω in Ω and v in $\mathcal{D}$.*
- *Given a set of smooth vector fields $v_1(x)$, $v_2(x)$, ..., $v_m(x)$ defined on U, the* distribution spanned by these vector fields *is*

$$\mathcal{D} := \text{span}\{v_1, v_2, \ldots, v_m\},$$

where the span is meant over the ring of smooth functions; that is, the elements of the span have the form

$$\alpha_1(x)v_1(x) + \cdots + \alpha_m(x)v_m(x),$$

where the α_i are smooth real-valued functions on U.
- *Given a set of smooth covector fields $\omega_1(x)$, $\omega_2(x)$, ..., $\omega_m(x)$ defined on U, the* codistribution spanned by these covector fields *is*

$$\Omega(x) := \text{span}\,\{\omega_1, \omega_2, \ldots, \omega_m\},$$

where the span is meant over the ring of smooth functions; that is, the elements of the span have the form

$$\beta_1(x)\omega_1(x) + \cdots + \beta_m(x)\omega_m(x),$$

where the β_i are smooth real-valued functions on U.

A smooth distribution need not have constant dimension. The point x_0 is a *regular point* of a distribution $\mathcal{D}$ if $\mathcal{D}$ has constant dimension in a neighborhood of x_0; that is, $\dim \mathcal{D}(x) = r$ for all x in some neighborhood of x_0.

It can be shown that if $\mathcal{D}$ is the span of m smooth vector fields and x_0 is a regular point of $\mathcal{D}$, then x_0 is also a regular point of the annihilator $\mathcal{D}^\perp$; moreover, the annihilator $\mathcal{D}^\perp$ is a smooth codistribution in some neighborhood of the regular point x_0.

Definition D.2 (Nonsingular Distribution, Dimension, Codimension)
Let U be an open set in $\mathbf{R}^n$. A distribution $\mathcal{D}$ is nonsingular *in U if* $\dim \mathcal{D}(x) = k$ *is constant in U, and then we write* $\dim \mathcal{D} = k$ *on U and say that $\mathcal{D}$ has* dimension k *on U. If $\mathcal{D}$ is nonsingular in $U \subset \mathbf{R}^n$ and* $\dim \mathcal{D} = k$ *on U, then we say $\mathcal{D}$ has* codimension $n - k$, *written* $\text{codim}\, \mathcal{D} = n - k$.

If $\mathcal{D}$ is a smooth nonsingular distribution on an open set U and $\dim \mathcal{D}(x) = k$ for all x, then it can be represented as the span of k smooth vector fields $v_1, \ldots, v_k$. By the remarks just before Definition D.2, the annihilator $\mathcal{D}^\perp$ is also a nonsingular (constant dimension) codistribution of dimension $n-k$, and thus $\mathcal{D}^\perp$ can be represented as the span of $n-k$ smooth covector fields $\omega_1, \ldots, \omega_{n-k}$. Hence,

$$\omega_j(x)v_i(x) = 0 \quad \text{for } 1 \le i \le k,\ 1 \le j \le n-k.$$

The Frobenius theorem answers the question of whether these covector fields are exact, that is, whether there exist smooth functions $\lambda_1, \ldots, \lambda_{n-k}$ such that $\omega_j = d\lambda_j = \frac{\partial \lambda_j}{\partial x}$ for $1 \le j \le n-k$. The question of exactness of the ω_j is equivalent to the solvability of the k partial differential equations

$$\frac{\partial \lambda_j}{\partial x}(x)[v_1(x) \cdots v_k(x)] = 0 \tag{D.1}$$

by $n-k$ independent smooth functions $\lambda_1, \ldots, \lambda_{n-k}$, where independence of the λ_j means that their differentials $d\lambda_j$, $1 \le j \le n-k$, are linearly independent at each x in U.

We recall from Definition 11.2 that the Lie bracket $[g_1, g_2]$ of two smooth vector fields g_1, g_2 is the vector field

$$[g_1, g_2](x) := \frac{\partial g_2}{\partial x}(x)g_1(x) - \frac{\partial g_1}{\partial x}(x)g_2(x) = Dg_2(x)g_1(x) - Dg_1(x)g_2(x).$$

Next is the definition of *involutivity* and *complete integrability* of a distribution.

Definition D.3 (Involutivity, Complete Integrability of a Distribution) *Let U be an open set in $\mathbf{R}^n$.*

- *A distribution $\mathcal{D}(x)$ is* involutive *in U if, for vector fields v_1 and v_2, and all $x \in U$,*

$$v_1(x), v_2(x) \in \mathcal{D}(x) \Longrightarrow [v_1, v_2](x) \in \mathcal{D}(x).$$

- *A nonsingular distribution $\mathcal{D}$ with $\dim \mathcal{D}(x) = k$ is* completely integrable *in U if for each point x_0 in U there is a neighborhood U_0 of x_0 and $n-k$ smooth real-valued functions λ_j, $1 \le j \le n-k$, defined on U_0, such that*

$$\text{span}\{d\lambda_j(x) : 1 \le j \le n-k\} = \mathcal{D}^\perp(x) \quad \textit{for all } x \in U_0,$$

or, equivalently,

$$\bigcap_{j=1}^{n-k} N(d\lambda_j(x)) = \mathcal{D}(x) \quad \textit{for all } x \in U_0.$$

If the right-hand side of $\dot{x} = f(x)$ has continuous partial derivatives of order r with respect to all components of x, then the solution $\phi_t(x) = \phi(t, x)$

also has continuous partial derivatives of order r with respect to t and all components of x; see [25]. (A C^r vector field has a C^r flow.)

In the next lemma, we assume a C^2 vector field so that we may interchange the order of differentiation in second order partial derivatives. We also write $G_*(x) = \frac{\partial G}{\partial x}(x)$ for the Jacobian at x of a smooth mapping G of an open set in $\mathbf{R}^n$ to $\mathbf{R}^n$. Thus, we may write $(\phi_t)_*(x) = \frac{\partial}{\partial x}\phi(t,x) = \frac{\partial \phi_t}{\partial x}(x)$.

Lemma D.1 *Let $f : U \to \mathbf{R}^n$ be a C^2 vector field and $\phi_t(x)$ its flow. Then for any C^1 vector field $\eta : U \to \mathbf{R}^n$,*

$$\frac{d}{dt}[(\phi_{-t})_*(\phi_t(x))\eta(\phi_t(x))] = (\phi_{-t})_*(\phi_t(x))[f,\eta](\phi_t(x)). \tag{D.2}$$

Proof. First, note that

$$\frac{d}{dt}(\phi_t)_*(x) = \frac{\partial}{\partial x}f(\phi_t(x)) = \frac{\partial f}{\partial x}(\phi_t(x))\frac{\partial \phi_t}{\partial x}(x) = \frac{\partial f}{\partial x}(\phi_t(x))(\phi_t)_*(x).$$

We have $(\phi_{-t})_*(\phi_t(x))(\phi_t)_*(x) = I$, since $\phi_{-t} \circ \phi_t(x) = x$ for all (t,x) for which the flow is defined. By the formula for the derivative with respect to t of the inverse of a smooth matrix function (Exercise 2.1 (b)), we have

$$\frac{d}{dt}(\phi_{-t})_*(\phi_t(x)) = -(\phi_{-t})_*(\phi_t(x))\frac{\partial f}{\partial x}(\phi_t(x)).$$

By the product rule, we have

$$\begin{aligned}\frac{d}{dt}[(\phi_{-t})_*(\phi_t(x))\eta(\phi_t(x))] = &-(\phi_{-t})_*(\phi_t(x))\frac{\partial f}{\partial x}(\phi_t(x))\eta(\phi_t(x))\\ &+(\phi_{-t})_*(\phi_t(x))\frac{\partial \eta}{\partial x}(\phi_t(x))f(\phi_t(x)).\end{aligned}$$

Hence, by definition of $[f,\eta](\phi_t(x))$, we have (D.2). □

We can now state and prove the Frobenius theorem.

Theorem D.4 (Frobenius Theorem)
A nonsingular distribution on $U \subset \mathbf{R}^n$ is completely integrable in U if and only if it is involutive in U.

Proof. We first prove that complete integrability implies involutivity. Thus, we may assume that the completely integrable distribution $\mathcal{D}$ has dimension k. Then we may write $\mathcal{D}^\perp = \operatorname{span}\{d\lambda_1, \ldots, d\lambda_{n-k}\}$ and, for each $1 \le j \le n-k$,

$$d\lambda_j v \equiv 0$$

for any vector field v in $\mathcal{D}$. An equivalent statement is that

$$L_v\lambda_j(x) \equiv 0 \quad \text{for } 1 \le j \le n-k$$

and any vector field v in $\mathcal{D}$. By the Jacobi identity, if v_1 and v_2 are any two vector fields in $\mathcal{D}$, we have

$$L_{[v_1,v_2]}\lambda_j = L_{v_1}L_{v_2}\lambda_j - L_{v_2}L_{v_1}\lambda_j.$$

Thus,

$$L_{[v_1,v_2]}\lambda_j \equiv 0.$$

Since $\mathcal{D}^\perp$ is the span of the differentials $d\lambda_j$, we have that the vector field $[v_1, v_2]$ is in $\mathcal{D}$. Since v_1 and v_2 were arbitrary vector fields in $\mathcal{D}$, this proves involutivity of $\mathcal{D}$.

We now show that involutivity implies complete integrability. Thus we assume that $\mathcal{D}$ is involvutive in U. First, note that we may write

$$\mathcal{D} = \text{span}\,\{v_1, \ldots, v_k\}.$$

The proof is a constructive proof for the required solutions of the partial differential equations (D.1). Let $\phi_t^{v_i}(x)$ denote the solution flow for the vector field v_i. Thus, $\phi_t^{v_i}(x)$ satisfies

$$\frac{d}{dt}\phi_t^{v_i}(x) = v_i(\phi_t^{v_i}(x)), \quad \phi_0^{v_i}(x) = x.$$

Let $v_{k+1}, \ldots, v_n$ be vector fields complementary to $v_1, \ldots, v_k$ in U, that is, such that

$$\text{span}\{v_1, \ldots, v_k, v_{k+1}, \ldots, v_n\} = \mathbf{R}^n.$$

Fix x_0. Consider the mapping $F(z)$ from a neighborhood of the origin to $\mathbf{R}^n$ defined by

$$F(z_1, \ldots, z_n) := \phi_{z_1}^{v_1} \circ \phi_{z_2}^{v_2} \circ \cdots \circ \phi_{z_n}^{v_n}(x_0),$$

where $\circ$ denotes composition of the time z_i flow maps with respect to the space argument x. We need two important facts about F:

(1) F is defined in an open set W about the origin, and F is a diffeomorphism onto its range $F(W)$, that is, F is smooth with a smooth inverse $F^{-1} : F(W) \to W$.
(2) The first k columns of the Jacobian matrix $[\frac{\partial F}{\partial z}(z)]$ are linearly independent vectors in $\mathcal{D}(F(z))$.

Before establishing these two properties, we show that they imply a construction of the required solutions of the partial differential equations (D.1). Thus, we assume properties (1) and (2) for the moment. By (1), since W is open, $F(W)$ is an open set by the basic existence theorem for ODEs, and $F(W)$ contains the origin since $F(0, \ldots, 0) = x_0$. By (1), F^{-1} is smooth on $F(W)$; write

$$F^{-1}(x) = \begin{bmatrix} \lambda_1(x) \\ \cdots \\ \lambda_n(x) \end{bmatrix},$$

where the λ_i are smooth real-valued functions defined on $F(W)$. We now show that the last $n-k$ component functions of F^{-1}, namely, $\lambda_{k+1}, \ldots, \lambda_n$, are independent solutions of (D.1). To see this, note that, by the definition of inverse and the chain rule, we have

$$\left[\frac{\partial F^{-1}}{\partial x}\right]_{x=F(z)} \left[\frac{\partial F}{\partial z}\right]_z = I \quad \text{for all } z \in W, x \in F(W),$$

where I is the $n \times n$ identity matrix. Thus, the last $n-k$ rows of the Jacobian of F^{-1} must annihilate the first k columns of the Jacobian of F. That is, the differentials $d\lambda_{k+1}, \ldots, d\lambda_n$ annihilate the first k columns of the Jacobian of F. By property (2), the first k columns of the Jacobian of F span the distribution $\mathcal{D}$. Since, by construction, these $n-k$ differentials are independent, we have the required solutions of (D.1) and therefore the distribution $\mathcal{D}$ is completely integrable.

To complete the proof of the theorem, we now show that properties (1) and (2) hold for the mapping F.

Proof of (1): By the basic existence theorem for ODEs, for each x, the flow maps $\phi_t^{v_i}(x)$ are well-defined for all t in some interval about the time 0. Thus, F is defined in some open set about the origin. We write $M_* = \frac{\partial M}{\partial x}$ for the differential mapping of a smooth mapping M of an open set in $\mathbf{R}^n$ to $\mathbf{R}^n$. By the chain rule, we have

$$\begin{aligned}\frac{\partial F}{\partial z_i}(z) &= (\phi_{z_1}^{v_1})_* \cdots (\phi_{z_{i-1}}^{v_{i-1}})_* \frac{\partial}{\partial z_i}(\phi_{z_i}^{v_i} \circ \cdots \circ \phi_{z_n}^{v_n}(x_0)) \\ &= (\phi_{z_1}^{v_1})_* \cdots (\phi_{z_{i-1}}^{v_{i-1}})_* v_i(\phi_{z_i}^{v_i} \circ \cdots \circ \phi_{z_n}^{v_n}(x_0)).\end{aligned} \tag{D.3}$$

Set $z = 0$ and use $F(0) = x_0$ to get

$$\frac{\partial F}{\partial z_i}(0) = v_i(x_0).$$

Since, by construction, the tangent vectors $v_1(x_0), \ldots, v_n(x_0)$ are linearly independent, the columns of $F_* = [\frac{\partial F}{\partial z}]$ are linearly independent. Hence, by the inverse function theorem, F has a local inverse mapping $F(W)$ to W and F^{-1} is as smooth as F.

Proof of (2): First, notice that, using the definition of the mapping F and the composition properties of flow maps, we can write

$$\phi_{z_i}^{v_i} \circ \cdots \circ \phi_{z_n}^{v_n}(x_0) = \phi_{-z_{i-1}}^{v_{i-1}} \circ \cdots \circ \phi_{-z_1}^{v_1}(F(z))$$

Thus, (D.3) may also be expressed as

$$\frac{\partial F}{\partial z_i}(z) = (\phi_{z_1}^{v_1})_* \cdots (\phi_{z_{i-1}}^{v_{i-1}})_* v_i(\phi_{-z_{i-1}}^{v_{i-1}} \circ \cdots \circ \phi_{-z_1}^{v_1}(x)), \tag{D.4}$$

where $x = F(z)$. By definition, the vector fields v_i are in $\mathcal{D}$. If we show that for all x in a neighborhood of x_0, for small $|t|$, and for any two vector fields

τ and θ belonging to $\mathcal{D}$,

$$(\phi_t^\theta)_* \tau \circ \phi_{-t}^\theta(x) \in \mathcal{D}(x),$$

so that $(\phi_t^\theta)_* \tau \circ \phi_{-t}^\theta(x)$ is a locally defined vector field belonging to $\mathcal{D}$, then repeated application of this fact in the composition (D.4) establishes item (2), namely, that the first k columns of the Jacobian $[\frac{\partial F}{\partial z}(z)]$ are vector fields belonging to $\mathcal{D}(F(z))$ for $x = F(z)$. We now proceed to show this. Let θ be a vector field belonging to $\mathcal{D}$, and define the functions

$$V_i(t) := (\phi_{-t}^\theta)_* v_i \circ \phi_t^\theta(x), \quad \text{for } 1 \le i \le k.$$

By Lemma D.1, the function $V_i(t)$ satisfies the differential equation

$$\dot{V}_i = (\phi_{-t}^\theta)_* [\theta, v_i] \circ \phi_t^\theta(x).$$

The vector fields θ and v_i belong to $\mathcal{D}$, and $\mathcal{D}$ is assumed involutive, so there exist smooth functions μ_{ij}, defined near x_0, such that

$$[\theta, v_i] = \sum_{j=1}^{k} \mu_{ij} v_j, \quad \text{for } 1 \le i \le k;$$

hence,

$$\dot{V}_i = (\phi_{-t}^\theta)_* \left(\sum_{j=1}^{k} \mu_{ij} v_j \right) \circ \phi_t^\theta = \sum_{j=1}^{k} \mu_{ij}(\phi_t^\theta(x)) V_j(t).$$

This is a system of k linear differential equations for the functions V_i; thus, there exists a fundamental matrix $X(t) \in \mathbf{R}^{k \times k}$, $X(0) = I_{k \times k}$, such that the general solution of the system is given by

$$[V_1(t) \quad \cdots \quad V_k(t)] = [V_1(0) \quad \cdots \quad V_k(0)] X(t). \tag{D.5}$$

Premultiply both sides of (D.5) by $(\phi_t^\theta)_*$ and use the fact that $V_i(0) = v_i(x)$ to get

$$[(\phi_t^\theta)_* V_1(t) \quad \cdots \quad (\phi_t^\theta)_* V_k(t)] = [(\phi_t^\theta)_* v_1(x) \quad \cdots \quad (\phi_t^\theta)_* v_k(x)] X(t).$$

Now use $(\phi_t^\theta)_* V_i(t) = v_i(\phi_t^\theta(x))$ to obtain

$$[v_1(\phi_t^\theta(x)) \quad \cdots \quad v_k(\phi_t^\theta(x))] = [(\phi_t^\theta)_* v_1(x) \quad \cdots \quad (\phi_t^\theta)_* v_k(x)] X(t).$$

Now, for small $|t|$, we may replace x in the last equation by $\phi_{-t}^\theta(x)$ to yield

$$[v_1(x) \quad \cdots \quad v_k(x)] = [(\phi_t^\theta)_* v_1 \circ \phi_{-t}^\theta(x) \quad \cdots \quad (\phi_t^\theta)_* v_k \circ \phi_{-t}^\theta(x)] X(t).$$

Since $X(t)$ is nonsingular, we multiply both sides of the last equation by $X^{-1}(t)$ on the right to get

$$[v_1(x) \quad \cdots \quad v_k(x)] X^{-1}(t) = [(\phi_t^\theta)_* v_1 \circ \phi_{-t}^\theta(x) \quad \cdots \quad (\phi_t^\theta)_* v_k \circ \phi_{-t}^\theta(x)],$$

and this equation says that each of the vector fields $(\phi_t^\theta)_* v_i \circ \phi_{-t}^\theta(x)$, for $1 \le i \le k$, is a (smooth) linear combination of the vector fields $v_1, \ldots, v_k$;

that is,

$$(\phi_t^\theta)_* v_i \circ \phi_{-t}^\theta(x) \in \text{span}\{v_1(x), \ldots, v_k(x)\} = \mathcal{D}(x), \quad 1 \le i \le k.$$

Finally, note that any vector field τ belonging to $\mathcal{D}$ can be written as

$$\tau = \sum_{i=1}^{k} c_i v_i$$

for some smooth functions c_i, and therefore the argument above completes the proof of (2). This completes the proof of the Frobenius theorem. □

In the text, the proof of Proposition 11.2 on the normal form applies the Frobenius theorem to a one-dimensional nonsingular distribution to invoke the existence of an $(n-1)$-dimensional annihilator codistribution. Another special case of the Frobenius theorem, when the distribution has the form

$$\mathcal{D}(x) = \text{span}\{g(x), ad_f g(x), \ldots, ad_f^{n-2} g(x)\}, \quad x \in U \subset \mathbf{R}^n,$$

and $\mathcal{D}$ has codimension one, is central to the feedback linearization result in Theorem 12.3 for the nonlinear single-input single-output system $\dot{x} = f(x) + g(x)u$.

The next example illustrates the Frobenius theorem.

Example D.7 Suppose the vector fields $v(x) = [\alpha_1(x) \quad \alpha_2(x) \quad \alpha_3(x)]^T$ and $w(x) = [\beta_1(x) \quad \beta_2(x) \quad \beta_3(x)]^T$ are linearly independent at each point x in some open set $U \subset \mathbf{R}^3$. Then the distribution $\mathcal{D}$ defined on U by

$$\mathcal{D}(x) = \text{span}\{v(x), w(x)\}$$

is nonsingular since $\dim \mathcal{D}(x) = 2$ at each x. Thus, by the Frobenius theorem, the system of first order partial differential equations

$$\begin{aligned} \alpha_1(x)\frac{\partial \lambda}{\partial x_1} + \alpha_2(x)\frac{\partial \lambda}{\partial x_2} + \alpha_3(x)\frac{\partial \lambda}{\partial x_3} &= 0, \\ \beta_1(x)\frac{\partial \lambda}{\partial x_1} + \beta_2(x)\frac{\partial \lambda}{\partial x_2} + \beta_3(x)\frac{\partial \lambda}{\partial x_3} &= 0, \end{aligned}$$

is solvable, in a neighborhood $\tilde{U} \subset U$ of each point x, for a smooth function $\lambda(x)$ defined on $\tilde{U}$, if and only if the distribution $\mathcal{D}$ is involutive in U, that is, if and only if

$$[v, w](x) = c_1(x)v(x) + c_2(x)w(x)$$

for some smooth functions c_1 and c_2 defined on U.

As a specific case, let $v(x) = [x_1 \quad x_2 x_3 \quad 0]^T$ and $w(x) = [0 \quad 1 \quad 0]^T$ be defined on all of $\mathbf{R}^3$. Note that

$$[v, w](x) = Dw(x) \cdot v(x) - Dv(x) \cdot w(x) = -x_3 w(x), \quad x \in \mathbf{R}^3.$$

Thus, the distribution $\mathcal{D} = \text{span}\,\{v, w\}$ is involutive in $\mathbf{R}^3$. By the Frobenius theorem, there are local solutions of the PDEs

$$x_1 \frac{\partial \lambda}{\partial x_1} + x_2 x_3 \frac{\partial \lambda}{\partial x_2} = 0,$$
$$\frac{\partial \lambda}{\partial x_2} = 0$$

around any point x in $\mathbf{R}^3$. By inspection, it is easy to see that any nonzero, globally defined function of x_3 yields a solution, which, in this case, is a global solution on $\mathbf{R}^3$. △

See Example 12.4 for an example of a nonsingular two-dimensional distribution $\mathcal{D}$ on $\mathbf{R}^3$ which is not involutive, and therefore not integrable, in any open set in $\mathbf{R}^3$.

It is easy to see that any one-dimensional distribution is involutive, since $[g, g](x) = Dg(x) \cdot g(x) - Dg(x) \cdot g(x) = 0$ for all x, and the zero vector field is certainly in the linear subspace span $\{g(x)\}$ for each x. If $g(x) \neq 0$ in U, then span $\{g(x)\}$ is nonsingular in U, and hence integrable in U, by the Frobenius Theorem, since it is involutive there. The next example illustrates how this is related to the idea of integrability of the vector field g in the sense of solving the equation $\dot{x} = g(x)$ for implicit local solutions.

Example D.8 Consider the first order partial differential equation

$$\sin x_1 \frac{\partial \lambda}{\partial x_2}(x) + \cos x_1 \frac{\partial \lambda}{\partial x_3}(x) = 0, \quad x = (x_1, x_2, x_3) \in \mathbf{R}^3,$$

which has the form $d\lambda(x) \cdot g(x) = 0$ with $g(x) = [0 \quad \sin x_1 \quad \cos x_1]^T$. The one-dimensional distribution $\mathcal{D}$ defined by

$$\mathcal{D}(x) = \text{span}\{g(x)\} = \text{span}\{[0 \quad \sin x_1 \quad \cos x_1]^T\}, \quad x \in \mathbf{R}^3,$$

is nonsingular and involutive. By inspection, one can find the solutions

$$\psi_1(x) = -x_2 \cos x_1 + x_3 \sin x_1, \qquad \psi_2(x) = x_1.$$

These solutions have linearly independent differentials $d\psi_1(x)$, $d\psi_2(x)$ at every point, since

$$\text{rank} \begin{bmatrix} d\psi_1(x) \\ d\psi_2(x) \end{bmatrix} = \text{rank} \begin{bmatrix} (x_2 \sin x_1 + x_3 \cos x_1) & \cos x_1 & \sin x_1 \\ 1 & 0 & 0 \end{bmatrix} = 2.$$

Thus, by the Frobenius theorem, $\mathcal{D}$ is integrable. The independent functions $\psi_1(x)$ and $\psi_2(x)$ must be constant along solutions of the system $\dot{x} = g(x)$; this fact is equivalent to the local existence theorem for the system $\dot{x} = g(x)$. △

The next example emphasizes the local nature of the complete integrability property of a distribution.

Example D.9 Consider the distribution in the plane defined by

$$\text{span}\{g\} := \text{span}\left\{\begin{bmatrix} x_1^2 \\ 1 \end{bmatrix}\right\}.$$

It is nonsingular and involutive in $\mathbf{R}^2$. The Frobenius theorem implies the integrability of this distribution, which implies that the partial differential equation

$$d\lambda(x) \cdot g(x) = x_1^2 \frac{\partial \lambda}{\partial x_1}(x_1, x_2) + \frac{\partial \lambda}{\partial x_2}(x_1, x_2) = 0$$

has at least a local solution in a neighborhood of each point (x_{10}, x_{20}) in the plane. It is straightforward to check that the function

$$\lambda(x_1, x_2) = \frac{x_1}{1 + x_1(x_2 - x_{20})} - x_{10}$$

is a solution in a neighborhood of the point (x_{10}, x_{20}). Note that λ is not globally defined in the plane, but only in some neighborhood of (x_{10}, x_{20}) such that $x_2 - x_{20} \neq -\frac{1}{x_1}$. Thus the functions whose existence is guaranteed by the integrability of a distribution need not be globally defined solutions of the requisite PDEs. △

For detail on the construction of the solution λ in Example D.9, see Example 12.7 of the text.

Appendix E

Comparison Functions and a Comparison Lemma

The comparison functions could be introduced in connection with the basic Lyapunov theorems if desired; however, their use in proving those results is included within Chapter 16. They provide a natural and convenient way to express the basic inequality estimates needed in Lyapunov theory.

Chapter 16 (Input-to-State Stability) is the only chapter in the book where comparison functions are used directly (except for a brief mention in the proof of Theorem 10.2 (d)). In the literature on input-to-state stability, comparison functions are part of the standard language. Thus the time invested in learning about them is definitely worthwhile. Besides the definitions, the main statements needed in Chapter 16 are Lemma E.1, Lemma E.2, Lemma E.4, and Lemma E.5. The main resources for the comparison functions and their basic properties are [32] and [58]. Lemma E.3 and Lemma E.4 are from [33].

E.1 DEFINITIONS AND BASIC PROPERTIES

We begin with the definitions of the important classes of comparison functions.

Definition E.1 (Class $\mathcal{K}$ and Class $\mathcal{K}_\infty$)
A continuous function $\alpha : [0, a) \to [0, \infty)$ belongs to class $\mathcal{K}$ *if* α *is strictly increasing and* $\alpha(0) = 0$. *If, in addition,* $a = \infty$ *and* $\alpha(r) \to \infty$ *as* $r \to \infty$, *then* α *is called a* class $\mathcal{K}_\infty$ function.

If α belongs to class $\mathcal{K}$, we may simply say that α is a class $\mathcal{K}$ function. Note that every class $\mathcal{K}$ function $\alpha : [0, a) \to [0, \infty)$ is invertible, since it is strictly increasing. We write the inverse as α^{-1}. Note that $\alpha^{-1} : [0, \beta) \to [0, a)$, where $\beta = \sup\{\alpha(r) : 0 \le r < a\}$.

Definition E.2 (Class $\mathcal{KL}$)
A continuous function $\beta : [0, a) \times [0, \infty) \to [0, \infty)$ belongs to class $\mathcal{KL}$ *if, for each fixed* s, *the mapping* $r \mapsto \beta(r, s)$ *is a class* $\mathcal{K}$ *function, and for each fixed* r, *the mapping* $s \mapsto \beta(r, s)$ *is decreasing in* s *and* $\beta(r, s) \to 0$ *as* $s \to \infty$.

If β belongs to class $\mathcal{KL}$, we also say that β is a class $\mathcal{KL}$ function. Here are some examples of class $\mathcal{KL}$ functions:

- $\beta(r,s) = \frac{r}{1+sr}, \quad r \geq 0, \ s \geq 0;$
- $\beta(r,s) = \frac{r}{\sqrt{2r^2s+1}}, \quad r \geq 0, \ s \geq 0.$

The reader will have no difficulty in verifying that these are class $\mathcal{KL}$ functions.

The next lemma gives some useful properties of comparison functions.

Lemma E.1 *Let α_1 and α_2 be class $\mathcal{K}$ functions on $[0,a)$, let α_3 and α_4 be class $\mathcal{K}_\infty$ functions, and let β be a class $\mathcal{KL}$ function. Then*

(a) *$\alpha_1 \circ \alpha_2$ belongs to class $\mathcal{K}$.*
(b) *$\alpha_3 \circ \alpha_4$ belongs to class $\mathcal{K}_\infty$.*
(c) *α_1^{-1} belongs to class $\mathcal{K}$.*
(d) *α_3^{-1} belongs to class $\mathcal{K}_\infty$.*
(e) *$\sigma(r,s) := \alpha_1(\beta(\alpha_2(r),s))$ belongs to class $\mathcal{KL}$.*

Proof. The proofs are elementary and a recommended exercise. □

Positive Definite Functions

Class $\mathcal{K}$ and class $\mathcal{K}_\infty$ functions provide a convenient way to express inequalities involving Lyapunov functions as inequalities involving vector norms. The basic result that allows this expression is given below in Lemma E.2. The lemma says that for small $\|x\|$, a positive definite function $V(x)$ can be bracketed between two strictly increasing functions of $\|x\|$. The bracketing functions of class $\mathcal{K}$ may be difficult to produce in specific cases. However, in most situations, it is important mainly that such functions exist. For example, consider the positive definite function

$$V(x) = \frac{x_1^2}{1+x_1^2} + x_2^2.$$

An upper bound for $V(x)$ is easy to obtain as

$$V(x) \leq x_1^2 + x_2^2,$$

and this upper bound is a class $\mathcal{K}$ function of $\|x\|$. Two lower bounds for $V(x)$ that are easy to obtain are given by

$$\frac{x_1^2}{1+x_1^2} \leq V(x) \quad \text{and} \quad x_2^2 \leq V(x);$$

however, neither of these lower bounds is a class $\mathcal{K}$ (strictly increasing) function of $\|x\|$. We might modify the first of these lower bounds to get

$$\frac{x_1^2+x_2^2}{1+x_1^2+x_1^2+2x_2^2} \leq \frac{x_1^2}{1+x_1^2} \leq V(x).$$

This new lower bound has the form $\|x\|_2^2/(1+2\|x\|_2^2)$, which is a class $\mathcal{K}$ function of $\|x\|_2$.

Exercise. Show that $\alpha(r) = r^2/(1+2r^2)$ is a class $\mathcal{K}$ function on $[0,\infty)$, but not a class $\mathcal{K}_\infty$ function.

Let D be a set containing the origin in $\mathbf{R}^n$. A function $V : D \to \mathbf{R}$ is *positive definite* if $V(0) = 0$ and $V(x) > 0$ for $x \in D$, $x \neq 0$.

Lemma E.2 (Class $\mathcal{K}$ Bounds)
Suppose $V : D \to \mathbf{R}$ is a continuous positive definite function defined on a set D that contains a neighborhood of the origin, and let $r > 0$ be such that $B_r(0) \subset D$. Then there exist class $\mathcal{K}$ functions α_1 and α_2, both defined on $[0,r]$, such that

$$\alpha_1(\|x\|) \le V(x) \le \alpha_2(\|x\|) \quad \text{for all } x \in B_r(0). \tag{E.1}$$

If $D = \mathbf{R}^n$ and $V(x)$ is radially unbounded, then there exist functions α_1 and α_2 of class $\mathcal{K}_\infty$ such that (E.1) *holds for all x in $\mathbf{R}^n$.*

Lemma E.2 is proved below. For a positive definite quadratic form $V(x) = x^T P x$, the inequality (E.1) follows from the standard estimate

$$\lambda_{\min}(P)\|x\|_2^2 \le x^T P x \le \lambda_{\max}(P)\|x\|_2^2,$$

where $\lambda_{\min}(P), \lambda_{\max}(P)$ are the minimum and maximum eigenvalues of P, respectively. (We are assuming that P is symmetric and therefore all eigenvalues of P are positive real numbers.) In this case, we may take $\alpha_1(r) = \lambda_{\min}(P)r^2$ and $\alpha_2(r) = \lambda_{\max}(P)r^2$. These are clearly $\mathcal{K}_\infty$ functions.

The next proposition is a simple but important consequence of Lemma E.2.

Proposition E.1 *Let D be an open set containing the origin, and $V : D \to \mathbf{R}$ a continuous positive definite function. For sufficiently small $c > 0$, the set*

$$\Omega_c = \{x : V(x) \le c\}$$

is compact.

Proof. By continuity of V, the set Ω_c is closed for every c. We must show that Ω_c is bounded for sufficiently small $c > 0$. Let $r > 0$, α_1, and α_2 be as in Lemma E.2, so that (E.1) holds for $x \in B_r(0)$. Choose any $c > 0$ such that $c < \alpha_1(r)$, and let $x \in \Omega_c$. Then we have

$$\alpha_1(\|x\|) \le V(x) \le c < \alpha_1(r),$$

and therefore $\|x\| < r$. The conclusion follows. □

In general it will not be true that Ω_c is bounded for all c. If $V(x)$ is radially unbounded, then Ω_c will be bounded, hence compact, for all c. (To see this, argue the contrapositive: if Ω_c is unbounded for some $c > 0$, then there is a sequence of points $x_k \in \Omega_c$ such that $\|x_k\| \to \infty$ while $V(x_k) \le c$, so V cannot be radially unbounded.) Consider again the function

$$V(x) = \frac{x_1^2}{1 + x_1^2} + x_2^2.$$

$V(x)$ is positive definite, but it is not radially unbounded. A plot of level curves shows that for small $c > 0$, the set $\{x : V(x) = c\} = V^{-1}(c)$ is a closed curve surrounding the origin, and Ω_c is the closed region consisting of the curve $V(x) = c$ and all points interior to that curve. On the other hand, for larger $c > 0$, the level curve $V(x) = c$ splits into two unbounded curves, and then Ω_c is an unbounded region with boundary defined by $V(x) = c$.

Proof of Lemma E.2. The function $V(x)$ is defined for $\|x\| \le r$. Let $\psi(s)$ be the function

$$\psi(s) := \inf_{s \le \|x\| \le r} V(x) \quad \text{for } 0 \le s \le r.$$

Then ψ is positive definite and continuous since V has those same properties, and ψ is increasing but may not be strictly increasing. We have

$$\psi(\|x\|) \le V(x) \quad \text{for } \|x\| \le r.$$

We may choose $0 < k < 1$ and a class $\mathcal{K}$ function α_1 such that $\alpha_1(s) \le k\psi(s)$ for $0 \le s \le r$. Then

$$\alpha_1(\|x\|) \le V(x) \quad \text{for } \|x\| \le r.$$

Now let $\phi(s)$ be the function

$$\phi(s) := \sup_{\|x\| \le s} V(x) \quad \text{for } 0 \le s \le r.$$

Then ϕ is positive definite and continuous since V has those properties, and ϕ is increasing but may not be strictly increasing. We have $V(x) \le \phi(\|x\|)$ for $\|x\| \le r$. We may choose $k > 1$ and a class $\mathcal{K}$ function α_2 such that $\alpha_2(s) \ge k\phi(s)$ for $0 \le s \le r$. Then

$$V(x) \le \alpha_2(\|x\|) \quad \text{for } \|x\| \le r.$$

This proves the first statement of Lemma E.2. For the final statement, if $V(x)$ is radially unbounded, then there are positive constants c and $\bar{r}$ such that $V(x) \ge c$ for $\|x\| > \bar{r}$. Thus, we may modify the definitions of ψ and ϕ to the following:

$$\psi(s) := \inf_{\|x\| \ge s} V(x), \quad \phi(s) := \sup_{\|x\| \le s} V(x) \quad \text{for all } s \ge 0.$$

Then $\psi(s)$ and $\phi(s)$ are positive definite, continuous, and increasing but not necessarily strictly increasing. Moreover, $\psi(s), \phi(s) \to \infty$ as $s \to \infty$ since V is radially unbounded. As before, we may choose class $\mathcal{K}$ functions $\alpha_1(s)$ and $\alpha_2(s)$ with $\alpha_1(s), \alpha_2(s) \to \infty$ as $s \to \infty$ such that

$$\alpha_1(\|x\|) \leq V(x) \leq \alpha_2(\|x\|) \quad \text{for all } \|x\|.$$

α_1, α_2 are of class $\mathcal{K}_\infty$. This completes the proof of Lemma E.2. □

E.2 DIFFERENTIAL INEQUALITY AND COMPARISON LEMMA

The material of this section is part of the standard toolbox for ordinary differential equations. However, it is referenced within the text only in Chapter 16.

We begin with the definition of the right-hand derivative of a function.

Definition E.3 (Right-Hand Derivative)
Let $v(t)$ be a real-valued function defined on an interval $[a, b)$ of real numbers. If the limit

$$\lim_{h \to 0^+} \frac{v(t+h) - v(t)}{h}$$

exists as a finite real number, then the limit is called the right-hand derivative *of v at t, and we write $D_r v(t)$ for this limit.*

In Definition E.3, the point t can be any point satisfying $a \leq t < b$. A similar definition applies to a vector function u of t, provided all component functions of u have a right-hand derivative at t.

Let $w(t, u)$ be a real-valued function of the real variables t, u, defined in an open connected set Ω. We want to establish some basic facts about the differential inequality

$$D_r v(t) \leq w(t, v(t)). \tag{E.2}$$

The function $v(t)$, $a \leq t < b$, is a solution of (E.2) on the interval $[a, b)$ if $v(t)$ is continuous on $[a, b)$ and v has a right-hand derivative on $[a, b)$ satisfying (E.2).

Lemma E.3 *If $x : [a, b] \to \mathbf{R}^n$ is a C^1 function, then $\|x(t)\|$ has a right-hand derivative, $D_r\|x(t)\|$, which exists on $[a, b)$ and satisfies*

$$|\, D_r(\|x(t)\|)| \leq \|\dot{x}(t)\|, \quad a \leq t < b.$$

Proof. First, note that for any two vectors x and u in $\mathbf{R}^n$, and $0 < \theta \leq 1$, $h > 0$, the triangle inequality implies that

$$\|x + \theta h u\| - \|\theta x + \theta h u\| \leq (1 - \theta)\|x\|.$$

Thus, by simple algebra

$$\frac{\|x+\theta hu\|-\|x\|}{\theta h} \le \frac{\|x+hu\|-\|x\|}{h},$$

which says that the difference quotient

$$\frac{\|x+hu\|-\|x\|}{h}$$

is a nondecreasing function of h. Since $\|x\| - \|\theta hu\| \le \|x+\theta hu\|$, this difference quotient is bounded below by $-\|u\|$. Thus,

$$\lim_{h\to 0^+} \frac{\|x+hu\|-\|x\|}{h}$$

exists for each vector u.

Let $t \in [a,b]$. If $\dot{x}(t)$ exists, then set $u=\dot{x}(t)$ in the previous limit statement, to see that

$$\lim_{h\to 0^+} \frac{\|x(t)+h\dot{x}(t)\|-\|x(t)\|}{h}$$

exists. Note that this is not the definition of $D_r(\|x(t)\|)$. However, we have

$$|(\|x(t+h)\|-\|x(t)\|)-(\|x(t)+h\dot{x}(t)\|-\|x(t)\|)| = \\ |\|x(t+h)\|-\|x(t)+h\dot{x}(t)\||,$$

and we have the estimate

$$\Big|\|x(t+h)\|-\|x+h\dot{x}(t)\|\Big| \le \|x(t+h)-x(t)-h\dot{x}(t)\| \\ = o(\|h\|) \quad (\text{as } h\to 0).$$

It now follows that $D_r\|x(t)\|$ exists, and

$$D_r\|x(t)\| = \lim_{h\to 0^+} \frac{\|x(t)+h\dot{x}(t)\|-\|x(t)\|}{h}.$$

The triangle inequality now implies that $|\,D_r(\|x(t)\|)\,| \le \|\dot{x}(t)\|$. □

Lemma E.4 (Comparison Lemma)
Suppose $w(t,u)$ is continuous on a connected open set Ω in $\mathbf{R}^2$ and the scalar equation

$$\dot{u} = w(t,u) \tag{E.3}$$

has a unique solution for each initial condition (t_0,u_0). If $u(t)$ is a solution of (E.3) *on $[a,b]$ and $v(t)$ is a solution of the differential inequality* (E.2) *on $[a,b)$ with $v(a)\le u(a)$, then*

$$v(t) \le u(t), \quad a \le t < b.$$

Proof. Let $u(t)$ be the unique solution of (E.3) defined on $[a, b]$ with initial condition $u(a)$. Consider the sequence of equations

$$\dot{u} = w(t, u) + \frac{1}{n}, \tag{E.4}$$

for positive integers n. Let $u_n(t)$ be the unique solution of (E.4) with initial condition $u_n(a) = u(a)$. By the continuous dependence of solutions on initial conditions and right-hand sides, for some n_0 sufficiently large the solutions $u_n(t)$ are all defined on the entire interval $[a, b]$. Moreover, the sequence u_n converges uniformly on $[a, b]$ to the solution u of (E.3) with initial condition $u(a)$. We would like to conclude that $v(t) \le u_n(t)$, $a \le t \le b$, for all $n \ge n_0$, because it would then follow immediately that $v(t) \le u(t)$, $a \le t < b$. Suppose, to the contrary, that there exist $\hat{n} \ge n_0$ and numbers $t_1 < t_2$ such that $v(t) > u_{\hat{n}}(t)$ for $t_1 < t \le t_2$, and $v(t_1) = u_{\hat{n}}(t_1)$, where we have used the continuity of solutions. Then

$$v(t) - v(t_1) > u_{\hat{n}}(t) - u_{\hat{n}}(t_1), \quad t_1 < t \le t_2.$$

Thus,

$$\begin{aligned} D_r v(t_1) \ge \dot{u}_{\hat{n}}(t_1) &= w(t_1, u_{\hat{n}}(t_1)) + \frac{1}{\hat{n}} \\ &= w(t_1, v(t_1)) + \frac{1}{\hat{n}} \\ &> w(t_1, v(t_1)), \end{aligned}$$

which is a contradiction of the definition of $v(t)$. Therefore $v(t) \le u_n(t)$, $a \le t \le b$, for all $n \ge n_0$. Then, since u_n converges uniformly to u on $[a, b]$, the lemma is proved. $\square$

In the text we reference the following very useful result.

Lemma E.5 *Consider the scalar differential equation*

$$\dot{y} = -\alpha(y), \quad y(t_0) = y_0,$$

where α is a locally Lipschitz class $\mathcal{K}$ function defined on $[0, a)$ and we may have $a = \infty$. For all $0 \le y_0 < a$, the initial value problem has a unique solution $y(t)$ defined for all $t \ge t_0$. Moreover,

$$y(t) = \sigma(y_0, t - t_0),$$

where σ is a class $\mathcal{KL}$ function defined on $[0, a) \times [0, \infty)$.

Proof. The unique solution for $y(t_0) = 0$ is $y(t) \equiv 0$ for all t. For each initial condition $y(t_0) = y_0 > 0$ there is a unique local solution $y(t)$. Since $\dot{y} < 0$ for $y > 0$, this solution remains bounded, in fact $y(t) \le y_0$, and thus can be extended for all $t \ge t_0$. Separate the variables in the differential equation to write

$$-\int_{y_0}^{y} \frac{1}{\alpha(x)}\,dx = \int_{t_0}^{t} d\tau.$$

Let b be any fixed number such that $0 < b < a$. Define the function $\eta(y)$ by

$$\eta(y) := -\int_{b}^{y} \frac{1}{\alpha(x)}\,dx. \tag{E.5}$$

Then $\eta(y)$ is a differentiable function defined for $0 < y < a$, and $\frac{d}{dy}\eta(y) = -\frac{1}{\alpha(y)}$. Also, η is strictly decreasing on the interval $(0,a)$. Note, in addition, that the integral on the right side of (E.5) must become unbounded as $y \to 0$, because $\alpha(\cdot)$ is Lipschitz. Thus, the Lipschitz condition implies that $\lim_{y\to 0^+} \eta(y) = \infty$.

Now consider the range of η. We have noted that $\lim_{y\to 0^+} \eta(y) = \infty$. Now define

$$c := -\lim_{y\to a} \eta(y),$$

so $c > 0$ and it is possible that $c = \infty$. Then the range of η is the interval $(-c,\infty)$. Since η is strictly decreasing on $(0,a)$, its inverse η^{-1} is welldefined on the interval $(-c,\infty)$.

From (E.5), we find that, for any $y_0 > 0$,

$$\eta(y(t)) - \eta(y_0) = -\int_{y_0}^{y(t)} \frac{1}{\alpha(x)}\,dx = t - t_0.$$

It follows that

$$y(t) = \eta^{-1}(\eta(y_0) + t - t_0), \quad y_0 > 0.$$

And, if $y_0 = 0$ then $y(t) \equiv 0$ since $y = 0$ is an equilibrium point. Thus, set

$$\sigma(r,s) = \begin{cases} \eta^{-1}(\eta(r)+s), & r > 0,\ s \geq 0, \\ 0, & r = 0. \end{cases}$$

Then $y(t) = \sigma(y_0, t - t_0)$ for all $y_0 \geq 0$ and all $t \geq t_0$. The function σ is continuous in (r,s) since both η and η^{-1} are continuous on their domains and

$$\lim_{x\to\infty} \eta^{-1}(x) = 0.$$

It remains to show that σ is a class $\mathcal{KL}$ function. From the fact that

$$\eta(\sigma(r,s)) = \eta(r) + s, \quad r > 0,$$

a calculation using the chain rule yields

$$\frac{\partial}{\partial r}\sigma(r,s) = \frac{\alpha(\sigma(r,s))}{\alpha(r)} > 0.$$

So $\sigma(r,s)$ is strictly increasing in r for each fixed s. Similarly,

$$\frac{\partial}{\partial s}\sigma(r,s) = -\alpha(\sigma(r,s)) < 0;$$

hence, $\sigma(r,s)$ is strictly decreasing in s for each fixed r. Moreover, we have

$$\lim_{s\to\infty} \sigma(r,s) = 0$$

for each r. Thus, $\sigma(r,s)$ is a class $\mathcal{KL}$ function. □

The example $\dot{y} = -\alpha(y) = -y^2$ shows the need for the restriction of the initial condition to $y_0 \geq 0$ in Lemma E.5. For this equation, $y(t) \to 0$ as $t \to \infty$ only for $y_0 \geq 0$.

Example E.1 Let $t_0 = 0$. By direct integration, the general solution of the equation $\dot{x} = -x^2$ for $x_0 = x(0) > 0$ is

$$x(t) = \frac{x_0}{1 + tx_0} \quad \text{for } t > -\frac{1}{x_0}.$$

For $x_0 \geq 0$, we have

$$x(t) = \beta(x_0, t),$$

where

$$\beta(r,s) = \frac{r}{1+sr}, \quad r \geq 0,\ s \geq 0.$$

△

Example E.2 Consider the equation $\dot{x} = -x^3$. Take the initial time to be $t_0 = 0$. By elementary integration, the general solution for $x_0 = x(0) \neq 0$ is

$$x(t) = \left(\frac{x_0^2}{2x_0^2 t + 1}\right)^{\frac{1}{2}} \quad \text{for } -\frac{1}{2x_0^2} < t < \infty.$$

Because of this formula, there is no particular reason to restrict attention only to $x_0 > 0$. In order to analyze asymptotic behavior, we may write

$$\begin{aligned} |x(t)| &= \frac{|x_0|}{\sqrt{2x_0^2 t + 1}} \\ &= \beta(|x_0|, t), \end{aligned}$$

where

$$\beta(r,s) := \frac{r}{\sqrt{2r^2 s + 1}}.$$

By differentiation, we find that

$$\frac{\partial}{\partial r}\beta(r,s) > 0 \quad \text{for } r > 0,$$

and $\beta(0,s) = 0$; hence, $\beta(r,s)$ is a class $\mathcal{K}$ function for each fixed s. Also,

$$\frac{\partial}{\partial s}\beta(r,s) < 0 \quad \text{for all } (r,s),$$

and $\beta(r,s) \to 0$ as $s \to \infty$ for each fixed r. Therefore $\beta(r,s)$ is a class $\mathcal{KL}$ function. If we take some other initial time t_0, then direct integration again gives the general result

$$|x(t)| = \frac{|x(t_0)|}{\sqrt{2[x(t_0)]^2(t-t_0)+1}} = \beta(|x(t_0)|, t-t_0)$$

for the same function $\beta(r, s)$, with $r = |x(t_0)|$ and $s = t - t_0$. △

The next example has no direct integrations of the differential equation.

Example E.3 Consider the equation $\dot{x} = -\tan^{-1}(x)$ for $x_0 = x(0) \geq 0$. Then Lemma E.5 applies. Without attempting a direct integration of this separable differential equation, we know that the solution has the form

$$x(t) = \sigma(x_0, t - t_0), \quad \text{with } x(t_0) = \sigma(x_0, 0) = x_0,$$

where σ is a class $\mathcal{KL}$ function defined on $[0, \infty) \times [0, \infty)$. Thus every solution with $x(t_0) = x_0 \geq 0$ satisfies $x(t) \to 0$ as $t \to \infty$. △

Appendix F

Hints and Solutions for Selected Exercises

Chapter 2: Mathematical Background

Exercise 2.2: (a) Linear independence of $v_1, \ldots, v_k$ over $\mathbf{C}$ clearly implies linear independence over $\mathbf{R}$. For the converse, suppose $v_1, \ldots, v_k$ are linearly independent over $\mathbf{R}$, and suppose $c_1 = a_1 + ib_1, \ldots, c_k = a_k + ib_k$ are complex numbers such that $c_1v_1 + \cdots + c_kv_k = 0$. Then $a_1v_1 + \cdots + a_kv_k = 0$ and $b_1v_1 + \cdots + b_kv_k = 0$, and by linear independence over $\mathbf{R}$, $a_1 = \cdots = a_k = 0$ and $b_1 = \cdots = b_k = 0$. Thus, $c_1 = \cdots = c_k = 0$, which shows linear independence of $v_1, \ldots, v_k$ over $\mathbf{C}$.
(b) Apply part (a) to the column vectors of A.

Exercise 2.3: Suppose $A(S) \subseteq S$. Let $w \in S^\perp$; we want to show that $A^T w \in S^\perp$. Let $v \in S$, and compute the product $(A^T w)^T v = w^T A v = 0$, since $Av \in S$ and $w \in S^\perp$. Since v was an arbitrary vector in S, we have $A^T w \in S^\perp$. Therefore $A^T(S^\perp) \subseteq S^\perp$. For the converse, apply the preceding argument, assuming that $A^T(S^\perp) \subseteq S^\perp$; since $(A^T)^T = A$ and $(S^\perp)^\perp = S$, the converse follows.

Exercise 2.7: We have $s_n = \sum_{k=1}^n a_k$ and $S_n = \sum_{k=1}^n \|a_k\|$. Suppose that $n > m > 0$. By the triangle inequality,

$$\|s_n - s_m\| = \|a_{m+1} + \cdots + a_n\| \le \|a_{m+1}\| + \cdots + \|a_n\| = |S_n - S_m|.$$

Absolute convergence implies $\{S_n\}$ is Cauchy, so given $\epsilon > 0$, there exists $N(\epsilon) > 0$ such that $n > m > N(\epsilon)$ implies $|S_n - S_m| < \epsilon$. Thus, $n > m > N(\epsilon)$ implies $\|s_n - s_m\| < \epsilon$ as well. Therefore $\{s_n\}$ is Cauchy in X. Since X is complete, $s_n \to s$ for some $s \in X$.

Chapter 3: Linear Systems and Stability

Exercise 3.2: Writing

$$P = \begin{bmatrix} p_1 & p_2 \\ p_2 & p_3 \end{bmatrix},$$

the required equations for the entries are $-2(p_1 + p_2) = -5$, $9p_1 - 2p_2 - p_3 = 0$, and $18p_2 - 2p_3 = -5$.

Exercise 3.16: (a) The unique solution is the symmetric matrix

$$P = \begin{bmatrix} \frac{1}{2} & -\frac{1}{4} \\ -\frac{1}{4} & \frac{1}{20} \end{bmatrix}.$$

(b) The solution is

$$P = \begin{bmatrix} -1 & \frac{3}{7} \\ 1 & -\frac{37}{63} \end{bmatrix}.$$

Chapter 4: Controllability of LTI Systems

Exercise 4.3: If $u_1 = \alpha$ for $0 \leq t \leq 2$, then

$$x_2(t) = -\frac{\alpha}{2}(1 - e^{2t}).$$

Then

$$x_1(t) = e^{2t} \int_0^t e^{-2s}[x_2(s) + u_2(s)]\, ds.$$

By choosing the constant control $u_2 = \beta$ for $0 \leq t \leq 2$, we have

$$x_1(t) = e^{2t} \int_0^t \left[\frac{\alpha}{2} - \frac{\alpha}{2}e^{-2s} + e^{-2s}\beta\right] ds.$$

Apply the final condition to determine unique constants α and β. (And there are many other possibilities for $u = [u_1(t)\ u_2(t)]^T$.)

Exercise 4.4: The system matrices are

$$A = \begin{bmatrix} -a & a & 0 & 0 \\ 0 & 0 & 1 & 0 \\ \omega^2 & -\omega^2 & 0 & 0 \\ c & 0 & 0 & 0 \end{bmatrix}, \quad B = \begin{bmatrix} 0 \\ 0 \\ b\omega^2 \\ 0 \end{bmatrix}.$$

The controllability matrix is

$$\mathcal{C}(A,B) = \begin{bmatrix} 0 & 0 & ab\omega^2 & -a^2b\omega^2 \\ 0 & b\omega^2 & 0 & -b\omega^2 \\ b\omega^2 & 0 & -b\omega^4 & ab\omega^4 \\ 0 & 0 & 0 & cab\omega^2 \end{bmatrix}.$$

Since a, b, c, ω are all positive, $\operatorname{rank} \mathcal{C}(A,B) = 4$.

Exercise 4.7: (b) Both b_1, b_3 nonzero, and note part (a).

Exercise 4.14: Let $\beta = e_2$, then $B\beta = e_3$. Let

$$K = \begin{bmatrix} 0 & 1 & 0 \\ 0 & 0 & 0 \end{bmatrix}.$$

Chapter 5: Observability and Duality

Exercise 5.5: (a) The flaw lies with the "if" implication. As one illustration, take A_1 and A_2 to be Jordan blocks with the same eigenvalue, and let C be partitioned conformably as shown:

$$A = \begin{bmatrix} 2 & 0 & 0 \\ 0 & 2 & 1 \\ 0 & 0 & 2 \end{bmatrix} = \begin{bmatrix} A_1 & 0 \\ 0 & A_2 \end{bmatrix}, \quad \text{where} \quad A_{22} \text{ is } 2 \times 2;$$

$$C = \begin{bmatrix} 1 & 1 & 0 \end{bmatrix} = \begin{bmatrix} C_1 & C_2 \end{bmatrix}, \quad \text{where} \quad C_2 \text{ is } 1 \times 2.$$

Then the pairs (C_1, A_1) and (C_2, A_2) are observable, but the PBH test shows that (C, A) is not observable. (b) Again the "if" statement is flawed; consider the dual system of the example just given.

Exercise 5.7: (a) The unique (and positive definite) solution of $A^T P + PA = -C^T C$ is

$$P = \begin{bmatrix} \frac{\sigma}{2} + \frac{1}{2\sigma} & \frac{1}{2} \\ \frac{1}{2} & \frac{1}{2\sigma} \end{bmatrix}.$$

(b) $\frac{dV}{dt}(x(t)) \equiv 0$ only when $x_1(t) \equiv 0$, and by observability, this occurs only for the zero solution.

(c) $\frac{dW}{dt}(x(t)) \equiv 0$ only when $x_2(t) \equiv 0$. The differential equations then imply that $x_1(t) \equiv 0$ as well.

Chapter 6: Stabilizability of LTI Systems

Exercise 6.2: (a) The controllability Gramian Q_β is

$$Q_\beta = \begin{bmatrix} \frac{1}{2}\beta - \frac{1}{4}\sin 2\beta & \frac{1}{2}\cos^2\beta - \frac{1}{2} \\ \frac{1}{2}\cos^2\beta - \frac{1}{2} & \frac{1}{2}\beta + \frac{1}{4}\sin 2\beta \end{bmatrix},$$

and

$$K_\beta = -B^T Q_\beta^{-1} = \begin{bmatrix} \frac{1}{2} - \frac{1}{2}\cos^2\beta & -\left(\frac{1}{2}\beta + \frac{1}{4}\sin 2\beta\right) \end{bmatrix}.$$

(b) We get velocity feedback from $K = -B^T Q^{-1}$ if and only if $\beta = n\pi$, where n is a positive integer. Oscillations cannot be avoided in the closed loop system with these particular velocity feedbacks.

Exercise 6.6: The eigenvalues of A are -1, 0, 1, 2.

(i) For $b = e_1$, the system is already in normal form. Not stabilizable.

(ii) For $b = e_2$, let $T = [b\ Ab\ A^2b\ e_3]$, $x = Tz$. A normal form is given by a pair

$$T^{-1}AT = \begin{bmatrix} A_{11} & 0 \\ 0 & -1 \end{bmatrix} \quad \text{and} \quad T^{-1}b = T^{-1}e_2 = e_1,$$

where A_{11} is 3×3. Eigenvalue -1 corresponds to uncontrollable modes. Stabilizable.

(iii) For $b = e_3$, let $T = [e_3\ v_1\ v_2\ v_3]$, $x = Tz$. A normal form is given by a pair

$$T^{-1}AT = \begin{bmatrix} -1 & * \\ 0 & A_{22} \end{bmatrix} \quad \text{and} \quad T^{-1}b = T^{-1}e_3 = e_1,$$

where A_{22} is 3×3. Eigenvalues 0, 1, 2 correspond to uncontrollable modes. Not stabilizable.

(iv) For $b = e_4$, let $T = [b\ Ab\ A^2b\ e_3]$, $x = Tz$. A normal form is given by a pair

$$T^{-1}AT = \begin{bmatrix} A_{11} & 0 \\ 0 & -1 \end{bmatrix}, \quad \text{and} \quad T^{-1}b = T^{-1}e_4 = e_1,$$

where A_{11} is 3×3. Eigenvalue -1 corresponds to uncontrollable modes. Stabilizable.

Chapter 7: Detectability and Duality

Exercise 7.2: (a) The eigenvalues are

$$\lambda = -\frac{1}{2} \pm \frac{1}{2}\sqrt{4K - 3}.$$

If $4K - 3 \le 0$, then the origin is a stable spiral point and the exponential decay rate is $-\frac{1}{2}$; if $0 < 4K - 3 < 1$, then the origin is a stable node and the exponential decay rate is greater than or equal to -1; if $4K - 3 > 1$, then the origin is an unstable saddle point.

Exercise 7.3: *Proof of* (b). Theorem 5.5 implies that by a linear change of coordinates, the system $\dot{x} = Ax$, $y = Cx$ may be written in the normal form

$$\begin{bmatrix} \dot{z}_1 \\ \dot{z}_2 \end{bmatrix} = \begin{bmatrix} A_{11} & 0 \\ A_{21} & A_{22} \end{bmatrix} \begin{bmatrix} z_1 \\ z_2 \end{bmatrix}, \quad y = \begin{bmatrix} C_1 & 0 \end{bmatrix} \begin{bmatrix} z_1 \\ z_2 \end{bmatrix},$$

where the pair (C_1, A_{11}) is observable. Note that, by the hypothesis applied to the normal form, the eigenvalues of A_{22} all have negative real part. Thus, the observability normal form in z coordinates is detectable. It follows from part (a) of this exercise that the original pair (C, A) is detectable.

Exercise 7.7: (a) The equations are linear in the second derivatives $\ddot{\xi}$, $\ddot{\theta}$. Using the state variables $x_1 = \xi$, $x_2 = \dot{\xi}$, $x_3 = \theta$, $x_4 = \dot{\theta}$, the state equations are

$$\begin{aligned}
\dot{x}_1 &= x_2, \\
\dot{x}_2 &= \frac{1}{M + m(\sin x_3)^2}(u + mlx_4^2 \sin x_3 - bx_2 - mg \cos x_3 \sin x_3), \\
\dot{x}_3 &= x_4, \\
\dot{x}_4 &= \frac{1}{l(M + m(\sin x_3)^2)} \big((M+m)g \sin x_3 - u \cos x_3 \\
&\quad + bx_2 \cos x_3 - mlx_4^2 \sin x_3 \cos x_3\big).
\end{aligned}$$

(b) For any choice of the constants, the linearization of the system at the origin $x = 0$ is given by

$$\dot{x} = \begin{bmatrix} 0 & 1 & 0 & 0 \\ 0 & -\frac{b}{M} & -\frac{mg}{M} & 0 \\ 0 & 0 & 0 & 1 \\ 0 & \frac{b}{lM} & \frac{g(m+M)}{lM} & 0 \end{bmatrix} x + \begin{bmatrix} 0 \\ \frac{1}{M} \\ 0 \\ -\frac{1}{lM} \end{bmatrix} u.$$

Exercise 7.8: The unique solution of the ARE is given by

$$P = \begin{bmatrix} \sqrt{2} & 1 \\ 1 & \sqrt{2} \end{bmatrix}; \quad \text{and} \quad A - BB^T P = \begin{bmatrix} 0 & 1 \\ -1 & -\sqrt{2} \end{bmatrix}.$$

Chapter 8: Lyapunov Stability Theory

Exercise 8.3: (b) Separate variables and use a partial fraction decomposition in order to integrate and obtain explicit solutions.

Exercise 8.9: (b) Positive definiteness of V is equivalent to $\alpha > 0$ and $\alpha\gamma - \frac{\beta^2}{2} > 0$. Requiring negative definiteness of $\dot{V}$ allows many choices for the coefficients, one of which is $\alpha = 5$, $\beta = 2$, $\gamma = 2$.

Chapter 9: Cascade Systems

Exercise 9.5: When $x_0 z_0 > 2$, the finite positive escape time equals $\hat{t}(x_0, z_0) = -\frac{1}{2} \ln \frac{x_0 z_0 - 2}{x_0 z_0}$.

Exercise 9.6: With $\eta = \frac{1}{x}$, the solution for η is

$$\eta(t) = e^t \eta_0 + \frac{z_0}{k+1} e^{-kt} - \frac{z_0}{k+1} e^t.$$

Consider initial conditions (x_0, z_0) with $x_0 \neq 0$, $z_0 \neq 0$. We have

$$x(t) = \frac{x_0(k+1)}{[(k+1) - x_0 z_0]e^t + x_0 z_0 e^{-kt}}.$$

Each branch of the hyperbola $xz = k + 1$ is a complete trajectory. If $x_0 z_0 > k + 1 > 0$, then the forward solution $(x(t), z(t))$ has a finite positive escape time given by

$$\hat{t}(x_0, z_0) = -\frac{1}{k+1} \ln \left(\frac{x_0 z_0 - (k+1)}{x_0 z_0} \right).$$

If $0 < x_0 z_0 < k + 1$, then $x(t)$ is defined on $[0, \infty)$, and $x(t) \to 0$ as $t \to \infty$. If $x_0 z_0 < 0$, then $x(t)$ is defined on $[0, \infty)$ and again $x(t) \to 0$ as $t \to \infty$. Therefore the basin can be described completely by the inequality $x_0 z_0 < k + 1$. Thus, the basin enlarges with increasing $k > 0$. Every compact set K in the plane is closed and bounded, hence contained in some ball $B_R(0)$. Thus, choose k with $k + 1 > R$ to include $B_R(0)$ in the basin.

Exercise 9.8: Assume that $x(s)$ exists on the interval $0 \le s \le t$ and obtain, by separation of variables and integration with respect to s over the interval $0 \le s \le t$, the expression

$$[x(t)]^2 = \frac{1}{\frac{1}{[x(0)]^2} + 2t + 2z_0 e^{-t} - 2z_0}, \quad \text{where} \quad z_0 = z(0).$$

Consider a fixed $x(0)$. The denominator $d(t)$ of $[x(t)]^2$ satisfies $d(0) = \frac{1}{[x(0)]^2} > 0$. Show that, if z_0 is sufficiently large positive, then $d(t)$ is zero at some (first) positive time $\hat{t}$, and thus $x(\hat{t})$ cannot exist.

Exercise 9.10: (a) Each of the matrices $A - iI$, $A + iI$ has rank 3. Stability of the origin for the full system cannot be deduced on the basis of linearization alone.

Chapter 10: Center Manifold Theory

Exercise 10.4: (a) Following the initial linear coordinate change which does not change z, the right-hand side of (10.36) takes the form $Az + p(T^{-1}w, z) =: Az + P(w_1, w_2, z)$. Consequently, the right-hand side of (10.39) becomes $Ak_1(w_1) + P(w_1, k_2(w_1), k_1(w_1))$.

Chapter 11: Zero Dynamics

Exercise 11.2: Start with the system equations

$$\dot{x} = f(x) + g(x)u,$$
$$y = h(x).$$

Invariance of the relative degree under coordinate transformations: Let $z = \Phi(x)$ be a local coordinate transformation defined near x_0. The system in z coordinates is described by

$$\dot{z} = \frac{\partial \Phi}{\partial x}(x) f(x) + \frac{\partial \Phi}{\partial x}(x) g(x) u,$$

where $x = \Phi^{-1}(z)$. Define

$$\bar{f}(z) := \left[\frac{\partial \Phi}{\partial x}(x) f(x) \right]_{x=\Phi^{-1}(z)},$$

$$\bar{g}(z) := \left[\frac{\partial \Phi}{\partial x}(x) g(x) \right]_{x=\Phi^{-1}(z)},$$

$$\bar{h}(z) := h(\Phi^{-1}(z)).$$

Then show by direct calculation that

$$L_{\bar{f}} \bar{h}(z) = [\, L_f h(x) \,]_{x=\Phi^{-1}(z)}$$

and, for $k \geq 1$,

$$L_{\bar{g}} L_{\bar{f}}^k \bar{h}(z) = [\, L_g L_f^k h(x) \,]_{x=\Phi^{-1}(z)}.$$

Conclude that the relative degree is invariant under the coordinate transformation $z = \Phi(x)$.

Invariance of the relative degree under a regular feedback transformation $u = \alpha(x) + \beta(x)v$, $\beta(x_0) \neq 0$: First, establish by induction the identities

$$L_{f+g\alpha}^k h(x) = L_f^k h(x), \quad \text{for } 0 \leq k \leq r-1.$$

Then show that these identities imply that

$$L_{g\beta} L_{f+g\alpha}^k h(x) = 0 \quad \text{for } 0 \leq k < r-1.$$

Use $\beta(x_0) \neq 0$ to deduce that

$$L_{g\beta} L_{f+g\alpha}^{r-1} h(x_0) \neq 0,$$

which completes the argument that the relative degree r is invariant under regular feedback transformations.

Exercise 11.4: (a) Let v and w be smooth vector fields, and λ a smooth function. We will write $d^2\lambda(x)$ for the matrix of second partial derivatives of λ at x. At any point x, we have

$$\begin{aligned} L_v L_w \lambda(x) - L_w L_v \lambda(x) &= L_v(d\lambda(x) \cdot w(x)) - L_w(d\lambda(x) \cdot v(x)) \\ &= \left(d^2\lambda(x) \cdot w(x) + d\lambda(x) \cdot \frac{\partial w}{\partial x}(x) \right) \cdot v(x) \\ &\quad - \left(d^2\lambda(x) \cdot v(x) + d\lambda(x) \cdot \frac{\partial v}{\partial x}(x) \right) \cdot w(x). \end{aligned}$$

Since $d^2\lambda$ is symmetric, we get the identity

$$L_v L_w \lambda - L_w L_v \lambda = d\lambda \cdot \left(\frac{\partial w}{\partial x} v - \frac{\partial v}{\partial x} w \right) = d\lambda \cdot [v, w] = L_{ad_v w} \lambda.$$

Exercise 11.13: *Hint*: The output $y = x_1 = \alpha$ still has relative degree three at the origin. Is the order of approximation of the center manifold still adequate? Reexamine (11.32) to ensure that the argument can proceed as before with the same parameter values for the k_j and for β.

Exercise 11.14: *Hint*: A transformation to normal form is given by

$$\begin{bmatrix} \xi \\ \eta \end{bmatrix} = \begin{bmatrix} c_1 & c_2 \\ c_2 & 0 \end{bmatrix} \begin{bmatrix} x_1 \\ x_2 \end{bmatrix}.$$

Obtain the normal form $\dot{\xi} = v$, $\dot{\eta} = -\xi + \frac{c_1}{c_2}\eta$. The zero dynamics is $\dot{\eta} = \frac{c_1}{c_2}\eta$, which is Lyapunov stable if $c_1 = 0$, and asymptotically stable if and only if c_1, c_2 have opposite sign.

Exercise 11.15: One expression of the answer is that the zero dynamics has Lyapunov stable origin if and only if all nonzero coefficients c_i have the same sign. More detail is given here, including cases of asymptotic stability:

(i) If all three c_i are nonzero, then $\operatorname{tr} Q = -c_2/c_3 \neq 0$ and $\det Q = c_1/c_3 \neq 0$. Thus the zero dynamics has asymptotically stable origin if and only if $c_1/c_3 > 0$ and $c_2/c_3 > 0$, that is, if and only if all three coefficients have the same sign. Output examples: $y = x_1 + 2x_2 + x_3$, $y = -x_1 - x_2 - 3x_3$.

(ii) If only c_1, c_3 are nonzero, then $\operatorname{tr} Q = -c_2/c_3 = 0$ and $\det Q = c_1/c_3 \neq 0$. Thus the origin of the zero dynamics is merely Lyapunov stable if and only if c_1, c_3 have the same sign, and then the solutions are oscillatory. Output examples: $y = x_1 + x_3$, $y = -x_1 - 2x_3$.

(iii) If only c_2, c_3 are nonzero, then $\operatorname{tr} Q = -c_2/c_3 \neq 0$ and $\det Q = c_1/c_3 = 0$. Thus the origin of the zero dynamics is merely Lyapunov stable if and only if c_2, c_3 have the same sign. In this case, there is a line of equilibria through the origin. Output examples: $y = x_2 + x_3$, $y = -x_2 - 3x_3$.

Chapter 12: Feedback Linearization

Exercise 12.2: *Hint*: By induction,

$$ad_f^k g(x) = (-1)^k A^k b + R_k(x), \quad \text{where } R_k(x_0) = 0, \text{ for } 0 \leq k \leq n-1.$$

Exercise 12.6: *Hints*: The matrix $[g(x)\ ad_f g(x)\ ad_f^2 g(x)]$ is lower-right triangular with nonzero skew-diagonal entries. If $\phi_3(x)$ is of class C^3, then $[g, ad_f g]$ is a C^1 vector field and span $\{g, ad_f g\}$ is involutive.

Chapter 13: Introduction to Damping Control

Exercise 13.1: *Proof.* Let $V(x) = \frac{1}{2}x^T x$. We have $\dot{V}(x) = L_f V(x) + L_g V(x) u$. By the assumption on the eigenvalues of A and the *Hint*, $L_f V(x) \le 0$. If $u = -L_g V(x)$, then $L_g V(x) u = -(L_g V(x))^2 \le 0$. Thus, $\dot{V}(x) = 0$ if and only if both $L_f V(x) = 0$ and $L_g V(x) = 0$. If $x(t)$ satisfies $\dot{V}(x(t)) \equiv 0$, then $x(t)$ must be a solution of the system $\dot{x} = Ax$; in fact, by the eigenvalue assumption, the solution $x(t)$ must lie within the center subspace of A, that is, within the sum of the eigenspaces associated with the zero eigenvalues (otherwise $V(x(t))$ is not constant, a contradiction). Thus, the invariance theorem applies and the origin is asymptotically stabilized by $u = -L_g V(x)$.

Exercise 13.2: (a) Bracket calculations yield

$$[g \quad ad_f g \quad ad_f^2 g] = \begin{bmatrix} x_2 & 0 & 0 \\ -x_1 & 0 & 0 \\ 1 & 2 & 4 \end{bmatrix}.$$

(b) Only the eigenvalues $\pm i$ of A need to be tested by the PBH test.
(c) The 3×4 Jacobian of $F(x,u) = Ax + g(x)u$ is

$$DF(x,u) = \begin{bmatrix} \frac{\partial F}{\partial x} & \frac{\partial F}{\partial u} \end{bmatrix} = \begin{bmatrix} 0 & (1+u) & 0 & x_2 \\ -(1+u) & 0 & 0 & -x_1 \\ 0 & 0 & -2 & 1 \end{bmatrix},$$

which has rank 3 at $(x,u) = (0,0)$. We conclude that the image of F contains an open neighborhood of the origin; this is Brockett's necessary condition for smooth stabilization (see Theorem 8.3).

Exercise 13.3: (a) *Hints*: (1) For the linear case with $F(x) = Ax$, $g(x) = b$, and $V(x) = x^T P x$, we have $L_g V(x) = x^T P b$; the rank condition on $h(x) = b^T P x$ implies that the pair $(b^T P, A)$ is observable. Use $V(x)$ as a Lyapunov function for the closed loop system when $u = -b^T P x$. (2) For the nonlinear case, note that the rank condition at 0 implies that the differentials $dh(x)$, $d(L_f h)(x)$, $\dots$, $d(L_f^{n-1} h)(x)$ are linearly independent throughout a neighborhood of the origin; and use the fact that $x = 0$ is stable for the unforced system.

(b) For Example 13.5: If $V(x)$ is the total energy for the unforced system and $h(x) = L_g V(x)$, then

$$\text{rank} \begin{bmatrix} dh(0) \\ d(L_f h)(0) \end{bmatrix} = \text{rank} \begin{bmatrix} 0 & 1 \\ -\cos 0 & 0 \end{bmatrix} = 2.$$

For Example 13.4: If $V(x)$ is the total energy for the unforced system and $h(x) = L_g V(x)$, then

$$\operatorname{rank}\begin{bmatrix} dh(0) \\ d(L_f h)(0) \end{bmatrix} = \operatorname{rank}\begin{bmatrix} 0 & 1 \\ 0 & 0 \end{bmatrix} = 1.$$

Part (a) does not apply; nevertheless, the feedback $u = -L_g V(x) = -x_2$ is globally stabilizing, by the invariance theorem.

Chapter 14: Passivity

Exercise 14.1: Using the solution of part (a), passivity would require that

$$0 \le S(x_1(T), x_2(T)) \le \int_0^T x_1(s)u(s)\,ds.$$

We may choose, for example, $\omega = 2$ and $T = 2\pi$, which gives

$$\int_0^T x_1(s)u(s)\,ds = -\int_0^{2\pi} \frac{1}{3}\sin^2 2s\,ds < 0$$

and violates (14.6).

Exercise 14.2: (1) If the system is passive with a C^1 storage function V, then the KYP conditions are $\frac{\partial V}{\partial x_1}(x_1, x_2)x_2 \le 0$ and $\frac{\partial V}{\partial x_2}(x_1, x_2) = x_1$. The second condition says that $V(x) = x_1 x_2 + \gamma(x_1)$ for some differentiable function $\gamma(x_1)$, and then the first condition implies that $x_2^2 + \gamma'(x_1)x_2 \le 0$. Fix an intial condition x_{10}, and then set $x_{20} = 1$ and $x_{20} = -1$ to get a contradiction.

Exercise 14.3: *Proof of Theorem 14.2.* If the system has the KYP property with quadratic function $V(x) = \frac{1}{2}x^T Px$, then $P \ge 0$ and P satisfies (14.13)–(14.14). By (14.13)–(14.14), along solutions $x(t)$ with input $u(t)$, we have $\frac{d}{dt}V(x(t)) \le y^T u$. Writing $x(0) = x_0$ and integrating over the interval $0 \le s \le t$ gives $V(x(t)) - V(x_0) = \int_0^t \frac{d}{ds}V(x(s))\,ds \le \int_0^t y^T(s)u(s)\,ds$, which is the dissipation inequality for storage V. Therefore the system is passive with quadratic storage $V(x) = \frac{1}{2}x^T Px$.

Conversely, if the system is passive with C^1 storage $V(x)$, then, in particular, $L_g V(x) = h^T(x)$. For the quadratic V and $g(x) = B$, this reads $(\nabla V(x))^T B = x^T C^T$ for all x. Since the right hand side is linear in x, $\nabla V(x)$ must be linear in x; hence V must be quadratic. Since $V(0) = 0$, we may write $V(x) = \frac{1}{2}x^T Px$ for some symmetric matrix P, with $P \ge 0$ since V is positive semi-definite. Then $\nabla V(x) = Px$, hence $x^T PB = x^T C^T$ for all x. Thus, $PB = C^T$ and therefore (14.14) holds. The first KYP condition (14.8) also holds, by hypothesis, so that P also satisfies (14.13). □

Exercise 14.5: Let V be a storage function. For any points $x_0 = x(0)$ and $x_1 = x(t_1)$ on a solution trajectory corresponding to input $u(\cdot)$,

$$V(x_1) - V(x_0) \le \int_0^{t_1} y^T(s)u(s)\,ds,$$

and therefore

$$V(x_0) \ge V(x_1) - \int_0^{t_1} y^T(s)u(s)\,ds.$$

Since $t_1 \ge 0$ is arbitrary and $V(x_1) = V(x(t_1)) \ge 0$, we have

$$V(x_0) \ge \sup_{u(\cdot)\in\mathcal{U},t_1\ge 0} \left\{ -\int_0^{t_1} y^T(s)u(s)\,ds \right\} = V_a(x_0),$$

which shows that $V_a(x_0)$ is finite. Since x_0 was arbitrary, $V_a(x)$ is finite for all x and $V_a(x) \le V(x)$ for any storage function V. Take $u = 0$ or $t = 0$ in the definition (14.17) to see that $V_a(x) \ge 0$ for every x. (Note that $V_a(0) \le V(0) = 0$ for any storage V, hence $V_a(0) = 0$.)

Exercise 14.8: (b) Not ZSD. (c) No.

Exercise 14.10: Show that the zero dynamics subsystem is $\dot{\eta} = -\frac{c_1}{c_2}\eta$.

Exercise 14.11: Use the algebraic catalog of linear planar systems in Example 2.15.

Exercise 14.12: *Hint*: Use a converse Lyapunov theorem.

Chapter 15: Partially Linear Cascade Systems

Exercise 15.1: Let $y = \xi_2$. The condition $Pb = c^T$ requires that P have the form

$$P = \begin{bmatrix} p_{11} & 0 \\ 0 & 1 \end{bmatrix}.$$

Take $p_{11} = 1$, $K = [k_1 \quad k_2] = [-1 \quad -1]$; then $A + BK$ is Hurwitz, while $(A+bK)^T P + P(A+bK) \le 0$ as required (although not negative definite).

Exercise 15.4: Let $y = \xi_1 + \xi_2$. The condition $Pb = c^T$ requires that P have the form

$$P = \begin{bmatrix} p_{11} & 1 \\ 1 & 1 \end{bmatrix}.$$

$P > 0$ requires $p_{11} > 1$. Take $p_{11} = 2$, $K = [k_1 \quad k_2] = [-1 \quad -1]$; then $A + BK$ is Hurwitz. Check that $(A + bK)^T P + P(A + bK) \le 0$ as required (although not negative definite).

Chapter 16: Input-to-State Stability

Exercise 16.2: Let γ be an ISS gain function. Suppose $u(t) \to 0$ as $t \to 0$. Take any initial condition $x(0) = x_0$ in $\mathbf{R}^n$ and consider the corresponding forward solution of $\dot{x} = f(x, u(t))$. Given $\epsilon > 0$, there exists $T_1(\epsilon) > 0$ such that

$$t \geq T_1(\epsilon) \Longrightarrow \|u(t)\| < \gamma^{-1}\left(\frac{\epsilon}{2}\right).$$

All solutions are bounded in forward time, so all solutions exist for all $t \geq 0$. Consider the forward trajectory of $x(t)$ for $t \geq T_1(\epsilon)$. By input-to-state stability, we have

$$\|x(t)\| \leq \beta(\|x(T_1)\|, t) + \frac{\epsilon}{2} \quad \text{for all } t \geq T_1.$$

Since $\beta(\cdot,\cdot)$ is a class $\mathcal{KL}$ function, there exists $T_2(\epsilon) > T_1(\epsilon)$ such that

$$\beta(\|x(T_1)\|, t) < \frac{\epsilon}{2} \quad \text{for all } t \geq T_1(\epsilon) + T_2(\epsilon).$$

Thus, for all $t \geq T_1(\epsilon) + T_2(\epsilon)$, we have

$$\|x(t)\| \leq \frac{\epsilon}{2} + \frac{\epsilon}{2} = \epsilon.$$

Hence, $\lim_{t\to\infty} x(t) = 0$.

Exercise 16.3:

(a) Not ISS; it is not BIBS
(b) Not ISS; it is not CICS
(c) ISS: Take $V(x) = \frac{1}{2}x^2$ and find that, for $0 < \theta < 1$, $\dot{V}(x) \leq -(1-\theta)x^4 - x^6$ if $|x| \geq \chi(r) := (\frac{|u|}{\theta})^{\frac{1}{3}}$.
(d) Not ISS; the origin is not GAS when $u = 0$
(e) Not ISS.
(f) ISS: Take $V(x) = \frac{1}{2}x^2$ and find that, for $0 < \theta < 1$, $\dot{V}(x) = -(1-\theta)x^4 - \theta x^4 + x^3 u$, so that $\dot{V}(x) \leq -(1-\theta)x^4$ if $|x| \geq \chi(r) := \frac{|u|}{\theta}$.

Exercise 16.4: (a) Not ISS; when $u = 0$ the origin is not GAS, (b) Not ISS; when $u = 0$ the origin is not GAS, (c) Not ISS, (d) ISS

Exercise 16.6: (a) ISS, by Theorem 16.4 (b), (b) Not ISS; in particular, it is not CICS, (c) Not ISS.

Bibliography

[1] R. Abraham. *Foundations of Mechanics.* W. A. Benjamin, New York, 1967.

[2] R. Abraham and J. Marsden. *Foundations of Mechanics.* Addison-Wesley, Reading, MA, second edition, 1978.

[3] J. Ackermann. Der entwurf linearer regelungssysteme im zustandsraum. *Regelungestecknik und Prozessedatenverarbeitung*, 7:297–300, 1972. See also J. Ackermann, *Abtastregulung*, Springer, Berlin, 1972.

[4] D. Aeyels. Stabilization of a class of nonlinear systems by a smooth feedback control. *Sys. Control Lett.*, 5:289–294, 1985.

[5] B. D. O. Anderson and J. B. Moore. *Optimal Control: Linear Quadratic Methods.* Prentice-Hall, Englewood Cliffs, NJ, 1990.

[6] P. J. Antsaklis and A. N. Michel. *Linear Systems.* Birkhauser, Boston, 2006.

[7] T. M. Apostol. *Mathematical Analysis.* Addison-Wesley, Reading, MA, second edition, 1974.

[8] E. A. Barbashin and N. N. Krasovskiĭ. On global stability of motion. *Doklady Akad. Nauk SSSR*, 86:453–456, 1952. In Russian.

[9] S. Barnett and R. G. Cameron. *Introduction to Mathematical Control Theory.* Clarendon Press, Oxford, second edition, 1985.

[10] J. S. Bay. *Fundamentals of Linear State Space Systems.* WCB/McGraw-Hill, New York, 1999.

[11] S. Behtash and S. Sastry. Stabilization of nonlinear systems with uncontrollable linearization. *IEEE Trans. Automat. Contr.*, 33:585–590, 1988.

[12] N. P. Bhatia and G. P. Szegő. *Stability Theory of Dynamical Systems.* Classics in Mathematics. Springer, Berlin, 2002. Reprint of the 1970 edition.

[13] A. M. Bloch with the collaboration of J. Baillieul, P. Crouch, and J. Marsden. *Nonholonomic Mechanics and Control.* Interdisciplinary Applied Mathematics: Systems and Control. Springer, New York, 2003.

[14] W. M. Boothby. *An Introduction to Differentiable Manifolds and Riemannian Geometry.* Academic Press, NewYork, second edition, 1986.

[15] F. Brauer and J. A. Nohel. *The Qualitative Theory of Ordinary Differential Equations: An Introduction.* W. A. Benjamin, New York, 1969. Reprinted by Dover, NewYork, 1989.

[16] R. W. Brockett. *Finite Dimensional Linear Systems.* John Wiley, New York, 1970.

[17] R. W. Brockett. Feedback invariants for nonlinear systems. In *Proceedings International Federation on Automatic Control Congress*, pp. 1115–1120, Helsinki, 1978. IFAC.

[18] F. Bullo and A. D. Lewis. *Geometric Control of Mechanical Systems.* Texts in Applied Mathematics 49. Springer, New York, 2005.

[19] C. I. Byrnes, A. Isidori, and J. C. Willems. Passivity, feedback equivalence, and the global stabilization of minimum phase nonlinear systems. *IEEE Trans. Automat. Control*, 36:1228–1240, 1991.

[20] C. I. Byrnes, F. Delli Priscoli, and A. Isidori. *Output Regulation of Uncertain Nonlinear Systems.* Systems and Control: Foundations and Applications. Birkhauser, Basel, 1997.

[21] S. L. Campbell and R. Haberman. *Introduction to Differential Equations with Boundary Value Problems.* Houghton Mifflin, Boston, 1996.

[22] S. L. Campbell and R. Nikoukhah. *Auxiliary Signal Design for Failure Detection.* Princeton Series in Applied Mathematics. Princeton University Press, Princeton, 2004.

[23] J. Carr. *Applications of Centre Manifold Theory.* Applied Mathematical Sciences 35. Springer-Verlag, New York, 1981.

[24] D. Cheng and C. Martin. Stabilization of nonlinear systems via designed center manifold. *IEEE Trans. Aut. Contr.*, 46:1372–1383, 2001.

[25] C. Chicone. *Ordinary Differential Equations with Applications.* Texts in Applied Mathematics 34. Springer, New York, 1999.

[26] M. J. Corless and A. E. Frazho. *Linear Systems and Control: An Operator Perspective.* Number 254 in Pure and Applied Mathematics. Marcel Dekker, New York, 2003.

[27] J-M. Coron, L. Praly, and A. Teel. Feedback stabilization of nonlinear systems: Sufficient conditions and lyapunov and input-output techniques. In A. Isidori, editor, *Trends in Control*, pp. 293–348. Springer Verlag, New York, 1995.

[28] G. E. Dellerud and F. Paganini. *A Course in Robust Control Theory: A Convex Approach.* Texts in Applied Mathematics 36. Springer, New York, 2000.

[29] E. Gilbert. Controllability and observability in multivariable control systems. *SIAM J. Control*, 1:128–151, 1963.

[30] J. Guckenheimer and P. Holmes. *Nonlinear Oscillations, Dynamical Systems, and Bifurcations of Vector Fields.* Applied Mathematical Sciences 42. Springer, New York, 1983.

[31] V. Guillemin and A. Pollack. *Differential Topology.* Prentice-Hall, Englewood Cliffs, NJ, 1974.

[32] W. Hahn. *Stability of Motion.* Springer, Berlin, 1967.

[33] J. K. Hale. *Ordinary Differential Equations.* Krieger, Malabar, second edition, 1980.

[34] J. Hale and H. Koçak. *Dynamics and Bifurcations.* Springer-Verlag, New York, 1991.

[35] M. L. J. Hautus. A simple proof of Heymann's lemma. *IEEE Trans. Autom. Control*, AC-22:885–886, 1977.

[36] R. Hermann. The differential geometry of foliations. *J. Math. Mech.*, 11:302–316, 1962.

[37] R. Hermann. On the accessibility problem in control theory. In J. P. Lasalle and S. Lefschetz, editors, *International Symposium on Nonlinear Differential Equations and Nonlinear Mechanics*, pages 325–332. Academic Press, New York, 1963.

[38] R. Hermann and A. J. Krener. Nonlinear controllability and observability. *IEEE Trans. Aut. Contr.*, AC-22:728–740, 1977.

[39] D. J. Hill and P. J. Moylan. Stability of nonlinear dissipative systems. *IEEE Trans. Aut. Contr.*, AC-21:708–711, 1976.

[40] M. Hirsch and S. Smale. *Differential Equations, Dynamical Systems, and Linear Algebra.* Academic Press, 1974.

[41] M. Hirsch, S. Smale, and R. L. Devaney. *Differential Equations, Dynamical Systems and an Introduction to Chaos.* Elsevier Academic Press, New York, 2004.

[42] R. A. Horn and C. R. Johnson. *Matrix Analysis.* Cambridge University Press, Cambridge, 1985.

[43] R. A. Horn and Johnson C. R. *Topics in Matrix Analysis.* Cambridge University Press, Cambridge, 1991.

[44] P. Horowitz and W. Hill. *The Art of Electronics.* Cambridge University Press, Cambridge, second edition, 1989.

[45] J. Huang. *Nonlinear Output Regulation: Theory and Applications.* Advances in Design and Control. SIAM, Philadelphia, 2004.

[46] W. Hurewicz. *Lectures on Ordinary Differential Equations.* MIT Press, Cambridge, MA, 1958.

[47] A. Isidori. *Nonlinear Control Systems.* Springer, New York, third edition, 1995.

[48] A. Isidori. *Nonlinear Control Systems II.* Springer, New York, 1999.

[49] B. Jacubczyk and W. Respondek. Feedback equivalence of planar systems and stabilizability. In M. A. Kaashoek, J. W. van Schuppen, and A. C. M. Ran, editors, *Robust Control of Linear Systems and Nonlinear Control*, pages 447–456. Birkhauser, Boston, 1990.

[50] D. W. Jordan and P. Smith. *Nonlinear Ordinary Differential Equations.* Oxford Applied Mathematics and Computing Science Series. Clarendon Press, Oxford, second edition, 1987.

[51] V. Jurdjevic. *Geometric Control Theory.* Cambridge Studies in Advanced Mathematics 51. Cambridge University Press, Cambridge, England, 1997.

[52] V. Jurdjevic and J. P. Quinn. Controllability and stability. *J. Diff. Eqs.*, 28:381–389, 1979.

[53] T. Kailath. *Linear Systems.* Prentice-Hall, Englewood Cliffs, NJ, 1980.

[54] R. E. Kalman. On the general theory of control systems. In *Proc. First IFAC Congress Automatic Control, Moscow*, volume 1, pp. 481–493, Butterworths, London, 1960.

[55] R. E. Kalman. When is a linear control system optimal? *Trans. ASME Ser. D: J. Basic Eng.*, 86:1–10, 1964.

[56] R. E. Kalman, P. L. Falb and M. A. Arbib. *Topics in Mathematical Control Theory.* McGraw-Hill, New York, 1969.

[57] W. Kelley and A. Peterson. *The Theory of Differential Equations: Classical and Qualitative.* Pearson Education, Upper Saddle River, NJ, 2004.

[58] H. K. Khalil. *Nonlinear Systems.* Prentice-Hall, Englewood Cliffs, NJ, third edition, 2002.

[59] P. V. Kokotović. Recent trends in feedback design. *Automatica*, 21:225–236, 1985.

[60] P. V. Kokotović. The joy of feedback: Nonlinear and adaptive. *IEEE Control Systems Magazine*, 12:7–17, June 1992. 1991 Bode Prize Lecture.

[61] P. V. Kokotović and H. J. Sussmann. A positive real condition for global stabilization of nonlinear systems. *Syst. Contr. Lett.*, 13:125–133, 1989.

[62] P. V. Kokotović and M. Arcak. Constructive nonlinear control: a historical perspective. *Automatica*, 37:637–662, 2001.

[63] A. J. Krener. A generalization of chow's theorem and the bang-bang theorem to nonlinear control problems. *SIAM., Series A. Control*, 12:43–52, 1974.

[64] A. J. Krener. Chapter 3: Feedback linearization. In J. Baillieul and J. C. Willems, editors, *Mathematical Control Theory*, pages 66–98. Springer, New York, 1999.

[65] M. Krstic, I. Kanellakopoulos, and P. Kokotović. *Nonlinear and Adaptive Control Design.* John Wiley, New York, 1995.

[66] Y. A. Kuznetsov. *Elements of Applied Bifurcation Theory.* Applied Mathematical Sciences 112. Springer, New York, third edition, 2004.

[67] S. Lang. *Linear Algebra.* Springer, New York, third edition, 1987.

[68] S. Lang. *Undergraduate Analysis.* Springer, New York, second edition, 1997.

[69] J. LaSalle, and S. Lefschetz. *Stability by Lyapunov's Direct Method with Applications.* Academic Press, New York, 1961.

[70] D. G. Luenberger. Observing the state of a linear system. *IEEE Trans. Mil. Electron.*, MIL-8:74–80, 1964.

[71] D. G. Luenberger. An introduction to observers. *IEEE Trans. Aut. Contr.*, AC-16:596–603, 1971.

[72] D. G. Luenberger. *Introduction to Dynamic Systems: Theory, Models, and Applications.* John Wiley, New York, 1979.

[73] A. M. Lyapunov. *The General Problem of the Stability of Motion.* Taylor and Francis, London-Bristol, PA, 1992. (A reprint of the March 1992 special issue of the *International Journal of Control*'s first English translation of Lyapunov's 1892 dissertation. A. T. Fuller, translator and editor of the volume. The work is also available in *Problème général de la stabilité du mouvement*, Ann. Math. Stud. 17, Princeton University Press, Princeton).

[74] J. L. Massera. Contributions to stability theory. *Annals of Math.*, 64:182–206, 1956. Erratum in *Annals of Math.* 68: 202, 1958.

[75] C. D. Meyer. *Matrix Analysis and Applied Linear Algebra.* SIAM, Philadelphia, 2000.

[76] P. J. Moylan. Implications of passivity in a class of nonlinear systems. *IEEE Trans. Aut. Control*, AC-19:373–381, 1974.

[77] J. Murdock. *Normal Forms and Unfoldings for Local Dynamical Systems.* Springer Monographs in Mathematics. Springer, New York, 2003.

[78] R. M. Murray, Z. Li and S. S. Sastry. *A Mathematical Introduction to Robotic Manipulation.* Texts in Applied Mathematics 26. CRC Press, Boca Raton, FL, 1994.

[79] H. Nijmeijer and A. J. van der Schaft. *Nonlinear Dynamical Control Systems.* Springer, New York, 1990.

[80] J. W. Polderman and J. C. Willems. *Introduction to Mathematical Systems Theory: A Behavioral Approach.* Texts in Applied Mathematics 26. Springer, New York, 1998.

[81] R. C. Robinson. *Dynamical Systems: Stability, Bifurcations, and Chaos.* CRC Press, Boca Raton, FL, first edition, 1994.

[82] W. J. Rugh. *Linear System Theory.* Prentice-Hall, Englewood Cliffs, NJ, second edition, 1996.

[83] A. Saberi, P. Kokotović and H. J. Sussmann. Global stabilization of partially linear composite systems. *SIAM J. Control Optim.*, 28:1491–1503, 1990.

[84] D. A. Sanchez. *Ordinary Differential Equations: A Brief Eclectic Tour.* Mathematical Association of America, Washington, DC, 2002.

[85] S. Sastry. *Nonlinear Systems: Analysis, Stability, and Control.* Springer, New York, 1999.

[86] R. Sepulchre, M. Jankovic and P. Kokotović. *Constructive Nonlinear Control.* Springer, London, 1997.

[87] E. D. Sontag. A concept of local observability. *Systems and Control Letters*, 5:41–47, 1984.

[88] E. D. Sontag. Controllability is harder to decide than accessibility. *SIAM J. Control Optim.*, 26(5):1106–1118, 1988.

[89] E. D. Sontag. Smooth stabilization implies coprime factorization. *IEEE Trans. Aut. Contr.*, AC-34:435–443, 1989.

[90] E. D. Sontag. Feedback stabilization of nonlinear systems. In M. A. Kaashoek, J. W. van Schuppen, and A. C. M. Ran, editors, *Robust Control of Linear Systems and Nonlinear Control*, pages 61–81. Birkhauser, Boston, 1990.

[91] E. D. Sontag. *Mathematical Control Theory: Deterministic Finite Dimensional Systems.* Springer, New York, second edition, 1998.

[92] H. J. Sussmann. Orbits of families of vector fields and integrability of distributions. *Trans. Am. Math. Soc.*, 180:171–188, 1973.

[93] H. J. Sussmann. A sufficient condition for local controllability. *SIAM J. Control Optim.*, 16(5):790–802, 1978.

[94] H. J. Sussmann. Lie brackets and local controllability: A sufficient condition for scalar-input systems. *SIAM J. Control Optim.*, 21(5):686–713, 1983.

[95] H. J. Sussmann. Lie brackets, real analyticity and geometric control. In R. W. Brockett, R. S. Millman and H. J. Sussman, editors, *Differential Geometric Control Theory*, pages 1–116. Birkhauser, Boston, 1983.

[96] H. J. Sussmann. A general theorem on local controllability. *SIAM J. Control Optim.*, 25(1):158–194, 1987.

[97] H. J. Sussmann and P. V. Kokotović. The peaking phenomenon and the global stabilization of nonlinear systems. *IEEE Trans. Aut. Contr.*, 36:424–439, 1991.

[98] H. J. Sussmann and V. Jurdjevic. Controllability of nonlinear systems. *J. Diff. Eq.*, 12:95–116, 1972.

[99] M. Vidyasagar. Decomposition techniques for large-scale systems with nonadditive interactions: stability and stabilizability. *IEEE Trans. Aut. Contr.*, AC-25:773–779, 1980.

[100] M. Vidyasagar. *Nonlinear Systems Analysis.* Classics in Applied Mathematics, CL 42. SIAM, Philadelphia, 2002.

[101] J. C. Willems. Dissipative dynamical systems part I: General theory. *Arch. Rati. Mech. Anal.*, 45:321–351, 1972.

[102] J. C. Willems. Dissipative dynamical systems part II: Linear systems with quadratic supply rates. *Arch. Rati. Mech. Anal.*, 45:352–393, 1972.

[103] H. R. Wilson. *Spikes, Decisions, and Actions: The Dynamical Foundations of Neuroscience.* Oxford University Press, Oxford, 1999.

[104] W. M. Wonham. *Linear Multivariable Control: A Geometric Approach.* Springer, New York, third edition, 1985.

[105] J. Zabczyk. *Mathematical Control Theory: An Introduction.* Birkhauser, Boston, 1992.

Index